AF411306

Energy Efficiency in Wireless
Networks

Energy Efficiency in Wireless Networks

Editors

R. Maheswar
M. Kathirvelu
K. Mohanasundaram

Basel • Beijing • Wuhan • Barcelona • Belgrade • Novi Sad • Cluj • Manchester

Editors
R. Maheswar
Department of ECE,
KPR Institute of Engineering
and Technology
Coimbatore, India

M. Kathirvelu
Department of ECE,
KPR Institute of Engineering
and Technology
Coimbatore, India

K. Mohanasundaram
Department of EEE,
KPR Institute of Engineering
and Technology
Coimbatore, India

Editorial Office
MDPI
St. Alban-Anlage 66
4052 Basel, Switzerland

This is a reprint of articles from the Special Issue published online in the open access journal *Energies* (ISSN 1996-1073) (available at: https://www.mdpi.com/journal/energies/special_issues/energy_efficiency_wireless_networks).

For citation purposes, cite each article independently as indicated on the article page online and as indicated below:

Lastname, A.A.; Lastname, B.B. Article Title. *Journal Name* **Year**, *Volume Number*, Page Range.

ISBN 978-3-0365-9311-1 (Hbk)
ISBN 978-3-0365-9310-4 (PDF)
doi.org/10.3390/books978-3-0365-9310-4

Contents

About the Editors

R. Maheswar

Dr. R. Maheswar completed his B.E. (ECE) from Madras University in the year 1999, M.E. (Applied Electronics) from Bharathiyar University in the year 2002, and Ph.D. in the field of Wireless Sensor Network from Anna University in the year 2012. He has 21 years of teaching experience at various levels, and his research interest includes Wireless Sensor Network, IoT, Queueing theory, and Performance Evaluation.

M. Kathirvelu

Dr. M. Kathirvelu received his B.E. degree in Electrical and Electronics Engineering from Madras University in the year 1998 and his M.E. degree in Applied Electronics from Anna University in the Year 2004. He received his Ph.D. degree from Anna University, Chennai, in the year 2013. He has 23 years of teaching experience in engineering colleges, and his research area includes low-power VLSI Design, Embedded System Design, and Soft Computing Techniques.

K. Mohanasundaram

Dr. K. Mohanasundaram received his B.E. degree in Electrical and Electronics Engineering from the University of Madras in 2000, his M. Tech degree in High Voltage Engineering from SASTRA University in 2002, and his Ph.D. degree from Anna University, India, in 2014. He has 21 years of teaching experience, and his research interests include intelligent controllers, power systems, embedded systems, and power electronics.

Preface

In recent years, the widespread use of wireless devices has seen significant growth in all sectors. Specifically, the post-COVID-19 situation has caused a huge revolution in the utilization of wireless devices across the globe. Advances are towards not only the number of devices in the network but also their applications, including sensors, Internet of Things (IoT) devices, mobile phones, and other wireless electronic gadgets, having a huge impact on maintaining global communication without any failure. It has been clearly witnessed that enormous power has been consumed by wireless devices when the entire world was using them during the pandemic. The varied utilization of wireless devices can be found not only among people working for industries but also among children and teachers working with schools and colleges and even beyond. As all wireless devices are battery-powered, energy is a critical issue, and it becomes essential to have the necessary energy management and control techniques and infrastructure in place to prolong the lifetime of both the individual device and the network. Hence, the present scenarios of a wireless environment have urged the research community to work more on energy efficiency issues on wireless devices. Hence, the articles in this reprint primarily target energy efficiency in wireless devices that focus on communication protocols, energy harvesting, energy management, device scheduling, edge computing, and so on for various wireless sensor, underwater, and IoT applications.

This reprint features 18 selected high-quality research articles spanning diverse domains in the field of energy-efficient wireless communication and networking. The contributions in this reprint collectively underscore the significance of energy optimization in wireless systems, illuminating innovative approaches, algorithms, and technologies that pave the way for sustainable and efficient wireless communication in various application scenarios.

Thus, this reprint brings together a rich tapestry of research that illuminates the ongoing efforts to design, optimize, and implement energy-efficient wireless communication and networking solutions. This reprint also encourages researchers to explore finer areas in the energy-efficient wireless network domain and provides a significant contribution to the research community to help further extend their areas of research.

R. Maheswar, M. Kathirvelu, and K. Mohanasundaram
Editors

Article

Key Technologies, Applications and Trends of Internet of Things for Energy-Efficient 6G Wireless Communication in Smart Cities

M. M. Kamruzzaman

Department of Computer Science, College of Computer and Information Sciences, Jouf University, Sakakah 72311, Al Jouf, Saudi Arabia; mmkamruzzaman@ju.edu.sa

Abstract: Smart cities can be made into super-smart cities through IoT devices' implication of energy-efficient 6G. IoT devices are expected to reach fifty billion, but limited information is available regarding the energy-efficient 6G wireless communication standard. This article highlights the key technologies, applications, and trends in the Internet of Things (IoT) for energy-efficient 6G wireless communication in smart cities. The systematic review helped to achieve the aim of the study by considering the 20 articles extracted from databases and Google that fell between 2015 and 2021 and are written in English. The findings identified that quantum communication, blockchain, visible light communication (VLC), 6G brain–computer interface (BCI), symbiotic radio, and others are the key technologies. The applications of IoT technologies and energy-efficient 6G are found in 15 Minute City, Industrial Town, Intelligent Transport systems and others. Furthermore, the trend of using 6G through IoT devices in smart cities is promising.

Keywords: 6G; energy-efficient; Internet of Things (IoT); key technologies; smart cities

Citation: Kamruzzaman, M.M. Key Technologies, Applications and Trends of Internet of Things for Energy-Efficient 6G Wireless Communication in Smart Cities. *Energies* **2022**, *15*, 5608. https://doi.org/10.3390/en15155608

Academic Editors: M. Kathirvelu, K. Mohanasundaram and R. Maheswar

Received: 21 June 2022
Accepted: 28 July 2022
Published: 2 August 2022

Publisher's Note: MDPI stays neutral with regard to jurisdictional claims in published maps and institutional affiliations.

1. Introduction

6G wireless communication is a successor of 5G communication. As per Akhtar et al. [1], it uses a higher frequency than any of its predecessors and provides a higher capacity and lower latency. This helps the technology to integrate communication over land, water, and air into one network with higher speed and reliability. This is a very important prerequisite for the large-scale adoption of the IoT. As stated by Ranger [2], the devices connected to IoT infrastructure are expected to reach billions, and this connection will ultimately depend on the speed of connectivity. It should also be noted that, at present, very little information is available on the standard of 6G wireless communication. However, given the rapid scale of the progression of communication from 2G to 5G in the space of a few years, the arrival of 6G communication is inevitable. This will eventually be a potential game-changer, as this technology would disrupt the current state of technology and intelligence. This is discussed by Zhao et al. [3], who point to the revolutionary nature of 6G communication, as it can have a wider frequency with greater communication transmission. Given the increasing role of artificial intelligence (AI), it is doubtless that 6G wireless communication will have a very important role to play in the future. This is because, unlike 5G, 6G does not require high power consumption and can be considered highly secure. Security is among the key issues that confront communication and IoT. This is because the current web is built on technology created over 50 years ago. Since then, there has been no major improvement in the web. However, as more devices are connected to the internet, a new challenge has been posed, as academicians and practitioners need to adapt to the changing role of technology. Furthermore, the main limitation of 5G and other previous networks is that they are insufficient to fulfill the connectivity requirements of 2030 and above. For example, 5G is limited to typical scenarios; therefore, villages and motorways are not well-covered, and a transition from classical technologies to 6G is required. Therefore, the main limitations of the previous networks are the security, coverage, and other concerns, which urge the need for 6G communication, especially in smart cities.

In addition, as the research and markets show [4], there is a need for a critical re-assessment of smart cities in the realm of energy-efficient 6G communication. This should consider IoT infrastructure, market factors that could lead to a faster adoption rate, and strategies that enable the operations of smart cities. Smart cities are built on the principle of increased connectivity; thus, the existing communication infrastructure cannot enable the growth and operation of smart cities. This is one of the core reasons that the world should move towards the faster communication rate provided by energy-efficient 6G communication channels. Hence, it can be concluded that the development of smart cities occurs in tandem with the development of energy-efficient 6G communication. It is also acknowledged in the research and markets [4] that global investment in the infrastructure development of smart cities will exceed USD 3.6B, making it one of the next big things. Under this backdrop, it is important to carve out the technologies, trends, and applications of 6G wireless communication, which will remain the objective of this study.

1.1. Research Aim and Objectives

This research paper aims to highlight the key technologies, applications, and trends in IoT for energy-efficient 6G wireless communication in smart cities. Three objectives have been developed to accomplish this research aim for 6G wireless communication in smart cities.

RO1—Craving out the key technologies of the IoT for 6G wireless communication in smart cities.

RO2—Identifying the popular key technology applications for 6G wireless communication in smart cities.

RO3—Highlighting the trends in IoT technologies.

1.2. Research Significance

The increased realm of 6G key technologies leads to faster adoption strategies that enable the operation of smart cities. Therefore, it is very important to highlight the key technologies, applications, and trends in the IoT for 6G in smart cities. Based on the research aim, the study results are highly significant for technology companies, government officials, scholars, and smart city management. The report results provide a holistic view of popular IoT 6G wireless communication technologies, which could enable smart cities to perform operations with high-speed connections. This allows technology companies to predict future trends and manufacture communication devices for smart cities accordingly. In addition, it helps government officials and smart city management to maintain the essence of building smart cities by adopting technologies that are relevant to 6G fastest wireless communication. Moreover, the result fills the gap in the literature, where limited information is available related to 6G wireless communication technologies, applications and trends. This shows the significance of our study.

1.3. IoT for 6G in Smart Cities

A range of technologies that need to be described for the full-scale development of the 6G communication network. The study of Liu et al. [5] highlighted different technologies that can be useful in 6G wireless communication. These technologies include the Internet of Things (IoT), virtual reality (VR), Terahertz (THz) communication, visible light communication, etc. IoT is also discussed by Liu et al. [5], who states that the current state of communication networks, i.e., 5G, is relatively slow at keeping the billions of devices connected to IoT systems, as they require massive data processing and communication in real-time. As a result, scientists are already looking forward to a faster form of communication that can open up a new era in terms of connectivity.

1.4. Trends

Regarding the recent trends in 6G communication, the study of Sher et al. [6] has provided significant insights. Intelligent connectivity is the leading trend in 6G communication. This is because the earlier systems of communication, such as 4G or 5G, do not possess the technical sophistication needed for the rapid development of intelligent connectivity. It has been a challenge for 5G networks to provide a single network platform that can enable enhanced connectivity. This is clearly unsupported by the current state of communication networks and requires high-speed networks such as 6G that rely on artificial intelligence for network services and edge computing. Another emerging trend discussed in the realm of 6G is the use of IoT systems. Wang et al. [7] stated that the major motivation behind the use of 6G in IoT systems is to effectively connect each and every device to the network. Based on the network speed of 6G connectivity, technologies based on IoT can enable the integration of many technologies and communication between devices that could not previously have been connected. It can also provide real-time data processing and communication and provide users with an excellent user experience in different IoT applications, such as smart homes and smart cities. However, this trend is largely undeveloped due to the limited research that has been conducted and high cost of implementation [8]. In future, given the rate of technological advances and the current interest shown by countries such as China and the US in the development of 6G, it is only a matter of time before 6G is widely applicable in IoT systems in smart cities. The aspect of security in the development of 6G communication also needs to be highlighted. As per Wang et al. [7], security is one of the key features leading to the advancement of a new communication paradigm, also known as 6G communication. In addition to the major aspects of the 6G network, such as real-time intelligent edge computing, distributed artificial intelligence, intelligent radio, and 3D intercoms, the issues of security and privacy must also be brought into the limelight. Through the advanced security system that is possible in 6G, the smart cities' IoT-based systems will be more secure, to keep up with the expectations of privacy and confidentiality. The study has illustrated the evolution of security under different networks; see Figure 1.

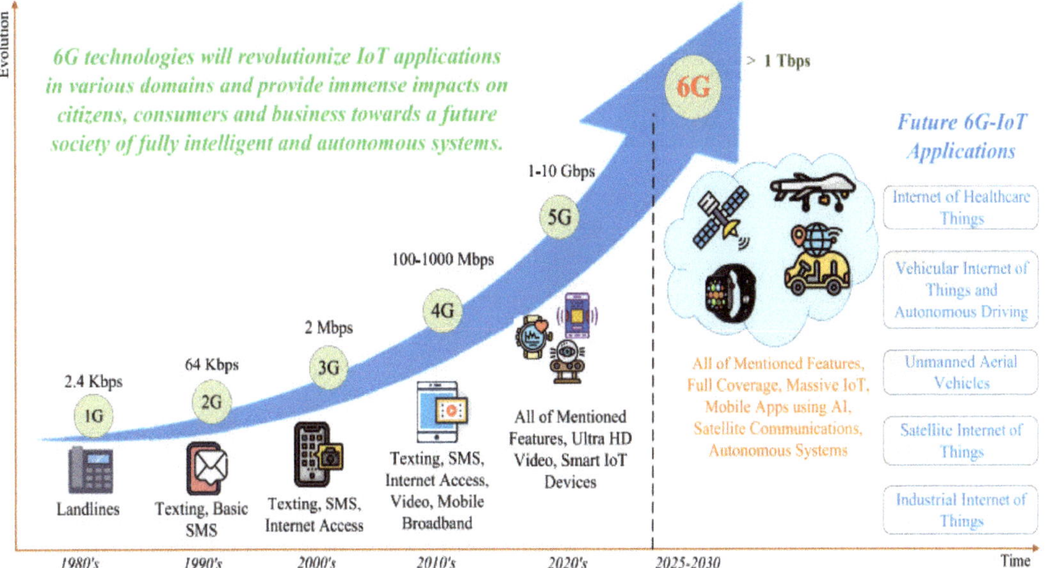

Figure 1. Phases in the development of networking technology source [9].

2. Methods and Materials

The main methodology used in our study is a systematic review. As discussed by Aromataris and Pearson [10], a systematic review is achieved through a critical assessment of different academic papers. This is essentially a review of the evidence and proven concepts regarding the formulated research objectives. In such a study, it is left to the researcher to select, screen, and critically analyze different studies. The study is dictated with the purpose of stating the summary and findings of the primary or secondary research conducted on the chosen topic. In this manner, a systematic review can also be referred to as secondary research, since it is based on a meticulous review of academic papers. The papers selected for this study are highly technical. They were obtained through searching for keywords such as "6G Network," "6G Network Applications," "Technologies," "Trends," and "Internet of Things (IoT)". The minimum number of papers that was reviewed in this study is 20. This was to make the study comprehensive and inclusive. However, it must also be noted that this study is not limited to summarizing different academic papers; the author's own discussion and analysis are also a key aspect of this study. The inclusion criteria were that the study only considers articles published from 2015 to 2022. The study excludes blogs and news to ensure the quality of content. The bias in the extraction of studies, selected using Preferred Reporting Items for Systematic Reviews and Meta-Analysis (PRISMA), is shown in Figure 2. The AMSTAR results for this study are presented in Table 1. In Appendix A, Table A1, shows the results of the systematic literature review conducted in the study.

Table 1. AMSTAR results.

Study Type ($n = 60$)		
	Number	Percentage
Overview	39	65%
Methodological	21	35%
Study designs ($n = 60$)		
Descriptive	27	45%
Experimental	33	55%
AMSTAR assessment ($n = 60$)		
Yes	19	32%
No	41	68%
Number of reviewers in total ($n = 60$)		
2	11	18%
>2	49	82%
IRR reported ($n = 60$)		
Yes	42	70%
No	18	30%
Publication Year > 2015 ($n = 60$)		
Yes	58	23%
No	2	77%

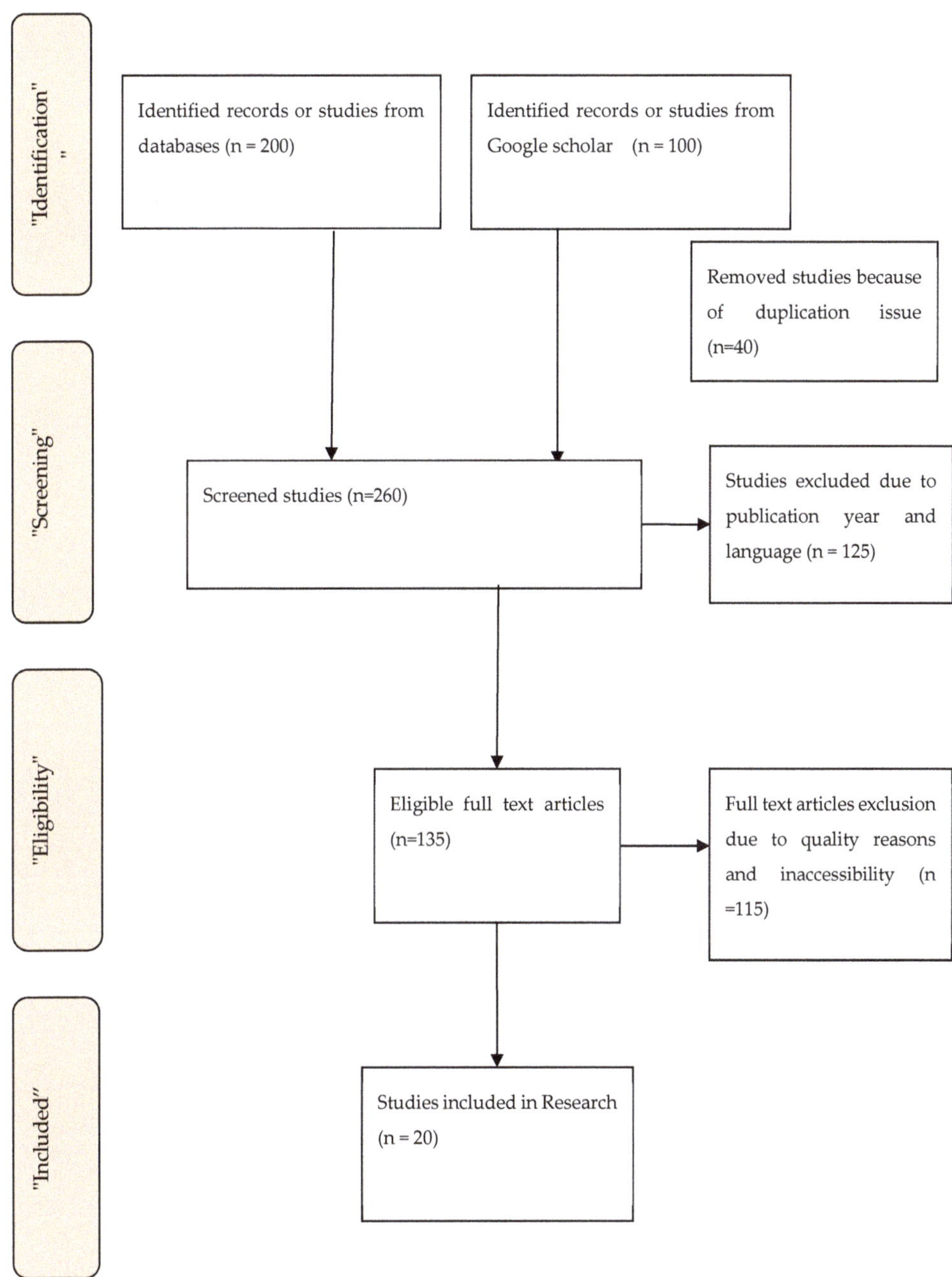

Figure 2. *Cont.*

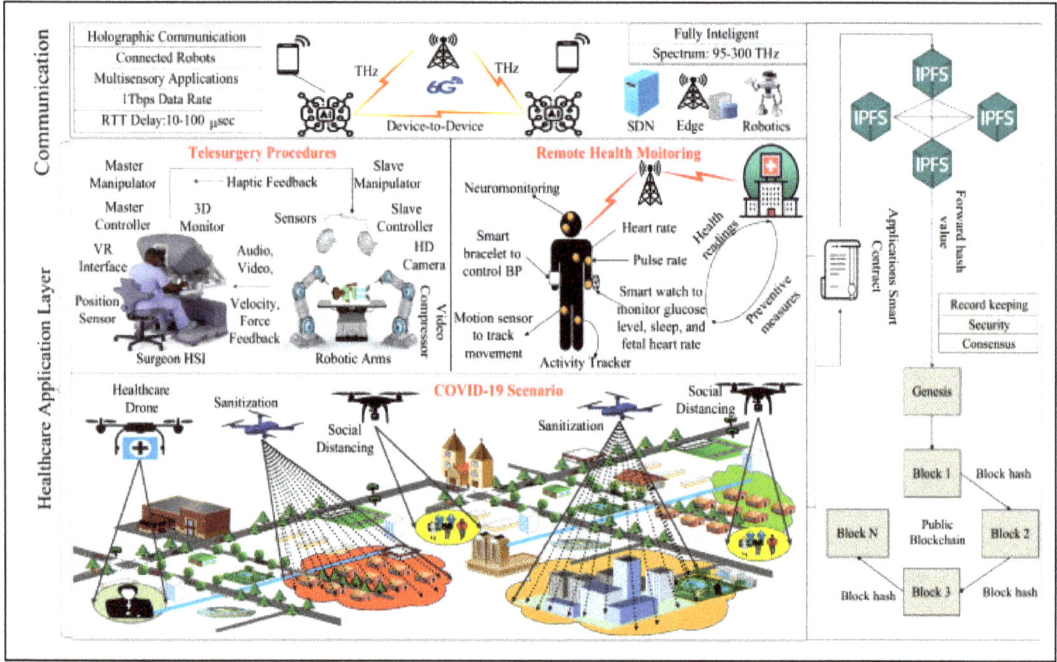

Figure 2. Blockchain-based 6G environment [11].

3. Results and Discussion

3.1. Identified Key Technologies of IoT for Energy-Efficient 6G in Smart Cities

The industrial revolution witnessed in the 18th century revolutionized cities around the globe. The research and markets [4] articulated that smart cities were built to enhance connectivity; however, the contemporary connection facilitations are unable to fulfill the demand of smart cities. Allam et al. [12] discussed the phrase '15-Minute City', referring to the smart urbanization in a post-pandemic era that offers better living standards and improved urban health facilities. The '15 Minute City' could only be made possible by identifying IoT key technologies related to 6G wireless communication in smart cities. However, no holistic study was found in the literature that provided a systematic review of key IoT technologies for 6G in smart cities [13]. Therefore, the current research fills this gap by systematically reviewing 60 peer-reviewed published articles. Ranger [2] proclaimed that IoT infrastructure is expected to reach a billion devices, dependent on connectivity. However, the connectivity provided by 2G–5G is not enough to meet the high demands for IoT devices, especially in smart cities. Therefore, it is highly significant to identify the key IoT technologies for 6G in smart cities that can be used to fulfill the high connectivity needs.

Machine learning and self-organization strategies can considerably improve the network level of 6G communication. 6G will not rely on typical AI operations but will instead require a collaborative AI that is capable of truly revolutionizing the world. According to Mohsan et al. [14], a new kind of communication should have real-time intelligence combined with extremely low latency. As a result, quantum communication technology will be required for the 6G network. Quantum communication has the potential to greatly improve the network's reliability and security. Zhang et al.'s [15] research has shed new light on quantum communication. This technology is, thus, very effective for IoT systems, as it can ensure that the data are processed and communicated whenever required. Additionally, through the use of 6G technology in IoT, big data can be handled with great efficiency. Moreover, high security and instant communication can be obtained using a model of direct transmission, carried out using a quantum channel. This sort of communication is also tied to the blockchain, which is a popular 6G communication technology. The blockchain, which is a combination of network decentralisation and general ledger technology, will be a crucial technology in 6G applications [16].

Blockchain, according to the researchers, is the most disruptive sort of technology, capable of enabling and ensuring the seamless operation of the 6G network. Intelligent resource management is one of the characteristics that blockchains can provide to 6G [17]. Spectrum sharing, orchestration, and decentralised operations, according to the researchers, cannot be made compatible with the current communication infrastructure. VLC is another sort of technology that cannot be overlooked when it comes to 6G. VLC transmits and receives data using visible light [18]. It belongs to the optical wireless communication category and can carry data at frequencies between 400 and 750 THz. The technology's capacity to transmit a large volume of data without major time delays is why it is considered a prerequisite for 6G. Given these characteristics, it is deemed necessary for 6G development to proceed without considering VLC. All these 6G technologies and features add various benefits to the IoT system, which can become highly competitive and optimized based on their speed and bandwidth.

The identified key technologies include quantum communication, blockchain, visible light communication (VLC), 6G BCI, symbiotic radio, and many others. This contradicts the research and markets' [4] findings of limited IoT technologies, as discussed in the previous literature. Zhao et al. [3] proposed that a comprehensive set of IoT key technologies for 6G wireless communication in smart cities is present in the literature. The authors shed light on cell-free massive MIMO technology, referred to as a device that does not centrically attach to any device; instead, they all serve the bases coherently. The IoT devices, i.e., massive cell-free MIMO, can tackle the low communication at the edge of the cell by using 6G in smart cities. In a similar context, Zhang et al. [19] invented the IoT key technology, such as the simultaneous wireless information and power transfer used in smart cities, incorporating the feature of utilizing 6G connectivity. The device is battery-free, and is pretty helpful in providing the fastest communication to members. In addition, Han et al. [20] shed light on the importance of an IoT that can provide internet for everything in the smart city. The authors proposed an intelligent hybrid random access scheme for smart cities that can serve as the IoT technology for 6G in smart cities [21–23]. Therefore, it is not wrong to articulate that the literature provided a holistic set of technologies invented by different scholars. However, a study that incorporates a systematic review of all these invented IoT key technologies for 6G in smart cities is lacking.

According to Kohli et al. [24], there are numerous technologies, such as 6G wireless communication in smart cities, that comprise a brain–computer interface. Since this is one of these emerging technologies, it used human consciousness more than external sources for better interaction. Based on the signals and information that monitor and control machines using sensible wearable headsets and devices, the 6G brain–computer interface (BCI) comprises five datasets, comprising features of human senses that are used for human interaction with the machine [24].

On the other hand, Liang et al. [25] claim that smart cities use symbiotic ration techniques to form cognitive radio (CR) and ambient backscattering communications (AmBC) for enhanced communication across the smart cities. Another technology, such as Reconfigurable Intelligent Surfaces (RIS), is used for the indoor windows of buildings located in smart cities. Ideally, the building receives the signals without interference, and communication among people is easily facilitated [26]. Simultaneous Wireless Information and Power Transfer (SWIPT) is used in smart cities with 6G wireless connectivity to increase communication among members. Typically, this technology enables sensors to use wireless connections for power transfer; therefore, battery-free devices are required in 6G networks [19]. Similarly, the Space–Air–Ground–Sea Integrated Network (SAGSIN) is also used in 6G wireless communication, mainly for global coverage. Using 6G wireless communication, the SAGSIN is formed, which can be helpful in communication in the sky

(10,000 km) and 20 nautical miles for the sea. The biggest advantage of employing SAGSIN is in its integrating air, land and sea to control traffic and communicate with users [27].

The previous literature demonstrated that scholars invented different IoT key technologies for 6G in smart cities. According to Khan et al. [16], blockchain is the key technology used when applying 6G. The rationale is that blockchain provides a transparent and trustful decentralized network that ensures users' data security in the 6G network [28]. As per Sodhro et al. [29], the combination of 6G and blockchain in smart cities could be made possible by using key technologies such as AI, RIS, and TeraHertz Communication (THz). Kumari et al. [11] formed a 6G architecture, as illustrated in Figure 2, which shows that IoT is used for various applications in smart cities, from monitoring traffic signals to managing emergency vehicle services.

As shown in Figure 2, the IoT- and blockchain-integrated system in smart cities may be useful in delivering flexible, secure and reliable communication to the users. This 6G environment comprises three layers, including an (i) application layer, (ii) communication layer and (iii) blockchain layer. Similarly, Hewa et al. [17] discussed the blockchain as a powerful technology that can support and ensure the smooth working of 6G. The authors posited that the current telecommunication infrastructure could not support virtual reality massive Machine Type Communication (mMTC). These technologies require blockchain and 6G to perform smart city operations effectively. Blockchain is significantly essential to IoT security. Therefore, the amalgamation of blockchain, IoT and 6G can allow smart cities to take advantage of the faster and safer connectivity. Moreover, Chi et al. [18] discussed the use of VLC in 6G. The authors propounded that the use VLC in 6G facilitates high transmission service through offline and real-time processing. The two types of VLC devices include PureLiFi and ByteLight and two light sources: LED and laser diode (LD). However, the main disadvantage of LED VLC is its limited bandwidth, which is compensated by LD. Hence, blockchain and VLC, as IoT devices, are related to 6G's implication in smart cities.

The emergence of COVID-19 transformed the lives of human beings. Imoize et al. [26] articulated that COVID-19 brought a new normal, working from the home, for people around the globe [30]. However, the 5G wireless communication network is unable to fulfill people's connectivity demands. Therefore, the authors discussed RIS, Ambient Backscatter Communication (ABC) system, UAVs, CubeSats, and others, which can be used as IoT key technologies for 6G in smart cities to fulfill the post-pandemic connectivity needs. Liang et al. [25] shed light on the ABC system and declared it to be an efficient communication tool. The device improves overall transmission because it forms a backscattering link, forming an effective and advanced communication system. In addition, Kohli et al. [24] posited that 6G BCI is an emerging device used to control or support sensible wearable headsets and embedded devices. Moreover, Khan et al. [31] discussed the use of RIS technologies. The systematic research analysis depicts that RIS technologies with AI, IoT, and blockchain can revolutionize communication network dynamics. Hence, IoT devices with 6G help people to fulfill their connectivity needs after humanitarian crises, i.e., COVID-19. The idea that using such devices in smart cities can benefit humanity is elaborated in the subheading below.

3.2. Applications Key IoT Technologies for Energy-Efficient 6G in Smart Cities

The implications of using 6G in key IoT technologies such as blockchain, VLC, and others rely on their main performance metrics. Mohsan et al. [14] proclaimed that the implementation of 6G wireless communication depends on specific performance metrics. These performance metrics include mobility, energy and spectrum efficiency, throughput, latency, and reliability. The devices and structures tried to meet such standards to implement 6G through IoT devices in smart cities. Ghorbani et al. [31] proposed that the application of 6G and IoT devices ensures unlimited networking possibilities for people in smart cities. According to Upadhyaya et al. [32], the number of smart IoT devices could reach up to fifty, making people lives easier. The application of IoT devices for 6G in smart cities provides a 100 Gbps data rate, <0.1 ms latency rate, up to 1000 km/h mobility rate,

100 bps/Hz spectral efficiency, and 1000 GHz frequency. Moreover, Kamruzzaman [33] discussed the implication of lightweight security modules such as 6G and IoT devices, which can improve the working of smart cities. As per the authors, Smart City Networking Model using Lightweight Security Module (SCNM-LSM) improves smart city living. The application of 6G in IoT devices facilitates e-health, intelligent transportation system, managing energy resources, and networking modeling [34–36]. This depicts that the implication of IoT devices for 6G in smart cities enhances the connectivity rate, which enables people to enjoy unlimited networking possibilities during a post-pandemic situation.

The application of IoT devices in 6G in smart cities can produce incredible benefits. Saad et al. [36] articulated that 6G wireless communication with the help of IoT devices helps to create super-smart cities. The rationale is that 6G connectivity enhances the communication and transmission rate, providing premium-quality features to smart city residents. However, Ghorbani et al. [31] claimed that the implementation of 6G through IoT devices in a smart city can give rise to data security risks, psychological abuse, cyber risks, and others. In contrast, Tariq et al. [37] articulated that the application of AI in 6G can help to control the system's security risks [38]. The collaboration of AI and 6G transform the healthcare system by providing precise medical solutions and health-monitoring. This allows for healthcare workers, patients, and hospital management to keep the details secure, enhancing their benefits to the community. In addition, Al-Turjman and Lemayian [39] discussed the intelligent transport system (ITS) that is implemented in smart cities, facilitating the use of smart logistics. The authors discussed the United Nations Population Funds' predictions that 60% of the world population in the next ten years will move to smart cities that imply the use of IoT devices with 6G [40,41]. Hence, the pattern of 6G being implemented through IoT devices has started in smart cities, which are the future of the global population.

It has demonstrated that 6G application through IoT devices can help people to receive unlimited benefits. This can be seen in Allam et al. [10] 's example, presenting a '15 Minute City' through 6G and IoT devices. According to Allam et al. [10], the '15 Minute City' is the new urban planning model presented by Franco-Colombian scientists, using intelligent devices with the fastest connectivity rate to enhance people's livelihood. As per the authors, the '15 Minute City' is the true example of applying 6G and IoT devices to build a super-smart city. In addition, Kumari et al. [11] highlighted the amalgamation of IoT, blockchain, and 6G in the smart city. As per the authors, the application of 6G wireless communication in the smart city can address latency and reliability issues. Blockchain technology can handle security and privacy risks, and IoT facilities play a significant role in security, privacy, latency, and reliability [42]. The authors provided different application examples, including ITS, smart grids, smart healthcare facilities and others. Hence, the application of 6G with IoT devices can help to formulate the '15 Minute City', ITS, smart grid, smart healthcare facilities and others.

Smart cities address people's issues and could lead to a revolution in people's lives. The application of 6G can facilitate connected robotics and autonomous systems, such as drone-delivery UAV systems. The drone-delivery UAV system can ensure the timely delivery of packages. In addition, the implication of 6G through IoT devices in smart cities can provide the necessary facilities for people to use self-driving cars that protect them from traffic accidents and hustle [42]. Moreover, Chimmanee and Jantavongso [43] have provided an example of the "Bangkok Metropolitan Region" (BMR) that entails a region known as an "Industrial Town", built as a smart and sustainable city. The industrial town ensures the deployment of the 6G facility with IoT devices, which can provide the fastest connectivity. The authors applied the practical mobile planning and optimization framework. They found that the implication of 6G through IoT devices in industrial towns allows for people to fulfill their techno-economic opportunities by focusing on sustainable goals. Moreover, Ibba et al. [44] discussed the use of IoT devices to create a better environment. The sensors from devices produce digital measurements that improve people's quality of life in every smart city. The authors shared an example of the 'CitySense System', which characterizes the quantities of different environmental elements such as carbon dioxide, methane, temperature, air, humidity, and light. The measured quantities regarding environmental pollution concerns provide solutions to maintain the environment of smart cities and ensure a healthy lifestyle for residents. In the context of smart cities, 6G wireless communication is extremely useful when building a super-smart society. The superior and premium features of 6G, enhancing communication and interaction in terms of quality of life, are remarkable when applied in smart cities [45]. AI-based Machine-to-Machine (M2M) communication is used to monitor environmental activities, ensure automation in super-smart societies and improve quality of life. Another application of smart cities is using extended reality and augmented reality (AR), virtual reality (VR) and mixed reality (MR) to form features, using 3D objects for effective communication. AR allows for individuals to interact with the physical world based on the 3D landscape. On the other hand, MR forms a real and virtual atmosphere, where the interaction takes place in real time. This is why MR is also recognized as a hybrid application of the technology, which creates a 3D environment for users [46]. Figure 3 shows that Nguyen et al. [9] have shown that fog computing is a real-time application, minimising time, computational cost, energy and storage issues when IoT is implemented in smart cities.

Fog computing is a cloud-computing combination of routers, storage data and computers with end-users in close proximity. Moreover, connected robotics and autonomous systems are also an application of the 6G wireless communication network. The 6G could be useful in the deployment of connected robots and autonomous systems, such as drone-delivery UAV systems. The 6G communication system employing robotics and autonomous systems promotes self-driving cars and improves traffic and driving patterns using the sensors. A wide range of sensors is used in 6G wireless communication, including Light Detection and Ranging (LiDAR), radar GPS, sonar and odometry and inertial measurement units. When using 6G wireless communication such as UAV, there is a system communication between the ground-based controller and systems communications [42]. Furthermore, the AI application of the 6G wireless communication can be useful in providing smart healthcare systems with health monitoring and precision medical treatment. It also supports privacy and a high-level protection and support system for healthcare workers, professionals and patients, thus maximizing the advantages of its application [37].

6G wireless communication can also provide haptic communication, as it is a nonverbal communication medium that employs the sense of touch. This is extremely useful for remote users, as they can experience haptic communication and interaction in the real world. The superior features of the 6G application are useful in haptic communication in smart cities among the members and individuals. Similarly, a 6G wireless communication network can be applied in industry for manufacturing and automation purposes. By providing full automation to provide secure communication and interaction mediums, 6G wireless communication can ensure high-data-quality transfer that is nearly error-free. Another significant aspect of employing a 6G system is data transfer without any data loss when transmitted and received by the end-user [42]. Hence, the application of IoT devices is revolutionizing the people's lives through the new state-of-the-art, i.e., 6G wireless communication in smart cities.

3.3. Key IoT Trends for Energy-Efficient 6G in Smart Cities

IoT has become highly popular globally, due to its ability to automate processes and make processes smarter. IoT has several applications that are very beneficial to the entire globe. One major application of IoT includes smart cities [47,48]. Therefore, a large amount of research and development is being conducted in this area to enhance the application and use of IoT in smart cities. One of the major issues faced by previous technologies in the application of IoT and smart cities is that they are unable to handle and process the large amount of data that are produced [6,49]. However, 6G technology can enhance the processing and communication of IoT systems by making them more efficient and faster. Therefore, the current research paper aims to highlight the key trends in the (IoT) for energy-efficient 6G wireless communication in smart cities.

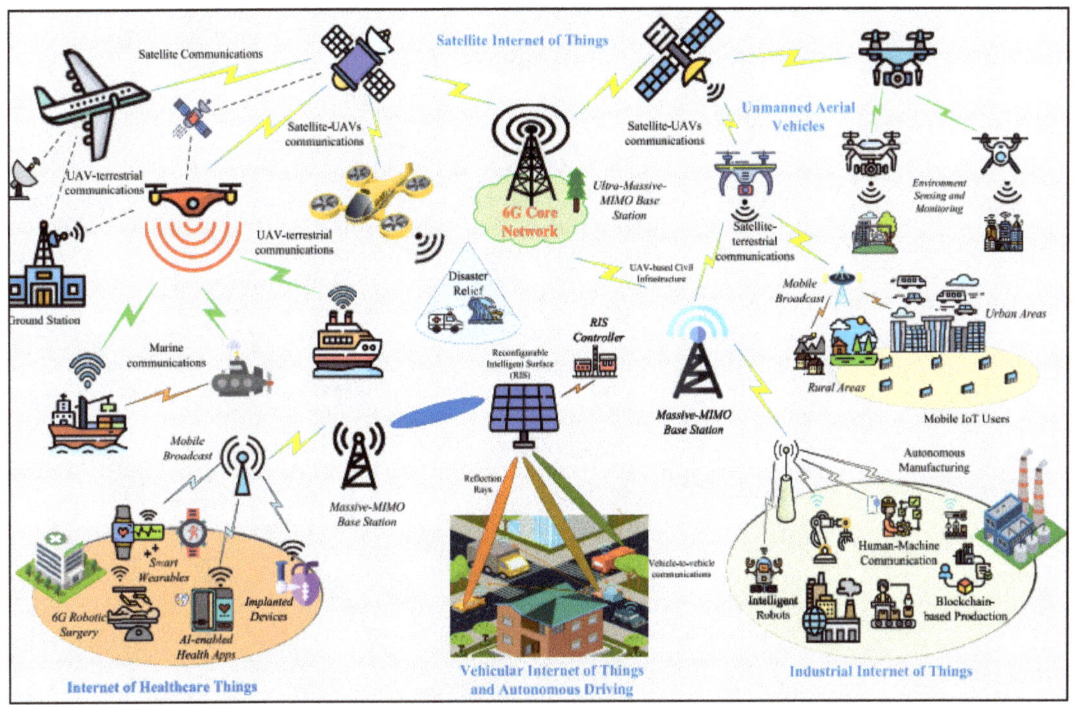

Figure 3. Smart cities and IoT devices. Source: Nguyen et al. [9].

A new trend that has been noted in the area of IoT is Holographic Connectivity. Technologies based on holography can enable the integration of various technologies and communication between devices that could not previously be connected, thanks to the network speed of 6G connectivity [7,49]. It may also deliver a real-time simulation experience and provide consumers with an amazing naked-eye simulation experience. This trend, like 6G, is highly underdeveloped and only exists in principle. However, considering the current rate of technological advancements and the present interest displayed by nations such as China and the United States in the development of AR holographic technology, it is only a matter of time before this technology becomes a reality [50,51]. Lu and Zheng [50] proposed that 6G can be referred to as holographic connectivity, intelligent connectivity, deep connectivity, and ubiquitous connectivity. 5G is the early stage of commercialization network connectivity function, which is not sustainable for the current era. Therefore, the future application of 6G technology can further enhance the IoT systems in smart cities through large-scale applications, as data transmission and processing speed will increase.

Other technological advancements have made intelligent monitoring devices in IoT, such as integrated circuits and wireless communications, smaller, lighter, and more energy-efficient [52]. In addition, another key area in IoT systems and 6G technology is data security. Wang et al. [7] and Chen and Okada [52] noted that the use of 6G technology could lead to more effective security solutions. The radio latency of a 6G network may be lowered to 0.1 ms, which is significantly lower than that of a 5G network. However, there are still security vulnerabilities in the 6G network, including in authentication, encryption, and communication [53]. Furthermore, a 6G wireless communication system is likely to be effective in reducing heterogeneous hardware restrictions in the coming years. As per [42] and [54], 6G connectivity will transform individual interactions in smart cities. Access points and mobile terminals that are not part of the hardware configuration may be required for quick and error-free communication. Additionally, the use of MIMO techniques can lead to the advancement of 6G architecture, making it more secure and effective [55,56]. Borges et al. [57] articulated that MIMO communication systems are based on the antenna array at the receiver and transmitter side, which can also be used in 5G technology. The antenna array's presence made the MIMO communication system a high-speed transmission technology with a minimum quality level. Therefore, the use of a 6G communication system makes MIMO as energy efficient, high-speed communication tool with the assurance of high-quality level transmission. Hence, the trends in 6G and IoT development show that, in the future, it is likely that smart cities will be safer in terms of privacy and confidentiality as the systems will be able to transfer and store data using strong security protocols [58]. This would also lead to an increase in the trust of the users and public, leading to the enhanced applicability of IoT systems based on 6G technology.

Moreover, it is also noted that the collaborative use of 6G technology and IoT system leads to the development of diverse architectures with the potential to gather large amounts of data. New driving trends, such as Smart Reflective Surfaces and Environments, may emerge as a result of the 6G wireless system. The smart reflecting surfaces, according to this concept, serve as walls, highways, entrances, and complete structures. When driving, this also aids in maintaining a clear line of sight and obtaining high-quality signals [26,59]. As a consequence, when a 6G communication network is used in the system, there will be little to no risk of loss or accidents on the roadways. Zhihan et al. [59] and Nguyen et al. [9] noted that the use of 6G technology could lead to the large-scale application of IoT systems with technologies such as AI and big data analytics. Through the development of these systems, IoT systems in smart cities could be developed to provide a more convenient and safer transportation infrastructure. Additionally, these systems are likely to enhance the security of smart cities, making people more comfortable. Another critical feature of 6G technology is AI-based computer vision using deep learning, which was recently worked on in many applications [60]. As per Mahmoud et al. [22] and Mucchi et al. [61], in smart cities, 6G wireless connectivity might be a new trend in wireless energy transfer and collection. Energy transfer for harvesting may be accomplished in smart cities utilizing 6G cellular networks, according to this approach. A highly positive effect on the environment can be noted through the development of such trends, as they will reduce the large number of carbon footprints. Hence, these benefits of 6G technology and IoT in smart cities are motivating researchers to find convenient, cost-effective, and efficient solutions that can easily be employed to enhance the infrastructure of smart cities [62,63].

Moreover, in the coming years, it is expected that a 6G wireless communication system will be useful in minimizing the heterogeneous hardware constraints. This implies that the 6G communication would revolutionize interactions among individuals in smart cities. Access points and mobile terminals may be needed for fast and error-free communication that will be entirely different from the hardware setting [42]. It is also expected that MIMO techniques can further be upgraded in the 6G architecture for advanced-level communication. A 6G wireless system may also give rise to new driving trends, such as Smart Reflective Surfaces and Environments. It even helps to maintain a line of sight and obtain high-quality signals when driving [26]. As a result, there would be minimal to no chances of loss or accidents on the roads when a 6G communication network is employed in the system. Likewise, Zhao et al. [63] claims that there is an emerging trend of 6G communication in the network layer. Within this domain, there is a traffic and mobility prediction that allows for smart cities to improve prediction accuracy in traffic. Moreover, 6G may even promote mobility and handover management intelligent network management, which could minimize accidents. According to Ho et al. [62], there is another emerging trend of employing 6G in the smart grid system, so that a number of IoT devices can be connected to ensure real-time remote monitoring of the system. Table 2 provides a comparison between the technologies being discussed with respect to smart cities and wireless communication.

Table 2. Comparison between the technologies with respect to smart cities and wireless communication.

Technologies	Application	Trend	Challenges
Quantum communication	The application of quantum communication with respect to wireless communication and smart cities protects information channels against eavesdropping by means of using quantum cryptography [61].	Among the key trends relating to quantum communication with respect to wireless communication and smart cities is the security of sensitive information, as the technology allows for strong encryption protocols to protect the communication messages [64]. Quantum communication has the ability to transform multiple markets, including many tech giants and startups, due to its unique abilities. Moreover, quantum communication is also expected to facilitate long-distance communication such as transmission across ocean, as the technology relies on satellite for the transfer of information [62].	The key challenges include maintaining a quantum communication network, transferring of quantum states, a secure infrastructure and encryption, along with the creation of public trust [63].
Blockchain	Blockchain technology can facilitate the interconnection of cities through its vertical services such as accessibility, mobility, transversal system and security benefits.	The technology has the ability to empower smart cities by allowing for data exchange with a higher degree of transparency and reliability without the need for a centralized administrator.	The challenges include a lack of awareness relating to the usage and working of the technology, which results in a waste of resources during the exploration process.
Visible light communication (VLC)	VLC is a wireless communication method, which allows for the transmission of information at a high speed with visible light. The information acquired through this technology is transmitted by modulating the intensity of light from the light source.	The key trends in VLC in smart cities include traffic management by facilitating vehicle-to-vehicle and vehicle-to-infrastructure communication. Moreover, streetlights that communicate with pedestrians' VLC-enabled smart phone devices can help to regulate traffic. The technology leads to possibilities for hybrid communication.	The key challenges in the technology include flicker, which is a major problem as light travelling at such a high speed can cause damage to the human eye. For indoor environments, dimming also is a challenge as the light needs to be adjusted as per the requirements.
6G BCI	The 6G applications include optical wireless communication, wireless power transfer and 3D networking. In addition, the technology can facilitate unmanned aerial vehicles, along with the advanced usage of AI in the smart cities.	Among the key trend of the technology include improved privacy in communication and transmission of information, improvement in global health, improved transportation, logistic facilities and security, all of which contribute to the building of smart cities.	The challenges relating to 6G communication include the attainment of a cost-effective approach for network deployment and expansion, data privacy enforcement, security concerns and the challenge of reducing the price of mobile communications.
Symbiotic radio	Symbiotic radio technology is a batteryless smart device that allows for the sharing of data with corresponding access points by employing Wi-Fi TV signals through backscattering waves.		The key challenges of the technology include the establishment of an accurate channel model to capture the essential behavior of backscattering channel and the management of the limited resources required to fulfill the wireless communication requirements. Moreover, privacy and security are also concerns in the wireless communication system; since the primary and secondary transmissions in symbiotic radio technology are connected, if an attacker disrupts primary transmission, the probability of affecting the secondary transmission becomes increasingly easy.

4. Conclusions

This study showed that IoT, along with energy-efficient 6G technology, has the potential to make smart cities more efficient and effective. The use of 6G by IoT devices can transform smart cities into super-smart cities. As a result, this research focused on IoT technologies, applications, and trends for energy-efficient 6G wireless communication in smart cities. The three goals emphasize the most important IoT technologies, applications, and trends in smart cities. The systematic review aided in accomplishing the study goal by analyzing 20 papers taken from online databases and Google Scholar, published between 2015 and 2021. The most relevant and reliable papers were selected and reviewed systematically. The findings of the paper noted that quantum communication, blockchain, VLC, 6G BCI, symbiotic radio, and other technologies are essential technologies. Through the use of these technologies, IoT systems in smart cities can work efficiently to provide high-quality services of traffic management, security, surveillance, etc. In addition to this, this study also examined the different applications of IoT systems based on 6G technology in smart cities. These applications include the 15 Minute City, Industrial Town, and ITS. Furthermore, more applications of IoT and 6G are likely to be researched and developed in the future, considering the high pace at which they have been adopted by researchers and experts in this field. Besides this, the trends in research and development in IoT and 6G technology show that future smart cities will be more efficient and effective, as they will be able to handle data at great speeds. These systems are likely to solve sustainability problems and make smart cities more environmentally friendly. In addition, the recent trends show the researchers' significant focus on security and data transmission speed and the performance of the systems, which is likely to make IoT and energy-efficient 6G-based smart cities highly effective. The application of IoT devices to 6G in smart cities provides a 100 Gbps data rate, <0.1 ms latency rate, up to 1000 km/h mobility rate, 100 bps/Hz spectral efficiency, and 1000 GHz frequency. This is most promising in resolving the issues of energy inefficiency and other concerns in classical communication networks. Conclusively, the use of energy-efficient 6G in smart cities via IoT devices is promising and is likely to solve various problems that are encountered by existing smart city systems.

Funding: This research received no external funding.

Informed Consent Statement: Not applicable.

Conflicts of Interest: The author of this publication declares that there is no conflict of interest associated with this publication.

Appendix A

The table below, Table A1, shows the results of the systematic literature review conducted in the study.

Table A1. Results of systematic literature reviews.

Serial No.	Author Name	Title	Year	Findings and Results
		Key Technologies		
1	Mohsan et al.	6G: Envisioning The Key Technologies, Applications and Challenges	2020	Real-time intelligence twinned with extremely low latency is enhancing communication through the 6G network. Quantum communication can also increase the reliability of communication.
2	Zhang et al.	Quantum Secure Direct Communication With Quantum Memory	2017	A direct communication model employing a quantum channel is useful for instant communication through a 6G network, such as blockchain technology.
3	Hewa et al.	The Role of Blockchain in 6G: Challenges, Opportunities and Research Directions	2020	Blockchain is a disruptive kind of technology that allows for 6G intelligent resource management for smooth functioning.
4	Chi et al.	Visible Light Communication in 6G: Advances, Challenges, and Prospects	2020	Visible Light Communication (VLC) enhances data transmission through the 6G network. With the ability to transfer data ranging up to 400 and 800 THz, VLC is a time-efficient technology for smart cities.

Table A1. *Cont.*

Serial No.	Author Name	Title	Year	Findings and Results
5	Kohli et al.	A Review of Virtual Reality and Augmented Reality Use-Cases of Brain-Computer Interface Based Applications For Smart Cities	2022	The brain–computer interface (BCI) allows for the exchange of information and signals through different controlling machines such as sensible wearable headsets and embedded devices.
6	Liang et al.	Symbiotic Radio: Cognitive Backscattering Communications For Future Wireless Networks	2020	Symbiotic radio technology is effective for smooth communication processes as it employs backscattering link signals for advanced level communication and interaction in smart cities.
7	Imoize et al.	6G Enabled Smart Infrastructure For A Sustainable Society: Opportunities, Challenges, and Research Roadmap	2021	Reconfigurable Intelligent Surface (RIS) is a technology that is used in the doors and windows of buildings in smart cities as it shares signals without any interference among individuals.
8	Zhao et al.	A Comprehensive Survey of 6G Wireless Communications	2020	MIMO technology in 6G wireless connection can operate without cells and provides spectral efficiency in a communication network.
9	Zhang et al.	Wireless Information and Power Transfer: From Scientific Hypothesis to Engineering Practice	2015	Simultaneous Wireless Information and Power Transfer (SWIPT), technology is able to detect sensors using wireless connections to improve the communication network.
10	Giordani and Zorzi	Satellite Communication At Millimeter Waves: A Key Enabler of The 6G Era	2020	the Space–Air–Ground–Sea Integrated Network (SAGSIN) integrates air, land and sea to control traffic and enhance communication.
11	Nguyen et al.	6G Internet of Things: A Comprehensive Survey	2021	The use of 6G technology enables the integration of various IoT technologies, such as edge intelligence, space–air–ground–underwater communications, reconfigurable intelligent surfaces, massive ultra-reliable, Terahertz communications, low-latency communications, and blockchain.
12	Allam et al.	Fundamentals of Smart Cities	2022	IoT-based 6G systems use technologies such as RFID and wireless sensor networks to process and transfer data at a high speed, which ensures their high performance.
Key Applications				
1	Saad et al.	A Vision of 6G Wireless Systems: Applications, Trends, Technologies, and Open Research Problems	2019	6G wireless communication can be employed in smart cities to develop a hybrid model of technology in a 3D setting.
2	Chowdhury et al.	6G Wireless Communication Systems: Applications, Requirements, Technologies, Challenges, and Research Directions	2020	6G wireless communication can be used for LiDAR, radar GPS, sonar and odometry and inertial measurement units when autonomous driving patterns are promoted. 6G wireless communication may improve ground-based controller communication when employed in robotics and the autonomous driving of vehicles. Moreover, the 6G wireless communication network can be used in the manufacturing industry with a high-data-quality transfer rate, with minimal chances of error. The 6G system also prevents data loss and ensures the safe transmission of data from the sender to the receiver. 6G is also applicable in haptic communication in smart cities.
3	Tariq et al.	A Speculative Study on 6G	2020	A smart healthcare system includes 6G precision medical treatment, providing high-level protection and supporting healthcare workers.

Table A1. *Cont.*

Serial No.	Author Name	Title	Year	Findings and Results
4	Xiaohu et al.	6G Internet of Things: A Comprehensive Survey	2020	6G has high potential in the following IoT applications: healthcare IoT, vehicular IoT autonomous driving, unmanned aerial vehicles, satellite IoT, and industrial IoT.
5	Atitallah et al.	Leveraging Deep Learning and Iot Big Data Analytics to Support The Smart Cities Development: Review and Future Directions	2020	The application of IoT based on 6G has the potential to handle big datasets and integrate cloud-based technologies to provide various services to smart cities.
6	Ye et al.	Big Data Analysis Technology For Electric Vehicle Networks in Smart Cities	2020	The application of 6G and IoT has enhanced the application of electric vehicles and autonomous cars.
7	Liu et al.	An Overview of Key Technologies and Challenges of 6G	2021	The use of 6G in IoT has widened the scope of applications, as various advanced technologies, such as VR and AR, can be implemented easily with great speed and security.
8	Ho et al.	Next-Generation Wireless Solutions For The Smart Factory, Smart Vehicles, The Smart Grid and Smart Cities	2019	IoT based on 6G has enhanced the application of smart grids with the potential to save various power losses and make energy systems more efficient.
9	Al-Turjman and Lemayian	Intelligence, Security, and Vehicular Sensor Networks in The Internet of Things (Iot)-Enabled Smart-Cities: An Overview	2020	IoT smart cities have taken the communication of information to another level due to the use of 6G, as it has very high speeds and ensures the fast travel of large data.
10	Wang et al.	Security and Privacy in 6G Networks: New Areas and New Challenges	2020	The use of 6G in IoT has also enhanced security applications, making smart cities safe from privacy and data breaches. This ensures large-scale applications and easy adoption by the public.
Trends				
1	Kamruzzaman	6G wireless communication assisted security management using cloud edge computing	2022	Security management is the process of identifying a company's assets (such as people, buildings, equipment, systems, and information assets) and then developing, documenting, and implementing policies and procedures to secure those assets. Meanwhile, artificial intelligence (AI) applications are flourishing thanks to advances in deep learning and numerous hardware architecture improvements based on cloud edge computing (CEC) issues are associated with the Internet of Things (IoT), including inadequate security measures, user ignorance, and the dreaded active monitoring.
2	Wang et al.	Security and Privacy in 6G Networks: New Areas and New Challenges	2020	6G is facilitating holographic connectivity, which implies that each and every device is connected to the network. Based on this integration of technologies, holographic connectivity may provide a strong connection between devices with a real-time simulation experience. In countries such as China and the US, augmented reality (AR) holographic technology with security features may be seen in the future. Thus, holographic technology in 6G may emerge as a new trend due to the security and privacy of the system.

Table A1. *Cont.*

Serial No.	Author Name	Title	Year	Findings and Results
3	Chowdhury et al.	6G Wireless Communication Systems: Applications, Requirements, Technologies, Challenges, and Research Directions	2020	6G communication systems may create access points and mobile terminals for fast and error-free communication. This can be helpful in reducing heterogeneous hardware constraints. MIMO techniques may advance the 6G architecture in the future.
4	Imoize et al.	6G Enabled Smart Infrastructure For a Sustainable Society: Opportunities, Challenges, and Research Roadmap	2021	6G wireless communication may be employed in smart reflective surfaces and environments in smart cities. This trend may increase when smart reflective surfaces are used in walls, roads and doors.
5	Luo	Machine Learning For Future Wireless Communications	2020	6G may help in network layers such as traffic and mobility prediction in smart cities, prediction and accuracy in traffic, promoting mobility and handover management to reduce the chance of accidents.
6	Mahmoud et al.	A Comprehensive Survey on Technologies, Applications, Challenges, and Research Problems	2021	6G wireless communication is a new trend in wireless energy transfer and harvesting in smart cities, allowing for energy transmission for harvesting to be possible in smart cities.
7	Sher et al.	An Overview of Key Technologies and Challenges of 6G	2021	Virtual reality (VR), Terahertz (THz) communication, visible light communication, radio stripes, quantum networks communication, and large intelligent surface (IRS) are the six different technologies that can be useful for 6G wireless communication.
8	Ji et al.	Amalgamation of Blockchain and Iot For Smart Cities Underlying 6G Communication: A Comprehensive Review	2021	The use IoT and 6G together increases efficiency in smart cities, but a single point of failure can negatively impact the lives of individuals. Therefore, blockchain can be effective in overcoming the challenges of implementing IoT and 6G in smart cities.
9	Elmeadawy and Shubair	Fundamentals of Smart Cities	2019	Smart cities include wireless sensor networks, IoT, RFID and 6G to increase intelligent urban facilities.
10	Han et al.	Role of Iot-Cloud Ecosystem in Smart Cities: Review and Challenges	2020	There is a limited computational capacity of end-devices in IoT infrastructure in smart cities. However, the future of smart cities may require both IoT and cloud infrastructure for better activities.
13	Fang et al.	Intelligence, Security, and Vehicular Sensor Networks in The Internet of Things (IoT)-Enabled Smart-Cities: An Overview	2021	The IoT smart cities paradigm is based on different design features such as reliability and robustness. However, 6G communication networks in smart cities may give rise to security concerns that must be addressed, and one way of doing this is to develop security standards for IoT devices.

References

1. Akhtar, M.W.; Hassan, S.A.; Ghaffar, R.; Jung, H.; Garg, S.; Hossain, M.S. The shift to 6G communications: Vision and requirements. *Hum. Cent. Comput. Inf. Sci.* **2020**, *10*, 1–27. [CrossRef]
2. Ranger, S. *What Is the IoT? Everything You Need to Know about the Internet of Things Right Now. ZDNet. Retrieved 16 August 2021;* Volume 16, p. 2021. Available online: https://www.zdnet.com/article/what-is-the-internet-of-things-everything-you-need-to-know-about-the-iot-right-now/ (accessed on 20 June 2022).

3. Zhao, Y.; Zhai, W.; Zhao, J.; Zhang, T.; Sun, S.; Niyato, D.; Lam, K.Y. A comprehensive survey of 6G wireless communications. *arXiv* **2020**, arXiv:2101.03889. [CrossRef]

4. *Research and Market (2021) 6G and Smart Cities: Transformation of Communications, Services, Content, and Commerce 2025–2030.* Report, Mind Commerce. 2021. Code: ASDR-572020. Available online: https://www.asdreports.com/market-research-related-572020/g-smart-cities-transformationcommunications-services-content-commerce (accessed on 20 June 2022).

5. Liu, Q.; Sarfraz, S.; Wang, S. An overview of key technologies and challenges of 6G. In *International Conference on Machine Learning for Cyber Security*; Springer: Cham, Switzerland, 2020; pp. 315–326. [CrossRef]

6. Sher, A.; Sohail, M.; Shah, S.B.H.; Koundal, D.; Hassan, M.A.; Abdollahi, A.; Khan, I.U. New trends and advancement in next generation mobile wireless communication (6G): A Survey. *Wirel. Commun. Mob. Comput.* **2021**, *2021*, 9614520.

7. Wang, M.; Zhu, T.; Zhang, T.; Zhang, J.; Yu, S.; Zhou, W. Security and privacy in 6G networks: New areas and new challenges. *Digit. Commun. Netw.* **2020**, *6*, 281–291. [CrossRef]

8. Ji, B.; Wang, Y.; Song, K.; Li, C.; Wen, H.; Menon, V.G.; Mumtaz, S. A survey of computational intelligence for 6G: Key technologies, applications and trends. *IEEE Trans. Ind. Inform.* **2021**, *17*, 7145–7154. [CrossRef]

9. Nguyen, D.C.; Ding, M.; Pathirana, P.N.; Seneviratne, A.; Li, J.; Niyato, D.; Dobre, O.; Poor, H.V. 6G Internet of Things: A comprehensive survey. *IEEE Internet Things J.* **2021**, *9*, 359–383. [CrossRef]

10. Aromataris, E.; Pearson, A. The systematic review: An overview. *Am. J. Nurs.* **2014**, *114*, 53–58. [CrossRef]

11. Kumari, A.; Gupta, R.; Tanwar, S. Amalgamation of blockchain and IoT for smart cities underlying 6G communication: A comprehensive review. *Comput. Commun.* **2021**, *172*, 102–118. [CrossRef]

12. Allam, Z.; Bibri, S.E.; Jones, D.S.; Chabaud, D.; Moreno, C. Unpacking the '15-minute city' via 6G, IoT, and digital twins: Towards a new narrative for increasing urban efficiency, resilience, and sustainability. *Sensors* **2022**, *22*, 1369. [CrossRef]

13. Kim, J.H. 6G and Internet of Things: A survey. *J. Manag. Anal.* **2021**, *8*, 316–332. [CrossRef]

14. Mohsan, S.A.H.; Mazinani, A.; Malik, W.; Younas, I.; Othman, N.Q.H.; Amjad, H.; Mahmood, A. 6G: Envisioning the key technologies applications and challenges. *Int. J. Adv. Comput. Sci. Appl.* **2020**, *11*, 14–23. [CrossRef]

15. Zhang, W.; Ding, D.S.; Sheng, Y.B.; Zhou, L.; Shi, B.S.; Guo, G.C. Quantum secure direct communication with quantum memory. *Phys. Rev. Lett.* **2017**, *118*, 220501. [CrossRef]

16. Khan, A.H.; Hassan, N.U.; Yuen, C.; Zhao, J.; Niyato, D.; Zhang, Y.; Poor, H.V. Blockchain and 6G: The future of secure and ubiquitous communication. *IEEE Wirel. Commun.* **2022**, *29*, 194–201. [CrossRef]

17. Hewa, T.; Gür, G.; Kalla, A.; Ylianttila, M.; Bracken, A.; Liyanage, M. The role of blockchain in 6G: Challenges, opportunities and research directions. In Proceedings of the 2020 2nd IEE 6G Wireless Summit (6G SUMMIT), Lapland, Finland, 20 March 2020; pp. 1–5.

18. Chi, N.; Zhou, Y.; Wei, Y.; Hu, F. Visible light communication in 6G: Advances, challenges, and prospects. *IEEE Veh. Technol. Mag.* **2020**, *15*, 93–102. [CrossRef]

19. Zhang, R.; Maunder, R.G.; Hanzo, L. Wireless information and power transfer: From scientific hypothesis to engineering practice. *IEEE Commun. Mag.* **2015**, *53*, 99–105. [CrossRef]

20. Han, H.; Zhao, J.; Zhai, W.; Xiong, Z.; Lu, W. Smart city enabled by 5G/6G networks: An intelligent hybrid random access scheme. *arXiv* **2021**, arXiv:2012.13537. [CrossRef]

21. Guo, F.; Yu, F.R.; Zhang, H.; Li, X.; Ji, H.; Leung, V.C. Enabling massive IoT toward 6G: A comprehensive survey. *IEEE Internet Things J.* **2021**, *8*, 11891–11915. [CrossRef]

22. Mahmoud, H.H.H.; Amer, A.A.; Ismail, T. 6G: A comprehensive survey on technologies, applications, challenges, and research problems. *Trans. Emerg. Telecommun. Technol.* **2021**, *32*, e4233. [CrossRef]

23. Kamruzzaman, M.M.; Alrashdi, I.; Alqazzaz, A. New Opportunities, Challenges, and Applications of Edge-AI for Connected Healthcare in Internet of Medical Things for Smart Cities. *J. Healthc. Eng.* **2022**, *2022*, 2950699. [CrossRef]

24. Kohli, V.; Tripathi, U.; Chamola, V.; Rout, B.K.; Kanhere, S.S. A review on Virtual Reality and Augmented Reality use-cases of Brain Computer Interface based applications for smart cities. *Microprocess. Microsyst.* **2022**, *88*, 104392. [CrossRef]

25. Liang, Y.C.; Zhang, Q.; Larsson, E.G.; Li, G.Y. Symbiotic radio: Cognitive backscattering communications for future wireless networks. *IEEE Trans. Cogn. Commun. Netw.* **2020**, *6*, 1242–1255. [CrossRef]

26. Imoize, A.L.; Adedeji, O.; Tandiya, N.; Shetty, S. 6G enabled smart infrastructure for sustainable society: Opportunities, challenges, and research roadmap. *Sensors* **2021**, *21*, 1709. [CrossRef] [PubMed]

27. Giordani, M.; Zorzi, M. Satellite communication at millimeter waves: A key enabler of the 6G era. In Proceedings of the IEEE 2020 International Conference on Computing, Networking and Communications (ICNC), Big Island, HI, USA, 17–20 February 2020; pp. 383–388.

28. Kamruzzaman, M.M.; Yan, B.; Sarker, M.N.I.; Alruwaili, O.; Wu, M.; Alrashdi, I. Blockchain and Fog Computing in IoT-Driven Healthcare Services for Smart Cities. *J. Healthc. Eng.* **2022**, *2022*, 9957888. [CrossRef] [PubMed]

29. Sodhro, A.H.; Pirbhulal, S.; Luo, Z.; Muhammad, K.; Zahid, N.Z. Toward 6G architecture for energy-efficient communication in IoT-enabled smart automation systems. *IEEE Internet Things J.* **2020**, *8*, 5141–5148. [CrossRef]

30. López, O.L.; Alves, H.; Souza, R.D.; Montejo-Sánchez, S.; Fernández, E.M.G.; Latva-Aho, M. Massive wireless energy transfer: Enabling sustainable IoT toward 6G era. *IEEE Internet Things J.* **2021**, *8*, 8816–8835. [CrossRef]

31. Ghorbani, H.; Mohammadzadeh, M.S.; Ahmadzadegan, M.H. Modeling for malicious traffic detection in 6G next generation networks. In Proceedings of the IEEE 2020 International Conference on Technology and Entrepreneurship-Virtual (ICTE-V), Big Island, HI, USA, 20–21 April 2020; pp. 1–6.
32. Prashant, U.; Suniti, D.; Ruchi, R.; Sheetal, U. 6G Communication: Next Generation Technology for IoT Applications. In Proceedings of the International Conference on Advances in Computing & Future Communication Technologies (ICACFCT-2021), Meerut, India, 16–17 December 2021; pp. 1–4.
33. Kamruzzaman, M.M. 6G-enabled smart city networking model using lightweight security module. *Spine* **2013**, *38*, 1790–1796. [CrossRef]
34. Xiaohu, Y.O.U.; Hao, Y.I.N.; Hequan, W.U. On 6G and wide-area IoT. *Chin. J. Internet Things* **2020**, *4*, 3–11.
35. Ye, N.; Yu, J.; Wang, A.; Zhang, R. Help from space: Grant-free massive access for satellite-based IoT in the 6G era. *Digit. Commun. Netw.* **2021**, *8*, 215–224. [CrossRef]
36. Saad, W.; Bennis, M.; Chen, M. A vision of 6G wireless systems: Applications, trends, technologies, and open research problems. *IEEE Netw.* **2019**, *34*, 134–142. [CrossRef]
37. Tariq, F.; Khandaker, M.R.; Wong, K.K.; Imran, M.A.; Bennis, M.; Debbah, M. A speculative study on 6G. *IEEE Wirel. Commun.* **2020**, *27*, 118–125. [CrossRef]
38. Barakat, B.; Taha, A.; Samson, R.; Steponenaite, A.; Ansari, S.; Langdon, P.M.; Keates, S. 6G opportunities arising from internet of things use cases: A review paper. *Future Internet* **2021**, *13*, 159. [CrossRef]
39. Al-Turjman, F.; Lemayian, J.P. Intelligence, security, and vehicular sensor networks in internet of things (IoT)-enabled smart-cities: An overview. *Comput. Electr. Eng.* **2020**, *87*, 106776. [CrossRef]
40. Fang, X.; Feng, W.; Wei, T.; Chen, Y.; Ge, N.; Wang, C.X. 5G embraces satellites for 6G ubiquitous IoT: Basic models for integrated satellite terrestrial networks. *IEEE Internet Things J.* **2021**, *8*, 14399–14417. [CrossRef]
41. Viswanathan, H.; Mogensen, P.E. Communications in the 6G era. *IEEE Access* **2020**, *8*, 57063–57074. [CrossRef]
42. Chowdhury, M.Z.; Shahjalal, M.; Ahmed, S.; Jang, Y.M. 6G wireless communication systems: Applications, requirements, technologies, challenges, and research directions. *IEEE Open J. Commun. Soc.* **2020**, *1*, 957–975. [CrossRef]
43. Chimmanee, K.; Jantavongso, S. Practical mobile network planning and optimization for Thai smart cities: Towards a more inclusive globalization. *Res. Glob.* **2021**, *3*, 100062. [CrossRef]
44. Ibba, S.; Pinna, A.; Seu, M.; Pani, F.E. CitySense: Blockchain-oriented smart cities. In Proceedings of the XP2017 Scientific Workshops, Cologne, Germany, 22–26 May 2017; pp. 1–5.
45. Verma, S.; Kaur, S.; Khan, M.A.; Sehdev, P.S. Toward green communication in 6G-enabled massive internet of things. *IEEE Internet Things J.* **2020**, *8*, 5408–5415. [CrossRef]
46. Atitallah, S.B.; Driss, M.; Boulila, W.; Ghézala, H.B. Leveraging Deep Learning and IoT big data analytics to support the smart cities development: Review and future directions. *Comput. Sci. Rev.* **2020**, *38*, 100303. [CrossRef]
47. Kamruzzaman, M.M. 6G wireless communication assisted security management using cloud edge computing. *Expert Syst.* **2022**, e13061. [CrossRef]
48. Tomkos, I.; Klonidis, D.; Pikasis, E.; Theodoridis, S. Toward the 6G network era: Opportunities and challenges. *IT Prof.* **2020**, *22*, 34–38. [CrossRef]
49. Elmeadawy, S.; Shubair, R.M. 6G wireless communications: Future technologies and research challenges. In Proceedings of the IEEE 2019 International Conference on Electrical and Computing Technologies and Applications (ICECTA), Ras Al Khaimah, United Arab Emirates, 19–21 November 2019; pp. 1–5.
50. Lu, Y.; Zheng, X. 6G: A survey on technologies, scenarios, challenges, and the related issues. *J. Ind. Inf. Integr.* **2020**, *19*, 100158. [CrossRef]
51. Sekaran, R.; Patan, R.; Raveendran, A.; Al-Turjman, F.; Ramachandran, M.; Mostarda, L. Survival study on blockchain based 6G-enabled mobile edge computation for IoT automation. *IEEE Access* **2020**, *8*, 143453–143463. [CrossRef]
52. Kamruzzaman, M.M.; Alruwaili, O. Energy Efficient Sustainable Wireless Body Area Network Design Using Network Optimization with Smart Grid and Renewable Energy Systems. *Energy Rep.* **2022**, *8*, 3780–3788. [CrossRef]
53. Chen, N.; Okada, M. Toward 6G Internet of Things and the Convergence with RoF System. *IEEE Internet Things J.* **2020**, *8*, 8719–8733. [CrossRef]
54. Deebak, B.D.; Al-Turjman, F. Drone of IoT in 6G wireless communications: Technology, challenges, and future aspects. In *Unmanned Aerial Vehicles in Smart Cities*; Springer: Cham, Switzerland, 2020; pp. 153–165.
55. Desai, A.; Kulkarni, J.; Kamruzzaman, M.M.; Hubálovský, S.; Hsu, H.-T.; Ibrahim, A.A. Interconnected CPW Fed Flexible 4-Port MIMO Antenna for UWB, X, and Ku Band Applications. *IEEE Access* **2022**, *10*, 57641–57654. [CrossRef]
56. Singh, M.; Kříž, J.; Kamruzzaman, M.M.; Dhasarathan, V.; Sharma, A.; Mottaleb, S.A.E. Design of a High-Speed OFDM-SAC-OCDMA-Based FSO System Using EDW Codes for Supporting 5G Data Services and Smart City Applications. *Front. Phys.* **2022**, *10*, 2022. [CrossRef]
57. Borges, D.; Montezuma, P.; Dinis, R.; Beko, M. Massive mimo techniques for 5g and beyond—opportunities and challenges. *Electronics* **2021**, *10*, 1667. [CrossRef]
58. Shahraki, A.; Abbasi, M.; Piran, M.; Taherkordi, A. A comprehensive survey on 6G networks: Applications, core services, enabling technologies, and future challenges. *arXiv* **2021**, arXiv:2101.12475. [CrossRef]

59. Zhihan, L.; Qiao, L.; Kumar Singh, A.; Wang, Q. AI-empowered IoT security for smart cities. *ACM Trans. Internet Technol.* **2021**, *21*, 1–21.
60. Kamruzzaman, M.; Alruwaili, O. AI-based computer vision using deep learning in 6G wireless networks. *Comput. Electr. Eng.* **2022**, *102*, 108233. [CrossRef]
61. Mucchi, L.; Jayousi, S.; Caputo, S.; Paoletti, E.; Zoppi, P.; Geli, S.; Dioniso, P. How 6G technology can change the future wireless healthcare. In Proceedings of the 2020 2nd IEEE 6G Wireless Summit (6G SUMMIT), Levi, Finland, 17–20 March 2020; pp. 1–6.
62. Ho, T.M.; Tran, T.D.; Nguyen, T.T.; Kazmi, S.M.; Le, L.B.; Hong, C.S.; Hanzo, L. Next-generation wireless solutions for the smart factory, smart vehicles, the smart grid and smart cities. *arXiv* **2019**, arXiv:1907.10102. [CrossRef]
63. Zhao, Z.; Feng, C.; Yang, H.H.; Luo, X. Federated-learning-enabled intelligent fog radio access networks: Fundamental theory, key techniques, and future trends. *IEEE Wirel. Commun.* **2020**, *27*, 22–28. [CrossRef]
64. You, X.; Wang, C.X.; Huang, J.; Gao, X.; Zhang, Z.; Wang, M.; Huang, Y.; Zhang, C.; Jiang, Y.; Wang, J.; et al. Towards 6G wireless communication networks: Vision, enabling technologies, and new paradigm shifts. *Sci. China Inf. Sci.* **2021**, *64*, 1–74. [CrossRef]

Article

Design of Power Location Coefficient System for 6G Downlink Cooperative NOMA Network

Mohamed Hassan [1], Manwinder Singh [1,*], Khalid Hamid [2], Rashid Saeed [3], Maha Abdelhaq [4] and Raed Alsaqour [5]

1   Department of Wireless Communication, Lovely Professional University, Phagwara 144001, Punjab, India
2   Department of Communication Systems Engineering, University of Science & Technology, Khartoum P.O. Box 30, Sudan
3   Department of Computer Engineering, College of Computers and Information Technology, Taif University, P.O. Box 11099, Taif 21944, Saudi Arabia
4   Department of Information Technology, College of Computer and Information Sciences, Princess Nourah Bint Abdulrahman University, P.O. Box 84428, Riyadh 11671, Saudi Arabia
5   Department of Information Technology, College of Computing and Informatics, Saudi Electronic University, P.O. Box 93499, Riyadh 93499, Saudi Arabia
*   Correspondence: manwinder.2523@gmail.com

Citation: Hassan, M.; Singh, M.; Hamid, K.; Saeed, R.; Abdelhaq, M.; Alsaqour, R. Design of Power Location Coefficient System for 6G Downlink Cooperative NOMA Network. *Energies* **2022**, *15*, 6996. https://doi.org/10.3390/en15196996

Academic Editors: R. Maheswar, M. Kathirvelu and K. Mohanasundaram

Received: 27 July 2022
Accepted: 20 September 2022
Published: 23 September 2022

Publisher's Note: MDPI stays neutral with regard to jurisdictional claims in published maps and institutional affiliations.

Abstract: Cooperative non-orthogonal multiple access (NOMA) is a technology that addresses many challenges in future wireless generation networks by delivering a large amount of connectivity and huge system capacity. The aim of this paper is to design the varied distances and power location coefficients for far users. In addition, this paper aims to evaluate the outage probability (OP) performance against a signal-to-noise ratio (SNR) for a 6G downlink (DL) NOMA power domain (PD) and DL cooperative NOMA PD networks. We combine a DL cooperative NOMA with a 16×16, a 32×23, and a 64×64 multiple-input multiple-output (MIMO) and a 128×128, a 256×256, and a 512×512 massive MIMO in an innovative method to enhance OP performance rate and mitigate the power location coefficient's effect for remote users. The results were obtained from Rayleigh fading channels using the MATLAB simulation software program. According to the outcomes, increasing the power location coefficients for the far user from 0.6 to 0.8 reduces the OP rate because increasing the power location coefficient for the far user decreases the power location coefficient for the near user, which results in less interference between them. In terms of the OP performance rate, the DL cooperative NOMA outperforms the NOMA. According to the findings, the DL cooperative NOMA OP rate outperforms the DL NOMA by a rate of $10^{-0.5}$. Whereas the 16×16 MIMO enhances the OP for the far user by 78.0×10^{-4}, the 32×32 MIMO increases the OP for the far user by 19.0×10^{-4}, and the 64×64 MIMO decreases the OP rate for the far user by 5.0×10^{-5}. At a SNR of 10 dB, the 128×128 massive MIMO improves the OP for the far user by 1.0×10^{-5}. The 256×256 massive MIMO decreases the OP for the far user by 43.0×10^{-5}, and the 512×512 massive MIMO enhances the OP for the far user by 8.0×10^{-6}. The MIMO techniques improve the OP performance, while the massive MIMO technology enhances the OP performance dramatically.

Keywords: non-orthogonal multiple access (NOMA); outage probability (OP); power domain (PD); multiple-input multiple-output (MIMO); 6G network; massive multiple-input multiple-output (MIMO)

1. Introduction

The NOMA has been used to increase the spectral efficiency of mobile next-generation networks [1]. It is one of the most promising technologies for future wireless networking. The primary idea behind NOMA is to serve several users in the same frequency band in the NOMA power domain (PD) [2] but at different power levels, as opposed to the typical orthogonal multiple access (OMA) solutions, such as time-division multiple access (TDMA). The NOMA technology takes advantage of a new dimension in the power field [3].

The NOMA employs successive interference cancellation (SIC), in which one user decodes the other's message from a superposition and then codes the incoming signal before decoding their own. When performing SIC, the near user decodes the information provided by the far user. This is a process that cannot be avoided. Regardless, the close user must decode the data of the far user [4].

Since the near user now has access to the far user's data, he can assist the far user by relaying that data. The close user's retransmission of his data will provide him with diversity because the far user's channel with the transmitting base station (BS) is weak [5]. To present it in another way, he will obtain two copies of the same message. One is from the BS, while the other is a relay from a close user. As a result, the chances of a far user outage should be reduced [6]. The term for this notion is "cooperative communication" or "relaying". Since the close user has access to the data of the far user, the NOMA naturally supports cooperative communication. After all, you are supposed to decode it [7].

When two lines are connected, they convey the same message, which benefits cooperative communication. Even if one connection is down, the other is very certainly operational. Compared to the risk of one link breaking, the chance of both failing simultaneously is extremely unlikely [8].

Due to the increasing importance of fast data transmission and the worldwide expansion of services, significant advances have been achieved in this study area. An alternative method for measuring system effectiveness is spectral efficiency (SE) [9]. One of the most efficient ways to achieve high spectral efficiency is to combine NOMA with MIMO communication, which is a crucial component in designing cellular communication systems. The massive MIMO is a key 6G enabler [10]. By placing many antennas and exploiting the space field to multiplex varied users, the massive-MIMO technology can reduce system latency and deliver incredible communication benefits [11]. A greater spectrum and conductivity improvements are gained when a 6G cooperative NOMA technology is used with a massive MIMO [12].

The system's performance was examined in [13] by analyzing the near–far relay cooperative NOMA system in aiding perfect and imperfect channel state information, imperfect with imperfect SIC over Rayleigh fading channel, but the system is limited in single user situations. The authors of [14] explored the OP in Nakagami-m fading channels and investigated a half-duplex cooperative MIMO NOMA system with incomplete channel state information and SIC. However, the results revealed that no matter how far the user was from the BS, the study treated them all with equal value.

We observe that all previous work has a small number of users and only employs one relay to transmit signals to another user. It is also essential to mention that NOMA systems constantly need power to be allocated to all users because they perform overlapping power domain signals. Since the power assignment correlation coefficients are anticipated to not match, the choice of relay and power allocation is critical for constructing real cooperative NOMA systems.

Motivated by the aforementioned reason, we will investigate the performance analysis of the downlink (DL) NOMA power domain (PD) and the DL cooperative NOMA PD networks. For the sake of simplicity, we simply examine the case of two users, with no regard for interferences from other NOMA users. Our major contributions are summarized as follows:

a.   The OPs of a two-user NOMA and a cooperative NOMA system are expressed approximately in closed form via theoretical analyses with different distances and power location coefficients. Additionally, through simulation, we show that the derived OP expressions are more precise than those in [13].

b.   We analyzed the findings of OP and impact power location coefficients in the cooperative NOMA system using a 16×16, a 32×23, and a 64×64 MIMO and compared them to our previous results and improvement calculations.

c. We calculated the improved OP performance rate and mitigated the power location coefficient's impact on far users by a cooperative NOMA combined with a 128×128, a 256×256, and a 512×512 massive MIMO.

The remainder of this paper is structured as follows: Section 2 presents the related work. Section 3 discusses the proposed system model. The simulation, parameters, results, and discussion are presented in Section 4. Finally, Section 5 concludes the paper and presents further future work.

2. Related Work

The accurate performance characteristics should be known before implementing a 6G network design to fulfill the system objectives. Several studies offer robust supporting evidence to enhance the transmission circumstances. For instance, in [15], the authors describe a new harmonized dynamic direct and relay detection technique (DD-CDRT) to improve the transmission reliability that uses fully available lateral information to avoid user interference using numerical data to back up the theoretical study and show how the DD-CDRT approach works [16]. At the same time, the proposed broad framework for analyzing NOMA system performance utilizing a two-relaying selection method and spatially random relays has been conducted in [17] and achieved notable outcomes.

Also, the high SNR impact of inadequate user channel gain on dropout performance introduced a new collaborative NOMA protocol for users' DL networks in [18,19]. Hence, according to the remote user's input, the protocol allows the source to adaptively switch between the NOMA direct and the NOMA cooperative transmission modes. The author in [20] applied a remarkable effort via focusing on the cooperative relay selection system with the NOMA's effective resource usage method. Whereas [21] investigated the security of two relay selection approaches for collaborative NOMA systems, resulting in new closed-form equations for the fine and convergent secrecy interruption probability equations [22].

The study in [23] investigated the multiuser detection process for NOMA, which is largely affected by the power distribution of the received signals via the IDMA; the system needs an FEC rate to work properly. Another investigation occurred in [24] and looked at NOMA in which the base station delivers two signals to destinations, obtains OP formulae for two users (close and remote), and emphasizes the role of the close user as a relay. In the same line, the impact of relay considering the direct link has been discussed in [25], but it is better to give more attention to maximizing the received signal.

The NOMA cooperative with simultaneous wireless data and power transfer radio is evaluated in [26]. However, the BS required more respect, as well as added attention to route relaying when transmitting data to two users. Another significant survey in [27] was the performance of a DL NOMA network over Nakagami-m fading channels to assess the OP; the final result demonstrated maximal throughput under varying factors, and the model could be considered to contribute to the development of NOMA systems [28].

3. Cooperative NOMA System Model

In the first scenario, the BS in the DL NOMA PD network with two users, one close to the BS with a strong channel and the other far from the BS with a weak channel, where the distances $(d1, d2)$ and power location coefficients $\left(\alpha_n, \alpha_f \right)$ are variables, is illustrated in Figure 1. For the DL cooperative NOMA PD network with two users, one near the BS with a strong channel and the other far from the BS with a weak channel, Figure 2 shows the various distances $(d1, d2,$ and $d12)$ and power location coefficients $\left(\alpha_n, \alpha_f, \alpha_{nf} \right)$.

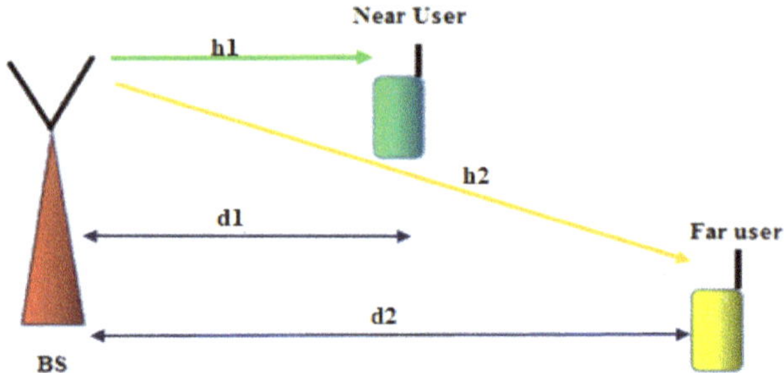

Figure 1. Downlink transmission for the NOMA network.

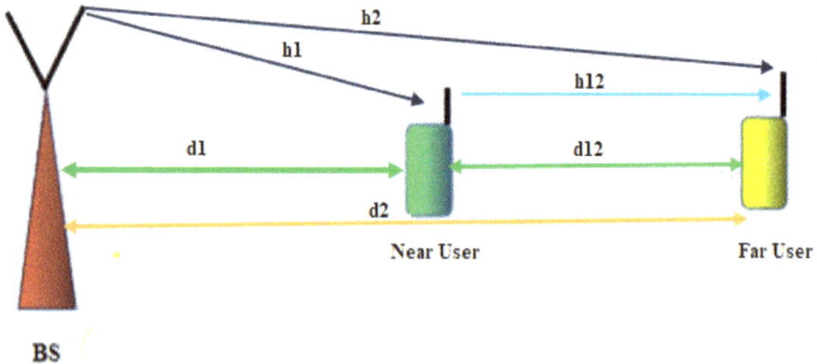

Figure 2. Downlink transmission for the cooperative NOMA network.

In the second case, the NOMA cooperative network is integrated with 16×16, 32×23, and 64×64 MIMO techniques. A similar distance and power location coefficients are used in the first scenario, as it is illustrated in Figure 3.

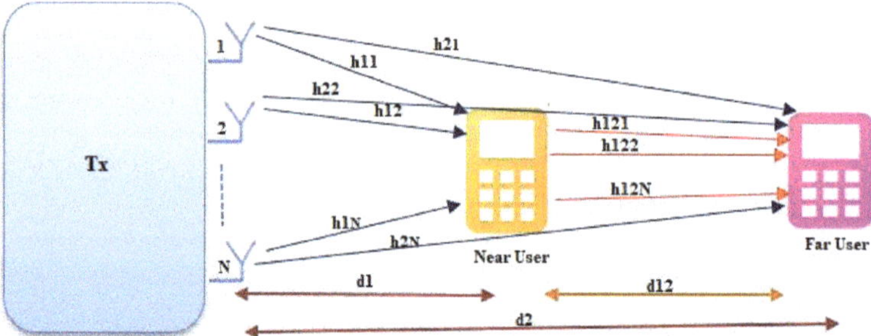

Figure 3. Downlink transmission for the cooperative NOMA network combined with a 16×16, a 32×23, and a 64×64 MIMO.

In the third scenario, the NOMA cooperative network is merged with massive MIMO techniques of 128×128, 256×256, and 512×512. As shown in Figure 4, the distance and power location coefficients employed in the first situation are the same.

In the DL cooperative NOMA, the transmission is divided into two slots [29]. The first slot is referred to as the direct transmission slot, while the second slot is referred to as the relay slot. These slots are used to calculate the total Rayleigh fading channel for each user.

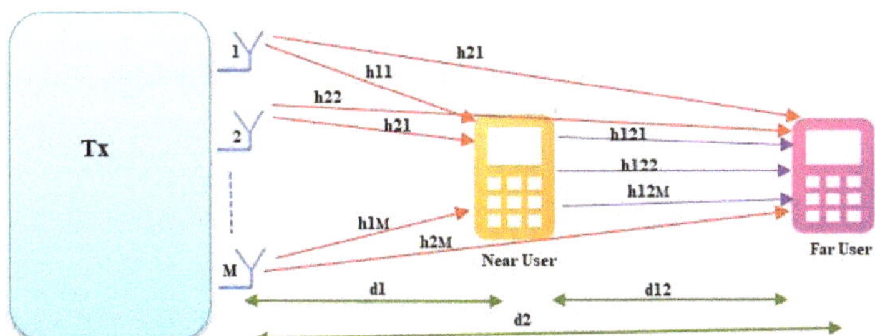

Figure 4. Downlink transmission for the cooperative NOMA network integrated with a 128×128, a 256×256, and a 512×512 MIMO.

The total Rayleigh fading channels for each user are given by [30] and are as follows:

$$h_{fN} = \sum_{f=1}^{N} h_{fN} \tag{1}$$

$$h_{nN} = \sum_{n=1}^{N} h_{nN} \tag{2}$$

where f denotes the far user, n represents the near user, and N indicates the number of antennas.

$N = 1$ is for the DL NOMA and for the DL cooperative NOMA. $N = 16$, 32, and 64 is for the MIMO DL 6G cooperative NOMA. $N = 128$, 256, and 512 is for the massive MIMO DL cooperative NOMA.

3.1. Direct Transmission Slot

The BS transmits data destined for the near user (x_n) and the far user $\left(x_f\right)$ in the direct transmission slot using the NOMA $\left(x_f\right)$. The near user uses SIC to decode the far user's data before decoding their own. The far user does only direct decoding. The possible data rates for the near and far users at the end of the direct transmission slot are given in [30] and are as follows:

$$R_n = \frac{1}{2} \log_2\left(1 + \alpha_n \rho |h_{nN}|^2\right) \tag{3}$$

$$R_{f,1} = \frac{1}{2} \log_2\left(1 + \frac{\alpha_f \rho \left|h_{fN}\right|^2}{\alpha_n \rho \left|h_{fN}\right|^2 + 1}\right) \tag{4}$$

where α_n is the power allocation coefficient for the near user, α_f is the power allocation coefficient for the far user, h_n is the channel between the BS and the near user, h_f is the channel between the BS and the far user. For $SNR = \rho / \sigma^2$, ρ is the transmit power and σ^2 is the noise variance. As usual, $\alpha_f > \alpha_n$, and $\alpha_n + \alpha_f = 1$. This is because there are time slots of equal duration; there is a factor of $1/2$ in front of the achievable rates and R_n, R_f are the achievable rates during the first time slot alone [31].

3.2. Relaying Slot

The relaying slot is the next half of the time slot. Since the near user decoded the data of the far user in the previous time slot, the near user already has it. The near user simply transmits this data to the far user during the relaying time slot [32]. The far user's achievable rate at the end of the relaying slot is as follows:

$$R_{f,2} = \frac{1}{2} \log_2 \left(1 + \alpha_n \rho \left| h_{n\,fN} \right|^2 \right) \tag{5}$$

The channel between the near and far users is denoted by h_{nfN}. $R_{f,2} > R_{f,1}$ because of the following two reasons: there is no interference from other transmissions and no fractional power allocation; the far user receives the absolute transmission power [33].

3.3. Diversity Combining

The far user now has two copies of the same information acquired over two distinct routes after the two-time intervals. The far user can now use a diversity-combining approach. For example, utilize selection combining to select the copy with the highest SNR. The far user's achievable rate after the selection combining would be as follows:

$$R_f = \frac{1}{2} \log_2 \left(1 + max \left(\frac{\alpha_f \rho \left| h_{fN} \right|^2}{\alpha_n \rho \left| h_{fN} \right|^2 + 1}, \; \rho \left| h_{n\,fN} \right|^2 \right) \right) \tag{6}$$

If cooperative relaying was not used, the feasible rate of the far user would be calculated as follows:

$$R_{f,noncoop} = \log_2 \left(1 + \frac{\alpha_f \rho \left| h_{fN} \right|^2}{\alpha_n \rho \left| h_{fN} \right|^2 + 1} \right) \tag{7}$$

4. Simulation Results and Discussions

After creating a channel gain and computing the outage probability for the far user DL (NOMA, cooperative NOMA, MIMO–NOMA cooperative, and massive MIMO–NOMA cooperative) versus the SNR [34], the system model and simulation parameters were applied in the MATLAB software. Table 1 displays the simulation settings.

Table 1. Simulation Parameters.

Parameters		Values
Distance		$d_2 = 2d_1$
SNR		0–25 dB
Slots		Direct Tx and Relaying slots
Channel		Rayleigh fading
Power allocation coefficients	α_f	0.9, 0.8, 0.7, and 0.6
	α_n	0.1, 0.2, 0.3, and 0.4
Path loss exponent		4
The No. of bits per symbol.		10^6
MIMO		$16 \times 16, 32 \times 32$ and 64×64
Massive MIMO		$128 \times 128, 256 \times 256$ and 512×512

Figure 5 shows the OP against the SNR for the two far DL cooperative NOMA PD users with distinct networks at 0.8 and 0.6 power location coefficients, with the findings demonstrating that the OP reduces as the SNR increases. As a result, the OP of the DL 6G cooperative NOMA for the user with a power location coefficient of 0.8 is better than the user with a power location coefficient of 0.6 because it achieves the lowest outage probability at a SNR of 44 dB. In contrast, the user with the lower power location coefficient is more susceptible to interference from the nearby user [35]. Figure 6 depicts the OP vs. the SNR for the DL NOMA PD at 0.8 and 0.6 power location coefficients. The results demonstrate that when the SNR improves, the OP decreases. For the DL NOMA with a power location coefficient of 0.8, the distant user's OP performance rate is identical to that of the 0.6 one until the 10 dB SNR is approached. The analysis results show that the level of performance achieved exceeds the level of the author Z. Ding in [36] by more than 30%.

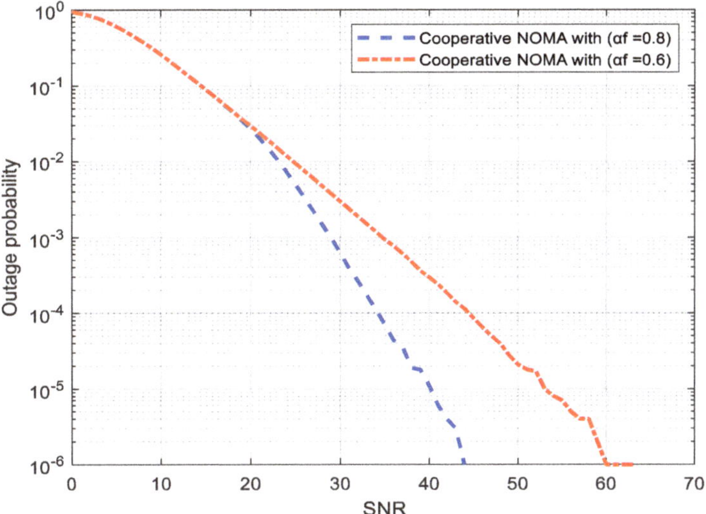

Figure 5. Outage probability against SNR for the two far users' DL cooperative NOMA PD.

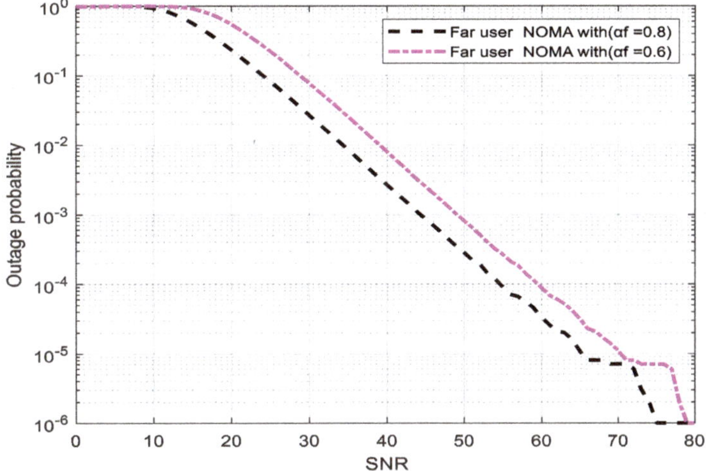

Figure 6. Outage probability against SNR for the two far users' DL NOMA PD.

Figure 7a,b illustrates the OP versus the SNR for the two far users' DL (6G cooperative NOMA and NOMA) PD at 0.9 and 0.7 power location coefficients, respectively. At a SNR of 40 dB, the OP performance at a power location coefficient of 0.7 for the DL cooperative NOMA users is 42.0×10^{-4} times better than the NOMA user. In contrast, the OP performance rate at a power location coefficient of 0.9 for the DL 6G cooperative NOMA user is 4.0×10^{-4} times better than the NOMA user. According to the observations, the DL cooperative NOMA outperforms the NOMA in terms of OP performance rate. Increasing the power location coefficient decreases the OP performance rate because increasing the power location coefficient of the far user decreases the power location coefficient of the near user, resulting in less interference between them. According to the data, the level of performance attained is 10% higher than the level attained in [13,14].

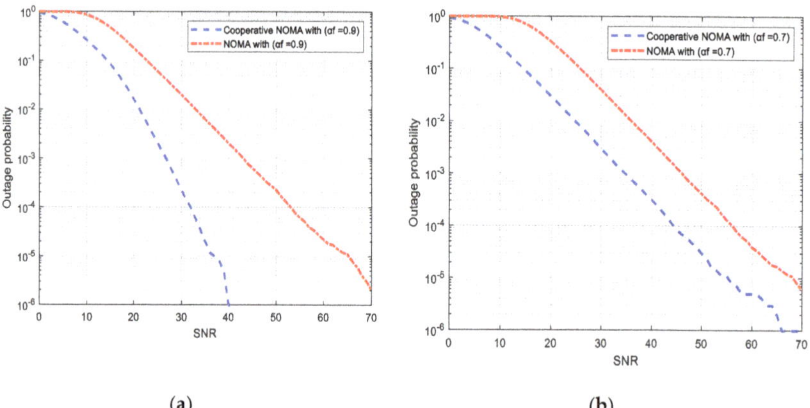

(a)   (b)

Figure 7. Outage probability against SNR for the two different users' DL NOMA and DL cooperative NOMA. (**a**) $\alpha_f = 0.9$ (**b**) $\alpha_f = 0.7$.

Figure 8 shows the OP vs. the SNR at 0.8 power location coefficients for the four varied far users of the cooperative NOMA, the 16×16 MIMO cooperative NOMA, the 32×32 MIMO cooperative NOMA, and the 64×64 MIMO cooperative NOMA [37]. At a SNR of 10 dB, the OP rate for the far user 64×64 MIMO cooperative NOMA is 5.0×10^{-4}. In contrast, the OP rate for the far user 32×32 MIMO cooperative NOMA is 19.0×10^{-4}. The OP rate for the user 32×32 MIMO cooperative NOMA is 78.0×10^{-4}, and the OP rate for the user 6G cooperative NOMA is 8644.0×10^{-4}. The rate of improvement in the OP by the best user using the cooperative 64×64 MIMO–NOMA versus the worst user using the cooperative NOMA is 8639.0×10^{-4}. The MIMO technique improves the overall OP performance; the obtained values are 4% higher than the values obtained in [38–40].

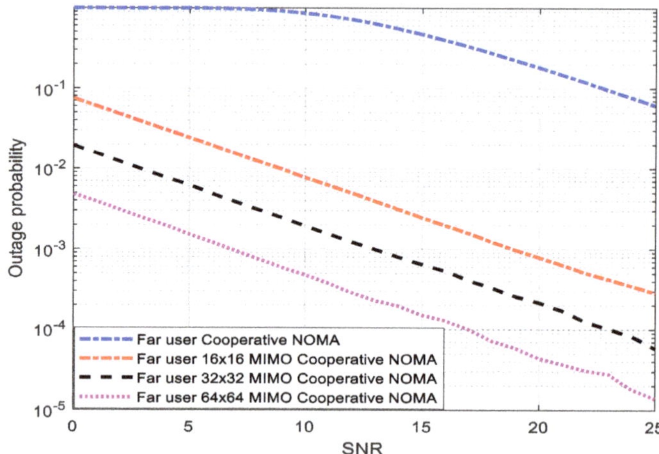

Figure 8. Outage probability against SNR for four different far users (cooperative NOMA and MIMO–cooperative NOMA).

Figure 9 shows the OP against the SNR at 0.8 power location coefficients for the four far users of the cooperative NOMA, the 128×128 massive MIMO cooperative NOMA, the

256 × 256 massive MIMO cooperative NOMA, and the 512 × 512 massive MIMO cooperative NOMA [41]. At a SNR of 14 dB, the OP for the far user 512 × 512 massive MIMO cooperative NOMA is 8.0×10^{-5}, while the OP for the user 256 × 256 massive MIMO cooperative NOMA is 43.0×10^{-5}. The OP for the user 128 × 128 massive MIMO cooperative NOMA is 1.0×10^{-5}, and the OP for the user cooperative NOMA is 8644.0×10^{-5}. Between the best user utilizing the cooperative 512 × 512 massive MIMO–NOMA and the worst user using the cooperative NOMA, the rate of improvement in the OP is 86432.0×10^{-5}. The massive MIMO technique significantly increases the OP's performance. The results show that the achieved performance is better than [42] by 15%.

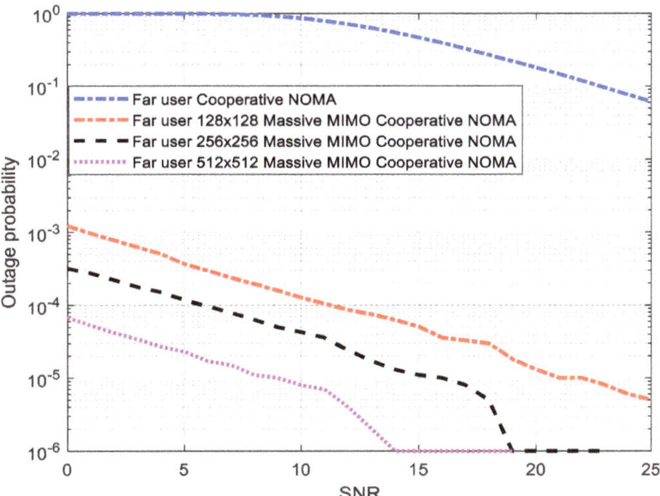

Figure 9. Outage probability against SNR for three different far users (cooperative NOMA, massive-MIMO–NOMA cooperative).

5. Conclusion and Future Work

The influence of distant users' power location coefficients on the DL NOMA PD and the DL cooperative NOMA PD concerning OP against SNR was investigated in this work. Furthermore, we designed and incorporated a MIMO, 16 × 16, 32 × 32, 64 × 64, and massive MIMO, 128 × 128, 256 × 256, and 512 × 512, into the DL cooperative NOMA PD system. The findings show that when the power location coefficients for the far user are increased, the OP performance rate goes down. This is because the power location coefficients for the near user are decreased; therefore, there is less interference between them.

The findings indicate that the OP rate of the DL 6G cooperative NOMA exceeds the DL NOMA by a range of 10–0.5. At a SNR of 10 dB, the 16 × 16 MIMO reduces the OP for the far user by 78.0×10^{-4}, the 32 × 32 MIMO decreases the OP for the far user by 16.0×10^{-4}, and the 64 × 64 MIMO improves the OP rate for the far user by 5.0×10^{-4}. In contrast, the 128 × 128 massive MIMO reduces the OP for the far user by 1.0×10^{-5}, the 256 × 256 massive MIMO enhances the OP for the far user by 43.0×10^{-5}, and the 512 × 512 massive MIMO improves the OP for the far user by 8.0×10^{-6}. The rate of improvement in the OP by the best user using the DL 512 × 512 massive MIMO cooperative NOMA versus the best user using the DL 64 × 64 cooperative NOMA is 492.0×10^{-6}.

The massive MIMO technology greatly improves the OP's performance, whereas the MIMO approach improves the OP's performance. In the future, researchers will explore combining the massive MIMO cooperative NOMA with a cognitive radio.

Author Contributions: Conceptualization, M.H.; methodology, M.H. and M.S.; software, M.H.; validation, M.H., M.S. and K.H.; formal analysis, M.H., M.S. and K.H.; investigation, R.S. and R.A.; resources, R.S., M.A. and R.A.; data curation, R.S., M.A. and R.A.; writing—original draft preparation, M.H.; writing—review and editing, M.A., R.S. and R.A.; visualization, M.H. and K.H.; supervision, M.S.; project administration, M.S.; funding acquisition, M.A. All authors have read and agreed to the published version of the manuscript.

Funding: This research was supported by the Princess Nourah bint Abdulrahman University Researchers Supporting Project Number (PNURSP2022R97), Princess Nourah bint Abdulrahman University, Riyadh, Saudi Arabia.

Data Availability Statement: Not applicable.

Acknowledgments: Princess Nourah bint Abdulrahman University Researchers Supporting Project Number (PNURSP2022R97), Princess Nourah bint Abdulrahman University, Riyadh, Saudi Arabia.

Conflicts of Interest: The authors declare no conflict of interest.

References

1. Li, A.; Lan, Y.; Chen, X.; Jiang, H. Non-orthogonal multiple access (NOMA) for future downlink radio access of 5G. *China Commun.* **2015**, *12*, 28–37. [CrossRef]
2. Mukhtar, A.M.; Saeed, R.A.; Mokhtar, R.A.; Ali, E.S.; Alhumyani, H. Performance Evaluation of Downlink Coordinated Multipoint Joint Transmission under Heavy IoT Traffic Load. *Wirel. Commun. Mob. Comput.* **2022**, *2022*, 6837780. [CrossRef]
3. Ahmed, M.A.; Mahmmod, K.F.; Azeez, M.M. On the performance of non-orthogonal multiple access (NOMA) using FPGA. *Int. J. Electr. Comput. Eng.* **2020**, *10*, 2151. [CrossRef]
4. Al-Adwany, M.A. Efficient power allocation method for non orthogonal multiple access 5G systems. *Int. J. Electr. Comput. Eng.* **2020**, *10*, 2139. [CrossRef]
5. Mokhtar, R.A.; Saeed, R.A.; Alhumyani, H. Cooperative Fusion Architecture-based Distributed Spectrum Sensing Under Rayleigh Fading Channel. *Wirel. Pers. Commun.* **2022**, *124*, 839–865. [CrossRef]
6. Budhiraja, I.; Kumar, N.; Tyagi, S.; Tanwar, S.; Han, Z.; Piran, M.J.; Suh, D.Y. A systematic review on NOMA variants for 5G and beyond. *IEEE Access* **2021**, *9*, 85573–85644. [CrossRef]
7. Do, D.-T.; Nguyen, T.-T.T. Impacts of relay and direct links at destinations in full-duplex non-orthogonal multiple access system. *Indones. J. Electr. Eng. Comput. Sci.* **2022**, *26*, 269–277. [CrossRef]
8. Hassan, M.B.; Ali, E.S.; Saeed, R.A. Ultra-Massive MIMO in THz Communications: Concepts, Challenges and Applications. In *Next Generation Wireless Terahertz Communication Networks*, 1st ed.; CRC Press: Boca Raton, FL, USA, 2021; pp. 267–297.
9. Oleiwi, H.; Saeed, N.; Al-Raweshidy, H. Cooperative SWIPT MIMO-NOMA for Reliable THz 6G Communications. *Netw. J.* **2022**, *2*, 257–269. [CrossRef]
10. Fang, F.; Zhang, H.; Cheng, J.; Roy, S.; Leung, V.C. Joint user scheduling and power allocation optimization for energy-efficient NOMA systems with imperfect CSI. *IEEE J. Sel. Areas Commun.* **2017**, *35*, 2874–2885. [CrossRef]
11. Satrya, G.B.; Shin, S.Y. Security enhancement to successive interference cancellation algorithm for non-orthogonal multiple access (NOMA). In Proceedings of the 2017 IEEE 28th Annual International Symposium on Personal Indoor, and Mobile Radio Communications (PIMRC), Montreal, QC, Canada, 8–13 October 2017; pp. 1–5.
12. Li, G.; Mishra, D.; Jiang, H. Cooperative NOMA with incremental relaying: Performance analysis and optimization. *IEEE Trans. Veh. Technol.* **2018**, *67*, 11291–11295. [CrossRef]
13. Singh, S.; Bansal, M. Performance analysis of non-orthogonal multiple access assisted cooperative relay system with channel estimation errors and imperfect successive interference cancellation. *Trans. Emerg. Telecommun. Technol.* **2022**, *32*, e4374. [CrossRef]
14. Srinivasarao, K.; Maruthu, S. Outage analysis of cooperative NOMA system with imperfect successive interference cancellation and channel state information over Rayleigh fading channel. *Int. J. Commun. Syst.* **2022**, in press. [CrossRef]
15. Jaber, Z.H.; Kadhim, D.J.; Al-Araji, A.S. Medium access control protocol design for wireless communications and networks review. *Int. J. Electr. Comput. Eng.* **2022**, *12*, 1711. [CrossRef]
16. Yang, M.; Chen, J.; Yang, L.; Lv, L.; He, B.; Liu, B. Design and performance analysis of cooperative NOMA with coordinated direct and relay transmission. *IEEE Access* **2019**, *7*, 73306–73323. [CrossRef]
17. Hassan, M.B.; Alsharif, S.; Alhumyani, H.; Ali, E.S.; Mokhtar, R.A.; Saeed, R.A. An enhanced cooperative communication scheme for physical uplink shared channel in NB-IoT. *Wirel. Pers. Commun.* **2021**, *120*, 2367–2386. [CrossRef]
18. Dahi, N.; Hamdi, N. Outage performance of generalized cooperative NOMA systems with SWIPT in nakagami-m fading. *J. Commun. Softw. Syst.* **2018**, *14*, 386–391. [CrossRef]
19. Kader, M.F.; Shahab, M.B.; Shin, S.Y. Exploiting non-orthogonal multiple access in cooperative relay sharing. *IEEE Commun. Lett.* **2017**, *21*, 1159–1162. [CrossRef]
20. Elfatih, N.M.; Hasan, M.K.; Kamal, Z.; Gupta, D.; Saeed, R.A.; Ali, E.S.; Hosain, M.S. Internet of vehicle's resource management in 5G networks using AI technologies: Current status and trends. *IET Commun.* **2022**, *16*, 400–420. [CrossRef]

21. Saeed, R.A.; Abbas, E.B. Performance evaluation of MIMO FSO communication with gamma-gamma turbulence channel using diversity techniques. In Proceedings of the 2018 International Conference on Computer, Control, Electrical, and Electronics Engineering (ICCCEEE), Khartoum, Sudan, 12–14 August 2018; pp. 1–5.

22. Han, S.; Chih-Lin, I.; Xu, Z.; Sun, Q. Energy efficiency and spectrum efficiency co-design: From NOMA to network NOMA. *IEEE COMSOC MMTC E-Lett.* **2014**, *9*, 21–24.

23. Lien, S.-Y.; Shieh, S.-L.; Huang, Y.; Su, B.; Hsu, Y.-L.; Wei, H.-Y. 5G new radio: Waveform, frame structure, multiple access, and initial access. *IEEE Commun. Mag.* **2017**, *55*, 64–71. [CrossRef]

24. Luo, S.; Teh, K.C. Adaptive transmission for cooperative NOMA system with buffer-aided relaying. *IEEE Commun. Lett.* **2017**, *21*, 937–940. [CrossRef]

25. Islam, S.R.; Avazov, N.; Dobre, O.A.; Kwak, K.-S. Power-domain non-orthogonal multiple access (NOMA) in 5G systems: Potentials and challenges. *IEEE Commun. Surv. Tutor.* **2016**, *19*, 721–742. [CrossRef]

26. Timotheou, S.; Krikidis, I. Fairness for non-orthogonal multiple access in 5G systems. *IEEE Signal Process. Lett.* **2015**, *22*, 1647–1651. [CrossRef]

27. Manglayev, T.; Kizilirmak, R.C.; Kho, Y.H.; Bazhayev, N.; Lebedev, I. NOMA with imperfect SIC implementation. In Proceedings of the IEEE EUROCON 2017—17th International Conference on Smart Technologies, Ohrid, Macedonia, 6–8 July 2017; pp. 22–25.

28. Manglayev, T.; Kizilirmak, R.C.; Kho, Y.H. Optimum power allocation for non-orthogonal multiple access (NOMA). In Proceedings of the 2016 IEEE 10th International Conference on Application of Information and Communication Technologies (AICT), Baku, Azerbaijan, 12–14 October 2016; pp. 1–4.

29. Nguyen, T.; Do, D.-T. Novel multiple access for cooperative networks with nakagami-m fading: System model and performance analysis. *Indones. J. Electr. Eng. Comput. Sci.* **2020**, *19*, 233–240.

30. Van, H.T.; Vo, T.A.; Le, D.H.; Phu, M.Q.; Nguyen, H.-S. Outage performance analysis of non-orthogonal multiple access systems with RF energy harvesting. *Int. J. Electr. Comput. Eng.* **2021**, *11*, 4135. [CrossRef]

31. Ding, Z.; Peng, M.; Poor, H.V. Cooperative non-orthogonal multiple access in 5G systems. *IEEE Commun. Lett.* **2015**, *19*, 1462–1465. [CrossRef]

32. Abdelrahman, Y.T.; Saeed, R.A.; El-Tahir, A. Multiple Physical Layer Pipes performance for DVB-T2. In Proceedings of the 2017 International Conference on Communication, Control, Computing and Electronics Engineering (ICCCCEE), Khartoum, Sudan, 16–18 January 2017; pp. 1–7.

33. Zhang, Y.; Wang, H.-M.; Yang, Q.; Ding, Z. Secrecy sum rate maximization in non-orthogonal multiple access. *IEEE Commun. Lett.* **2016**, *20*, 930–933. [CrossRef]

34. Zhang, Y.; Wang, H.-M.; Zheng, T.-X.; Yang, Q. Energy-efficient transmission design in non-orthogonal multiple access. *IEEE Trans. Veh. Technol.* **2016**, *66*, 2852–2857. [CrossRef]

35. Ding, Z.; Poor, H.V. Design of massive-MIMO-NOMA with limited feedback. *IEEE Signal Processing Lett.* **2016**, *23*, 629–633. [CrossRef]

36. Ding, Z.; Liu, Y.; Choi, J.; Sun, Q.; Elkashlan, M.; Chih-Lin, I.; Poor, H.V. Application of non-orthogonal multiple access in LTE and 5G networks. *IEEE Commun. Mag.* **2017**, *55*, 185–191. [CrossRef]

37. Liu, Y.; Ding, Z.; Elkashlan, M.; Poor, H.V. Cooperative non-orthogonal multiple access with simultaneous wireless information and power transfer. *IEEE J. Sel. Areas Commun.* **2016**, *34*, 938–953. [CrossRef]

38. Ding, Z.; Fan, P.; Poor, H.V. Random beamforming in millimeter-wave NOMA networks. *IEEE Access* **2017**, *5*, 7667–7681. [CrossRef]

39. Aldababsa, M.; Kucur, O. Outage and ergodic sum-rate performance of cooperative MIMO-NOMA with imperfect CSI and SIC. *Int. J. Commun. Syst.* **2020**, *33*, e4405. [CrossRef]

40. Sanjay, M.; Roy, D.; Kundu, S. Outage analysis for NOMA-based energy harvesting relay network with imperfect CSI and transmit antenna selection. *IET Commun.* **2020**, *14*, 2240–2249. [CrossRef]

41. Wang, Z.; Peng, Z. Secrecy performance analysis of relay selection in cooperative NOMA systems. *IEEE Access* **2019**, *7*, 86274–86287. [CrossRef]

42. Shashank, S.; Muralikrishnan, S.; Sheetal, K. Outage Probability of Uplink Cell-Free Massive MIMO Network with Imperfect CSI Using Dimension-Reduction Method. *arXiv* **2021**. [CrossRef]

Article

Modeling of NOMA-MIMO-Based Power Domain for 5G Network under Selective Rayleigh Fading Channels

Mohamed Hassan [1], **Manwinder Singh** [1,*], **Khalid Hamid** [2], **Rashid Saeed** [3], **Maha Abdelhaq** [4] and **Raed Alsaqour** [5]

[1] Department of Wireless Communication, Lovely Professional University, Phagwara 144001, Punjab, India
[2] Department Communication Systems Engineering, University of Science & Technology, Omdurman P.O. Box 30, Sudan
[3] Department of Computer Engineering, College of Computers and Information Technology, Taif University, P.O. Box 11099, Taif 21944, Saudi Arabia
[4] Department of Information Technology, College of Computer and Information Sciences, Princess Nourah bint Abdulrahman University, P.O. Box 84428, Riyadh 11671, Saudi Arabia
[5] Department of Information Technology, College of Computing and Informatics, Saudi Electronic University, Riyadh 93499, Saudi Arabia
* Correspondence: manwinder.25231@lpu.co.in

Abstract: The integration of multiple-input multiple-output (MIMO) and non-orthogonal multiple access (NOMA) technologies is a hybrid technology that overcomes a myriad of problems in the 5G cellular system and beyond, including massive connectivity, low latency, and high dependability. The goal of this paper is to improve and reassess the bit error rate (BER), spectrum efficiency (SE) of the downlink (DL), average capacity rate, and outage probability (OP) of the uplink (UL) in a 5G network using MIMO. The proposed model utilizes QPSK modulation, four users with different power location coefficients, SNR, transmit power, and two contrasting bandwidths 80 and 200 MHz under selective frequency Rayleigh fading channels. The proposed model's performance is evaluated using the MATLAB software program. The DL results found that the BER and SE against transmitted power showed the MIMO-NOMA enhanced the BER performance for the best user U4 from $10^{-1.7}$ to $10^{-5.2}$ at 80 MHz bandwidth (BW), and from $10^{-1.5}$ to 10^{-5} at 200 MHz for transmitting power of 40 dBm. In contrast, the SE performance for the best user U4 is enhanced from 24×10^{-3} to 25×10^{-3} bits/second/Hz at 80 MHz BW and from 19.8×10^{-3} to 20×10^{-3} bps/Hz at 200 MHz BW. Although the outcomes for the UL were obtained in terms of average capacity rate and OP versus SNR at 80, and 200 MHz BW, the MIMO-NOMA result showed that the average capacity rate for the best user U4 performance improves by 12 bps/Hz for 1 dB SNR and the OP is reduced by 15×10^{-3} for 80 MHz BW and by 12×10^{-3} for 200 MHz BW at an SNR of 0.17 dB. As the BW increased the BER, the average capacity rate increased while the SE and OP decreased. For both DL/UL NOMA with and without MIMO, closed-form expressions for BER, SE, average capacity rate, and OP were obtained. All users' performance, even those whose connections were affected by interference or Rayleigh fading channels significantly improved, when MIMO-NOMA was implemented.

Keywords: non-orthogonal multiple access (NOMA); bit error rate (BER); spectrum efficiency (SE); outage probability (OP); multiple-input multiple-output (MIMO)

Citation: Hassan, M.; Singh, M.; Hamid, K.; Saeed, R.; Abdelhaq, M.; Alsaqour, R. Modeling of NOMA-MIMO-Based Power Domain for 5G Network under Selective Rayleigh Fading Channels. *Energies* **2022**, *15*, 5668. https://doi.org/10.3390/en15155668

Academic Editors: R. Maheswar, M. Kathirvelu and K. Mohanasundaram

Received: 12 July 2022
Accepted: 1 August 2022
Published: 4 August 2022

Publisher's Note: MDPI stays neutral with regard to jurisdictional claims in published maps and institutional affiliations.

1. Introduction

Due to the compelling multimedia applications and the popularity of smart mobile devices, wireless communication has developed at a breakneck pace over the last decade [1,2]. Non-orthogonal multiple access (NOMA) has been offered as a possible strategy for achieving mass accessibility with keeping spectral efficiency [3,4]. NOMA achieves multiple access by modifying the power level of overlay user signals at the transmitter and receiving the signal using successive interference cancellation (SIC) in receivers, with noise-limiting user rates for better channels and bandwidth-limiting user rates for bad channels [5–8].

SIC is implemented at the power user level because NOMA is sequential interference cancellation, allowing the powerful user to detect and discard messages from users with weaker channel conditions. Data from users with better channel conditions is considered interference by the weaker user [9]. In the NOMA downlink (DL) system, multiple users share the same time, coding, and frequency resources due to the multiplexing of the power field. The base station (BS) sends an overlay signal to each user, which includes a signal for all users [10]. Because users no longer have to wait for an orthogonal resource block to become available, NOMA can accomplish huge connections while drastically lowering transmission delay [11]. As a result, NOMA with SIC is a promising multiple-access approach in the next-generation communication system [12]. NOMA is a wireless technology that can meet the demands of today's wireless environment [13]. The analysis of different access technologies is still in its early phases [14]. The main leading research group is working on determining the spectrum's efficiency because all features are constantly generated [15,16].

In uplink (UL) NOMA, a group of users simultaneously sends their signals to their associated BS [17,18]. Intra-cluster interference, as a result, impacts a user's received signal, which is determined statically by other users' channel data [19]. To reduce interference, the BS could use SIC to decode signals. Separate message signals with an adequate power variance must arrive at the receiver BS to correctly utilize the SIC technique. This is commonly handled in the DL by employing different scales at the transmitter. Furthermore, such values are superfluous because the UL channel gains already offer enough separation between the received signals. The standard UL transmits power control, which is supposed to balance the received signal levels of users, is not recommended for UL NOMA transmissions because it may reduce channel distinctness [20–24].

Using several antennas in both the transmitter and receiver can considerably increase the capacity of a radio communication channel. That is, many separate channels can be managed in the same bandwidth using multiple-input multiple-output (MIMO) technology with these antennas, but only if the propagation environment is sufficiently rich [25,26]. Although the use of MIMO techniques adds another dimension to enhance efficiency, research into combining MIMO and NOMA has recently a lot of interest [26,27].

The additive white Gaussian noise (AWGN) and Rayleigh fading channels were investigated for the accurate expression of the BER rate generated in a closed form for BPSK modulation in the perfect and deficient SIC states for the DL NOMA network. However, it did not include parameters that influence the BER, such as distance and power location coefficients in [28]. Three power assignment strategies are proposed to improve NOMA SE by optimizing the given power to each NOMA in [29,30]. The impact of interference on users of the energy allocation process, nevertheless, has not been investigated or examined. Investigate the attainable rates of UP NOMA with a synchronized transmission in [31] and estimate both the upper and lower bounds of a synchronized NOMA system's possible rates, illustrating that the measured lower bound is approximately the rate achievable by conventional synchronized NOMA systems. Nonetheless, each user is provided with a relatively limited number of transmitted symbols, which may lead to inaccurate findings. The ergodic rate achievable in [23,24] is monitored and checked to ensure optimal performance of the encoder and pre-detection systems that simplify decoding in the dual receive channels used for DL and UL transmissions. Nevertheless, the number of users is insufficient to confirm the optimal situation, and there has been no research into the method's impact on OP or BER, for example.

In this paper, we examine the impact of varying the bandwidth and the number of antennas in a 5G network on the bit error rate (BER) and spectrum efficiency (SE) of the downlink, and the average capacity rate and outage probability (OP) of the uplink in a network subject to Rayleigh fading. During the analytical process, the integral expressions of the BER, SE, capacity rate, and OP were generated. In addition, modeling is used to verify all the potential configurations of the system. The following are some of the research's most significant contributions:

- Two different bandwidths (BWs) for the NOMA system over a Rayleigh fading channel are proposed and investigated;
- Improvements in the system have been examined when NOMA and MIMO are used together to handle four users.

The article addresses the NOMA technique which is considered one of the main bullets of 5G technology. Our main innovative idea in the article is remodeling NOMA-MIMO for the power domain for a higher data rate, capacity, and throughput. This has been performed by proposing a new power domain scheme for NOMA-MIMO.

The remainder of the paper is structured as follows: Section 2 contains previous and related works. Section 3 discusses the proposed system mathematical model. Simulation and simulation parameters are presented in Section 4. The results and discussions are presented in Section 5 and, finally, Section 6 concludes the paper and presents further future work.

2. Related Work

Multiple beams forming with a single carrier are utilized in NOMA systems to accommodate numerous users as a two-stage beam forming solution for modular beam forming vectors, according to the author in [32]. A reduced total transmission packet shaping issue is built to identify both users' packet-shaping vectors and power.

The author in [33], established successful precoding and detecting procedures to produce a considerable difference between users' effective channel gains, allowing NOMA's potential to be achieved even when the users' initial channel conditions are comparable. The author investigated the performance of MIMO-NOMA when numerous users are aggregated into a group, finding that MIMO-NOMA outperforms MIMO-OMA in terms of total channel capacity and total practical capacity [34].

Using statistical channel state information at the transmitter, the ergodic capacity maximization problem for selective Rayleigh fading MIMO-NOMA systems was investigated in [30]. The MIMO-NOMA schemes greatly outperform the conventional OMA scheme, according to numerical results.

Following a review of the concept of integrating NOMA downlink with MIMO, an experiment was carried out to assess the performance of NOMA downlink combined with MIMO under realistic settings in [35]. In UL, the user connection. The author investigated NOMA in [36], considering numerous specified power allocation techniques. It has been demonstrated that NOMA with the suggested user pair technique outperforms NOMA with the previously described signal realignment.

The author looked in [37] at many NOMA DL and UL user power field-based communication systems with different fading bindings for all users who can follow one of the many conceivable distributions. At high SNRs, analytical expressions of the OP for the NOMA DL and UL systems were derived.

An unmanned aerial vehicle-assisted NOMA network with UL and DL transmissions was explored, and analytical expressions of OP as the major measure were derived in [38]. The author explores a novel UL/DL NOMA system with a uniform relay and set decode order that involves the use of statistical channel state information, resulting in enhanced fairness and applicability [39]. The author evaluated the potential of UL and DL resource utilization, adaptive control, and power control for wireless communications systems under the assumptions of in-band full-duplex BSs, NOMA operation, and queue stability limitations.

A method is proposed for solving by finding a correlation similarity in [40]. The efficacy of different NOMA plots over the tapping delay line channel in both normal and fast UE speed and correlation-level modeling was explored by the author in [40–42]. With UE's normal and fast speed, NOMA methods work differently.

3. System Model

3.1. DL Scenario

Consider a wireless network with four DL NOMA users and a 64×64 MIMO system (as depicted in Figure 1). $U1$, $U2$, $U3$, and $U4$ are the four users with different bandwidths of 80 and

200 MHz [43]. Let $d1$, $d2$, $d3$, and $d4$ represent their various BS distances, $d1 > d2 > d3 > d4$ indicating the preferred order. Depending on the distance, $U1$ is the weak/far user while $U4$ is the strong/near user from BS. Let h_{T1}, h_{T2}, h_{T3}, and h_{T4} identify which selective Rayleigh fading coefficients they correspond to $|h_{T1}|^2 < |h_{T2}|^2 < |h_{T3}|^2 < |h_{T4}|^2$.

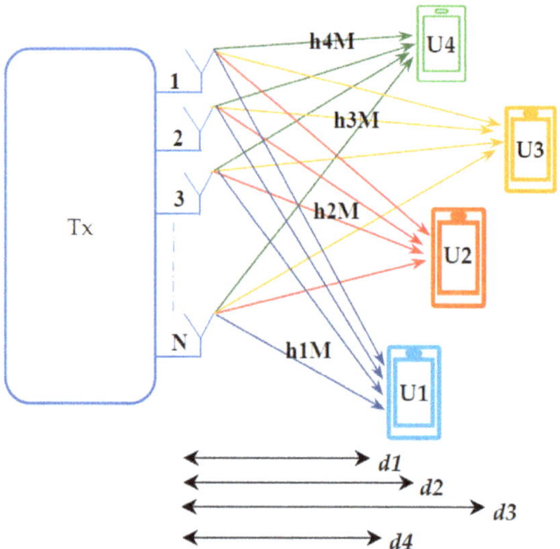

Figure 1. Illustrates the wireless network consisting of 4 users 64×64 MIMO-DL-NOMA (power domain).

The total Rayleigh fading channel for each user is given by [44]:

$$h_{Ti} = \sum_{i=1}^{M} h_{Ti} \tag{1}$$

where, $i = 1, 2, 3, 4$ is the number of users, and $M = 64$ is the number of channels. α_1, α_2, α_3 and α_4 show their respective power coefficients. According to the NOMA (power domain) principles, the lower user must have more power and the better user should have less power [45,46]. As a result, the power coefficients must be modified as $\alpha_1 > \alpha_2 > \alpha_3 > \alpha_4$. Let x_1, x_2, x_3 and x_4 be the QPSK-formed messages to send to BS $U1$, $U2$, $U3$, *and* $U4$. The BS's encoded overlay signal is then given by as in [47].

$$x = \sqrt{p}(\sqrt{\alpha_1}x_1 + \sqrt{\alpha_2}x_2 + \sqrt{\alpha_3}x_3 + \sqrt{\alpha_4}x_4) \tag{2}$$

$U1$ decodes y_1 directly when it has maximum power, interfering with the 2nd, 3rd, and 4th U signals. As a result, the first possible U rate is

$$R_1 = \log_2\left(1 + \frac{\alpha_1 P|h_{T1}|^2}{\alpha_2 P|h_{T1}|^2 + \alpha_3 P|h_{T1}|^2 + \alpha_4 P|h_{T1}|^2 + \sigma^2}\right) \tag{3}$$

The obtained rate is for $U2$ after SIC eliminated $U1$ data.

$$R_2 = \log_2\left(1 + \frac{\alpha_2 P|h_{T2}|^2}{\alpha_3 P|h_{T2}|^2 + \alpha_4 P|h_{T2}|^2 + \sigma^2}\right) \tag{4}$$

The achieved rate is for $U3$ after SIC deleted $U1$ and $U2$ data.

$$R_3 = \log_2\left(1 + \frac{\alpha_3 P|h_{T3}|^2}{\alpha_4 P|h_{T3}|^2 + \sigma^2}\right) \tag{5}$$

The acquired rate is for $U4$ after SIC deleted $U1$ data, $U2$ data, and $U3$ data.

$$R_4 = \log_2\left(1 + \frac{\alpha_4 P |h_{T4}|^2}{\sigma^2}\right) \tag{6}$$

To calculate the spectrum efficiency.

$$SE = \frac{Th}{BW} \tag{7}$$

where the SE is spectrum efficiency, Th is the throughput and BW is the bandwidth.

3.2. UL Scenario

The power domain multiplexing for uplink NOMA is almost entirely different. In downlink NOMA, the BS employed superposition coding to offer power domain multiplexing; however, the user's transmit power is limited only by their battery capacity in the uplink. That is, both users can transmit at full strength. Changes in the users' channel gains cause variation in the power domain at the receiver side of BS.

Let x_1, x_2, x_3 and x_4 represent the messages that will be sent by four UL NOMA users $U1$, $U2$, $U3$, and $U4$, accordingly. Suppose that both users' signals have the same strength and consider the 64×64 MIMO system and BW equal to 80 MHz on a wireless network (as depicted in Figure 2). Let $d1 > d2 > d3 > d4$ denote the various BS distances, with $d1 > d2 > d3 > d4$ being the preferred order. $U1$ from BS is the weak/far user, while $U4$ is the strong/near user, depending on the distance. Let h_{T1}, h_{T2}, h_{T3}, and h_{T4} determine which selective Rayleigh fading coefficients they relate to $|h_{T1}|^2 < |h_{T2}|^2 < |h_{T3}|^2 < |h_{T4}|^2$.

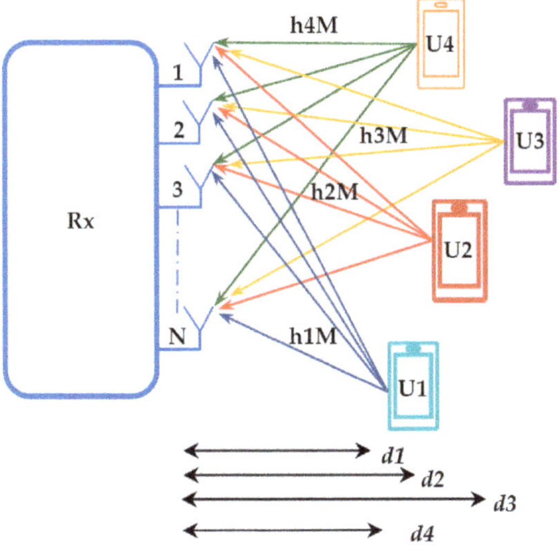

Figure 2. Illustrates the wireless network consisting of 4 users 64×64 MIMO-UL-NOMA (power domain).

The total Rayleigh fading channel for each user is given by:

$$h_{jT} = \sum_{j=1}^{N} h_{jT} \tag{8}$$

where $j = 1, 2, 3, 4$ is the number of users and $N = 64$ is the number of channels.

The signal was received at the BS.

$$y = \sqrt{P_{x1}} h_{1T} + \sqrt{P_{x2}} h_{2T} + \sqrt{P_{x3}} h_{3T} + \sqrt{P_{x4}} h_{4T} + w \tag{9}$$

where w is the noise power.

3.2.1. Capacity Rates Achievable of Four Users UL NOMA

The signal from the close user is decoded first, with the signal from the distant users being treated as interference. Therefore, the rate at which the BS can decode the data of a nearby user is, according to [48,49].

$$R_{U4} = \log_2\left(1 + \frac{P|h_{4T}|^2}{P|h_{1T}|^2 + P|h_{2T}|^2 + P|h_{3T}|^2 + \sigma^2}\right) \tag{10}$$

After the SIC has been calculated, the maximum rate $U3$ can be obtained.

$$R_{U3} = \log_2\left(1 + \frac{P|h_{3T}|^2}{P|h_{1T}|^2 + P|h_{2T}|^2 + \sigma^2}\right) \tag{11}$$

After the SIC has been calculated, the maximum rate $U2$ that can be accomplished

$$R_{U2} = \log_2\left(1 + \frac{P|h_{2T}|^2}{P|h_{1T}|^2 + \sigma^2}\right) \tag{12}$$

After the SIC has been calculated, the maximum rate of U1 can be achieved.

$$R_{U1} = \log_2\left(1 + \frac{P|h_{1T}|^2}{\sigma^2}\right) \tag{13}$$

3.2.2. OP of Four Users UL NOMA

Consider that the four users have different target rates.

$$r_1 = 1, r_2 = 2, r_3 = 3, r_4 = 4$$

The capacity $U4$ is calculated as follows:

$$C_4 = \sum_{i=1}^{N} \log_2\left(1 + \frac{P|h_4|}{P|h_1| + P|h_2| + P|h_3| + N_4}\right) \tag{14}$$

$U3's$ capacity is calculated as follows:

$$C_3 = \sum_{i=1}^{N} \log_2\left(1 + \frac{P|h_3|}{P|h_1| + P|h_2| + N_3}\right) \tag{15}$$

$U2's$ capacity is calculated as follows:

$$C_2 = \sum_{i=1}^{N} \log_2\left(1 + \frac{P|h_2|}{P|h_1| + N_2}\right) \tag{16}$$

$U1's$ capacity is calculated as follows:

$$C_1 = \sum_{i=1}^{N} \log_2\left(1 + \frac{P|h_1|}{N_1}\right) \tag{17}$$

For $U1$, the OP condition is:

$$P_r\ (C_1(k)) < r_1)\ ||\ P_r\ (C_2(k)\ < r_2)\ ||\ P_r\ (C_3(k)\ \langle\ r_3||\ P_r\ (C_4(k)\ <\ r_4)) < r$$

The OP of $U1$:

$$P_r(U1) = \left(\sum_{i=1}^{N} P_r(C_1(k)) < r_1 \right) \| P_r(C_2(k) < r_2) \| P_r(C_3(k) \langle r_3 \| P_r(C_4(k) < r_4)) / N \tag{18}$$

For $U2$, the OP condition is:

$$P_r \ (C_2(k) \ < r_2) \ \| \ P_r \ (C_3(k) \ \langle \ r_3 \| \ P_r \ (C_4(k) \ < \ r_4)) < r$$

The OP of $U2$:

$$Pr(U2) = \left(\sum_{i=1}^{N} P_r \ (C_2(k) \ < r_2) \ \right\| \ P_r \ (C_3(k) \ \langle \ r_3 \| \ P_r \ (C_4(k) \ < \ r_4)) / N \tag{19}$$

For $U3$, the OP condition is:

$$P_r \ (C_3(k) \ \langle \ r_3 \| \ P_r \ (C_4(k) \ < \ r_4)) < r$$

The OP of $U3$:

$$Pr(U3) = \left(\sum_{i=1}^{N} \ P_r \ (C_3(k) \ \langle \ r_3 \| \ P_r \ (C_4(k) \ < \ r_4)) / N \tag{20}$$

For $U4$, the OP condition is:

$$P_r \ (C_4(k) \ < \ r_4)) < r_4$$

The OP of $U4$:

$$Pr(U4) = \left(\sum_{i=1}^{N} \ P_r \ (C_4(k) \ < \ r_4)) / N \tag{21}$$

where N is the number of transferred samples.

4. Simulation Parameters

The system model and simulator parameters for the DL and UL NOMA power domains in 5G networks with and without MIMO were implemented using the MATLAB software program. Tables 1 and 2 show the simulation parameters that are properly considered in the simulation model.

Table 1. Simulator parameters for the DL scenario.

No.	Parameters	Values		
1.	Number of users		4 users	
2.	Transmit power		0 to 40 dBm	
3.	Bandwidth	BW 1		80 MHz
		BW 2		200 MHz
4.	Distances	User 1		800 m
		User 2		600 m
		User 3		300 m
		User 4		100 m
5.	Power coefficients	User 1		0.75
		User 2		0.188
		User 3		0.047
		User 4		0.011
6.	Path loss exponent		4	
7.	MIMO		64 × 64	
8.	Modulation		QPSK	

Table 2. Simulator parameters for the UL scenario.

No.	Parameters	Values	
1.	Number of users	4 users	
2.	Transmit power	-30 to 30 dBm	
3.	Bandwidth	BW1	80 MHz
		BW2	200 MHz
4.	Distances	User1	800 m
		User 2	600 m
		User 3	300 m
		User 4	100 m
6.	Path loss exponent	4	
7.	MIMO	64×64	

5. Results and Discussions

5.1. The Outcomes of the DL Scenario

The DL NOMA system results showed that using 64×64 MIMO improved the SE and BER performance. The near–far user's problem is also resolved, where the performance of all users becomes close to each other's for different power location coefficients, transmitted power, and distance parameters when compared without MIMO DL NOMA performance. Figure 3 depicts the DL NOMA BER performance versus transmitted power at 80 MHz BW. The findings indicate that the BER performance decreases as transmitted power increases. As an outcome, the $U4$ BER performance is best for all users, because $U4$ is the nearest one. At a transmitter power of 25 dBm, the BER rate for $U1$, $U2$, $U3$, and $U4$ is found to be 20%, 28%, 22%, and 8%, respectively. Figure 4 shows the DL NOMA BER performance against transmitted power at 200 MHz BW; the findings show that the BER performance decreases as transmitted power increases. As a result, the U4 BER performance is best when compared with all users because $U4$ is the nearest one. At a transmit power of 25 dBm, the BER rates for $U1$, $U2$, $U3$, and $U4$ are found to be 27%, 36%, 31%, and 13%, respectively. The 64×64 MIMO DL NOMA enhances the performance of BER for the best user U4 from $10^{-1.7}$ to $10^{-5.2}$ at 80 MHz then, from $10^{-1.5}$ to 10^{-5} at 200 MHz BW at a transmitter power of 40 dBm in Figure 4. In contrast, the SE performance for the best user $U4$ is improved by 8×10^{-3} bps/Hz for 80 MHz BW and by 10^{-2} bps/Hz for 200 MHz BW at a transmitter power of 40 dBm. The UL NOMA systems results obtained using 64×64 MIMO enhanced the average capacity rate performance by 12 bps/Hz, reduced the OP by 15×10^{-3} for 80 MHz BW at SNR of 1 dB.

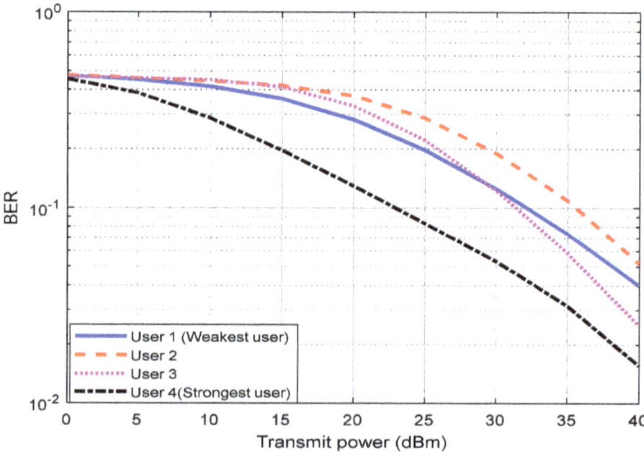

Figure 3. BER vs. transmit power for four users with varying distances and power coefficients, for DL NOMA at BW (80 MHz).

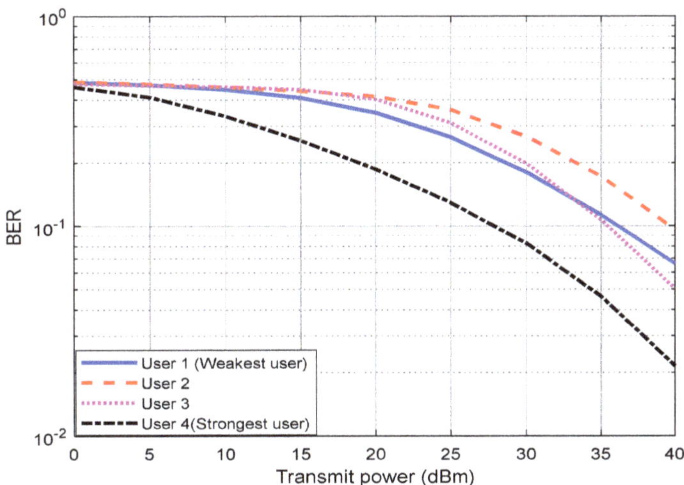

Figure 4. BER against transmitting power for four users with varying distances and power coefficients, for DL NOMA at BW (200 MHz).

At 80 MHz BW and 64×64 MIMO, Figure 5 shows the DL NOMA BER performance versus transmitted power. When transmitted power is 20 dBm, the BER rate for U1, U2, U3, and U4 is found to be 19×10^{-4}, 18×10^{-4}, 8×10^{-4}, and 5×10^{-4}, respectively. Figure 6 shows the DL NOMA BER performance against transmitted power at 200 MHz BW and 64×64 MIMO, at a transmitted power of 25 dBm, the BER rates for U1, U2, U3, and U4 are found to be 46×10^{-4}, 43×10^{-4}, 19×10^{-4}, and 7×10^{-4}, respectively. The MIMO system reduces the BER performance.

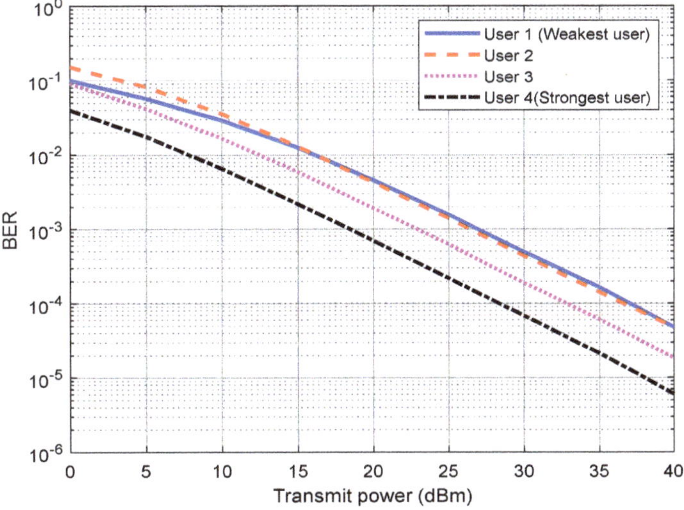

Figure 5. BER vs. transmitting power for four users with varying distances and power coefficients, for DL NOMA at BW (80 MHz with 64×64 MIMO for DL NOMA.

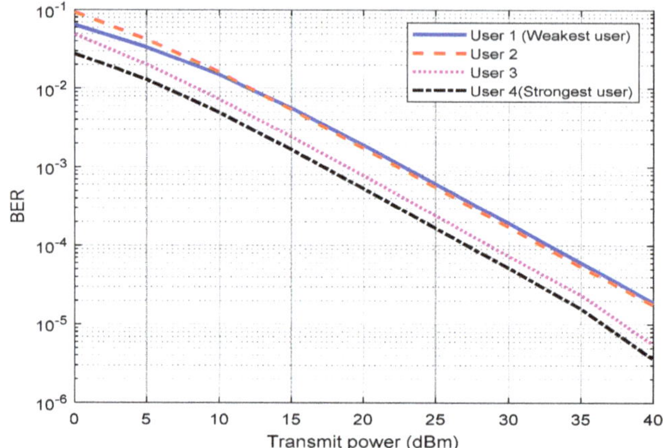

Figure 6. BER against transmitting power for four users with varying distances and power coefficients for DL NOMA at BW 200 MHz with 64×64 MIMO for DL NOMA.

Figure 7 shows the performance of the DL NOMA SE vs. transmitted power at 80 MHz BW, with the outcomes demonstrating that SE performance improves as transmitted power increases. As an outcome, the U4 BER performance is best for all users, because U4 is the nearest one. There is a clear separation of SE performance for all users from one another until the transmitted power reaches 5 dBm. Figure 8 depicts the DL NOMA SE performance versus transmitted power at 200 MHz BW, with the results indicating that increasing transmitted power improves SE performance. The U4 SE performance is best when compared with all users because U4 is the nearest one. The outcomes are superior to those of the best U2 users in [38], with an improvement rate of $10^{-2.3}$ in the BER.

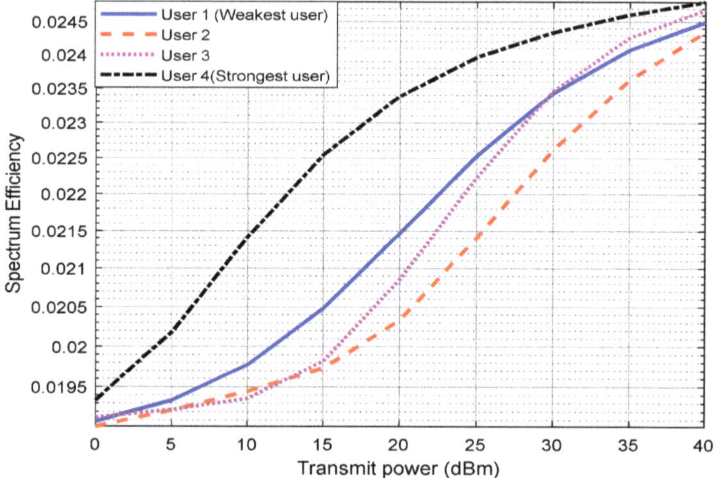

Figure 7. SE vs. transmit power for four users with varying distances, with power coefficients and BW 80 MHz for DL NOMA.

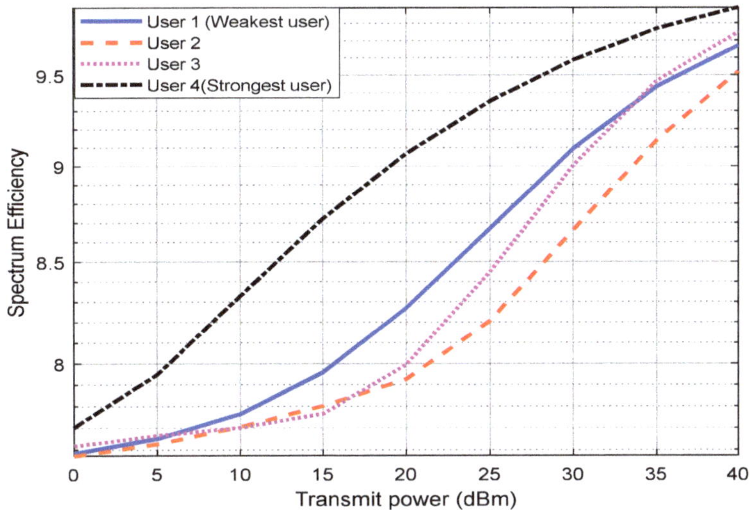

Figure 8. SE against transmitting power for four users with varying distances, with power coefficients and BW 200 MHz for DL NOMA.

The performance of the DL NOMA SE in terms of transmitted power at 80 MHz BW and 64×64 MIMO is shown in Figure 9. At the transmitting power of 5 dBm, the SE for all users is relatively close. Figure 10 depicts the DL NOMA SE performance versus transmit power at 200 MHz BW and 64×64 MIMO. At a transmitter power of 10 dBm, the SE for all users is relatively close. The MIMO improved SE performance.

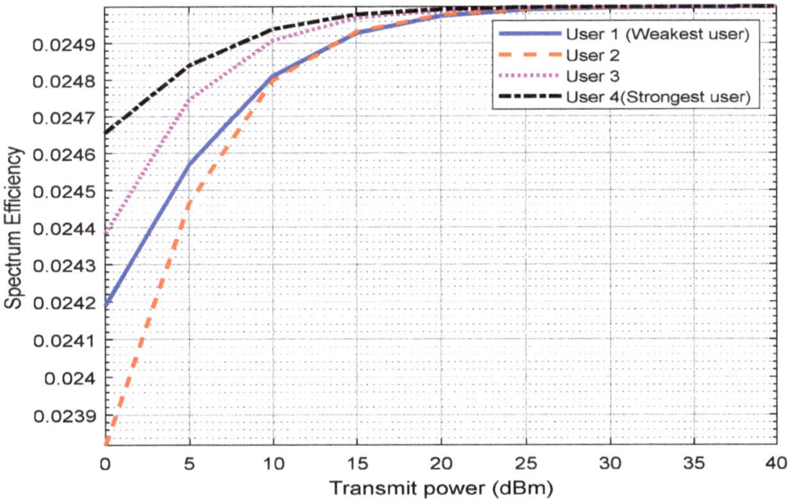

Figure 9. SE vs. transmitting power for four users with varying distances, with power coefficients, with the 64×64 MIMO and BW 80 MHz for DL NOMA.

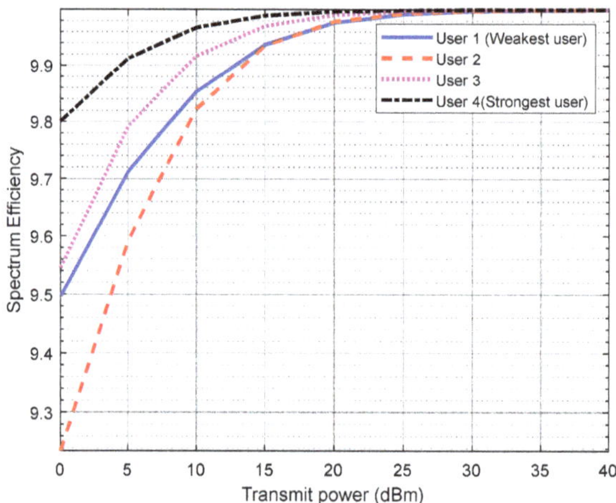

Figure 10. SE against transmitting power for four users with varying distances, with power coefficients with the 64×64 MIMO and BW 200 MHz for DL NOMA.

5.2. The Outcomes of the UL Scenario

The UL NOMA average capacity rate vs. SNR at 80 MHz BW is depicted in Figure 11. The result shows that the average capacity rate for U4 is best for all users because U4 is the closest. At SNR of 1 dB, the average capacity rate for U1, U2, U3, and U4 is found to be 1.6873, 2.8718, 6.4960, and 12.7814, respectively. Figure 12 shows the UL average capacity rate against SNR at 200 MHz BW. At SNR of 1 dB, the average capacity rate for U1, U2, U3, and U4 is found to be 2.6015, 3.9841, 7.7910, and 14.1068, respectively. The results reveal that when an SNR increases, the average capacity rate performance rises as well. The 64×64 MIMO improved the performance of the capacity average rate by 12 bps/Hz and reduced the OP by 15×10^{-3} for 80 MHz BW at SNR of 1 dB; it enhanced the performance capacity average rate by 12 bps/Hz, and decreased the OP by 12×10^{-3} for 200 MHz BW at 0.17 dB SNR for the user U4. In general, an increase in BW increases the capacity average rate and BER while decreasing OP and SE. MIMO significantly enhances the throughput of all users.

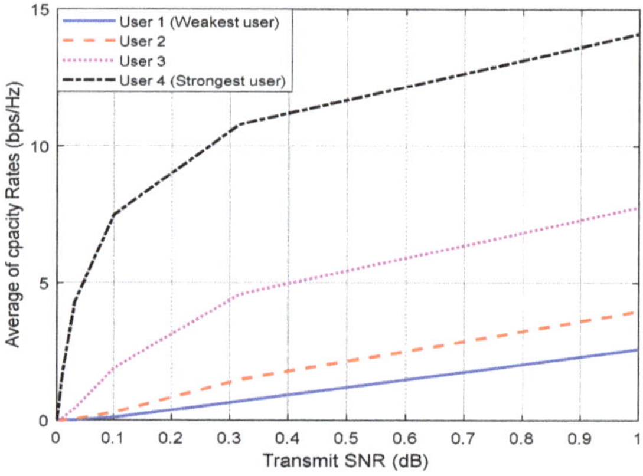

Figure 11. Average capacity rate vs. SNR for four users with varying distances and BW 80 MHz for UL NOMA.

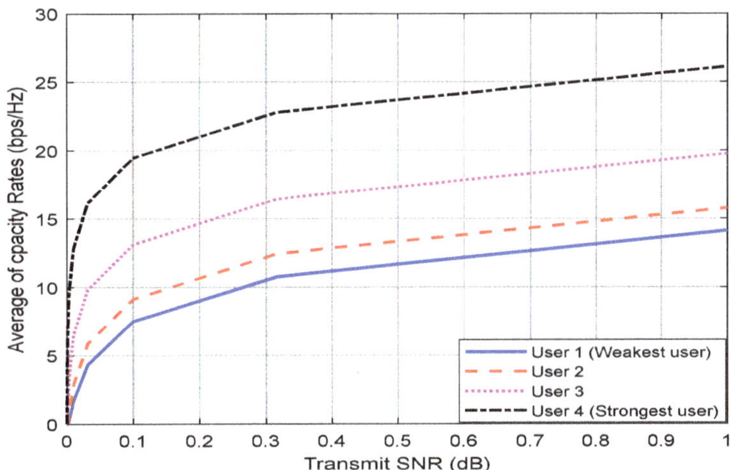

Figure 12. Average capacity rate against SNR for four users with varying distances and BW 200 MHz for UL NOMA.

The average capacity rate performance for UL NOMA versus SNR at 80 MHz BW and 64 × 64 MIMO is obtained in Figure 13. The outcomes were achieved for four users 12.7881, 14.4423, 18.4489, and 24.7815, respectively. Figure 14 shows the average capacity rate performance versus SNR for UL NOMA at 200 MHz BW and 64 × 64 MIMO. The results show that the average capacity rate performance improves as the SNR increases. The average capacity rate performance for U4 is the best according to the data obtained for four users 14.1110, 15.7659, 19.7693, and 26.1040 at the SNR of 1 dB. Users' performance has improved considerably.

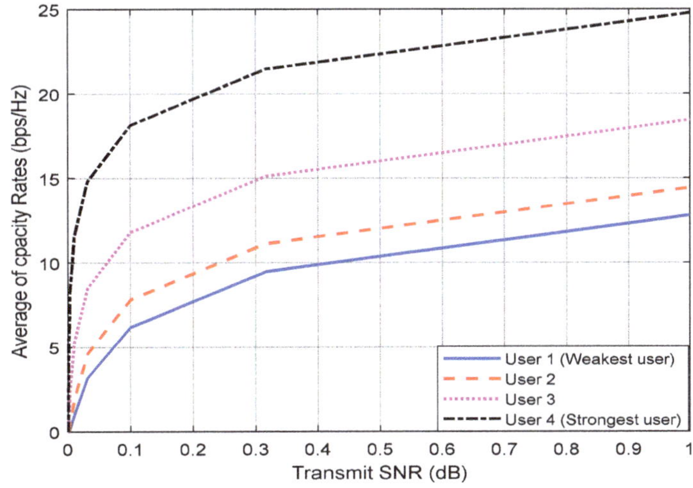

Figure 13. Average capacity rate vs. SNR for four users with varying distances, with 64 × 64 MIMO and BW 80 MHz for UL NOMA.

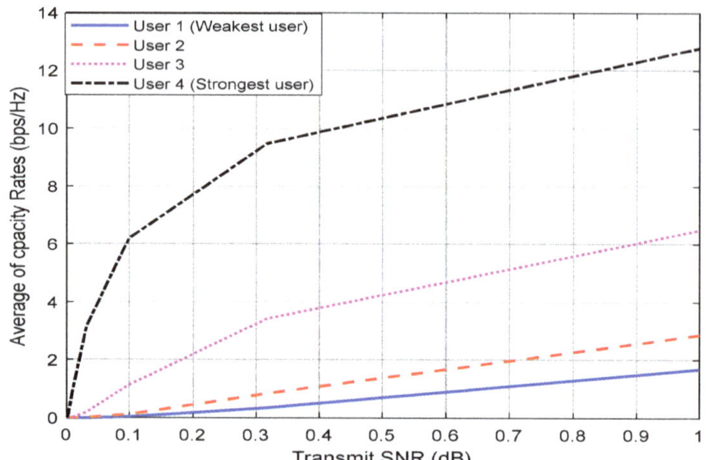

Figure 14. Average capacity rate against SNR for four users with varying distances, with 64×64 MIMO and BW 200 MHz for UL NOMA.

The BW and average capacity rate have a positive relationship, with an increase in BW leading to an increase in average capacity rate. The average capacity rate increases dramatically when the system is enhanced using the MIMO scheme.

The UL NOMA of OP vs. SNR correlation is shown in Figure 15 at 80 MHz BW. When SNR is 0.17 dB, the results for U1, U2, U3, and U4 are 99.9×10^{-2}, 98.9×10^{-2}, 44.3×10^{-2}, and 15×10^{-3}, respectively. Figure 16 depicts the UL NOMA of OP versus the SNR at 200 MHz BW. At an SNR of 0.17 dB, the results for U1, U2, U3, and U4 are 0.9989, 0.9715, 0.3709, and 0.0120, respectively. The findings show that as the SNR improves, the OP performance decreases. Results achieved have an improvement in average capacity rate and are superior to those of the best U2 users in [35].

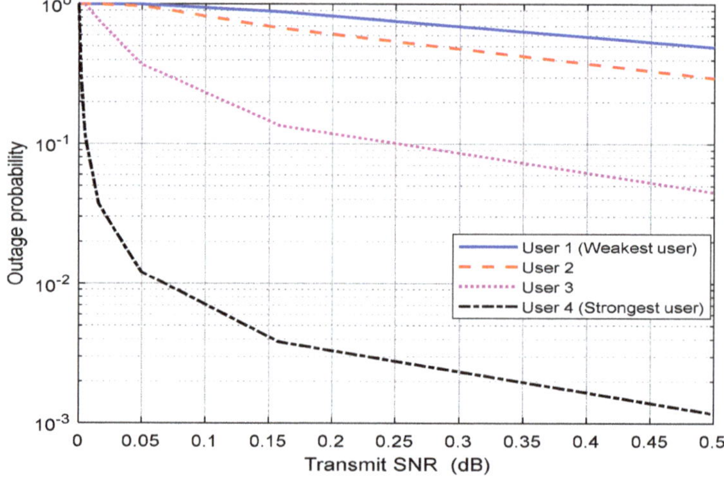

Figure 15. OP vs. SNR for four users with varying distances and BW 80 MHz for UL NOMA.

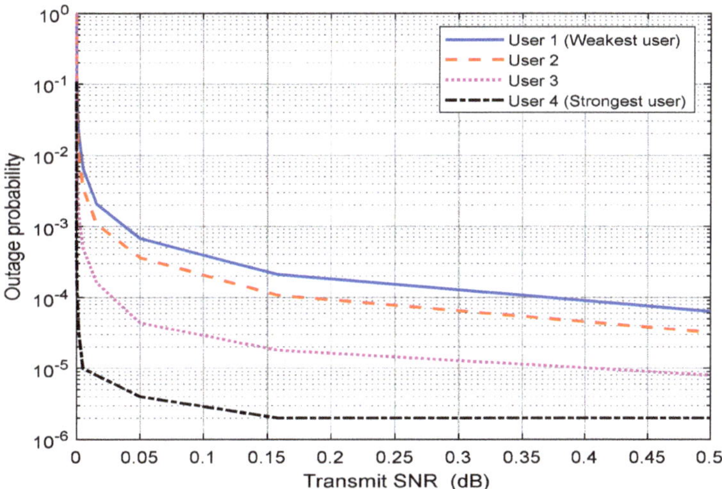

Figure 16. OP against SNR for four users with varying distances and BW 200 MHz for UL NOMA.

At 80 MHz BW with 64×64 MIMO, Figure 17 depicts the UL NOMA of OP vs. SNR. At an SNR of 0.17 dB, the outcomes for U1, U2, U3, and U4 are 0.0053, 0.0027, 0.0003, and 0.0000. At 200 MHz BW and 64×64 MIMO, Figure 18 depicts the UL NOMA of OP vs. the SNR. At an SNR of 0.17 dB, the results for U1, U2, U3, and U4 are 21×10^{-4}, 10^{-4}, 10^{-4}, and 10^{-5}. The data show that when the SNR increases, the performance of the OP decreases [39–41].

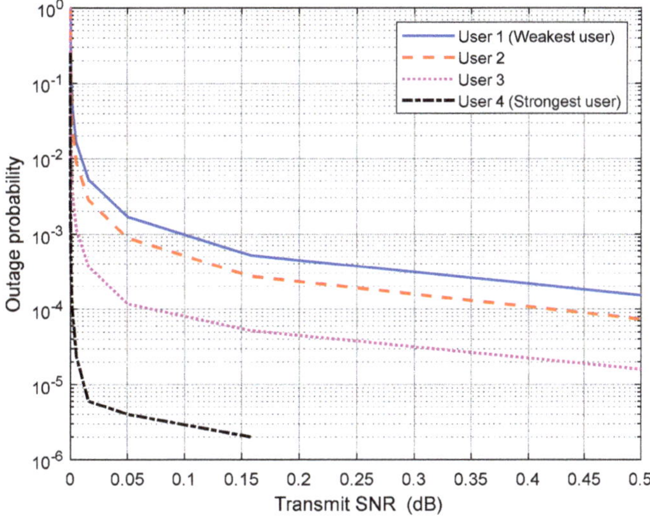

Figure 17. OP vs. SNR for four users with varying distances, with 64×64 MIMO and BW 80 MHz for UL NOMA.

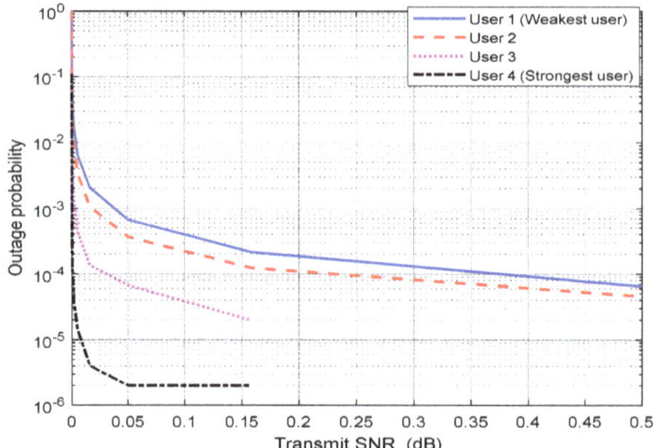

Figure 18. OP against SNR for four users with varying distances, with 64×64 MIMO and BW 200 MHz for UL NOMA.

The BW and OP have an inverse connection, with an increase in BW resulting in a drop in OP. The OP drops dramatically when the system is optimized utilizing the MIMO technique. With an improvement rate of $10^{-1.9}$ in OP, the results are superior to those of the top U2 users in [38].

6. Conclusions and Future Work

This paper demonstrated the performance of DL and UL NOMA PD in a 5G network with and without 64×64 MIMO technologies. The BER and SE performance of DL NOMA was investigated and analyzed for various distances, power location coefficients, transmitted power, and BW, whereas the average capacity rate and OP performance of UL NOMA were examined for various distances, SNR, and BW. The DL NOMA system results showed that using 64×64 MIMO enhanced the performance of BER, and SE, and solved the near–far user's problem, where the performance of all users becomes close to each other's for different transmitted power, distance, and power location coefficients parameters when compared without MIMO DL NOMA performance. The results demonstrated that the 64×64 MIMO DL NOMA enhances the BER performance for the best user U4 from $10^{-1.7}$ to $10^{-5.2}$ at 80 MHz BW, and from $10^{-1.5}$ to 10^{-5} at 200 MHz BW at a transmitter power of 40 dBm. In contrast, the SE performance for the best user U4 is improved by 0.8% bps/Hz for 80 MHz BW and by 1.01% bps/Hz for 200 MHz BW at a transmitter power of 40 dBm. The UL NOMA systems results obtained using 64×64 MIMO enhanced the average capacity rate performance by 12 bps/Hz, reduced the OP by 0.0150 for 80 MHz BW at SNR of 1 dB, improved the average capacity rate performance by 12 bps/Hz, and decreased the OP by 0.0120 for 200 MHz BW at SNR of 0.17 dB for the best user U4. In general, an increase in BW increases BER and average capacity rate while decreasing SE and OP. MIMO significantly improves the performance of all users. In the future, it will be looked into how MIMO cooperative NOMA and cognitive radio work together.

Author Contributions: Conceptualization, M.H.; methodology, M.H. and M.S.; software, M.H.; validation, M.H., M.S. and K.H.; formal analysis, M.H., M.S. and K.H.; investigation, R.S. and R.A.; resources, R.S., M.A. and R.A.; data curation, R.S., M.A. and R.A.; writing—original draft preparation, M.H.; writing—review and editing, M.A., R.S. and R.A.; visualization, M.H. and K.H.; supervision, M.S.; project administration, M.S.; funding acquisition, M.A. All authors have read and agreed to the published version of the manuscript.

Funding: This research was supported by Princess Nourah bint Abdulrahman University Researchers Supporting Project Number (PNURSP2022R97), Princess Nourah bint Abdulrahman University, Riyadh, Saudi Arabia.

Institutional Review Board Statement: Not applicable.

Informed Consent Statement: Not applicable.

Data Availability Statement: Not applicable.

Acknowledgments: Princess Nourah bint Abdulrahman University Researchers Supporting Project Number (PNURSP2022R97), Princess Nourah bint Abdulrahman University, Riyadh, Saudi Arabia.

Conflicts of Interest: The authors declare no conflict of interest.

References

1. Dai, L.; Wang, B.; Yuan, Y.; Han, S.; Chih-Lin, I.; Wang, Z. Non-orthogonal multiple access for 5G: Solutions, challenges, opportunities, and future research trends. *IEEE Commun. Mag.* **2015**, *53*, 74–81. [CrossRef]
2. Timotheou, S.; Krikidis, I. Fairness for non-orthogonal multiple access in 5G systems. *IEEE Signal Processing Lett.* **2015**, *22*, 1647–1651. [CrossRef]
3. Chen, Z.; Ding, Z.; Dai, X.; Zhang, R. An optimization perspective of the superiority of NOMA compared to conventional OMA. *IEEE Trans. Signal Processing* **2017**, *65*, 5191–5202. [CrossRef]
4. Feng, D. Performance comparison on NOMA schemes in high speed scenario. In Proceedings of the 2019 IEEE 2nd International Conference on Electronics Technology (ICET), Chengdu, China, 10–13 May 2019; pp. 112–116.
5. Bello, M.; Chorti, A.; Fijalkow, I.; Yu, W.; Musavian, L. Asymptotic performance analysis of NOMA uplink networks under statistical QoS delay constraints. *IEEE Open J. Commun. Soc.* **2020**, *1*, 1691–1706. [CrossRef]
6. Maatouk, A.; Assaad, M.; Ephremides, A. Minimizing the age of information: NOMA or OMA? In Proceedings of the IEEE INFOCOM 2019-IEEE Conference on Computer Communications Workshops (INFOCOM WKSHPS), Paris, France, 29 April–2 May 2019; pp. 102–108.
7. Wei, Z.; Yang, L.; Ng, D.W.K.; Yuan, J.; Hanzo, L. On the performance gain of NOMA over OMA in uplink communication systems. *IEEE Trans. Commun.* **2019**, *68*, 536–568. [CrossRef]
8. Ding, Z.; Zhao, Z.; Peng, M.; Poor, H.V. On the spectral efficiency and security enhancements of NOMA assisted multicast-unicast streaming. *IEEE Trans. Commun.* **2017**, *65*, 3151–3163. [CrossRef]
9. Hassan, M.; Singh, M.; Hamid, K. Survey on NOMA and Spectrum Sharing Techniques in 5G. In Proceedings of the 2021 IEEE International Conference on Smart Information Systems and Technologies (SIST), Nur-Sultan, Kazakhstan, 28–30 April 2021; pp. 1–4.
10. Makki, B.; Chitti, K.; Behravan, A.; Alouini, M.-S. A survey of NOMA: Current status and open research challenges. *IEEE Open J. Commun. Soc.* **2020**, *1*, 179–189. [CrossRef]
11. Shahab, M.B.; Johnson, S.J.; Shirvanimoghaddam, M.; Chafii, M.; Basar, E.; Dohler, M. Index modulation aided uplink NOMA for massive machine type communications. *IEEE Wirel. Commun. Lett.* **2020**, *9*, 2159–2162. [CrossRef]
12. Cejudo, E.C.; Zhu, H.; Alluhaibi, O. On the power allocation and constellation selection in downlink NOMA. In Proceedings of the 2017 IEEE 86th Vehicular Technology Conference (VTC-Fall), Toronto, ON, Canada, 24–27 September 2017; pp. 1–5.
13. Lei, L.; Yuan, D.; Ho, C.K.; Sun, S. Power and channel allocation for non-orthogonal multiple access in 5G systems: Tractability and computation. *IEEE Trans. Wirel. Commun.* **2016**, *15*, 8580–8594. [CrossRef]
14. Chandrasekhar, R.; Navya, R.; Kumari, P.K.; Kausal, K.; Bharathi, V.; Singh, P. Performance evaluation of MIMO-NOMA for the next generation wireless communications. In Proceedings of the 2021 3rd International Conference on Signal Processing and Communication (ICPSC), Coimbatore, India, 13–14 May 2021; pp. 631–636.
15. Saetan, W.; Thipchaksurat, S. Application of deep learning to energy-efficient power allocation scheme for 5G SC-NOMA system with imperfect SIC. In Proceedings of the 2019 16th International Conference on Electrical Engineering/Electronics, Computer, Telecommunications and Information Technology (ECTI-CON), Pattaya, Thailand, 10–13 July 2019; pp. 661–664.
16. Tweed, D.; Le-Ngoc, T. Dynamic resource allocation for uplink MIMO NOMA VWN with imperfect SIC. In Proceedings of the 2018 IEEE International Conference on Communications (ICC), Kansas City, MO, USA, 20–24 May 2018; pp. 1–6.
17. Krishnamoorthy, A.; Huang, M.; Schober, R. Precoder design and power allocation for downlink MIMO-NOMA via simultaneous triangularization. In Proceedings of the 2021 IEEE Wireless Communications and Networking Conference (WCNC), Nanjing, China, 29 March–1 April 2021; pp. 1–6.
18. Hua, Y.; Wang, N.; Zhao, K. Simultaneous unknown input and state estimation for the linear system with a rank-deficient distribution matrix. *Math. Probl. Eng.* **2021**, *2021*, 6693690. [CrossRef]
19. Sun, H.; Sun, J.; Zhao, K.; Wang, L.; Wang, K. Data-Driven ICA-Bi-LSTM-Combined Lithium Battery SOH Estimation. *Math. Probl. Eng.* **2022**, *2022*, 9645892. [CrossRef]
20. Rehman, B.U.; Babar, M.I.; Ahmad, A.W.; Alhumyani, H.; Abdel Azim, G.; Saeed, R.A.; Abdel Khalek, S. Joint power control and user grouping for uplink power domain non-orthogonal multiple access. *Int. J. Distrib. Sens. Netw.* **2021**, *17*, 15501477211057443. [CrossRef]
21. Shieh, S.-L.; Lin, C.-H.; Huang, Y.-C.; Wang, C.-L. On gray labeling for downlink non-orthogonal multiple access without SIC. *IEEE Commun. Lett.* **2016**, *20*, 1721–1724. [CrossRef]

22.	Al Rabee, F.; Davaslioglu, K.; Gitlin, R. The optimum received power levels of uplink non-orthogonal multiple access (NOMA) signals. In Proceedings of the 2017 IEEE 18th Wireless and Microwave Technology Conference (WAMICON), Cocoa Beach, FL, USA, 24–25 April 2017; pp. 1–4.

23.	Tweed, D.; Derakhshani, M.; Parsaeefard, S.; Le-Ngoc, T. Outage-constrained resource allocation in uplink NOMA for critical applications. *IEEE Access* **2017**, *5*, 27636–27648. [CrossRef]

24.	Ding, Z.; Lei, X.; Karagiannidis, G.K.; Schober, R.; Yuan, J.; Bhargava, V.K. A survey on non-orthogonal multiple access for 5G networks: Research challenges and future trends. *IEEE J. Sel. Areas Commun.* **2017**, *35*, 2181–2195. [CrossRef]

25.	Moriyama, M.; Kurosawa, A.; Matsuda, T.; Matsumura, T. A Study of Parallel Interference Cancellation Combined with Successive Interference Cancellation for UL-NOMA Systems. In Proceedings of the 2021 24th International Symposium on Wireless Personal Multimedia Communications (WPMC), Okayama, Japan, 14–16 December 2021; pp. 1–6.

26.	Hassan, M.B.; Ali, E.S.; Saeed, R.A. Ultra-Massive MIMO in THz Communications: Concepts, Challenges and Applications. In *Next Generation Wireless Terahertz Communication Networks*, 1st ed.; CRC Press: Boca Raton, FL, USA, 2021; Chapter 10, pp. 267–297.

27.	Budhiraja, I.; Kumar, N.; Tyagi, S.; Tanwar, S.; Han, Z.; Piran, M.J.; Suh, D.Y. A systematic review on NOMA variants for 5G and beyond. *IEEE Access* **2021**, *9*, 85573–85644. [CrossRef]

28.	Celik, A.; Al-Qahtani, F.S.; Radaydeh, R.M.; Alouini, M.-S. Cluster formation and joint power-bandwidth allocation for imperfect NOMA in DL-HetNets. In Proceedings of the GLOBECOM 2017-2017 IEEE Global Communications Conference, Singapore, 4–8 December 2017; pp. 1–6.

29.	Zeng, J.; Lv, T.; Liu, R.P.; Su, X.; Peng, M.; Wang, C.; Mei, J. Investigation on evolving single-carrier NOMA into multi-carrier NOMA in 5G. *IEEE Access* **2018**, *6*, 48268–48288. [CrossRef]

30.	Islam, S.R.; Avazov, N.; Dobre, O.A.; Kwak, K.-S. Power-domain non-orthogonal multiple access (NOMA) in 5G systems: Potentials and challenges. *IEEE Commun. Surv. Tutor.* **2016**, *19*, 721–742. [CrossRef]

31.	Alsaqour, R.; Ali, E.S.; Mokhtar, R.A.; Saeed, R.A.; Alhumyani, H.; Abdelhaq, M. Efficient Energy Mechanism in Heterogeneous WSNs for Underground Mining Monitoring Applications. *IEEE Access* **2022**, *10*, 72907–72924. [CrossRef]

32.	Aldababsa, M.; Göztepe, C.; Kurt, G.K.; Kucur, O. Bit error rate for NOMA network. *IEEE Commun. Lett.* **2020**, *24*, 1188–1191. [CrossRef]

33.	Al-Abbasi, Z.Q.; Khamis, M.A. Spectral efficiency (SE) enhancement of NOMA system through iterative power assignment. *Wirel. Netw.* **2021**, *27*, 1309–1317. [CrossRef]

34.	Li, S.; Wei, Z.; Yuan, W.; Yuan, J.; Bai, B.; Ng, D.W.K. On the achievable rates of uplink NOMA with asynchronized transmission. In Proceedings of the 2021 IEEE Wireless Communications and Networking Conference (WCNC), Nanjing, China, 29 March–1 April 2021; pp. 1–7.

35.	Choi, J. Minimum power multicast beamforming with superposition coding for multiresolution broadcast and application to NOMA systems. *IEEE Trans. Commun.* **2015**, *63*, 791–800. [CrossRef]

36.	Liu, F.; Petrova, M. Proportional fair scheduling for downlink single-carrier NOMA systems. In Proceedings of the GLOBECOM 2017-2017 IEEE Global Communications Conference, Singapore, 4–8 December 2017; pp. 1–7.

37.	Saeed, R.A.; Abbas, E.B. Performance evaluation of MIMO FSO communication with gamma-gamma turbulence channel using diversity techniques. In Proceedings of the 2018 International Conference on Computer, Control, Electrical, and Electronics Engineering (ICCCEEE), Khartoum, Sudan, 12–14 August 2018; pp. 1–5.

38.	Shen, D.; Wei, C.; Zhou, X.; Wang, L.; Xu, C. Photon Counting Based Iterative Quantum Non-Orthogonal Multiple Access with Spatial Coupling. In Proceedings of the 2018 IEEE Global Communications Conference (GLOBECOM), Abu Dhabi, United Arab Emirates, 9–13 December 2018; pp. 1–6.

39.	Mokhtar, R.A.; Saeed, R.A.; Alhumyani, H. Cooperative Fusion Architecture-based Distributed Spectrum Sensing Under Rayleigh Fading Channel. *Wirel. Pers. Commun.* **2022**, *124*, 839–865. [CrossRef]

40.	Lo, S.-H.; Chen, Y.-F. Subcarrier Allocation for Rate Maximization in Multiuser OFDM NOMA Systems on Downlink Beamforming. In Proceedings of the 2020 6th International Conference on Applied System Innovation (ICASI), Taitung, Taiwan, 5–8 November 2020; pp. 56–61.

41.	Abdelrahman, Y.T.; Saeed, R.A.; El-Tahir, A. Multiple Physical Layer Pipes performance for DVB-T2. In Proceedings of the 2017 International Conference on Communication, Control, Computing and Electronics Engineering (ICCCCEE), Khartoum, Sudan, 16–18 January 2017; pp. 1–7.

42.	Sedaghat, M.A.; Müller, R.R. On user pairing in uplink NOMA. *IEEE Trans. Wirel. Commun.* **2018**, *17*, 3474–3486. [CrossRef]

43.	Agarwal, A.; Chaurasiya, R.; Rai, S.; Jagannatham, A.K. Outage probability analysis for NOMA downlink and uplink communication systems with generalized fading channels. *IEEE Access* **2020**, *8*, 220461–220481. [CrossRef]

44.	Mukhtar, A.M.; Saeed, R.A.; Mokhtar, R.A.; Ali, E.S.; Alhumyani, H. Performance Evaluation of Downlink Coordinated Multipoint Joint Transmission under Heavy IoT Traffic Load. *Wirel. Commun. Mob. Comput.* **2022**, *2022*, 6837780. [CrossRef]

45.	Do, D.-T.; Nguyen, T.-T.T.; Nguyen, T.N.; Li, X.; Voznak, M. Uplink and downlink NOMA transmission using full-duplex UAV. *IEEE Access* **2020**, *8*, 164347–164364. [CrossRef]

46.	Krishnamoorthy, A.; Schober, R. Uplink and downlink MIMO-NOMA with simultaneous triangularization. *IEEE Trans. Wirel. Commun.* **2021**, *20*, 3381–3396. [CrossRef]

47.	Do, D.-T.; Nguyen, T.-L.; Ekin, S.; Kaleem, Z.; Voznak, M. Joint user grouping and decoding order in uplink/downlink MISO/SIMO-NOMA. *IEEE Access* **2020**, *8*, 143632–143643. [CrossRef]

48. Elbamby, M.S.; Bennis, M.; Saad, W.; Debbah, M.; Latva-Aho, M. Resource optimization and power allocation in in-band full duplex-enabled non-orthogonal multiple access networks. *IEEE J. Sel. Areas Commun.* **2017**, *35*, 2860–2873. [CrossRef]
49. Elfatih, N.M.; Hasan, M.K.; Kamal, Z.; Gupta, D.; Saeed, R.A.; Ali, E.S.; Hosain, M.S. Internet of vehicle's resource management in 5G networks using AI technologies: Current status and trends. *IET Commun.* **2022**, *16*, 400–420. [CrossRef]

51

Article

Multiagent Reinforcement Learning Based on Fusion-Multiactor-Attention-Critic for Multiple-Unmanned-Aerial-Vehicle Navigation Control

Sangwoo Jeon [1], Hoeun Lee [1], Vishnu Kumar Kaliappan [2,*], Tuan Anh Nguyen [2], Hyungeun Jo [1], Hyeonseo Cho [1] and Dugki Min [1,*]

1   Department of Computer Science and Engineering, Konkuk University, Seoul 05029, Korea
2   Konkuk Aerospace Design-Airworthiness Research Institute, Konkuk University, Seoul 05029, Korea
*   Correspondence: vishnudms@gmail.com (V.K.K.); dkmin@konkuk.ac.kr (D.M.)

Citation: Jeon, S.; Lee, H.; Kaliappan, V.K.; Nguyen, T.A.; Jo, H.; Cho, H.; Min, D. Multiagent Reinforcement Learning Based on Fusion-Multiactor-Attention-Critic for Multiple-Unmanned-Aerial-Vehicle Navigation Control. *Energies* 2022, 15, 7426. https://doi.org/10.3390/en15197426

Academic Editors: R. Maheswar, M. Kathirvelu and K. Mohanasundaram

Received: 2 September 2022
Accepted: 5 October 2022
Published: 10 October 2022

Abstract: The proliferation of unmanned aerial vehicles (UAVs) has spawned a variety of intelligent services, where efficient coordination plays a significant role in increasing the effectiveness of cooperative execution. However, due to the limited operational time and range of UAVs, achieving highly efficient coordinated actions is difficult, particularly in unknown dynamic environments. This paper proposes a multiagent deep reinforcement learning (MADRL)-based fusion-multiactor-attention-critic (F-MAAC) model for multiple UAVs' energy-efficient cooperative navigation control. The proposed model is built on the multiactor-attention-critic (MAAC) model, which offers two significant advances. The first is the sensor fusion layer, which enables the actor network to utilize all required sensor information effectively. Next, a layer that computes the dissimilarity weights of different agents is added to compensate for the information lost through the attention layer of the MAAC model. We utilize the UAV LDS (logistic delivery service) environment created by the Unity engine to train the proposed model and verify its energy efficiency. The feature that measures the total distance traveled by the UAVs is incorporated with the UAV LDS environment to validate the energy efficiency. To demonstrate the performance of the proposed model, the F-MAAC model is compared with several conventional reinforcement learning models with two use cases. First, we compare the F-MAAC model to the DDPG, MADDPG, and MAAC models based on the mean episode rewards for 20k episodes of training. The two top-performing models (F-MAAC and MAAC) are then chosen and retrained for 150k episodes. Our study determines the total amount of deliveries done within the same period and the total amount done within the same distance to represent energy efficiency. According to our simulation results, the F-MAAC model outperforms the MAAC model, making 38% more deliveries in 3000 time steps and 30% more deliveries per 1000 m of distance traveled.

Keywords: air logistics; multiagent reinforcement learning; actor-attention-critic; sensor fusion; multiple UAV

1. Introduction

In recent years the usage of unmanned aerial vehicles (UAVs) for various applications has increased spontaneously. Multiple UAVs are deployed for cooperative missions such as passenger transportation, logistics delivery, and surveillance [1]. In order to successfully carry out the mission in limited resources and time, an energy-efficient multiple-UAV navigation control is needed for the cooperative task. Since the energy consumption of a UAV is proportional to operating time, the UAV's energy efficiency is directly related to a high performance [2]. To develop an energy-efficient multiple-UAV control model, control complexity is a typical problem that needs to be resolved. When UAVs perform cooperative missions together, the decision of one UAV affects the decision of other UAVs. Moreover, complexity increases exponentially as the number of UAVs increases [3]. Consequently, there are clear limitations in solving such problems with existing conventional heuristic-based search algorithms.

Multiagent deep reinforcement learning (MADRL) is a novel model that enables each agent to perform cooperative tasks by interacting with other agents through their own decisions. MADRL is a suitable model compared to a conventional model, which can be applied to various environments where multiple agents exist, such as multirobot controls, multiplayer games, and multiple-UAV control, etc. [4,5]. Unlike a ground vehicle that moves on a 2D plane, the range of a UAV's motion is much broader. As a result, the movement strategy for mission performance is more diverse. Furthermore, UAVs must make appropriate decisions by using their own sensor information and the information retrieved by other UAVs. For these reasons, a suitable MADRL model must be selected for efficient navigation control.

There has been considerable research carried out in reinforcement learning (RL) based on UAV navigation and its application. G. Muñoz et al. [6] developed a DQN-based model applied to a single UAV for navigation with obstacle avoidance. The Airsim-based realistic simulated 3D environment was utilized for training the agent. The author evaluated and demonstrated that the proposed model outperformed other DQN-based algorithms. Similarly, H. Qie et al. [7] proposed a multiagent deep deterministic policy gradient (MADDPG)-based model for multiple-UAV target assignment and path planning. The results showed that agents could be assigned to their targets at a relatively close distance with a clear behavior for avoiding threat areas. Linfei Feng [8] introduced the policy gradient (PG) model, which could be applied to optimize the logistics distribution routes of a single UAV. The results showed that the UAV arranged delivery routes to multiple destinations with the shortest path. Ory Walker et al. [9] developed a framework based on the combination of proximal policy optimization (PPO) and adaptive belief tree (ABT) for multiple-UAV exploration and target finding. The proposed algorithm was verified in both 2D and 3D environments with the physically simulated UAVs using the PX4 software stack. W.J. Yoon et al. [10] utilized the QMIX model for eVTOL mobility in drone taxi applications. The proposed QMIX-based algorithm showed optimal performance when compared with independent DQN (I-DQN) and a random walk in the drone taxi service scenario. Zhou W. et al. [11] proposed a reciprocal-reward multiagent actor-critic (MAAC-R) method and applied it for learning cooperative tracking policies for UAV swarms. The training results demonstrated that the proposed model performed better than the MAAC model in terms of cooperative tracking behaviors of UAV swarms. D. Xu et al. [12] improved the MADDPG-based algorithm and applied it for the autonomous and cooperative control of UAV clusters in combat missions. The proposed algorithm was tested by performing two conventional combat missions. The result showed that the learning efficiency and the operational safety factor were improved when compared with the original MADDPG algorithm. Similarly, Guang Zhan et al. [13] applied multiagent proximal policy optimization (MAPPO) in a Unity based 3D-simulated air combat environment. The proposed algorithm was trained with a Ray based distributed training framework. In the experiment, MAPPO outperformed COMA and BiCNet in average accumulate reward. Table 1 shows a detailed comparison of research activity conducted utilizing MADRL and RL.

Table 1. Comparison of RL-based UAV application.

Name of the Research	Year	Baseline	Actor Critic	Single/ Multiagent	Centralized/ Decentralized	Applications	Simulated Environment
Multiple-UAV Reinforcement Learning Algorithm Based on Improved PPO in Ray Framework [13]	2022	MAPPO [14]	Yes	Multiagent	Centralized critic with decentralized actor	Distributed decision-making and complete cooperation task	Unity collaborative combat environment 3D
Autonomous and cooperative control of UAV cluster with multi-agent reinforcement learning [12]	2022	MADDPG [15]	Yes	Multiagent	Centralized critic with decentralized actor	Autonomous and cooperative control of UAV clusters	Conventional combat environment
Improving multi-target cooperative tracking guidance for UAV swarms using multi-agent reinforcement learning [11]	2021	MAAC [16]	Yes	Multiagent	Centralized critic with decentralized actor	Tracking the perceived targets and searching the unknown targets	Coordinate plane 2D
Distributed deep reinforcement learning for autonomous aerial eVTOL mobility in drone taxi applications [10]	2021	QMIX [17]	Yes	Multiagent	Centralized critic with decentralized actor	Computing the optimal passenger transportation routes	200-by-200 grid map 2D
A Framework for Multi-Agent UAV Exploration and Target-Finding in GPS-Denied and Partially Observable Environments [9]	2020	ABT + PPO [18]	Yes	Multiagent	Decentralized actor and critic	Multiple-UAV exploration and target finding	Occupancy map with OpenAI Gym 2D + 3DR Iris and 3DR Solo with Gazebo 3D
Reinforcement Learning to Optimize the Logistics Distribution Routes of Unmanned Aerial Vehicle [8]	2020	PG [19]	Yes	Single Agent	-	Path planning for UAVs in complex surroundings	Coordinate plane 2D
Joint Optimization of Multi-UAV Target Assignment and Path Planning Based on Multi-Agent Reinforcement Learning [7]	2019	MADDPG [15]	Yes	Multiagent	Centralized critic with decentralized actor	Multiple-UAV target assignment and path planning	OpenAI's platform 2D
Deep reinforcement learning for drone delivery [6]	2019	DDQN [20]	No	Single Agent	-	Navigation with obstacle avoidance in realistic environment	Realistic neighborhood environment on AirSim 3D

From Table 1, most of the research was carried out using actor-critic-based models. Additionally, based on the previous research related to MADRL, we conclude that centralized training with a decentralized execution methodology is more suitable for real-world situations. In real-world execution, it is difficult for one UAV to obtain data from all other UAVs in real time. A decentralized actor network can be used to infer the action in such a partially observable environment. We paid attention to the multi-actor-attention-critic (MAAC) model, which showed optimal performance among algorithms based on a centralized critic and a decentralized policy, which can be used in environments where information exchange between agents is not guaranteed [16].

This study makes the following significant contributions.

- The development of an MAAC-based model with two significant improvements by applying a sensor fusion layer in the actor network and a dissimilarity layer in the critic network.
- A new feature to calculate the energy efficiency of UAVs is incorporated with the previously developed UAV LDS simulation environment.
- The performance of the existing RL and MADRL models are compared with two energy efficiency indicators.

In this research, we focus on optimizing learning efficiency by efficiently processing the observations of multiple UAVs by adding two features to the MAAC model. First, we introduce a sensor fusion layer in the actor network to extract features from various sensors such as a ray-cast sensor for preventing collision with adjacent obstacles, an inertial navigation system (INS) for the self-awareness of flight status, and a radio detection and ranging (RADAR) system for collecting location data from other UAVs. Second, in the critic network, a dissimilarity layer is added to provide more weight to the information of agents with fewer similarities. By implementing these functions, the efficiency of information processing is increased, and we prove through experiments that it plays a decisive role in achieving the goal of energy-efficient UAV navigation control.

To experiment and validate our proposed MADRL model, the logistic delivery service virtual test bed is adopted from our previous research [21]. The test bed is customized by adding an energy efficiency module for multiple-UAV cooperation specifically for logistic delivery. To find out whether UAVs can cooperatively perform missions well, the environment includes a scenario in which two UAVs cooperate for transport logistics. A function to measure the total travel distance of UAVs has been added to validate the energy efficiency of the UAVs. Our proposed model shows the highest performance in terms of energy efficiency compared to conventional RL algorithms. We measure energy efficiency with the number of trips carried out during the same time, and the number of cargos carried out during the same distance traveled. Our model shows superiority in both indicators.

Our work is structured as follows. Section 2 covers the general background of the RL and MADRL algorithms. In Section 3, we expound on the proposed fusion-MAAC (F-MAAC) method. Section 4, the test bed for the training and evaluation is described in detail. Section 5 shows the results and discusses performance evaluation. Finally, the study concludes with future directions in Section 6.

2. Background

RL is a field that has recently been spotlighted in the field of machine learning. It is a technology that learns a model through the trial and error of an agent in a given environment without any data. RL can be described as a learning process that develops a behavior through trial and error to maximize the cumulative reward in a sequential decision-making problem. The Markov decision process (MDP) can be expressed as a sequential decision-making problem. RL is being utilized in various fields and situations expressed as sequential decision-making problems, such as stock investment, driving, and games.

2.1. Markov Decision Process (MDP)

RL is an optimization method for solving sequential decision-making problems using the Markov decision process (MDP). The MDP is defined as follows.

$$(S, A, P, R, \gamma) \tag{1}$$

Here, S stands for state space and A stands for action space. P is the probability distribution of the next state s' when the agent chooses the action $a \in A$ from the state $s \in S$, and R means the reward received in the next state s'. For the cumulative reward, the future reward is depreciated using the discount rate γ. This reflects future uncertainty and prevents the divergence of the cumulative reward so that learning can be performed stably. Figure 1a exemplifies the basic concept of MDP. When an agent chooses action a, the environments proceed to the next step by action a and return the next state s and reward r.

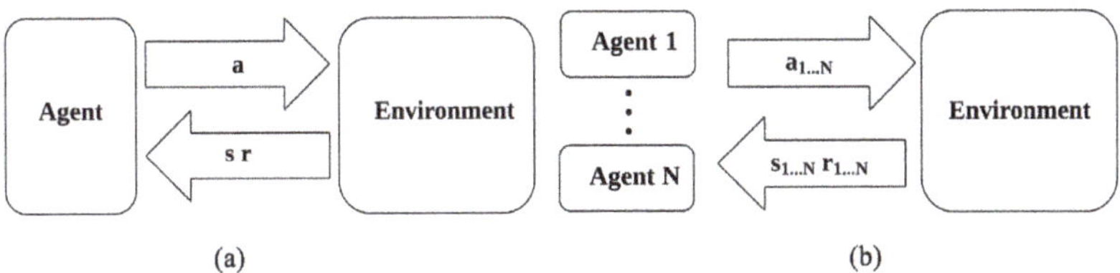

Figure 1. Conceptual diagram: (**a**) Markov decision process and (**b**) Markov game (a—action, s—state, r—reward).

A Markov game is a multiagent extension of the MDP [22]. A Markov game is defined as a set of states and actions for N agents. A probability distribution for the next state is given through the current state and action of each agent. The reward function for each agent depends on the global state and action of all agents. Observation O_i is a partial state that agent i can observe and includes some information of the global state. Each agent learns the policy $\pi : O_i \rightarrow P(A_i)$ that maximizes the expected sum of rewards. Figure 1b shows the multiple-agent interaction with the environment to update the rewards. Multiple agents $\{A_1 \ldots A_N\}$ send the action command $a_N = \{A_1\{a_1\}, A_2\{a_2\}, \ldots, A_N\{a_N\}\}$ to the environment. The environment returns the following set of state $s_N = \{A_1\{s_1\}, A_2\{s_2\}, \ldots, A_N\{s_N\}\}$ and reward $r_N = \{A_1\{r_1\}, A_2\{r_2\}, \ldots, A_N\{r_N\}\}$.

2.2. Bellman Equation

Solving the MDP is divided into prediction and control problems. Prediction is the problem of evaluating the value of each state given a policy. Control is the problem of finding the optimal policy. The policy and value need to be expressed through the Bellman equation to solve these problems. Bellman's equation is defined using the recursive relationship between the present time step t and the next time step t + 1. The value function $V(S)$ and the action value function $Q(S, A)$ can be expressed as the Bellman expectation equation and the Bellman optimal equation [23].

The expected reward G_t is derived using the following equation:

$$G_t = R_{t+1} + R_{t+2} + \ldots + R_T \tag{2}$$

where R is the reward, G_t is the sum of the rewards received from time step t + 1 to the final time step T.

Since immediate rewards are more important than the future reward, the discount factor γ is multiplied by Equation (2) to redefine G_t.

$$G_t = R_{t+1} + \gamma R_{t+2} + \gamma^2 R_{t+3} + \ldots = R_{t+1} + \gamma G_{t+1} \tag{3}$$

- *Bellman's expectation equation*—The value function $v_\pi(s_t)$ is calculated using the expected value G_t in Equation (4).

$$v_\pi(s) = E_\pi[G_t|S_t = s] \tag{4}$$

where E_π is the expectation when following the policy π.
Now to derive the value function $v_\pi(s_t)$, Equation (3) is substituted in Equation (4).

$$v_\pi(s_t) = E_\pi(R_{t+1} + \gamma v_\pi(s_{t+1})] \tag{5}$$

The action value function $q_\pi(s, a)$ is calculated using the expected value $E_\pi(G_t)$.

$$q_\pi(s, a) = E_\pi[G_t|S_t = s, A_t = a] \tag{6}$$

Now, to action value function $q_\pi(s_t, a_t)$, Equation (3) is substituted in Equation (6).

$$q_\pi(s_t, a_t) = E_\pi[R_{t+1} + \gamma q_\pi(s_{t+1}, a_{t+1})] \tag{7}$$

- *Bellman's optimal equation*—The optimal value $v_*(s)$ and $q_*(s, a)$ is calculated as follows:

$$v_*(s) = \max_\pi v_\pi(s_t) = \max_a E[R_{t+1} + \gamma v_\pi(s_{t+1})] \tag{8}$$

where $\max_\pi$ is the maximum cumulative rewards and $\max_{a'}$ is the best action a' out of all actions a_{t+1} that provides a maximum reward.

$$q_*(s, a) = \max_\pi q_\pi(s_t, a_t) = E[R_{t+1} + \gamma \max_{a'} q_*(s_{t+1}, a')] \tag{9}$$

2.3. Multiagent Deep Reinforcement Learning

Multiagent Deep Reinforcement Learning (MADRL) is one of the most popular and effective models for solving more complex problems where multiple agents collaborate to perform specific tasks. For example, playing soccer games with multiple robots where the team of robots collaborates to achieve the mission. One of the key challenges in such an environment is that the environment is more dynamic to the perceptive of each agent, which may affect the individual learning rate as a team.

- *Multiagent deep deterministic policy gradient (MADDPG)*— MADDPG [15] is a multiagent extension of DPG [24], which combines DDPG [25] with DQN [26] approaches such as replay buffer and target separation. Each agent has its own actor and critic. In the MADDPG method, centralized training with a decentralized execution approach is used. The architecture of the MADDPG model is shown in Figure 2. A centralized critic network $Q_{1...N}$ is used for centralized training with observations $o_{1...N}$ and actions $a_{1...N}$ from all other agents as input. In the decentralized execution, agents use an actor network $\pi_{1...N}$ to choose an action by only using local information. By this approach, the MADDPG model can be applied even in a partially observable environment where communication between agents is limited.
- *Multiactor-attention-critic (MAAC)*—MAAC was developed by [16] and adopted from the MADRL model. The model trains the decentralized policies in multiagent environments by utilizing centrally computed critics with an attention mechanism. It chooses relevant information for each agent at every time step. The multiattention head layer consists of multiple attention heads. The attention function in the attention head can be described as mapping a query and a set of key–value pairs to an output [27]. The attention function is calculated as Equation (10), where query Q has the corresponding key K and value V and d_k is a scaling factor. As shown in Figure 3, encodings of the agent's state and action denotes the state action encodings (SAE_i) are the key and value. The encodings of the other agent's state encoder (SE_j), $j \in \setminus i$ are the query. In

each attention head N, different attention head values (AHVs) are derived according to the influence of the query, key, and value extractors. The final output attention value (AV) is achieved with the combination of AHVs. The final output $Q_i(o,a)$ is derived through fully connected layers FC_1 and FC_2 with the input of AV and SE_i. In the multiattention head layer, the agent updates the weighted value which is more similar to other agents. This attention mechanism enables a more effective and flexible learning in complex multiagent environments compared to MADDPG.

$$AttentionFuntion(Q,K,V) = softmax\left(\frac{QK^T}{\sqrt{d_k}}\right)V \tag{10}$$

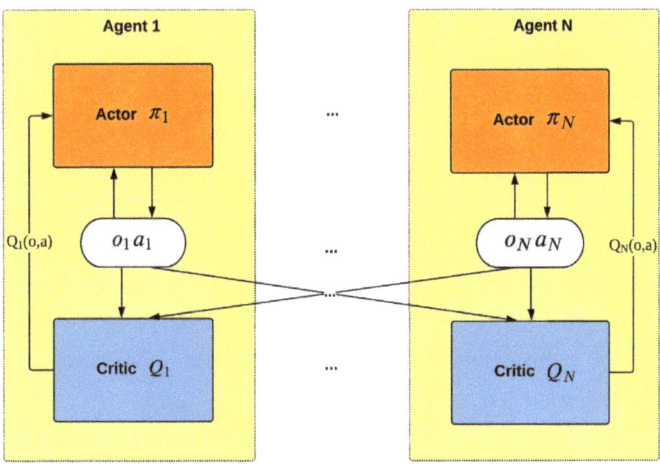

Figure 2. Overall architecture of DDPG.

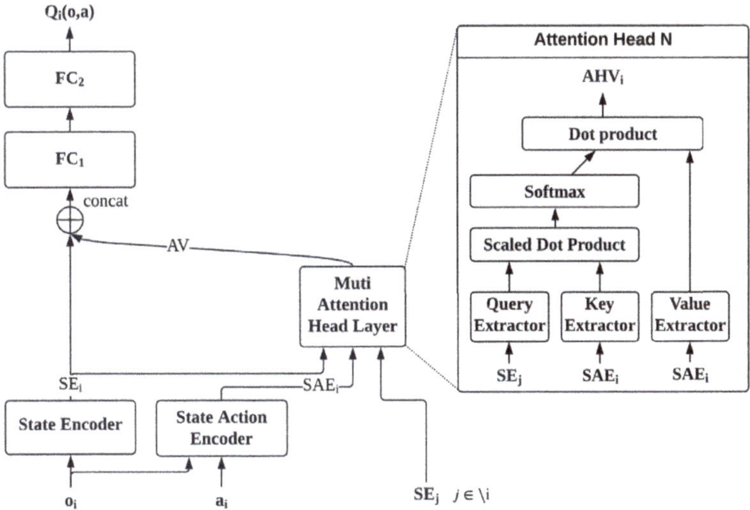

Figure 3. Critic network of MAAC.

3. Fusion-Multiactor-Attention-Critic (F-MAAC) Model

In this section, the F-MAAC model is discussed for the application of multiple-UAV cooperative navigation. To increase the learning efficiency of the agent, we used a new sensor fusion layer with MAAC. The sensor fusion layer was used for the UAV's local observation, and another layer named cosine dissimilarity was added to utilize global information obtained by other UAVs efficiently. The overall architecture of the proposed F-MAAC model is exemplified in Figure 4.

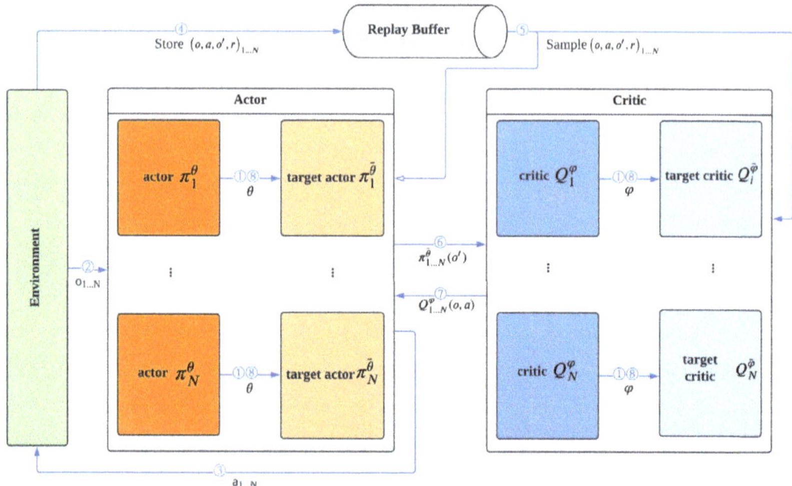

Figure 4. F-MAAC model's overall architecture with training flow.

The overall flow of the F-MAAC model follows the basis of the MAAC model, including a loss function and the gradients of the objective function. Each agent has its own independent actors and critics following a centralized training with a decentralized execution. In the training phase, all agents' observations are entered as inputs of each agent's critic network. In the execution phase, the decentralized actor network is used to choose the action as inference by using only its own observation for input data. This general F-MAAC model can be applied to N agents equipped with M types of sensors. The step-by-step training procedure of the F-MAAC model is as follows:

Step 1: Initialize the critic network $Q_{1...N}^\psi$ and actor network $\pi_{1...N}^\theta$ with random parameters and synchronize the parameters of target critics $Q_{1..N}^{\bar\psi}$ with critics $Q_{1..N}^\varphi$, and target actors $\pi_{1...N}^{\bar\theta}$ with actors $\pi_{1..N}^\theta$.

Step2: Get observation $o_{1...N}$ from the environment, feed-forward to actors $\pi_{1...N}^\theta(o)$, and select action $a_{1...N}$.

Step 3: Proceed to the next time step with actions $a_{1...N}$ and get the next observations $o'_{1...N}$ and rewards $r_{1...N}$ from the environment.

Step 4: Push the obtained set of $(o,a,o',r)_{1...N}$ to the replay buffer.

Step 5: Repeat step 2 to 4 until the number of E data is collected.

Step 6: Sample $B = (o,a,o',r)_{1...N}$ from the replay buffer,

Step 7: Perform a gradient descent by using B to minimize the loss function in Equation (11) with respect to the network parameter φ

$$L_Q(\varphi) = \sum_{i=1}^N E[(Q_i^\varphi(o,a) - y_i^2)] \tag{11}$$

where $y_i = r_i + \gamma E_{a'\sim\pi_\theta(o')}\left[Q_i^{\bar\psi}(o',a') - \alpha \log\left(\pi_{\theta_i}(a_i' \mid o_i')\right)\right]$

Step 8: Perform a gradient ascent by using $o_{1...N}$ in B to maximize the gradient of the objective function in Equation (12) with respect to the network parameter θ

$$\nabla_{\theta_i} J(\pi_\theta) = E_{o\sim D, a\sim\pi}\nabla_{\theta_i}\log\pi_{\theta_i}(a_i|o_i)\left(-\alpha\log\left(\pi_{\theta_i}(a_i|o_i)\right) + A_i(o,a)\right) \tag{12}$$

Step 9: Update the parameters of target critics $Q_{1...N}^{\bar{\psi}}(o,a)$ with Equation (13) and target actors $\pi_{1...N}^{\bar{\theta}}(a,s)$ with Equation (14) using an update rate of $\tau = 0.005$

$$\bar{\psi} = \bar{\psi} * (1.0 - \tau) + \psi * \tau \tag{13}$$

$$\bar{\theta} = \bar{\theta} * (1.0 - \tau) + \theta * \tau \tag{14}$$

Step 10: Steps 2 to 9 should be repeated until the end of the episode.

3.1. Deep Fusion Layer in Actor Network

As illustrated in Figure 5, we propose a deep fusion layer in the actor network to increase efficiency. Observations are separated into the M types of sensors to extract features from each sensor. For instance, three different types of sensors are used for UAVs in our virtual UAV LDS environment: a ray-cast sensor for preventing collision with surrounding obstacles, an INS for the self-awareness of flight status, and a RADAR for retrieving coordinates of other UAVs and hubs. Each sensor's data pass through the sensor encoder. The encoded sensor data are concatenated and pass through two fully connected layers. The output of the deep fusion layer can be expressed by Equation (15). FC_1, FC_2, and sensor encoders($SNE_{1...3}$) are fully connected layers.

$$Output = FC_2(FC_1(Concat(SNE_1(sensor_1),\ SNE_2(sensor_2),\ SNE_3(sensor_3)))) \tag{15}$$

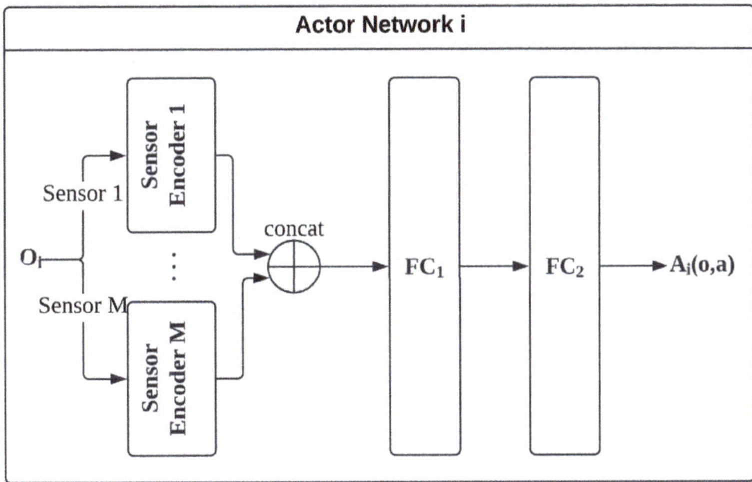

Figure 5. Actor network of F-MAAC model.

3.2. Dissimilarity Layer in Critic Network

In the critic network, state encodings ($SE_{1...N}$) are shared with other agents, as shown in Figure 6. The attention head in the multiattention head layer selects relevant information from other agents' observations. The attention head is constructed with a scaled dot product [27] which calculates the degree of similarity between encoded observations of agent i (SE_i, SAE_i) and the encoded observations of the other agents $j \in \backslash i$ (SE_j). The UAVs at adjacent distances will have similar observation data. When more weights are provided to similar observations, the UAVs will have a wider field of view and less chance of colliding with each other.

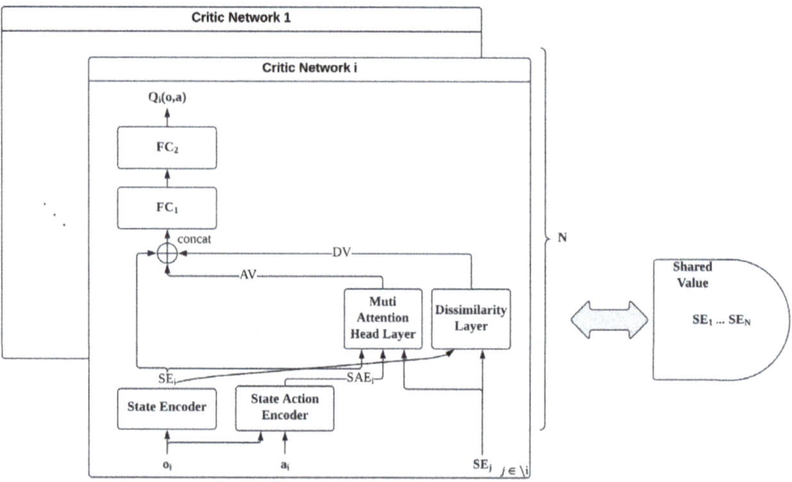

Figure 6. Critic network of F-MAAC model.

However, there are also drawbacks derived from the multiattention head layer. For example, when an agent's observation at a distance that is dissimilar from the current agent's observation plays an essential role in performing its mission, it can lead to serious performance degradation. More specifically, the observation from an agent at long distances near a target point may provide helpful information. For these reasons, a dissimilarity layer was added to prevent performance degradation due to attention and to improve learning stability. In the previous study, we verified the effect of adding a dissimilarity layer to the MAAC model in a simple 2D cooperative navigation environment [28].

Cosine similarity refers to the similarity between two vectors obtained by using the cosine angle between the two vectors. The additional use of the observation multiplied by the dissimilarity value may offset the effect of attention. The dissimilarity value is calculated with the encoded observations of agent i (SE_i) and the encoded observations of other agents $j \in \setminus i$ (SE_j). The value passed through the dissimilarity layer's dissimilarity value (DV) is concatenated with the value from the multiattention head layer's attention value (AV) and SE_i. Then, the concatenated value is sent to the fully connected layers FC_1 and FC_2 to calculate the critic value Q_i.

Figure 7 shows the detailed process of the dissimilarity layer. The dissimilarity weight between the agent's observations is calculated by multiplying the cosine similarity value by a negative number as in Equation (16).

$$CosineDisimilarity(SE_i, SE_n) = -1 \cdot \frac{SE_i \cdot SE_n}{max(\| SE_i \|_2 \cdot \| SE_n \|_2, \varepsilon)} \tag{16}$$

where, $\varepsilon = 1 \times 10^{-8}$.

The negative dissimilarity values are replaced with 0 to focus on the agents' information with different patterns. Then, the observations of each agent are multiplied by the cosine dissimilarity weight and concatenated. The concatenated value is entered as the input value of the fully connected layer. The output value DV from the dissimilarity layer, the output value AV from the multiattention head layer, and the encoded value SE_i are concatenated as an input of the fully connected layers.

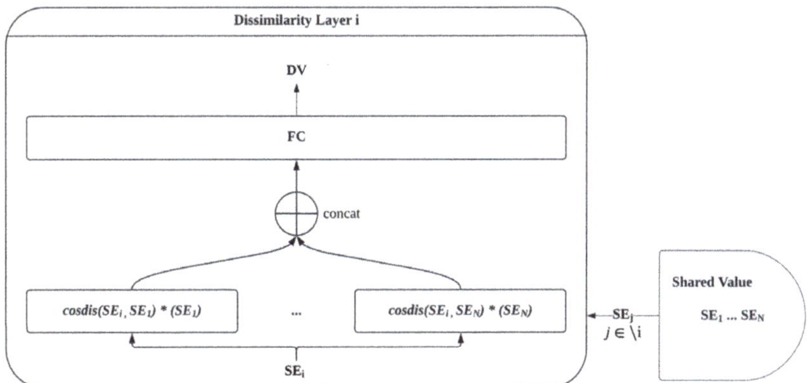

Figure 7. Dissimilarity layer in critic network.

4. Test bed

4.1. UAV LDS Environment

In our previous work [21], we developed a UAV logistic delivery service (UAV LDS) environment for evaluating MADRL-based models. To calculate energy efficiency for this research, we added the feature of calculating the total movement of all agents. The UAV LDS environment is a virtual environment designed to reflect simplified logistics delivery scenarios in the real world and implemented through the Unity platform equipped with the 3D physics engine. The modified source was updated in the following repository (https://github.com/leehe228/LogisticsEnv, accessed on 3 October 2022). The environment follows the Open AI Gym API [29] design which provides standard communication between learning algorithms and environments. In LDS, multiple UAVs act as an air transportation system which is used to carry cargo in the three-dimensional city sky that connects land and air. To implement this as a simulated environment, we constructed blocks representing obstacles such as buildings, warehouses, and cargo to be transported. In the scenario, UAVs delivered big cargo and small cargo from hubs to the destination. What was unique about this environment was that two UAVs had to collaborate to move a big cargo. The reason for including this scenario was that it was possible to check whether the cooperation of UAVs worked well directly. In addition, such cooperative situations could occur any time in the real world, such as when multiple UAVs need to move together to load multiple cargos. Figure 8 shows the UAV carrying cargo in the UAV LDS environment. The gray box indicates the buildings in the real world, the blue box is the small cargo, and the red box is the big cargo. Cargos are generated from the hubs, colored blue on the ground. The destination of the big cargo is colored pink and that of the small cargo is colored green.

4.2. Observation, Action, and Reward Design

This section describes the state, action, and reward of the environment which are essential elements of MDP. The state is the observation received by the agent, the action is the type of movement that can be selected, and the reward is the compensation according to the UAV's action.

- *Observation*—The UAVs received three different sensor data such as ray-cast for preventing collision with adjacent obstacles, INS for the self-awareness of flight status, and RADAR used to find the location of the other UAVs and hubs. In Table 2, a detailed description of the sensor data is provided.
- *Actions*—The UAVs could perform seven types of actions: ascend, descend, forward, backward, left, right, and not move.
- *Driving reward* —To make the UAVs deliver cargo in the shortest path, a driving reward was given at every step. The reward was calculated with the difference between the

distance of the previous time step d_{pre} and the distance of the current time step d_{curr}. Each distance was calculated with the distance to the target point. Before picking up the cargo, the nearest cargo was the target point. After picking up the cargo, the delivery point was the target point. If the UAV was not closer to the target point in the current time step than in the previous time step, a negative reward was given as $(d_{pre} - d_{curr}) \times 0.5$.

- *Delivery rewards*—The values in Table 3 were designed to make UAVs deliver cargo efficiently. For training numerous UAVs to work together to carry cargos, we delicately designed the rewards related to the delivery.
- *Collision penalty*—The UAVs must avoid buildings and other UAVs with ray-cast observations. A negative reward of -10 was given when a collision occurred.

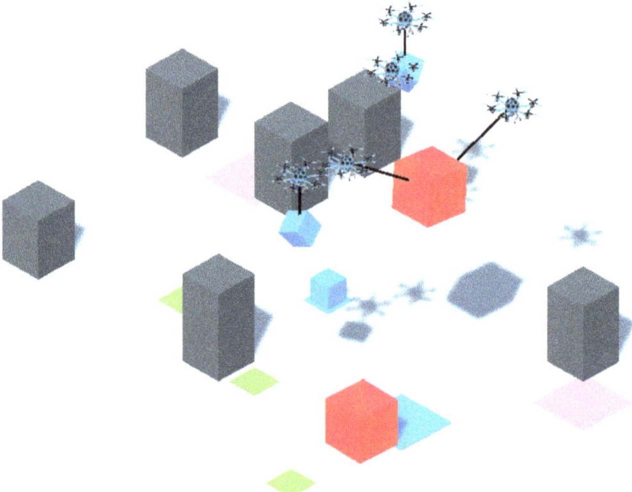

Figure 8. UAV logistic delivery service virtual environment.

Table 2. Summary of observations.

Sensor Type	Size	Description
Ray-cast	1×9	Distance of 9 directions of ray-cast sensor
	2×9	One-hot encoding of the detected object (nothing, building) of 9 direction of ray-cast sensor
INS	3	(x, y, z)—coordinates of UAV_i.
	3	(x, y, z)—velocity of UAV_i.
	3	One-hot encoded cargo type (not holding, small cargo, and big cargo).
RADAR	6	(x, y, z, x, y, z)—coordinates of a big cargo hub and a small cargo hub.
	2	Distance from UAV to big and small cargo hubs.
	6	(x, y, z, x, y, z)—each nearest big and small cargo coordinates.
	2	Distances from UAV_i to the nearest big and small cargos.
	4	(x, y, z, d) if UAV_i holds any cargo, the coordinates and distance of the destination are given.
	7×4	Coordinates of UAV_j (size 3), cargo type of UAV_j (size 3), and distance from UAV_i to UAV_j (size 1). *

* UAV_i is the current, and UAV_j are the rest of all UAVs except UAV_i.

Table 3. Summary of delivery rewards.

Action	Collaborative	First UAV	Second UAV
UAV picks up a small cargo	No	+20.0	-
Small cargo delivery completed	No	+20.0	-
First UAV picks up a big cargo	Yes	+10.0	-
The second UAV picks up a big cargo	Yes	+ 10.0	+20.0
Big cargo delivery completed	Yes	+30.0	+30.0
First UAV drops a big cargo	Yes	−8.0	-
Both UAVs drop a big cargo	Yes	−15.0	−15.0

4.3. Environmental Setup

The UAV environment provided custom settings for the environmental setup. In this research we used the default values in Table 4 for training and evaluation.

Table 4. Summary of environmental setup.

Parameter	Description	Default (Training)	Default (Execution)
NumAgent	Total number of UAVs	5	5
width	Width of the Unity window	480 pixels	1280 pixels
height	Height of the Unity window	270 pixels	720 pixels
timescale	The multiplier for the time	20×	1×
mapsize	Size of the map	13 m	13 m
numbuilding	Number of buildings	3 units	3 units
MaxSmallbox	Total number of small cargos that can be generated	100 units	100 units
MaxBigbox	Total number of big cargos that can be generated	100 units	100 units

5. Experimental Simulation and Results

The proposed F-MAAC model was validated using the environment proposed in Section 4. For more efficient evaluations of the proposed F-MAAC model, we first compared the mean episode rewards of the MAAC, MADDPG, and DDPG models with a training of 20k episodes. Then, the two models with the highest performance, F-MAAC and MAAC, were selected for the training of 150k episodes. To evaluate the trained model to achieve a meaningful scale length, the episode length was replaced with 3000 from 1000 time steps. The timescale of the environment was decreased in the evaluation phase to observe and analyze the strategies of UAVs. The total number of deliveries during one episode and the same distance traveled were evaluated to verify the energy efficiency.

The hyperparameters for training the RL models are shown in Table 5.

Table 5. Hyperparameter settings of RL models.

	DDPG	MADDPG	MAAC	F-MAAC
Number of episodes	1000	1000	1000	1000
Steps per update	100	100	250	250
Batch size	1024	1024	1024	1024
Number of attention heads	-	-	4	4
Policy hidden dimension	128	128	128	128
Learning rate of critic	0.01	0.01	0.001	0.001
Learning rate of policy	0.01	0.01	0.001	0.001

5.1. Comparison of Performance of RL Models

Two MADRL models (MAAC, MADDPG) and one single agent RL model (DDPG) were compared with the proposed F-MAAC model. Each model was trained for 20k episodes with 1000 steps per episode in the proposed UAV LDS simulation environment.

According to Figure 9, the DDPG's mean episode reward value showed the worst performance because it did not increase significantly tableuntil 20k episodes. Although it rose slightly higher from 5k episodes to 20k episodes when compared to DDPG, the increase in the mean episode reward value in the MADDPG model was also minor. The F-MAAC and MAAC models, on the other hand, displayed an impressive performance and successfully conveyed some quantities of both large and small cargos. Between 10k and 20k training episodes, the F-MAAC model demonstrated a more significant value than the MAAC model in the mean episode reward. At the end of 20k training sessions, the F-MAAC model demonstrated greater mean episode rewards than the MAAC model by more than 30%.

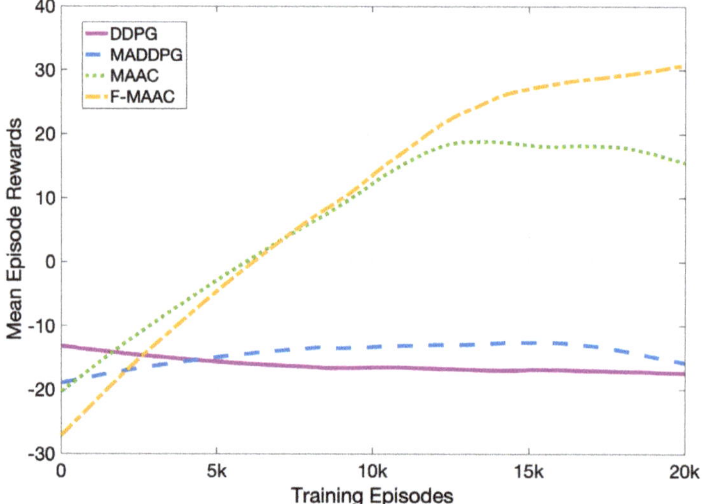

Figure 9. Mean episode rewards comparison of different models for 20k episodes.

5.2. Comparison of Performance between F-MAAC and MAAC Models

We retrained the F-MAAC and MAAC models with 150k episodes, which took about six days with two GPU machines. The detailed specifications of the machine are listed below in Table 6.

Table 6. Specifications and environmental setup of the GPU machine.

CPU	Intel i7 8700 k
GPU	Nvidia RTX 3080
RAM	64 GB
OS	Ubuntu 20.04 LTS
Deep Learning Framework	Pytorch 1.8.2

Figure 10 shows the mean episode rewards of the MAAC and F-MAAC models for 150k training episodes. The mean episode reward value of the F-MAAC and MAAC models increased noticeably in this experiment compared to the previous Section 5.1. The difference between them with training episodes until 40k was unnoticeable. After the training of 40k episodes, the F-MAAC model started to outperform the MAAC model. From 80k to 150k, the mean episode reward of the MAAC model decreased while that of the F-MAAC model constantly increased. At the end of the training, the F-MAAC model obtained 50% more rewards than the MAAC model. The randomness and instability of the complex 3D environment produced different learning patterns compared with the previous training, since the maps of the UAV LDS environment were generated randomly for every episode. However, both results showed that the F-MAAC model outperformed the MAAC model. The result of this experiment showed a more reliable comparison since it was trained longer, until 150k episodes.

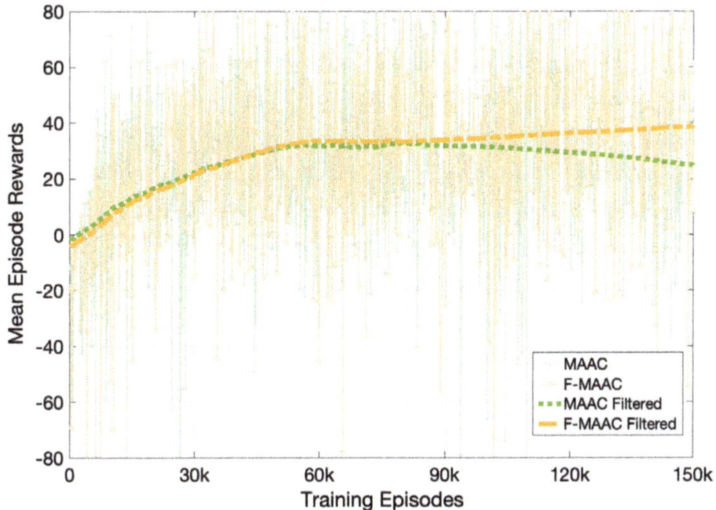

Figure 10. Mean episode rewards of the MAAC and F-MAAC models for 150k episodes.

5.3. Comparison of Energy Efficiency between F-MAAC and MAAC Models

For energy efficiency evaluation, we executed the trained model of F-MAAC and MAAC with 150k episodes. Each model was executed for 100 episodes with 3000 time steps of each episode. The average performance per episode is shown with a box plot in Figure 11. We show the number of successful deliveries of small cargo and big cargo. Furthermore, the total performance was evaluated with $Score = NumberOfSmallCargo + 1.5 * NumberOfBigCargo$. The weight of 1.5 was multiplied by the number of big cargos since we gave 50% more rewards to the big cargos in the training phase.

The result showed that the number of deliveries in both small and big cargos with the F-MAAC model was higher than in the MAAC model. Table 7 shows that the score of the F-MAAC model was 38% higher than that of the MAAC model during one episode, indicating that the F-MAAC model was more energy efficient.

Table 7. Overall comparison of MAAC and F-MAAC models.

	MAAC	**F-MAAC**
Score	13.29	18.31
Movement	1200 m	1270 m
Collision	9.2	8.4
Score_movement	11.08	14.42

We also provided the energy efficiency with Score_movement, which is the performance per 1000 m distance moved. We recorded the total movements of the UAV during execution. The Score_movement was calculated with $\frac{Score}{Movement} \times 1000$. The results showed that the F-MAAC model was 30% more efficient compared to the MAAC model.

In addition, the number of collisions of the F-MAAC model was about 9% less than that of the MAAC model. The improvement of the F-MAAC model's sensor processing efficiency can be interpreted as having a positive effect on the obstacle avoidance performance of the UAV.

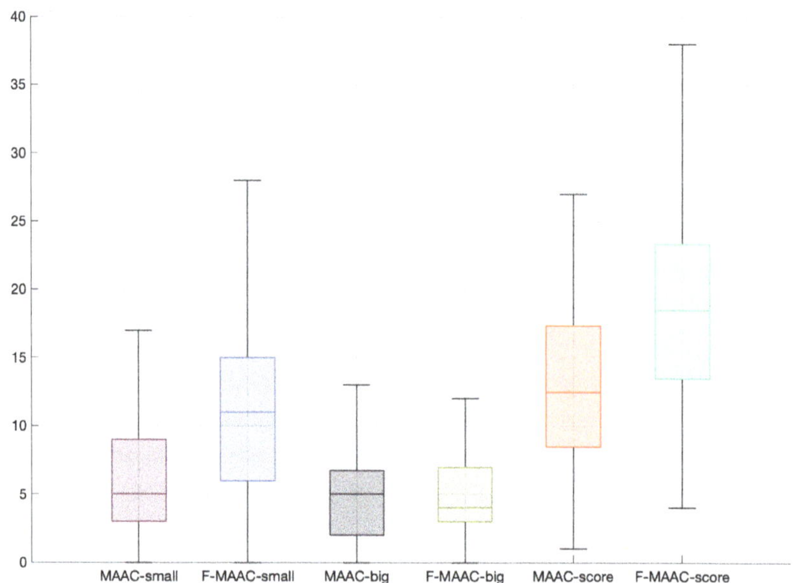

Figure 11. Comparison of delivery performance.

6. Conclusions

This study proposed an MAAC-based multiple-UAV navigation control model that improved energy efficiency through efficient data processing of the UAVs. The following significant findings were obtained.

(a) In the proposed model, the sensor fusion layer was adapted in the actor network, and the dissimilarity layer was utilized for the critic network. When applied to the UAV LDS simulation environment, it outperformed the conventional RL model in terms of energy efficiency.

(b) The sensor fusion layer extracted features from each sensor enabling the UAVs to use various sensor data efficiently. The dissimilarity layer compensated for the loss derived from the attention layer by providing data with high dissimilarity to other agents.

(c) The F-MAAC-applied UAVs transported more cargo than the MAAC in the same amount of time and distance with greater cooperation and fewer collisions.

The feature of measuring the total movement of UAVs was added to the existing UAV LDS environment to calculate energy efficiency. We provided two indicators that calculated the energy efficiency of UAVs. The proposed model showed the best performance in both types of energy efficiency indicators out of various RL models, including the original MAAC model. In future studies, further verification and development are needed for the model in a more sophisticated environment, including realistic sensors and dynamic flight models. Furthermore, the scalability should be verified in a broader environment where more agents exist.

Author Contributions: Conceptualization, S.J.; Investigation, S.J. and H.C.; Methodology, S.J. and H.J.; Project administration, V.K.K. and D.M.; Software, H.L.; Supervision, V.K.K. and D.M.; Validation, V.K.K. and T.A.N.; Visualization, S.J.; Writing—original draft, S.J.; Writing—review & editing, V.K.K. and T.A.N. All authors have read and agreed to the published version of the manuscript.

Energies **2022**, *15*, 7426

Funding: This research was supported by the Basic Science Research Program through the National Research Foundation of Korea (NRF) funded by the Ministry of Education (no. 2020R1A6A1A03046811). This work was supported by the National Foundation of Korea (NRF) grant funded by the Korea government (Ministry of Science and ICT (MIST)) (no. 2021R1A2C209494311).

Institutional Review Board Statement: Not applicable.

Informed Consent Statement: Not applicable.

Data Availability Statement: Not applicable.

Conflicts of Interest: The authors declare no conflict of interest.

Abbreviations

UAV	Unmanned aerial vehicle
RL	Reinforcement learning
MADRL	Multiagent reinforcement learning
LDS	Logistic delivery service
MAAC	Multiactor-attention-critic
F-MAAC	Fusion-multiactor-attention-critic
DDPG	Deep deterministic policy gradient
MADDPG	Multiagent deep deterministic gradient
MDP	Markov decision process
INS	Inertial navigation system
RADAR	Radio detection and ranging

References

1. Roldán, J.J.; Cerro, J.D.; Barrientos, A. A proposal of methodology for multi-UAV mission modeling. In Proceedings of the 2015 23rd Mediterranean Conference on Control and Automation (MED), Torremolinos, Spain, 16–19 June 2015; pp. 1–7.
2. Abeywickrama, H.V.; Jayawickrama, B.A.; He, Y.; Dutkiewicz, E. Comprehensive energy consumption model for unmanned aerial vehicles, based on empirical studies of battery performance. *IEEE Access* **2018**, *6*, 58383–58394. [CrossRef]
3. Zhang, J.; Jiahao, X.I.N.G. Cooperative task assignment of multi-UAV system. *Chin. J. Aeronaut.* **2020**, *33*, 2825–2827. [CrossRef]
4. Nguyen, T.T.; Nguyen, N.D.; Nahavandi, S. Deep reinforcement learning for multiagent systems: A review of challenges, solutions, and applications. *IEEE Trans. Cybern.* **2020**, *50*, 3826–3839. [CrossRef] [PubMed]
5. Chang, H.; Chen, Y.; Zhang, B.; Doermann, D. Multi-UAV mobile edge computing and path planning platform based on reinforcement learning. *IEEE Trans. Emerg. Top. Comput. Intell.* **2021**, *6*, 489–498. [CrossRef]
6. Muñoz, G.; Barrado, C.; Çetin, E.; Salami, E. Deep reinforcement learning for drone delivery. *Drones* **2019**, *3*, 72. [CrossRef]
7. Qie, H.; Shi, D.; Shen, T.; Xu, X.; Li, Y.; Wang, L. Joint optimization of multi-UAV target assignment and path planning based on multi-agent reinforcement learning. *IEEE Access* **2019**, *7*, 146264–146272. [CrossRef]
8. Feng, L. Reinforcement learning to optimize the logistics distribution routes of unmanned aerial vehicle. *arXiv* **2020**, arXiv:2004.09864.
9. Walker, O.; Vanegas, F.; Gonzalez, F. A framework for multi-agent UAV exploration and target-finding in GPS-denied and partially observable environments. *Sensors* **2020**, *20*, 4739. [CrossRef] [PubMed]
10. Yun, W.J.; Jung, S.; Kim, J.; Kim, J.H. Distributed deep reinforcement learning for autonomous aerial eVTOL mobility in drone taxi applications. *ICT Express* **2021**, *7*, 1–4. [CrossRef]
11. Zhou, W.; Li, J.; Liu, Z.; Shen, L. Improving multi-target cooperative tracking guidance for UAV swarms using multi-agent reinforcement learning. *Chin. J. Aeronaut.* **2022**, *35*, 100–112. [CrossRef]
12. Xu, D.; Chen, G. Autonomous and cooperative control of UAV cluster with multi-agent reinforcement learning. *Aeronaut. J.* **2022**, *126*, 932–951. [CrossRef]
13. Zhan, G.; Zhang, X.; Li, Z.; Xu, L.; Zhou, D.; Yang, Z. Multiple-UAV Reinforcement Learning Algorithm Based on Improved PPO in Ray Framework. *Drones* **2022**, *6*, 166. [CrossRef]
14. Yu, C.; Velu, A.; Vinitsky, E.; Wang, Y.; Bayen, A.; Wu, Y. The surprising effectiveness of ppo in cooperative, multi-agent games. *arXiv* **2021**, arXiv:2103.01955.
15. Lowe, R.; Wu, Y.I.; Tamar, A.; Harb, J.; Abbeel, O.P.; Mordatch, I. Multi-agent actor-critic for mixed cooperative-competitive environments. In Proceedings of the Advances in Neural Information Processing Systems 30 (NIPS 2017), Long Beach, CA, USA, 4–9 December 2017; 30p.
16. Iqbal, S.; Sha, F. Actor-attention-critic for multi-agent reinforcement learning. In Proceedings of the International Conference on Machine Learning, Long Beach, CA, USA, 10–15 June 2019; pp. 2961–2970.

17. Rashid, T.; Samvelyan, M.; Schroeder, C.; Farquhar, G.; Foerster, J.; Whiteson, S. Qmix: Monotonic value function factorisation for deep multi-agent reinforcement learning. In Proceedings of the International Conference on Machine Learning, Stockholm, Sweden, 10–15 July 2018; pp. 4295–4304.

18. Schulman, J.; Wolski, F.; Dhariwal, P.; Radford, A.; Klimov, O. Proximal policy optimization algorithms. *arXiv* **2017**, arXiv:1707.06347.

19. Sutton, R.S.; McAllester, D.; Singh, S.; Mansour, Y. Policy gradient methods for reinforcement learning with function approximation. In Proceedings of the Advances in Neural Information Processing Systems 12 (NIPS 1999), Denver, CO, USA, 29 November–4 December 1999; Volume 12.

20. Hasselt, H.V.; Guez, A.; Silver, D. Deep reinforcement learning with double q-learning. In Proceedings of the AAAI Conference on Artificial Intelligence, Phoenix, AZ, USA, 12–17 February 2016; Volume 30.

21. Jo, H.; Lee, H.; Jeon, S.; Kaliappan, V.K.; Nguyen, T.A.; Min, D.; Lee, J.W. Multi-Agent Reinforcement Learning-based UAS Control for Logistics Environments. In Proceedings of the Asia-Pacific International Symposium on Aerospace Technology, Jeju, Korea, 15–17 November 2021.

22. Littman, M.L. Markov games as a framework for multi-agent reinforcement learning. In *Machine Learning Proceedings*; Morgan Kaufmann: Waltham, MA, USA, 1994; pp. 157–163.

23. Glorennec, P.Y. Reinforcement learning: An overview. In Proceedings of the European Symposium on Intelligent Techniques (ESIT-00), Aachen, Germany, 14–15 September 2000; pp. 14–15.

24. Silver, D.; Lever, G.; Heess, N.; Degris, T.; Wierstra, D.; Riedmiller, M. Deterministic policy gradient algorithms. In Proceedings of the International Conference on Machine Learning, Beijing, China, 21–26 June 2014; pp. 387–395.

25. Lillicrap, T.P.; Hunt, J.J.; Pritzel, A.; Heess, N.; Erez, T.; Tassa, Y.; Silver, D.; Wierstra, D. Continuous control with deep reinforcement learning. *arXiv* **2015**, arXiv:1509.02971.

26. Mnih, V.; Kavukcuoglu, K.; Silver, D.; Rusu, A.A.; Veness, J.; Bellemare, M.G.; Hassabis, D. Human-level control through deep reinforcement learning. *Nature* **2015**, *518*, 529–533. [CrossRef]

27. Vaswani, A.; Shazeer, N.; Parmar, N.; Uszkoreit, J.; Jones, L.; Gomez, A.N.; Kaiser, L.; Polosukhin, I. Attention is all you need. In Proceedings of the Advances in Neural Information Processing Systems 30 (NIPS 2017), Long Beach, CA, USA, 4–9 December 2017; 30p.

28. Jeon, S.; Kaliappan, V.K. Dissimilarity Multi Actor Attention Critic based model for robot navigations in cooperative disaster recovery applications. In Proceedings of the International Virtual Conference on Industry 4.0, Amsterdam, The Netherlands, 22–24 September 2022.

29. OpenAi Gym. Available online: https://github.com/openai/gym (accessed on 3 October 2022).

Article

High-Secured Data Communication for Cloud Enabled Secure Docker Image Sharing Technique Using Blockchain-Based Homomorphic Encryption

Vishnu Kumar Kaliappan [1], **Seungjin Yu** [2], **Rajasoundaran Soundararajan** [3], **Sangwoo Jeon** [2], **Dugki Min** [2,*] **and Eunmi Choi** [4]

1   Konkuk Aerospace Design-Airworthiness Research Institute (KADA), Konkuk University, Seoul 05029, Korea; vishnudms@gmail.com
2   Department of Computer Science and Engineering, Konkuk University, Seoul 05029, Korea; seunggin.yu@gmail.com (S.Y.); ndrw5580@gmail.com (S.J.)
3   School of Computing Science and Engineering, VIT Bhopal University, Bhopal 466114, India; rajasoundaransraja@gmail.com
4   Department of Computer Science and Engineering, Kookmin University, Seoul 05029, Korea; emchoi@kookmin.ac.kr
*   Correspondence: dkmin@konkuk.ac.kr

Abstract: In recent years, container-based virtualization technology for edge and cloud computing has advanced dramatically. Virtualization solutions based on Docker Containers provide a more lightweight and efficient virtual environment for Edge and cloud-based applications. Because their use is growing on its own and is still in its early phases, these technologies will face a slew of security issues. Vulnerabilities and malware in Docker container images are two serious security concerns. The risk of privilege escalation is increased because Docker containers share the Linux kernel. This study presents a distributed system framework called Safe Docker Image Sharing with Homomorphic Encryption and Blockchain (SeDIS-HEB). Through homomorphic encryption, authentication, and access management, SeDIS-HEB provides secure docker image sharing. The SeDIS-HEB framework prioritizes the following three major functions: (1) secure docker image upload, (2) secure docker image sharing, and (3) secure docker image download. The proposed framework was evaluated using the InterPlanetary File System (IPFS). Secure Docker images were uploaded using IPFS, preventing unauthorized users from accessing the data contained within the secure Docker images. The SeDIS-HEB results were transparent and ensured the quality of blockchain data access control authentication, docker image metadata denial-of-service protection, and docker image availability.

Keywords: cloud computing; secure communication; docker security; homomorphic encryption; virtualization; blockchain; docker image sharing

Citation: Kaliappan, V.K.; Yu, S.; Soundararajan, R.; Jeon, S.; Min, D.; Choi, E. High-Secured Data Communication for Cloud Enabled Secure Docker Image Sharing Technique Using Blockchain-Based Homomorphic Encryption. *Energies* **2022**, *15*, 5544. https://doi.org/10.3390/en15155544

Academic Editors: R. Maheswar, M. Kathirvelu and K. Mohanasundaram

Received: 23 June 2022
Accepted: 26 July 2022
Published: 30 July 2022

1. Introduction

Over the past decade, a boom in virtualization technologies has made it possible to divide a computer system into many discrete virtual environments. Technology has progressed quickly and offers many advantages. The virtualization of servers in data centers is one of the most popular uses of virtualization technology. An administrator can set up one or more virtual system instances on a single server using server virtualization. These virtual servers may be hired on a subscription basis and function similarly to physical servers. The growing adoption of virtualization technologies drives the desire for a virtualization solution that can offer dense, scalable, and secure user environments. There are various virtualization products on the market right now. They can be divided into the following two primary categories: container-based virtualization and hypervisor-based virtualization. Container-based virtualization is also referred to as operating system virtualization since it allows the virtualization layer to run as an application along with the operating systems [1].

Containers emerged as mini virtual computers, which are lighter and more efficient because a single operating system can manage all hardware. This means that ten times more virtual environments can be run on a physical server than in hypervisor-based virtualization [2]. Containers are considered to be the future of virtual machines [3].

It is typical practice to use docker containers for provision across a shared physical host. A docker image repository simplifies the publishing and shares high-quality docker images generated by docker. Although the rapid deployment and sharing of docker images have the significant advantage of allowing developers to share a wide range of real-time apps [4,5], because there is no specific mechanism for dealing with docker image vulnerabilities, the docker platform is easily vulnerable to numerous security threats. The older packages containing vulnerabilities in the Docker image can be vulnerable to attacks such as DoS, acquire privilege, and so on [6,7].

The distribution of bitcoin mining software is one example of an attack using Docker images; this attack demonstrates that a variety of attacks may be carried out using Docker images irrespective of the target Operating System (OS). As a result, one of the critical difficulties in establishing a reliable Docker environment is the security of Docker images. The fundamental cause of this Docker image vulnerability issue is users failing to perform a separate security verification task on downloaded docker images or docker images that are about to be submitted to a docker image repository.

In light of the preceding factors, we proposed a secure docker image sharing framework with homomorphic encryption and blockchain (SeDIS-HEB) in this paper to secure the docker images. The proposed framework uses homomorphic encryption to offer authentication and access control to metadata for secure docker image sharing. During the uploading step, the SeDIS-HEB framework scans the docker image, extracts the CBE list of the docker, encrypts the matching docker image, and uploads it. Each value is called ImageBCID and is based on hash values, symmetric keys, and user information. The image is passed through user authentication during the sharing procedure via ImageBCID. Later, the system was designed so that it could only be imported via ImageBCID authentication by encrypting and recording the metadata to obtain docker images and CVE lists in the blockchain. The uploader proceeds with the sharing procedure using the ImageBCID issued in the subsequent sharing process. The uploader initially authenticates the user with authorization to the related docker image using the issued ImageBCID and personal information. It then provides access to docker metadata on top of the blockchain for users who pass the authentication. Finally, it generates a new ImageBCID by fusing metadata and homomorphically encrypted data from users that the uploader wishes to share with. The shared user receives the shared docker image via their personal information and ImageBCID.

This study makes the following significant contributions.

- A new paradigm for improving security in docker image sharing is proposed. The Docker Image Sharing with Homomorphic Encryption and Blockchain (SeDIS-HEB) platform enables metadata authentication and access control. The following three key features were developed to ensure authentication and access control: secure docker image upload, secure docker image sharing, and secure docker image download.
- The system structure for secure docker image sharing was implemented for the docker image, ensuring integrity via IFPS.
- The proposed framework was validated with the Proof of concept.

The remainder of the paper is organized as follows: Section 2 elaborates on the background and current research findings of security measures for docker images. Section 3 describes the potential threads in docker images. Section 4 presents the proposed Secure Docker Image Sharing with Homomorphic Encryption and Blockchain (SeDIS-HEB) framework. Section 5 presents the implementation and results analysis, and Section 6 explores the conclusion and future directions.

2. Background

2.1. Docker Image Systems

Docker image—The combination of program source code and library required for service operation is called a Docker image [8]. The service is launched by placing the docker image on top of the docker image container. Docker images must be chosen as the default option, and Docker base images are read-only files. The base image is typically comprised of operating system docker images such as Ubuntu and Debian. Users can then use different environments by layering the packages they want to use on top of the corresponding base image. Docker images can be built using docker build. Docker images can be built in the Advanced Union Mount File System format (AUFS). The AUFS file system supports the mounting of multiple image files using containers. Docker images save much space on these systems by reusing the images used to develop other concepts, allowing for a faster generation.

InterPlanetary File System (IPFS)—Docker images are usually heavy due to their nature. Hence, storage for them in the blockchain is limited. It takes longer at the start to save data in the blockchain since it mines blocks first and then saves data in them. Furthermore, while storing data in blocks, there is a problem with expanding the size of the block. We used IPFS to efficiently separate data storage [9]. A distributed P2P file system that connects all computers is known as an interplanetary File System (IPFS). When the files are shared, IPFS assist in partitioning and distributing them to nodes in the IPFS network. The IPFS hash is returned to the user based on the uploaded file's contents, which will be recognized as the same file if it is kept on the network under a different name. The distributed file can be accessed using the IPFS hash when a user requires a file.

2.2. Secure Docker Image Systems

CVE (Common Vulnerabilities and Exposures)—CVE is an abbreviation for a publicly available list of computer security issues. It is typically referred to as the CVE ID number allocated to a security problem. CVE allows professionals to collaborate to prioritize and address vulnerabilities to manage safer computer systems. CVE is overseen by MITRE Corporation [10], which receives funding from the US Department of Homeland Security's Cybersecurity and Infrastructure Security Agency. When a new CVE is reported, the CVE database registers and identifies each new CVE, allowing developers to stay up to date with the latest information by locating and continuously updating the system contained in their approach. CVE also provides a CVSS score [11] for security, allowing users to determine how secure the data is.

Docker Scan—Docker scanning can generate a list of CVEs for docker images. First and foremost, because the docker image is composed of layers, as previously mentioned, the docker scan examines the packages in each layer. Once verification is completed, the NVD dataset [11] is checked for security issues. In this procedure, we cross-check the CVE list and return the stated information to the user. Anchore [12], Clair [13], and Snyk Engine [14] are examples of standard Docker image scanning software. In this work, we utilized Snyk Engine docker scans to scan images that users download from the hub or their docker images.

Blockchain—Satoshi Nakamoto invented blockchain, a distributed book system, in 2008 [15]. Blockchain technology is primarily kept in a chain-based distributed storage system based on peer-to-peer (P2P) networks of small data units known as "blocks". It is a distributed book storage format that allows arbitrary users to modify the blockchain and anyone to inspect transactions or modifications through the blockchain. This research proposes a distributed system that stores metadata using blockchain as a database via a distributed book storage structure [16]. Each node must be mined to add data to the blockchain. A consensus technique known as proof-of-work [17] is required for each node in the mining process for each new block.

Furthermore, denaturing the data necessitates modification of the majority of the node's books, which is difficult. Blockchain was previously made available to anybody via public blockchain but currently employs private blockchain for usage by a specific organization or hybrid blockchain for a combination of the two. In this research, we used the Ethereum public blockchain.

Smart contract—Ethereum is a public blockchain that may be used to build decentralized applications [18]. The smart contract is a characteristic that distinguishes the Ethereum blockchain from others. An intelligent contract [19], a notion of Ethereum founder Vitalik Buterin, outlines the process that must be executed in advance, and when requested by the contract, it acts as described in the smart contract. The benefit of this method is that if the requester meets the contract requirements without the assistance of a third-party intermediary, the requester provides the desired data and returns it, and if this does not satisfy the criteria, the contract is stopped. Smart contracts are written in their own languages, which can be written in Solidity [20], which is similar to object-oriented programming languages such as JAVA and C++.

Symmetric, asymmetric encryption—In this work, we used various encryption methods to ensure the confidentiality of the data. The encryption method employs symmetric, asymmetric, and homomorphic keys. "Symmetric key encryption" refers to algorithms that enable encryption and decryption using the same key. Symmetric encryption offers the advantage of swiftly encrypting any data. This paper proposed an encryption procedure for datasets containing a large amount of data. The AES256 technique generates a key and encrypts the data using the generated key.

In asymmetric encryption [21], the user has a public and a private key, unlike symmetric keys. Data encrypted with a public key can be decrypted with a private key and vice versa. Asymmetric encryption has the advantage that symmetrically encrypted keys can only be utilized by users who encrypted keys with public keys for that symmetric encryption. Furthermore, asymmetric encryption is used for data registered in the blockchain to maintain the confidentiality of the data transmitted.

Homomorphic Encryption—Homomorphic encryption is a cryptographic method designed to allow operations to be carried out with encrypted data [22]. There are two types of homomorphic encryption: partial homomorphic encryption and fully homomorphic encryption. In this work, partial homomorphic encryption, also known as Pailier methods [23], was used to improve efficiency with less computation and high security. In addition, partial homomorphic encryption is used for access control and authentication.

3. Related Work

Several studies were conducted in order to avoid malicious docker container image distribution as well as malicious forgery or alteration of images in the repository. The relevant studies are described below.

Q. Xu et al. propose a blockchain-based Decentralized Content Trust [24] with the possibility of extending studies on notary services. Once the docker image is verified, the signature is converted into a transaction, and the transaction is uploaded to bitcoin using the Carbon chain's unique library. The purpose was to ensure signed data denial blocking to provide decentralized services. However, while the service validates the signature of the docker image, no scans of the docker image are performed, making it difficult to determine whether the docker image is malicious.

J. Sun et al. [25] developed a Blockchain-based automatic container cloud security by performing a vulnerability check on the docker image and registering it on the Ethereum blockchain. The vulnerability checks on the docker image are unambiguous. The user's information and other data are also recorded, leading to flaws in the docker image and its signature. This determines the CVE linked with the vulnerability and provides precise information about docker images. In this study, the limitation is that personal information could be revealed when data were saved in the blockchain. Similarly, Y. Zheng et al. [26] introduced ZeroDVS, a secure container image based on inheritance graphs. This research aimed to find the vulnerabilities of the docker image through the layer of the docker image. The public container images were checked for inherited images, the relationship was examined, and the user was notified of any vulnerabilities. There is no mechanism for secure sharing in the proposed solution.

To secure applications, S.H. Han et al. [27] developed a container image access control architecture. Unauthorized users are prevented from accessing the container with basic user information. For ID authorization, each time the user ID is retrieved after downloading the container image, it is registered with the kernel policy administrator to grant ID-based privileges. The author claims that by using this strategy, unauthorized users will be unable to utilize the Docker image without authentication.

The Docker Image Vulnerability Diagnostic System (DIVDS) was developed by Soonhong Kwon et al. [11] to allow users to identify the vulnerability level of each docker image through the vulnerability evaluation process. The evaluation process is based on the combined relationship of vulnerable software packages and vulnerability information in a docker image. DIVDS assures a reliable docker-based application build environment by preventing users from downloading or uploading vulnerable docker images in a docker image repository. Similarly, Manish K A et al. [28] proposed architecture to ensure security for docker containers. The proposed architecture includes a plugin that uses a CI/CD pipeline to deploy the application and to ensure the security of the application. The architecture ensures security from the starting stage of application development to the deployment stage including plugin for docker build. The application is bundled in the form of images along with required libraries, pushing the images to a docker registry. The vulnerabilities in both static and dynamic resource allocations are validated. The entire implementation is automated without any manual interventions. The docker security levels' application mechanism was proposed by Vipin Jain et al. [29] to test the different platforms and measure the security attribute to access the resource.

Blockchain-based lightweight security framework named BlockchainBus was proposed by Joseph Doyle et al. [30] to secure virtual machines when migrating from one cloud to another. BlockchainBus is implemented using the HyperLedger solution and deployed on the Microsoft Azure cloud platform. The proposed framework was validated by comparing with the overall VM migration time and was determined as 2.36 s, whereas the overall VM migration time was 5 s. Similarly, a general call policy restriction method for the domestic operating system based on a docker container was proposed by Xu Xin et al. [31]. The proposed scheme provides practical security protection reinforcement for containers. An active defence mechanism was developed to filter and isolate container threats. The following security measures were considered in a study that Xu Xin et al. conducted: filtering system call, authorized installation, real-time signature verification, monitoring, resource control, access control and boundary protection.

Applying Zero Trust Containers Architecture (ZTCA) [32] to secure docker containers was proposed by Darragh Leahy et al. [33]. Initially, the authors investigated the security state of a default deployment of the docker container engine on Linux and analyzed how the Zero Trust containers Architecture can be extended beyond the networking to secure docker deployments. The author validated the ZTCA with the following five test scenarios: (a) grant user access to remote API, (b) prevent Container network attack, (c) review a docker image, (d) review an existing container and (e) deploy a privileged container. Similarly, Robail Yasrab [7] developed a mitigation strategy to secure two types of docker attacks, namely insider and outside attacks. The proposed solution was based on an access control methodology to ensure appropriate access management. In this method, the image maintainers ship the SELinux policy module and its images to the host platform. The policy models for an image are stated in the docker file and placed in the image metadata during creation of the image. The author claims that correct configuration and security policy ensure the greater security of a docker container.

Ferdinand Brasser et al. [34] developed a novel container security architecture named Trusted Container Extension (TCX). The proposed architecture ensured integrity and confidentiality for containers executed in the untrusted cloud. The TCX provided an extended integration to the docker container which was based on AMD Secure Encrypted

Virtualization (SEV) [35] and the kata container project [36]. TCX acts as a secure channel for communication so that the docker cannot distinguish between locally or remotely executed containers. The proposed architecture was validated with three benchmarks, namely (a) computational intensive workloads, (b) network intensive workloads and (c) memory intensive workloads. The author claims a performance impact of 5.77% and network throughput overhead of 22.1% for the NGINX webserver and overhead of 13.36% for the Apache webserver.

According to the literature, no framework specifies how to share the Docker image safely. This allows unauthorized users to edit the docker image and hence update the identity, allowing the attack to proceed through the docker image. This paper examined suitable user authentication for Docker images in order to safeguard their sharing, allowing only authorized users to view and share docker image data. To prevent successive docker images from being lost, the Secure Docker Image Sharing with Homomorphic Encryption and Blockchain (SeDIS-HEB) architecture was developed in this research which accounted for data stability, availability, and integrity.

4. Blockchain-Based Homomorphic Encryption (SeDIS-HEB Framework)

The proposed architecture uses Homomorphic encryption, authentication, and access control to provide docker images securely. The functionality implementation of a docker image with modifications made using the SeDIS-HEB system's existing docker architecture is shown in Figure 1. The SeDIS-HEB system consists of the following four components: the SeDIS-HEB Docker client, the Docker Engine and Docker hub, the IPFS peer-to-peer network, and the Ethereum blockchain peer-to-peer network. The existing Docker client module was modified to include homomorphic encryption authentication. SeDIS-HEB is made up of several submodules that work together to ensure the security of docker image upload, image share, and image download at the DI customer's request.

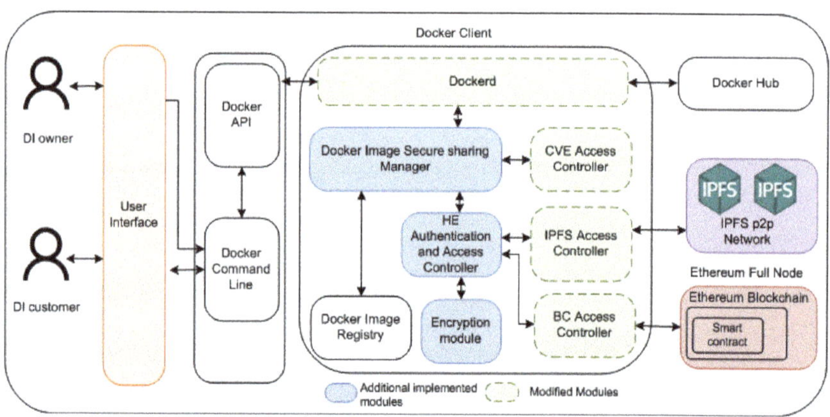

Figure 1. SeDIS-HEB System Architecture.

4.1. SeDIS-HEB Docker Client

The security components are integrated into the existing docker container design. The source code used is from the docker repository [37]. The modified source is updated in the following repository (https://github.com/seunggin/SDIS accessed on 25 July 2022). The following provides a detailed explanation of the three newly integrated modules:

Docker Image Secure Sharing Manager—The docker image secure manager ensures the following three logics: secure docker image upload for secure uploading of docker images, secure docker image share for sharing uploaded docker images, and secure docker images image download for download by the recipient. To make all the three logics possible, the docker manager ensures each controller, encryption module, and registry can be requested to perform the functions and pass on the resulting values.

HE Authentication & Access Controller—Homomorphic encryption, authentication, and access control are provided by the HE authentication and access controller. Upon receipt of essential information such as user data, docker image, CVE lists, HE authentication, and access controller module, the data are cross-checked and sent to the encryption module, where encryption is completed, and the image BCID is received. The image BCID controls access to the IPFS access controller.

Encryption Module—Paillier Homomorphic Encryption (PHE) [23], AES-256 symmetric encryption [38], and RSA-256 Asymmetric Encryption [21] were the three methods of encryption used in this module. The authentication and control access to the individual data were ensured using the Paillier homomorphic encryption technique. Before posting docker images and CVE lists to IPFS, AES-256 was used to encrypt them. To preserve the security of symmetric key data, RSA-256 asymmetric encryption was utilized to decrypt the docker image and CVE list hash values uploaded on top of the blockchain, as well as the encrypted metadata.

4.2. Docker Engine and Docker Hub

This paper used the existing Docker Engine and Docker Hub. Docker Engine is in charge of publishing docker images and managing containers, whereas Docker Hub [39] is in charge of publicly storing docker images.

4.3. IPFS P2P Network

The proposed architecture securely used IPFS to store the docker images and CVE lists. The user-requested files are sent to the user-registered node. The IPFS peer-to-peer (P2P) network is linked and connected with other peer nodes simultaneously to ensure the smooth transfer of the requested data.

4.4. Ethereum Blockchain P2P Network

Through the Smart Contract capability for data registration and invocation, the Ethereum blockchain P2P network grants access control to the docker image metadata.

4.5. Homomorphic Authentication via ImageBCID

Fowler_Noll-Vo function process—The Fowler Noll Vo (FNV) hash function varies substantially with a small change in value. We can see in the above ImageBCID generating procedure that we first processed FNV hash functions before proceeding to homomorphic encryption. This solution requires the following four types of data: user information, hashes obtained after uploading the docker to IPFS, hashes obtained after uploading the CVE list to IPFS, and symmetric keys that encrypt the docker image and CVE list using the AES-256 algorithm. After string processing, the four values were transformed into integers using the FNV hash algorithm. Converting these values to integers allows them to be used in homomorphic encryption. The three docker image-related data cannot be modified during this operation; however, the user information provided by the user can only be submitted by those who know the information accurately. Even if this value is significantly changed, the FNV hash method yields a different output.

4.6. Homomorphic Encryption Authentication and Access Control with ImageBCID

Docker image sharing and downloading require authentication processes in order to gain the required docker image rights. The certification process produced two results. This occurs when authentication fails for the incorrect user but passes for the correct user. When a user fails to authenticate, all functions, regardless of process, are terminated, and the user is notified of the failure. It also restricts access to blockchain data, which is required to obtain IPFS docker images. We provided access control over actions via data for docker image calls on top of the blockchain for users that are correctly logged in. The DI Owner can share docker images with others during this process, and the DI Consumer can decrypt and use docker images that are encrypted and downloaded from the shared docker images.

Use Case1: Secure docker Image Upload—The first use case is to upload the docker images securely. The docker image owner utilizes this functionality to check for vulnerabilities before publishing the docker image. Then, the docker image and CVE list are uploaded after the docker image has been analyzed. The ImageBCID key value will then be returned. This is the most critical value for future Docker image downloads or sharing.

Each CVSS value is generated separately to create the CVE list. The corresponding values are then gathered, and access is limited based on CVSS ratings. As seen in Figure 2, the user initially requests that the Docker image be uploaded. When the command is received, the Docker Command Line performs the Docker Image Secure Sharing Manager's Secure Docker Image Upload logic (DISSM). The DISSM uses the Docker API to request a scan of the Docker image and then uses the CVE Access Controller to obtain the CVSS score of the CVE list. If the CVVS rating exceeds the threshold, the CVE access controller prevents the Docker image from being uploaded. Otherwise, just the necessary docker image and a list of CVEs are returned. The safe docker image upload pseudocode is depicted in code snippet 1.

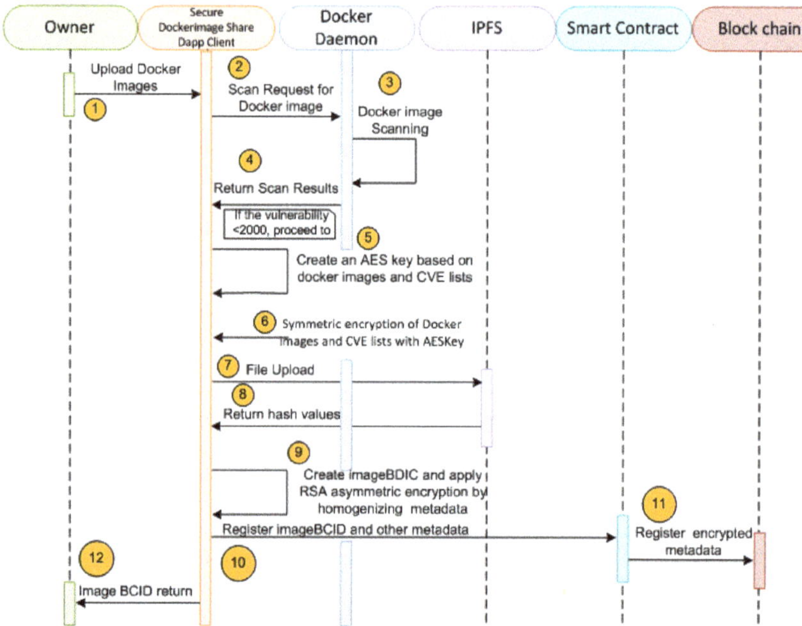

Figure 2. Sequence flow of the secure docker image upload.

Once the vulnerability is identified using the CVE list, the docker image is symmetrically encrypted using AES-256 symmetric encryption and the CVE list. Using IPFS Access Controller, the encrypted picture is then uploaded to IPFS (IPFS-AC). The IPFS-AC generates the corresponding hash value and returns it to the SeID client. As previously shown, the user can submit a Docker image and create an ImageBCID containing all the data. Because user data are hashed with FNV and consists of an ID and password, there is no risk of the value leaking even if it is later recovered. Prior to Homomorphic encryption, the FNV function was used to generate hashes and Homomorphic encryption. This approach creates value using the Paillier Homomorphic encryption technique, which allows encryption with the public key. The imageBCID is formed by combining the values produced via homomorphic encryption. The value is then used as the blockchain's key value in order to obtain further data and authenticate users. The ImageBCID is then registered on the blockchain with key values using BC Access Controller smart contracts. After that, the ImageBCID is returned to the user.

```
1  if docker score = assigned score then
2  deliver the corresponding doker image and CVE list
3  else
4  return Null
5  end if
6  return CVVS score
7  end function
8  CVEList  = CVE_AccessControl (Cvelist.json)
9  AESKey = sha256 (marshalJSON (ownerInfo + CVEList)) //
      Encrypting docker image and CVE Lists
10 enDimage, enCVE = AES256Encryption (AESKey, Docker_Image,
      CVElist) //file upload to IPFS
11 homoCVE, homodimage, homoAESKey, homoOwnerInfo =
      homoRSAEncryption (homoPublicKey, FNVHash (CVE_IPFSHash),
          FNVHash (Dimage_IPFSHash), FNVHash(AESKey), FNVHash(
      ownerInfo)) //return the hash value
12 ImageBCID = makeBCID_by_homoAdd (homoCVE + homoDimage +
      homoAESKey + homoOwnerInfo) // Image BCID creation and
      blockchain registration
13 rsaCVE_IPFSHash, rsaDimage_IPFSHash, rsaAESKey = RSAEncryption
              (ownerPublickey, CVE_IPFSHash, Dimage_IPFSHash,
      AESKey)
14 registertoBlockchain (ImageBCID, rsaCVE_IPFSHash,
      rsaDimage_IPFSHash, rsaAESKey)
```

Code Snippet 1: Secure docker image upload (Input: Address, Docker Image name, CVEs. Output: upload secure docker image, check for vulnerabilities, CVE list, ImageBCID key value.

4.7. Use Case2: Secure Docker Image Share

Secure Docker Image Share presupposes that a docker image is uploaded as indicated in Figure 3, in which the owner of the docker image has a corresponding ImageBCID, and the user who wishes to share the docker image has homomorphically encrypted it. This technique enables docker image owners to share the docker images with requested users safely. The new ImageBCID is generated, and the same is shared with the user. As illustrated in Figure 3, the metadata are retrieved from the docker image through blockchain via the BC access controller and image BCID. The retrieved metadata will be used for authentication using FNV hash functions and homomorphic encryption. Once the data are encrypted, the image is authenticated using the personal password of the docker image owner. Further decryption is carried out if the two values are matched. The safe docker image share pseudo code is depicted in code snippet 2.

To share these values again, it undergoes asymmetric encryption using the DI Consumer's public key. New values are prepared to be registered in the blockchain as a result of this. The data are then registered with the new New Image BCID as the key value by the BC Access Controller. It then returns the user the New Image BCID.

4.8. Use Case3: Secure Docker Image Download

Secure Docker Image Download works on the assumption that a docker image has been uploaded and that the image BCID is known. Figure 4 depicts the full process of the DI Owner downloading the image from the DI Consumer after it has been shared. To begin, the ImageBCID is transmitted to request the download of the corresponding docker image. The SeDIS-HEB Docker client imports Docker image metadata that matches the ImageBCID. Once the matching process is carried out, the docker metadata is decrypted by the DI consumer with the private key transmitted through the FNC. Like the previous homomorphic encryption authentication, the downloader combines the rest of the metadata with their password and proceeds with the authentication procedure. Users who complete the authentication process will be sent encrypted docker images and CVE lists via decrypted data. The docker image and CVE list are then re-symmetrized using asymmetric decrypted symmetric keys. It then returns the user the docker image and CVE list. The pseudocode is illustrated in code snippet 3.

```
1  CVE_IPFSHash, Dimage_IPFSHash, AESKey = RSADecryption (
       ownerPrivateKey, rsaCVE_IPFSHash, rsaDimage_IPFSHash,
       rsaAESKey) // CVE hash, image hash, AES key
2  homoCVE, homoDimage, homoAESKey, homoOwnerInfo =
       homoRSAEncryption     (homoOwnerPublicKey, FNVHash (
       CVE_IPFSHash), FNVHash (Dimage_IPFSHash), FNVHash (AESKey),
       FNVHash (ownerInfo)) // Verification with access control
3  function Verification Result (FHE Access control) Verification
       Result = FHE_AccessControl (imageBCID, homoCVE, homoDimage
       , homoAESKey, homoOwnerInfo) // Verification with access
       control
4  function Verification Result (FHE Access control) Verification
       Result = FHE_AccessControl (imageBCID, homoCVE, homoDimage
       , homoAESKey, homoOwnerInfo)
5  If verification Result == false then
6  return Null
7  else
8  newImageBCID = makeImageBCID (homoConsumerInfo, homoCVE,
       homoDimage, homoAESKey)
9  end if
10 return verification Result
11 end function
12 rsaConCVE_IPFSHash, rsaConDimage_IPFSHash, rsaConAESKey =
       RSAEncryption (consumerPublicKey, CVE_IPFSHash,
       Dimage_IPFSHash, AESKey) // register the newly generated
       BCID with blockchain
13 registertoBlockchain (newImageBCID, rsaConCVE_IPFSHash,
       rsaConDimage_IPFSHash , rsaConAESKey)
```

Code Snippet 2: Secure docker image share (Input: Request docker metadata. Output: New image BCID).

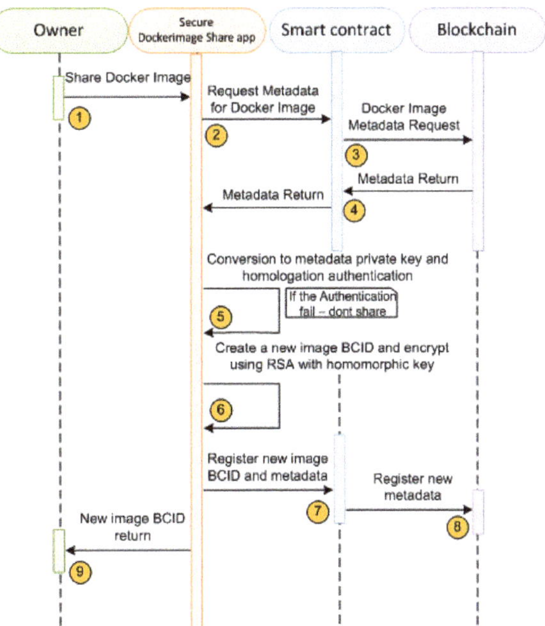

Figure 3. Sequence flow of the secure docker image share.

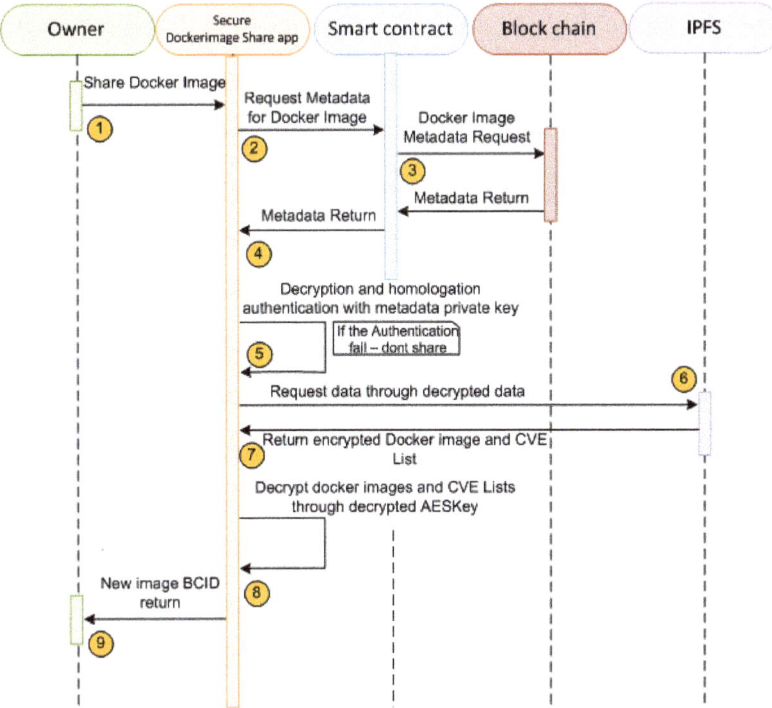

Figure 4. Sequence flow of the secure dockerimage download.

```
1   request Docker Image Metadata (newImageBCID)
2   homoCVE, homoDimage, homoAESKey, homoConsumerInfo =
       homoRSAEncryption (homoConPublicKey, FNVHash (CVE_IPFSHash)
       ,     FNVHash (Dimage_IPFSHash), FNVHash (AESKey), FNVHash(
       ownerInfo)) //homomorphic encryption
3   VerificationResult = FHE_AccessControl (newImageBCID, homoCVE,
        homoDimage, homoAESKey, homoConsumerInfo) //Verification
        of results
4   function Verification Result (FHE Access control)
5   Verification Result = FHE_AccessControl (newImageBCID, homoCVE
        ,
6   homoDimage, homoAESKey, homoConsumerInfo)
7   If verification Result == false then
8   return Null
9   else
10  enCVE, enImage = downloadFromIPFS (CVE_IPFSHash,
        Dimage_IPFSHash)
11  end if
12  return verification Result
13  end function
14  CVElist, Dimage = AES256Decrypt (AESKey, enCVE, enDimage) //
        decryption using AES
```

Code Snippet 3: Secure docker image download (Input: Request docker metadata. Output: Docker image, CVE List).

5. Verification and Validation

Verification and Validation are carried out using the docker that does not provide security guarantees for vulnerable software packages installed within the docker image. This paper applied the proposed SDIS HEB to the docker environment when uploading, sharing, or downloading docker images. To demonstrate the applicability of the SDIS HEB, we verified and validated the SDIS HEB works correctly. SDIS HEB security was ensured with the following quality properties.

5.1. Quality Properties

The quality properties are studied in this section to verify and validate the proposed security mechanism to ensure the docker image is securely shared among the user.

Homomorphic-based authentication—The docker image is not shared with anyone who is not the Data Owner of that docker image or who is not allowed to share it. By collecting the docker image provided by other attackers, this authentication stops the attacker from polluting other people's docker containers. This research provides authentication using ImageBCID, created via homomorphic encryption. Even if an unauthorized user attempts to download with ImageBCID, the download will fail since ImageBCID is not produced using the personal information they provided during the authentication process.

Docker Image Content-Based Access Control with Homomorphic Encryption—In this research, we used ImageBCID to control metadata access for Docker images stored on the blockchain. The docker image stored on IPFS cannot be accessible if no metadata is imported. The role of ImageBCID is to offer access control to blockchain data activities. ImageBCID contains metadata from Docker images as well as personal passwords that are encrypted after the hash, granting them access to data in the blockchain via that value. This document stops users from computing their data if they fail to authenticate in an Encryption module. In other words, the blockchain's access control fails to retrieve docker images and CVE lists from IPFS.

Confidentiality of data uploaded to blockchain and IPFS—The system then uses symmetric and asymmetric encryption to encrypt data written on the blockchain. Symmetric encryption was employed during the encryption operation for CVE lists and docker images with large data volumes. Metadata is intended to discourage users from using IPFS, even if they download it without authorization. Symmetric keys for decrypting metadata and hash values for retrieving data from IPFS are also encrypted and stored on the blockchain using asymmetric keys. As a result, even if the data are obtained by someone else, they cannot be used. During the ImageBCID creation process, the user sends a homomorphically encrypted value to the data owner instead of the original password of the Data consumer and then constructs a new ImageBCID using the encrypted value.

Denial containment of Docker image metadata—If the docker image metadata are corrupted, obtaining the related docker image is generally difficult. On the other hand, this approach prevents metadata deterioration since it maintains metadata about Docker images in the blockchain. Because of its structure, blockchain involves modifying the books of most nodes to corrupt data, which is essentially impossible. If an image posted to this system is infected, it is hard to dispute that the docker image was uploaded.

Stability for Docker Image—Other prior studies identified a lack of research on docker image storage. Therefore, this effort provided availability for this component through distributed storage. Because encrypted docker images and CVE lists are distributed over IPFS, local repositories utilizing existing containers may overcome the disadvantages of disappearing and data disappearing simultaneously, ensuring stability.

Integrity to Docker Image—Because IPFS returns hash values based on the file's contents, the hash values of the file change as the file's content changes. After the encryption procedure in this system, the hash values are placed on the blockchain. Following that, the encrypted hash values are re-imported, assuring the integrity of the docker image and the CVE list.

5.2. Comparative Analysis of the Proposed Work

As shown in Table 1, five related studies and the systems provided in this paper were divided and compared based on six items. We show that there were deficiencies in each related study and that the system structure proposed in this paper provides each of the quality attributes discussed in Section 3. Each property lacking in existing relevant studies is guaranteed through the proposed architecture. We provided non-repudiation for docker image metadata via blockchain, docker image vulnerability check with docker scan, user authentication for docker images, confidentiality guarantees for each docker image and metadata, access control for docker images, and stability in storing docker images reliably.

Table 1. Comparative analysis of SDIS HEB with landmark methods.

Core Security Essentials	Notary	Decentralized Content Trust	Security System	Zero DVS	Container Image Access Control	SDIS HEB
Blockade and Docker Image Information	NO	YES (bitcoin)	YES (Ethereum)	NO	NO	YES (Ethereum)
Vulnerability checking for docker image	NO	NO	YES	YES (image check inheritance)	NO	YES (Docker Scan)
Authentication procedures for docker image unique	NO	NO	NO	NO	NO	YES (ImageBCID)
Ensure confidentiality of docker information	NO	NO	NO	NO	NO	YES (AES, RSA)
Docker image data access control	NO	NO	NO	NO	YES (Container Access control)	YES (ImageBCID)
Stability for docker image	NO	NO	NO	NO	NO	YES (IPFS)

5.3. Evaluation and System Performance Analysis

We can observe that only metadata computations have been performed, with no actual file uploads or downloads. Figures 5–7 depicts the time spent in the sharing procedure.

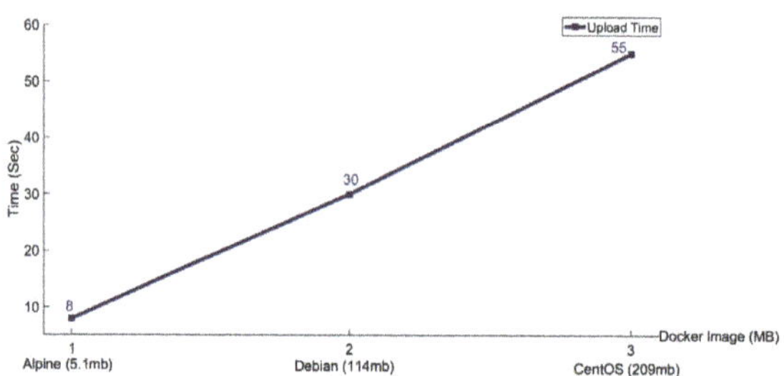

Figure 5. Comparison of upload time with Alpine, Debian and CentOS.

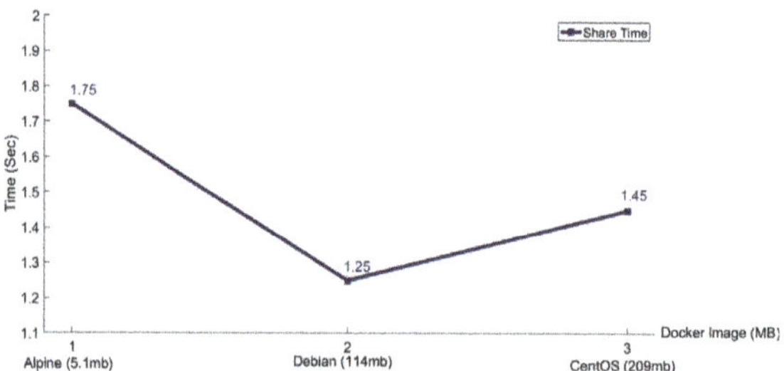

Figure 6. Comparison of shared time with Alpine, Debian and CentOS.

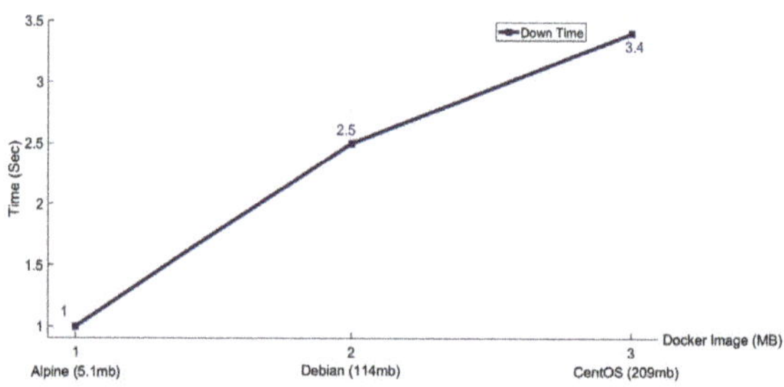

Figure 7. Comparison of download time with Alpine, Debian and CentOS.

The results show that alpine, Debian, and centos were uploaded from the official repository. However, Ubuntu had multiple flaws in the image, causing the upload to fail. Furthermore, ImageBCID was issued for all images using homomorphic encryption in a relatively short period of time using the paillier method. Likewise, wrongly supplied user information can be noticed to keep the user away from the data.

While reusing the docker image, the problem we face is that when the user downloads the data randomly and while sharing with the other user, our proposed work does not block the data flow. This leads to a lack of prevention of the unauthorized reuse of docker images. Currently, we are developing a Digital Right Management (DRM)-based prevention mechanism to prevent the unauthorized reuse of docker images.

6. Conclusions

This study proposed a distributed framework named Safe Docker Image Sharing with Homomorphic Encryption and Blockchain (SeDIS-HEB) for container image security to solve the security issues in one of the most extensively used infrastructures in cloud services. The following major findings were obtained:

(a) In the proposed frameworks, the information about the Docker image owner and docker image consumer users is accessed only by authorized users. This is achieved by providing a control scheme approach through authentication procedures and blockchain data through homomorphic encryption.

(b) While storing the docker image in IPFS, the confidentiality and integrity of data are ensured through asymmetric encryption.

(c) Homomorphic encryption allows users to secure their privacy by using features that can be computed by encrypting DI Consumer's privacy during docker image sharing.

The SeDIS-HEB architecture combines cutting-edge decentralized technologies such as IFPS and blockchain as core kernels. A secure scan will check for docker images on image upload to discover vulnerabilities in docker images, ensuring docker image metadata denial-of-service protection. We also provided authentication and access control using ImageBCID, which was constructed using homomorphic encryption, and privacy for the remaining data using various asymmetric and symmetric encryption methods. Furthermore, we are introducing the unauthorized reuse of docker images utilizing Digital Right Management (DRM) technology to improve the security of the docker image.

Author Contributions: Conceptualization, V.K.K., S.Y. and D.M.; methodology, D.M., S.Y.; software, S.Y.; validation, V.K.K., S.J., D.M. and R.S.; formal analysis, V.K.K.; investigation, D.M.; resources, E.C.; data curation, E.C.; writing—original draft preparation, V.K.K. and S.Y.; writing—review and editing, V.K.K. and S.Y.; visualization, D.M. and E.C.; supervision, D.M.; project administration, D.M.; funding acquisition, D.M. All authors have read and agreed to the published version of the manuscript.

Funding: This research was supported by the Basic Science Research Program through the National Research Foundation of Korea (NRF) funded by the Ministry of Education (No. 2020R1A6A 1A03046811). This work was supported by the National Foundation of Korea (NRF) grant funded by the Korea government (Ministry of Science and ICT (MIST)) (No. 2021R1A2C209494311).

Institutional Review Board Statement: Not applicable.

Informed Consent Statement: Not applicable.

Data Availability Statement: Not applicable.

Conflicts of Interest: The authors declare no conflict of interest.

References

1. Bernstein, D. Containers and cloud: From lxc to docker to kubernetes. *IEEE Cloud Comput.* **2014**, *1*, 81–84. [CrossRef]
2. Burniske, C. Containers: The Next Generation of Virtualization? 2016. Available online: https://ark-invest.com/articles/analyst-research/containers-virtualization/ (accessed on 25 July 2022).
3. Rodriguez, M.A.; Buyya, R. Container-based cluster orchestration systems: A taxonomy and future directions. *Softw. Pract. Exp.* **2019**, *49*, 698–719. [CrossRef]
4. Merkel, D. Docker: lightweight linux containers for consistent development and deployment. *Linux J.* **2014**, *239*, 2.
5. Boettiger, C. An introduction to Docker for reproducible research. *ACM SIGOPS Oper. Syst. Rev.* **2015**, *49*, 71–79. [CrossRef]
6. Tunde-Onadele, O.; He, J.; Dai, T.; Gu, X. A study on container vulnerability exploit detection. In Proceedings of the 2019 IEEE International Conference on Cloud Engineering (IC2E), Prague, Czech Republic, 24–27 June 2019; pp. 121–127.
7. Yasrab, R. Mitigating docker security issues. *arXiv* **2018**, arXiv:1804.05039.
8. Rad, B.B.; Bhatti, H.J.; Ahmadi, M. An introduction to docker and analysis of its performance. *Int. J. Comput. Sci. Netw. Secur. (IJCSNS)* **2017**, *17*, 228.
9. Rajalakshmi, A.; Lakshmy, K.; Sindhu, M.; Amritha, P. A blockchain and ipfs based framework for secure research record keeping. *Int. J. Pure Appl. Math.* **2018**, *119*, 1437–1442.
10. MITRE. CVE Records. Available online: https://www.cve.org/ResourcesSupport/Resources (accessed on 17 June 2022).
11. Kwon, S.; Lee, J.H. Divds: Docker image vulnerability diagnostic system. *IEEE Access* **2020**, *8*, 42666–42673. [CrossRef]
12. Anchore. *Docker Image Security*; Anchore: Santa Barbara, CA, USA, 2022.
13. Clair. Clair—Static Analysis of Vulnerabilities. 2020. Available online: https://github.com/quay/clair (accessed on 23 June 2022).

14. Snyk. Snyk Engine. 2019. Available online: https://snyk.io/product/open-source-security-management/ (accessed on 12 March 2019).

15. Nakamoto, S. Bitcoin: A peer-to-peer electronic cash system. *Decentralized Bus. Rev.* **2008**, 21260. Available online: https://www.researchgate.net/publication/228640975_Bitcoin_A_Peer-to-Peer_Electronic_Cash_System (accessed on 12 March 2019).

16. Naz, M.; Al-zahrani, F.A.; Khalid, R.; Javaid, N.; Qamar, A.M.; Afzal, M.K.; Shafiq, M. A secure data sharing platform using blockchain and interplanetary file system. *Sustainability* **2019**, *11*, 7054. [CrossRef]

17. Mohanta, B.K.; Panda, S.S.; Jena, D. An overview of smart contract and use cases in blockchain technology. In Proceedings of the 2018 9th International Conference on Computing, Communication and Networking Technologies (ICCCNT), Bengaluru, India, 10–12 July 2018; pp. 1–4.

18. Vujičić, D.; Jagodić, D.; Ranđić, S. Blockchain technology, bitcoin, and Ethereum: A brief overview. In Proceedings of the 2018 17th International Symposium Infoteh-jahorina (Infoteh), East Sarajevo, Bosnia and Herzegovina, 21–23 March 2018; pp. 1–6.

19. Buterin, V. *A Next-Generation Smart Contract and Decentralized Application Platform*; White Paper; nft2x.com: New York, NY, USA, 2014; Volume 3, pp. 1–2.

20. Solidity. Object-Oriented, High-Level Language. 2016. Available online: https://docs.soliditylang.org/en/v0.8.11/ (accessed on 2 May 2022).

21. Simmons, G.J. Symmetric and asymmetric encryption. *ACM Comput. Surv. (CSUR)* **1979**, *11*, 305–330. [CrossRef]

22. Ogburn, M.; Turner, C.; Dahal, P. Homomorphic encryption. *Procedia Comput. Sci.* **2013**, *20*, 502–509. [CrossRef]

23. Koç, Ç.K.; Özdemir, F.; Özger, Z.Ö. Paillier Algorithm. In *Partially Homomorphic Encryption*; Springer: Cham, Switzerland, 2021; Volume 20, pp. 95–105.

24. Xu, Q.; Jin, C.; Rasid, M.F.B.M.; Veeravalli, B.; Aung, K.M.M. Blockchain-based decentralized content trust for docker images. *Multimed. Tools Appl.* **2018**, *77*, 18223–18248. [CrossRef]

25. Sun, J.; Wu, C.; Ye, J. Blockchain-based Automated Container Cloud Security Enhancement System. In Proceedings of the 2020 IEEE International Conference on Smart Cloud, Washington, DC, USA, 6–8 November 2020; pp. 1–6.

26. Zheng, Y.; Dong, W.; Zhao, J. ZeroDVS: Trace-ability and security detection of container image based on inheritance graph. In Proceedings of the IEEE 5th International Conference on Cryptography, Security and Privacy, CSP 2021, Zhuhai, China, 8–10 January 2021; pp. 186–192.

27. Han, S.H.; Lee, H.K.; Lee, S.T.; Kim, S.J.; Jang, W.J. Container Image Access Control Architecture to Protect Applications. *IEEE Access* **2020**, *8*, 162012–162021. [CrossRef]

28. Abhishek, M.K.; Rao, D.R. Framework to Secure Docker Containers. In Proceedings of the 2021 Fifth World Conference on Smart Trends in Systems Security and Sustainability (WorldS4), London, UK, 29–30 July 2021; pp. 152–156.

29. Jain, V.; Singh, B.; Choudhary, N. Audit and Analysis of Docker Tools for Vulnerability Detection and Tasks Execution in Secure Environment. In Proceedings of the International Conference on Emerging Technologies in Computer Engineering, Jaipur, India, 4–5 February 2022; pp. 654–665.

30. Doyle, J.; Golec, M.; Gill, S.S. Blockchainbus: A lightweight framework for secure virtual machine migration in cloud federations using blockchain. *Secur. Priv.* **2022**, *5*, e197. [CrossRef]

31. Xu, X.; Zhang, Y.; Hao, Y.; Jiang, Y.; Geng, M. Research of Container Security Reinforcement Multi-Service APP Deployment for New Power System on Substation. In Proceedings of the 2022 4th Asia Energy and Electrical Engineering Symposium (AEEES), Chengdu, China, 25–28 March 2022; pp. 945–949.

32. Kindervag, J.; Balaouras, S. No more chewy centers: Introducing the zero trust model of information security. *Forrester Res.* **2016**, *3*, 1–15.

33. Leahy, D.; Thorpe, C. Zero Trust Container Architecture (ZTCA): A Framework for Applying Zero Trust Principals to Docker Containers. In Proceedings of the International Conference on Cyber Warfare and Security, Albany, NY, USA, 17–18 March 2022; Volume 17, pp. 111–120.

34. Brasser, F.; Jauernig, P.; Pustelnik, F.; Sadeghi, A.R.; Stapf, E. Trusted Container Extensions for Container-based Confidential Computing. *arXiv* **2022**, arXiv:2205.05747.

35. Kaplan, D. *Protecting VM Register State with SEV-ES*; White Paper; 2017. Available online: www.amd.com (accessed on 25 July 2022).

36. Kata. Kata Containers. 2017. Available online: https://katacontainers.io/ (accessed on 25 July 2022).

37. Docker. Docker. 2013. Available online: https://github.com/docker/docker.github.io (accessed on 25 July 2022).

38. Abdullah, A.M. Advanced encryption standard (AES) algorithm to encrypt and decrypt data. *Cryptogr. Netw. Secur.* **2017**, *16*, 1–11.

39. Yadav, S.P.; Agrawal, K.K.; Bhati, B.S.; Al-Turjman, F.; Mostarda, L. Blockchain-based cryptocurrency regulation: An overview. *Comput. Econ.* **2020**, *59*, 1659–1675. [CrossRef]

Article

Energy-Efficient Offloading Based on Efficient Cognitive Energy Management Scheme in Edge Computing Device with Energy Optimization

Vishnu Kumar Kaliappan [1], Aravind Babu Lalpet Ranganathan [2], Selvaraju Periasamy [3], Padmapriya Thirumalai [4], Tuan Anh Nguyen [1], Sangwoo Jeon [5], Dugki Min [5,*] and Enumi Choi [6]

[1] Konkuk Aerospace Design Airworthiness Institute, Koknkuk University, Seoul 05029, Korea
[2] Department of Computer and Information Science, Annamalai University, Chidambaram 608002, India
[3] Department of Mathematics, Rajalakshmi Institute of Technology, Chennai 600124, India
[4] Melange Academic Research Associates, Puducherry 605004, India
[5] Department of Computer Science and Engineering, Konkuk University, Seoul 05029, Korea
[6] Department of Computer Science and Engineering, Kookmin University, Seoul 05029, Korea
* Correspondence: dkmin@konkuk.ac.kr

Abstract: Edge devices and their associated computing techniques require energy efficiency to improve sustainability over time. The operating edge devices are timed to swap between different states to achieve stabilized energy efficiency. This article introduces a Cognitive Energy Management Scheme (CEMS) by considering the offloading and computational states for energy efficacy. The proposed scheme employs state learning for swapping the computing intervals for scheduling or offloading depending on the load. The edge devices are distributed at the time of scheduling and organized for first come, first serve for offloading features. In state learning, the reward is allocated for successful scheduling over offloading to prevent device exhaustion. The computation is therefore swapped for energy-reserved scheduling or offloading based on the previous computed reward. This cognitive management induces device allocation based on energy availability and computing time to prevent energy convergence. Cognitive management is limited in recent works due to non-linear swapping and missing features. The proposed CEMS addresses this issue through precise scheduling and earlier device exhaustion identification. The convergence issue is addressed using rewards assigned to post the state transitions. In the transition process, multiple device energy levels are considered. This consideration prevents early detection of exhaustive devices, unlike conventional wireless networks. The proposed scheme's performance is compared using the metrics computing rate and time, energy efficacy, offloading ratio, and scheduling failures. The experimental results show that this scheme improves the computing rate and energy efficacy by 7.2% and 9.32%, respectively, for the varying edge devices. It reduces the offloading ratio, scheduling failures, and computing time by 14.97%, 7.27%, and 14.48%, respectively.

Keywords: edge computing; energy efficiency; reward function; state learning

Citation: Kaliappan, V.K.; Lalpet Ranganathan, A.B.; Periasamy, S.; Thirumalai, P.; Nguyen, T.A.; Jeon, S.; Min, D.; Choi, E. Energy-Efficient Offloading Based on Efficient Cognitive Energy Management Scheme in Edge Computing Device with Energy Optimization. *Energies* **2022**, *15*, 8273. https://doi.org/10.3390/en15218273

Academic Editors: R. Maheswar, M. Kathirvelu and K. Mohanasundaram

Received: 14 September 2022
Accepted: 2 November 2022
Published: 5 November 2022

1. Introduction

An edge computing device is a type of hardware that drives edge computing applications in various industries. Edge computing devices are mainly used to accomplish specific tasks given by software applications [1]. The primary purpose of an edge computing device is to manage the application closer, which prevents unwanted data loss [2]. Edge devices are also used to control data flow that occurs among functions and operations. Energy efficiency is the level of energy that is used to perform a task. The energy efficiency level is an important variable in every application and system. Edge computing devices mostly reduce the overall energy consumption rate in a network, which enhances the efficiency of the system [3]. Edge computing devices provide better privacy-preserving policies that reduce energy consumption rates in the authentication process. A key agreement scheme is mostly used in edge computing devices that identify the actual users of devices [4]. A key agreement or value reduces the time consumption rate in the identification process, which improves the effectiveness and reliability of an application. Energy efficiency levels in edge computing devices are high, which reduces the error rate and latency rate in providing services for the users [5].

An edge network is a place where local networks and devices are interfaced with a network connection. An internet connection plays a major role in edge networks. Energy efficiency analysis is performed in the edge network [6]. The efficiency analysis process analyzes the datasets that are available in the database, and provides the necessary set of data for the edge network to perform particular tasks. The energy efficiency analysis process also analyzes the content and time needed to perform a certain task in a network [7]. The energy efficiency analysis process reduces the overall energy consumption rate in the computation process, enhancing the edge network's sustainability and reliability. The energy efficiency analysis process provides an appropriate set of data to provide accurate services for the users [8]. The efficiency analysis process reduces the latency rate in the computation process and improves the speed level of the process. Energy efficiency computing creates a safe path to obtain datasets that reduce the latency rate in the searching and identification process. The computing system enhances the quality of service and information in the edge network. Energy efficiency computing is mainly used in edge networks to improve the privacy and security of users' data and protect them from attackers [9,10].

Machine learning (ML) techniques are primarily used in various fields for prediction, recognition, and detection. ML techniques improve the overall accuracy rate in the prediction process, providing an accurate dataset for multiple functions. ML techniques are used in edge devices to reduce the computation cost and latency rate in the computation process [11]. Edge networks are mostly used in industries and smart devices to provide proper services for users. Convolutional neural network (CNN)-based wireless sensors are used in edge devices [12]. CNN reduces the time consumption rate in the identification process and improves the accuracy rate in the edge node detection process [13]. CNN also increases edge devices' energy efficiency rate, enhancing device efficiency. The deep reinforcement learning (DRL) algorithm is also used for energy efficiency-based edge devices [14]. DRL predicts the exact resources required to perform tasks that reduce the energy consumption rate in the allocation and classification process. DRL improves edge devices' performance and sustainability rate, providing necessary data for further processes in an application and system [15]. This study makes the following significant contributions:

- Designing a cognitive energy management scheme for edge devices by assimilating the machine learning paradigm.
- Introducing a joint scheduling and offloading process for preventing energy convergence across distributed edge devices.
- Performing a comparative analysis study for verifying the proposed scheme's performance with precise metrics and existing methods.

The paper's concept is designed for managing energy efficacy of edge devices through precise selection of scheduling and offloading concepts. The energy efficacy is predominant over increasing the user/application density. After the energy harvesting techniques, fundamental conservation is required for balancing the edge device operations over prolonged time.

The rest of the paper is structured as follows: Section 2 expands on the findings of related work on AI-based approaches for energy management schemes. Section 3 describes the proposed machine learning-based cognitive energy management scheme. Section 4 discusses implementation and results analysis, while Section 5 delves into the conclusion and future directions.

2. Related Works

Ali et al. [16] introduced a deep learning (DL)-based resource allocation approach for Mobile Edge Computing (MEC) systems. The power Migration Expand (powMigExpand) algorithm is used here for the resource allocation process that identifies the critical set of data. The powMigExpand algorithm provides the necessary resources for the allocation process by analyzing the exact device requirement. The proposed method achieves a high accuracy in the allocation process that increases the quality of service (QoS) rate of the MEC system.

Ale et al. [17] proposed a deep reinforcement learning (DRL) algorithm-based energy-efficiency computation offloading method for MEC systems. DRL is mainly used here to determine the exact need and requirements for the computation process. DRL identifies the exact edge nodes and servers for offloading and resource allocation. The proposed method reduces the computation process's time and energy consumption rates. The proposed approach also improves the performance and reliability of the system.

Irtija et al. [18] designed an energy-efficient edge computing system for multi-access edge computing. The proposed method is also used for a fully autonomous aerial system (FAAS) by using a deep neural network (DNN). The DNN-based approach identifies the area of interest (AOI) that provides the necessary set of data for the satisfaction process. From the AOI, the energy information is observed for leveraging deployment. These deployment issues are addressed by projecting energy-efficient devices across the AOI. Therefore, the energy utilization and improvements are linear for the available computing through multiple accesses (V1). The proposed method achieved a high accuracy rate, enhancing the efficiency and sustainability of the MEC and FASS systems.

Lu et al. [19] introduced a scheduling algorithm named the energy-aware double-fitness particle swarm optimization (EA-DFPSO) method for MEC. The PSO algorithm identifies both computation and edge nodes, providing an optimal dataset for the computation process. PSO reduces the energy consumption rate in the computation process, enhancing the MEC reliability and stability. The proposed EA-DEPSO method reduces the latency rate in the computation process and improves the energy efficiency rate of the MEC system.

Wen et al. [20] developed a cluster-based wireless sensor network (WSN) for edge computing. The main aim of the proposed method is to develop an energy-efficient task allocation process. The genetic algorithm (GA) algorithm is used here to identify the requirements necessary for the load balancing process. The proposed WSN method reduces the computation process's energy and time consumption rate, enhancing the edge network's load balancing level.

Xie et al. [21] introduced a minimal retention energy harvesting (MREH) method for edge devices. The proposed method is mainly used for Internet of Things (IoT)-based devices and applications. The MREH method focuses on swapping edge nodes that are needed for energy harvesting. MREH identifies the edge nodes and provides a possible dataset for the computation process. This supplied data equips multiple observation sequences for identifying the swapping instances. Considering the available edge nodes, IoT computations are required for enhancing the EH. The proposed MREH method improves energy efficiency.

Dai et al. [22] proposed a deep reinforcement learning (DRL) algorithm-based partition approach for an edge computing system. Game theory and Deep Neural Network (DNN) approaches are used here to identify the resource required for resource allocation. DNN improves the accuracy rate in partitioning, where it provides possible services for users. The Dai approach reduces the computation processes, time, and energy consumption. Added to it, DRL improves the scalability and efficiency of an edge computing system.

Zhang et al. [23] developed a dynamic programming-based energy-saving offloading (DPESO) for the mobile edge computing offloading (MECO) system. Identifying an offloading decision problem is a complicated task to perform in an edge computing system. The proposed method is mainly used to determine the offloading decision problem presented in a computing system. DPESO minimizes the latency rate in the computation process, improving the MECO system's effectiveness and scalability. The proposed DPESO increases the energy efficiency rate in an edge computing system.

Alsubhi et al. [24] designed a Mobile Energy Augmentation for smart devices using Cloud Computing (MEACC). MEACC is used to find the inefficient energy consumption source and produce a proper dataset for further process. Smart devices utilize more energy to perform specific tasks. MEACC reduces the communication cost and latency rate in the offloading process. The proposed MEACC method reduces the overall energy consumption rate in the computation process, providing a better load balancing level for an edge computing system.

Li et al. [25] introduced a multi-edge collaborative computation offloading strategy for MEC systems. The proposed method calculates the execution time to perform a specific task in the MEC system. A migration strategy is used here that analyzes the datasets that are required for the computation process. The mitigation strategy identifies the previous state utilization and energy drains for preventing failures. The computations are restricted to the energy availability of the devices through multiple remaining energy metrics. Due to the overheads in migration, the overloading tasks are confined using MEC offloading. The proposed method reduces the computation energy demands.

Zhou et al. [26] proposed an edge intelligent energy-efficient model (ECMS) for MEC systems. When compared with other methods, the proposed method improves the performance and feasibility of the MEC system. The Elman neural network (ENN) algorithm is used here that identifies the energy consumption rate to perform tasks in MEC. ENN predicts accurate energy consumption. This energy consumption feature is estimated across different intervals where the communication is either consistent or varying, provided the energy exhaustions are identified.

Liu et al. [27] introduced an energy-aware allocation method for MEC systems. An MEC server identifies the access points (AP) presented in MEC that provide necessary data for the offloading process. The proposed method reduces the complexity and time consumption rate in the computation process. The complexity of the varying multiple access switchovers is addressed using offloading procedures. This process reduces the computations over different intervals to prevent energy drops. The proposed method provides high-quality services for users and reduces the energy consumption rate in the computation process.

Xie et al. [28] proposed an energy-efficient collaborative computation method for MEC networks. The block coordinate descent (BCD) method is used to find the coordinates available in the MEC network. The proposed method is used primarily for reconfigurable intelligent surface (RIS)-assisted MEC networks. In the reconfiguration process, the remaining energy-based allocations are provided across multiple coordinates, preventing MEC delegations. Therefore, energy utilization is reformed in order to avoid various device utilizations. In this process, the variations are provided using available devices. The proposed method improves the energy efficiency level by reducing the energy consumption rate in the computation process.

The methods discussed above rely on assisted networks as in [20,24,28] for energy conservation by incorporating the conventional WSN strategies. Independent processes such as in [18,19,23] provide optimization alongside learning paradigms that increase the complexity during iterated training. The proposed energy management scheme steers between the offloading and scheduling decisions using different state models. In particular, the state model is incorporated into this work due to its action and reinforcement strategies. The action represents the allocation, scheduling, and offloading computed using the available energy and drain. The transitions between the states are required for limited intervals, i.e., before offloading and scheduling. Therefore, recurrent training machine learning is less required for this energy management scheme.

The aforementioned methods and techniques optimize the energy efficiency through harvesting, conservation, and device changeovers. These methods are conventional for improving the current transmissions across multiple shared devices in an edge computing scenario. However, the knowledge of service demands is unknown for the varying intervals, due to which the allocation and prolonged device operations are mandatory. As the energy drain is accepted, the convergence, device failure, or recharging intervals are frequent. Therefore, the device and task swapping instances are confined through pre-exhaustion identification and proper device selection. Considering the energy levels and tasks, the balancing takes place with considerable learning procedures in this proposed scheme. This prevents scheduling failures due to periodic device switching based on energy complexities.

3. Proposed Energy Management Scheme

The edge computing device-based energy efficiency processes specific tasks that different users and software applications observe. The edge computing device model maintains the application closer and requires data storage and a computation process to perform a certain task and enhance the system's efficiency. The challenges in operating the edge devices, such as scheduling, offloading, and allocation, are considered factors for improving the energy efficacy of edge computing to satisfy user needs and demands. The edge computing devices are highly competitive in achieving stabilized energy efficiency between different states considering their energy efficiency. The edge devices and their associated computing are used in many industries for improving energy conservation and distribution, along with allocating time for scheduling and offloading based on the sustainability of the proposed scheme to improve energy availability. However, addressing energy convergence at the time of cognitive management generates the edge device allocation that relies on carbon emissions, deforestation, pollution, and enormous energy consumption. These impacts augment concerns for swapping the edge computing time intervals for offloading or scheduling depending on the load needed to reduce failures. Offloading is one of the solutions used to identify the offloading features when scheduling edge devices through state learning. Offloading is identified in edge computing and requires diverse performances to reallocate the offloading features to save the minimum energy. An enormous amount of user needs and demands are available in energy management, which is feasible in scheduling failures, and computation of device exhaustion is a paramount consideration in edge computing. Figure 1 portrays the proposed CEMS in an edge environment.

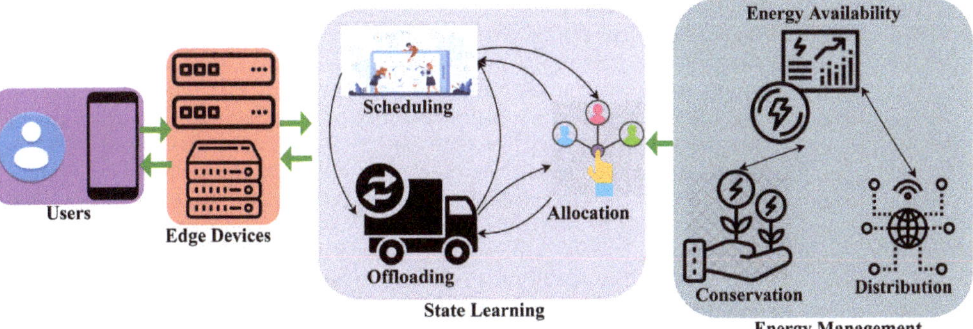

Figure 1. Proposed CEMS in edge environment.

The proposed energy management scheme mainly focuses on consideration by identifying rewards for all successful scheduling over the offloading in edge computing through Q-learning. In this scheme, edge computing is administered to control the data flow when performing functions and operations. The proposed scheme reduces the edge computing latency rate and augments the process's speed (refer to Figure 1). Energy efficiency computing generates a safe path to obtain datasets from edge devices. It prevents latency in identifying and searching processes. In this proposed scheme, the edge computing systems increase the quality of information, and services in the edge network are analyzed to determine any existing edge computing modifications to augment the privacy and security of users' datasets from attackers. Machine learning is a combination of performing the prediction analysis that requires a set of data for various operations in edge computing. State learning is used in this proposed scheme to identify the accurate edge nodes that are aided in performing tasks. Learning reduces the time consumption in the identification process and increases the overall accuracy of the edge node detection process. The energy availability is classified as conservation, and distribution depends on available datasets. In this proposed scheme, the edge computing devices reduce the energy consumption in an edge network, which enhances the sustainability and reliability of the edge network.

The edge devices and their associated techniques improve the speed level of the edge computing process depending on the user's demands through state learning. The energy efficiency analysis is based on the learning process content and computing time to perform a certain task in an edge network. The offloading and computational states for energy efficacy consideration are made. The energy efficiency analysis is based on the edge devices (E_d) in that network. Therefore, the scheduling or offloading based on the load is designed into three segments: scheduling, offloading, and allocating content and time to perform a task. The energy management scheme varies based on users' needs or demands to handle energy availability in that edge network. The initial function of edge device computing is keen on maintaining the quality of service and information, relying on the objective as in Equation (1).

$$\left.\begin{array}{c} \underset{N \in tk}{maximize}\ E_d(c)\ \forall\ Scl = Ofl = Alloc \\ and \\ \underset{N \in E_d(c)}{maximize}\ D_L \forall(\dfrac{Scl}{Ofl}) \end{array}\right\} \tag{1}$$

where:

$$D_L = E_d(c)(C_T - Id_P] \tag{2}$$

$$D_L = \underset{n \in i}{minimize} E_d(c)_{tk}\ \forall\ N \in Scl_{tk} \tag{3}$$

As per Equations (1)–(3), the variables $E_d(c)$, Scl, Ofl, and $Alloc$ represent the edge computing process of performing N tasks depending on scheduling, offloading, and allocation, respectively. In the following next edge device computing, the variables D_L, C_T, and Id_P are used to denote the unwanted data loss, time consumption rate, and identification process, respectively. The third objective is to minimize the data flow that occurs in functions and operations using the $E_d(C)_{tk}\ \forall\ N \in Scl_{tk}$ condition. If $U^S = (1, 2, \ldots, u^s\}$ represents the set of users in edge computing devices, the overall energy consumption in the computation process that enhances the reliability and sustainability of the edge network is based on $E_d(c) \times C_T$, whereas an appropriate set of data provides service for the users of $u^s \times E_d(c)$. The overall energy efficiency analysis is based on $u^s \times E_d(c)$ and $E_d(c) \times C_T$ for energy availability. The control of data flow and offloading is used to perform a certain task. The scheduling of offloading relies on the load to provide better privacy-preserving policies that reduce energy consumption in the authentication process. In this edge computing, energy efficiency analysis is essential to identify data flow, time slots, and devices in that network. The user needs or demands are based on improving the sustainability (st_b) of performing N tasks. The remaining energy consumption time for scheduling and organizing for FCFS for available offloading features relies on energy efficacy for improving device exhaustion. The cognitive management for the available N tasks is performed using machine learning. Later, the edge devices are distributed at the time of scheduling; the edge computing analysis is the improving factor to provide accurate services for the users. From this scheduling or offloading based on the edge, devices are the prevailing instance for the required computation process. The reliability and sustainability level of the system predicting the energy consumption rate for considering the energy-aware allocation are essential in the following section. The offloading process is illustrated in Figure 2.

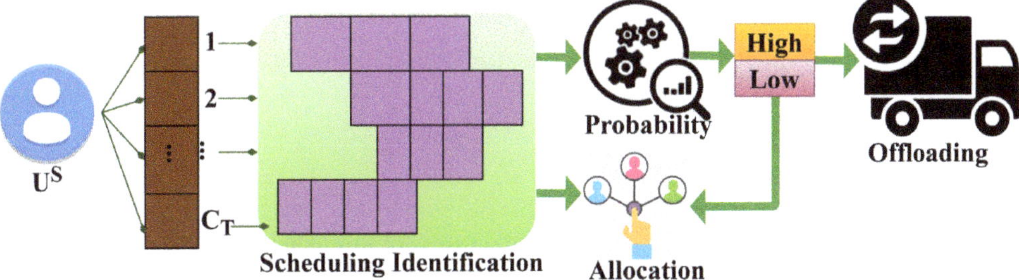

Figure 2. Offloading illustration.

The user tasks (requests) are streamlined from C_T for identifying scheduling probability. If the allocation does not fit the identified C_T (i.e.,) $\frac{N}{C_T} > Alloc$, then the offloading probability is high. Therefore, an edge device (s) is identified as satisfying energy management conditions for offloading. In the offloading process, the energy conditions defined in Equations (1) and (2) are to be satisfied to prevent D_L (refer to Figure 2). The identification process in edge device computing for performing a particular task, which relies on the state learning of $(E_d(c) \times C_T)$ and is computed for improving energy consumption for all N tasks on the basis of sustainability overtime, is the considering factor. The probability of offloading $\left(\rho_{Of} \right)$ in an edge device computation is given as:

$$\rho_{Of} = \left(1 - \rho_{E_d(c)} \right)^{C_T - 1}, \, N \in C_T \tag{4}$$

where:

$$\rho_{E_d(c)} = \left(1 - \frac{E_d(c) \in N}{E_d(c) \in C_T} \right) \tag{5}$$

From Equations (4) and (5), the continuous energy consumption in edge computing relies on the offloading and computational states of N tasks. Therefore, the remaining tasks are performed to swap between different states; hence, the scheduling time is substituted as in Equation (1). Therefore, the offloading computation for $\rho_{E_d(c)}$ follows:

$$Of(N(tk)] = \frac{1}{(Scl + Ofl - Alloc|} \cdot \left(\rho_{E_d(c)} \right)_{C_T}, \, N \in C_T \tag{6}$$

In Equation (6), the edge computing for the N load depending on the scheduling or offloading as in Equation (6) is to satisfy both the condition of $u^s \times E_d(c)$ and $E_d(c) \times C_T$, improving the energy efficiency of edge devices. The offloading process in edge computing is processed using state learning to assign different states for time consumption to reduce the impact of the data flow and reallocation for generating minimum energy based on $(u^s \times E_d(c)) > (E_d(c) \times C_T)$, and the computational states for energy efficacy descriptively use machine learning. Therefore, the successful scheduling over the offloading, which follows $u^s > C_T$ and $\rho_{E_d(c)}$ for minimal energy consumption, is to satisfy Equation (1). The various states depend on $\rho_{E_d(c)}$ and hence the edge computing, resulting in an offloading process for reallocation.

In an edge device computing scenario, the data flow and identification process are performed for the condition $u^s \times E_d(c)$ to maximize energy availability, and the time consumption rate and scheduling failures are invariant. The maximum and minimum scheduling in edge computing identify the offloading along with the reward allocated through Q-learning of N tasks, and the reward allocation is the considering factor here. The probability of reward allocation (ρ_{Rw}) is computed as:

$$\rho_{Rw} = \frac{\rho_{Of}.\, D_L(N(tk)].\left((Scl - Ofl) * \rho_{E_d(c)} - \left(\frac{Scl - Ofl}{N}\right)\right]}{F(S_L).N} \tag{7}$$

where:

$$F(Q_L) = \sum_{N=1} \left(\frac{(Scl - Ofl) * \rho_{E_d(c)} * \rho_{D_L}}{D_L(N)}\right) \tag{8}$$

Based on Equations (7) and (8), the variable $F(Q_L)$ represents the function of Q-learning at different time intervals. For all of the edge device computing processes, the sustainability over time of energy availability is analyzed for N tasks required for allocating the reward. As in Equation (1), the data flow identification requires more energy efficiency. The state models with reward allocations are illustrated in Figure 3.

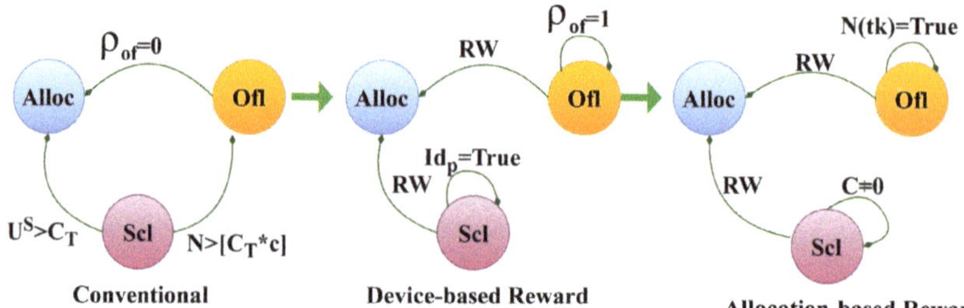

Figure 3. State models with reward allocation.

The Q-learning states are defined for "*Alloc*", "*Ofl*", and "*Scl*" based on different conditions satisfying Equations (1) and (2). This state modeling is distinguished as device-based and allocation-based. The RW is estimated for $\rho_{Of} = 1$ and $Id_p = True$ conditions is the device-based one; the $N(tk)$ and $c \neq 0$ validations are performed in the allocation-based method. The ρ_{RW} is used for computing the feasibility that prevents D_L and high C_T. This state modeling is induced for U^s until resource allocation is made (Figure 3). The continuous edge computing analysis, the energy conservation, and distribution outcomes depend on identifying the minimum or maximum time consumption for swapping the computing intervals for scheduling or offloading of $u^s > C_T$, and N task performance and computing time are the considered metrics. These metrics are addressable using Q-learning and energy management to mitigate the data flows through reward allocation. The decision to allocate rewards to scheduling relies on FCFS for offloading features. The following section represents the energy management scheme for edge computing to reduce edge devices' data flows and time consumption.

Energy Management: Energy availability is classified as conservation and distribution based on performing certain tasks. This scheme is used to control the economy evaluation time for both sequential and individual factors. The energy efficiency analysis is computed to identify the data flows and reallocates tasks in edge devices using machine learning. The edge computing process relies on energy management to reduce the data flows and time consumption when performing a task. Therefore, the condition for energy availability is different for each task in edge devices that follows individual computation processes for enhancing the sustainability of the edge device. The learning process is used for computing the time consumption rate for the N tasks and available offloading features. The first energy availability relies on maximum edge device computing $(E_d(c)_{ea})$ and $F(Q_L)$ is computed as:

$$F(Q_L, E_d(c)_{ea}) = \left(\left(\frac{\rho_{E_d(c)}}{\rho_{e_c} + \rho_{e_d}} \right) \times \frac{1}{n} \right] - D_L(n) + 1 \tag{9}$$

In Equation (9), the variables ea, e_c, and e_d represent the energy availability, energy conservation, and energy distribution in edge computing, respectively, depending on energy efficiency analysis as in $\rho_{E_d(c)}$ and $D_L(n)$ for swapping the states between edge devices. The state learning and reward estimation for ea, ec, and ed are illustrated in Figure 4.

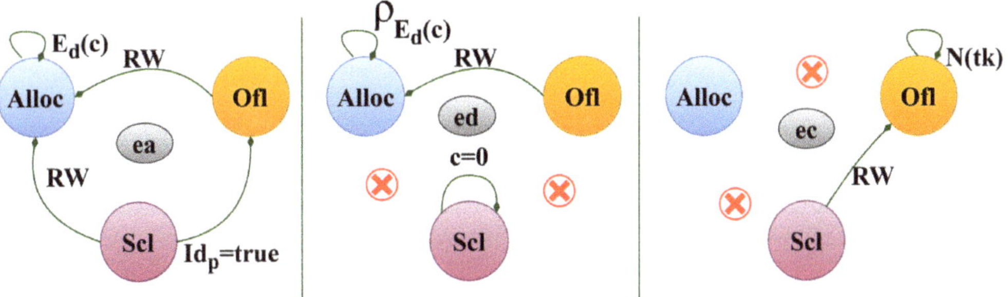

Figure 4. State learning and reward estimation for ea, ec, and ed.

The energy management processes for state modeling and ρ_{RW} are different from that of the device-based process. This model limits the "*Alloc*" based on the *RW* and ec phases. For the ea model, $E_d(c)$ is alone validated for $\rho_{ED}(c) = max$; this induces offloading fewer computations. In the ed phase, if $c = 0$, then *RW* for "*Ofl*" is estimated $\forall\ 0\ \rho_{ED}(c) < 1$ such that "*Scl*" is temporarily halted. Considering the ec, the ea is determined by identifying $N(tk)$ and *RW* for "*Scl*" to "*Ofl*", preventing any new "*Alloc*" (refer to Figure 4). Here, the chances of device exhaustion through previous reward achieving continuous energy distribution is computed as:

$$\rho_{e_d} = \frac{1}{\sqrt{2N}} experssion \left(\frac{Scl - \rho_{E_d(c)}}{N} \right) \tag{10}$$

In Equation (10), the probability of edge computing, the objective is to balance users and time is to minimize the data flow; hence, the actual energy distribution in that edge network is computed as:

$$Scl = max \left(\frac{\rho_{E_d(c)}}{D_L(N) - \rho_{e_c}} \right) \tag{11}$$

Therefore, the energy distribution in the edge device is validated as $\left(1 - \left(\frac{\rho_{E_d(c)}}{D_L(N) - \rho_{e_c}} \right) \right)$ and the time consumption of the energy management process in the device allocation instance depends on the scheduling process. The exceeding time slot and edge devices require different states and hence the energy availability is demandingly improved. The energy availability, conservation, and distribution probability are considered in edge computing. The offloading occurred in the edge network for the condition (Scl, Ofl) is differentiated based on ρ_{e_d} for $F(Q_L)$, and is given as:

$$D_L(N) = \begin{cases} \dfrac{N - (\rho_{ESG} * E_d)}{n + \left(\rho_{R_f}\right)} & \forall\ Scl = E_d(c) \\[4mm] \dfrac{N - \rho_{E_d(c)}}{N + \left(\rho_{e_d} + \rho_{e_c} - \rho_{D_L}\right)} & \forall\ Scl < E_d(c) \end{cases} \tag{12}$$

In Equation (12), the edge computing process of $\left(\rho_{e_d} + \rho_{e_c} - \rho_{D_L}\right)$ is the idle probability for energy conservation, and energy distribution is performed based on the edge network using state learning through $D_L(N)$ analysis. Finally, the differentiation scheduling is presented in Figure 5.

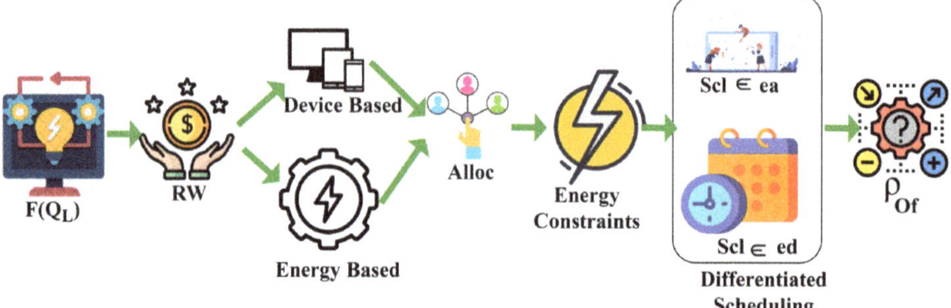

Figure 5. Differentiation scheduling.

The $F(Q_L)$ generates RW for device- and energy-based allocations. These allocations are discussed in Figures 3 and 4; the energy constraints determine the RW for different state models post the ec phase. Therefore, leaving out ec (due to no "$Alloc$"), $Scl \in ea$ and $Scl \in ed$ are performed. It is to be noted that "Ofl" is active $\forall\ Scl \in ed$ alone where device- and energy-dependent RW are computed. This increases the ρ_{RW} and ed, augmenting $F(Q_L, E_d(c)_{ea})$ (refer to Figure 5). Therefore, the energy availability performs the remaining user demands for energy-reserved scheduling or offloading, i.e., the remaining reward, identifying until the device exhaustion is prevented. Therefore, the remaining edge device computing is processed in this continuous manner, reducing the data flows and time consumption in the edge network.

4. Results and Discussion

The results and discussion section presents the self and comparative analysis of the metrics used in the CEMS. First, the scheme was experimentally verified using a Contiki Cooja environment with 110 edge devices and $N = 1200$. The scheduling interval was varied between 5 and 60 min, with a maximum time-out of 12 s. In this scenario, six resource servers were used for computing and allocating requests and resources for the U^s.

Figure 6 presents the self-analysis on $D_L(\%)$ for the varying N and ρ_{of} and $\rho_{ED}(c)$. The proposed scheme relies on two distinct RW for device and energy for performing allocations. For confining D_L, the Id_p in C_T is performed using ρ_{of} probability. In the process, $c \neq 0$ and $N > (C_T * c]$ conditions are validated for concurrent allocations. The allocations are performed with the Rw maximization $\forall\ (tk) = true$ and $\rho_{of} = 1\ (max)$ conditions. Therefore, the D_L facing conditions are suppressed through increasing ρ_{of} and hence allocations are maximum. The continuous computation and energy allocations are performed using the ea and ed phases. In particular, RW is estimated to ec for preventing failures. This means that the $Id_p = True$ is satisfied using "Scl" or "Ofl" and hence the $u^s > C_T$ is handled. Therefore, the continuous allocations endure the available edge devices for N tasks. The energy distribution and allocation are concurrent with preventing D_L in these state models, achieving fair consumption. The ec phase is instigated after this allocation and "Scl". In Figure 7, the energy and Scl analysis for the varying ρ_{Rw} are presented.

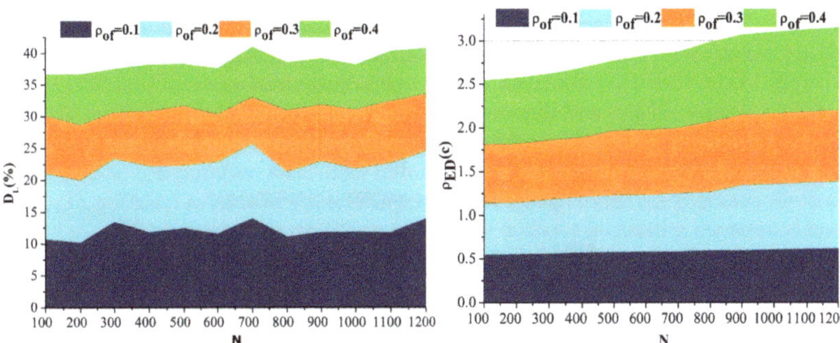

Figure 6. Analysis of D_L and $\rho_{E_d(c)}$ for the varying N.

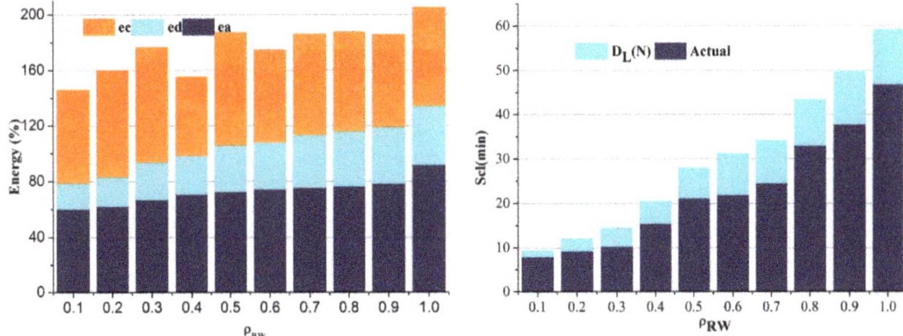

Figure 7. Analysis of energy (%) and Scl (min) for the varying ρ_{Rw}.

The self-analysis for energy (%) and "Scl" for the varying ρ_{RW} is presented in Figure 7. The energies (%) $\forall$ *ea*, *ed*, and *ec* are analyzed in this analysis, and the *ea* is high for the varying rewards due to $\rho_{ED}(c)$ and $N \in C_T$. In the "*Alloc*"-based state model, the "*Scl*" and "*Ofl*" are induced for an increasing $(N * C_T)$ rate. Therefore, the *ed* increases; the *ec* case is different from the other two phases. Depending on the allocation, less $C_T ec$ is performed. This computation determines the devices and remaining N for allocation. Therefore, as the need for *ec* arises to a mid *ed* and *ea*, it is unstable throughout C_T. The "*Scl*" time demands are varying based on "*Alloc*"; if $\rho_{ED}(c)$ is high, then "*Scl*" is high. In particular, the $D_L(N)$ is increased post the *ec* observed. This is intermittent for the distinguishable "*Scl*".

4.1. Comparative Analysis

Depending on the experimental information, the metrics of computing rate, energy efficacy, offloading ratio, scheduling failure, and computing time are comparatively analyzed. The methods DPESO [23], E2E_DRL [17], and EA-DFPSO [18] are used alongside the proposed CEMS in the comparative analysis.

4.2. Computing Rate

As shown in Figure 8, edge device computing performs certain tasks depending on the time consumption rate of energy efficiency and did not control data flow between the different states to swap the allocated timing. The offloading and computational state requires user needs and demands used for identifying the scheduling failure and computing time. The analysis of scheduling and offloading is performed to augment edge devices' sustainability to control functions and operations for energy efficacy using the state learning process. This offloading problem is identified for different edge devices in that network depending on the load for the condition $u^s \times E_d(c)$ and $E_d(c) \times C_T$ used for performing remaining tasks. The continuous edge devices distributed in the network achieve successive

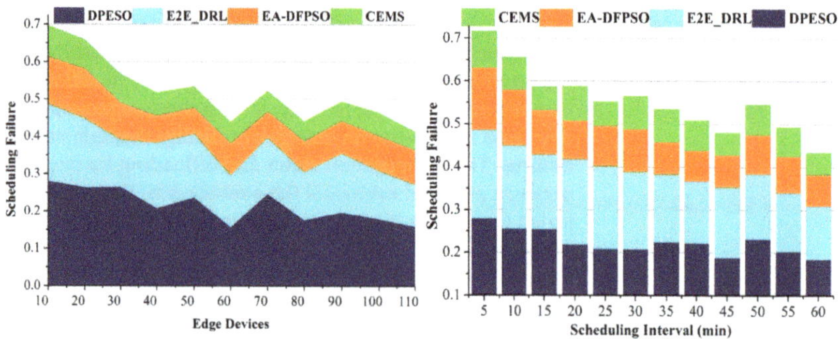

Figure 11. Scheduling failure analysis.

4.6. Computing Time

As shown in Figure 12, the probability of edge device computing performed for energy efficiency analysis using energy management is processed for operating edge devices in that network and does not swap different states at allocated time intervals. The scheduling failure identification is organized for performing certain tasks from the previous reward computed and the offloading ratio is considered for improving computing time. Based on the scheduling, offloading, and allocation for energy management based on the condition $\left(\frac{Scl-\rho_{E_d(c)}}{N}\right)$, the probability is analyzed in a consecutive manner of improving energy efficiency. This scheduling failure identification is computed from the edge devices using the state learning paradigm in current scheduling and relies on allocated time and analysis, preventing offloading. The energy conservation and distribution are analyzed and processed based on the scheduling over the offloading, and the reward is allocated through the Q-learning process at different intervals for maximizing energy. The device allocation relies on energy availability considering the successful scheduling in energy management for which the proposed scheme requires less computing time. The above analysis's summary is tabulated with the findings in Tables 1 and 2 for the varying edge devices and scheduling intervals.

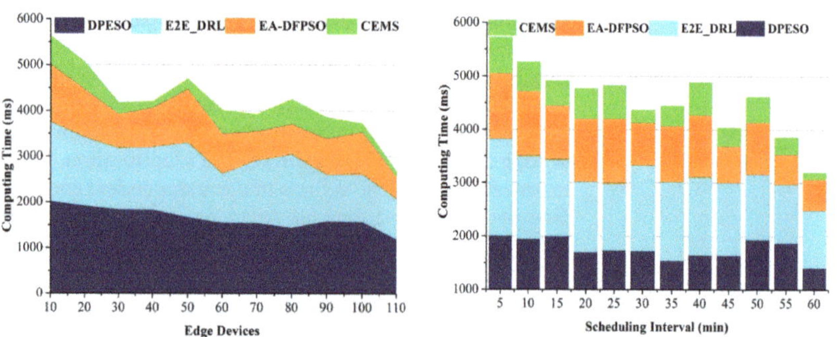

Figure 12. Computing time analysis.

This scheme improves the computing rate and energy efficacy by 7.2% and 9.32%, respectively. It reduces the offloading ratio, scheduling failures, and computing time by 14.97%, 7.27%, and 14.48% respectively.

This scheme improves the computing rate and energy efficacy by 7.43% and 9.36%, respectively. It reduces the offloading ratio, scheduling failures, and computing time by 15.81%, 7.81%, and 14.93% respectively.

Table 1. Analysis summary for edge devices.

Metrics	DPESO	E2E_DRL	EA-DFPSO	CEMS
Computing Rate (Scheduling/Devices)	22	41	51	67
Energy Efficiency	54.96	64.78	73.66	83.11
Offloading Ratio	43.25	32.36	27.52	19.406
Scheduling Failures	0.158	0.113	0.094	0.049
Computing Time (ms)	1193.1	879.2	491.2	112.21

Table 2. Analysis summary for scheduling interval.

Metrics	DPESO	E2E_DRL	EA-DFPSO	CEMS
Computing Rate (Scheduling/Devices)	22	39	52	68
Energy Efficiency	55.34	65.25	74.47	83.746
Offloading Ratio	42.69	33.79	27.78	18.947
Scheduling Failures	0.185	0.125	0.074	0.0499
Computing Time (ms)	1414.1	1072.8	596.4	107.29

5. Conclusions

This article introduces a cognitive management scheme to improve edge devices' computation and energy efficacy. This scheme performs differentiated task scheduling and cognitive offloading using state learning. The scheduling is based on a first-come, first-serve process wherein the offloading is performed using energy allocation feasibility. The continuous energy allocation, distribution, and computation factors are analyzed using independent state models. The models increase the chances for energy conservation amid the distribution and allocation phases. In the energy-conserved phase, the allocations are prevented, preventing data losses and hence early energy exhaustion. The reward function is used for identifying the offloading/scheduling-required intervals. This facilitates the decision on energy distribution or allocation for the pending and new tasks. In the concurrent allocation intervals, device availability and energy conservation features are estimated using the current states in maximizing energy efficacy. This scheme improves the computing rate and energy efficacy for varying intervals by 7.43% and 9.36%, respectively. It reduces the offloading ratio, scheduling failures, and computing time by 15.81%, 7.81%, and 14.93%, respectively.

Author Contributions: Conceptualization, V.K.K., and D.M.; methodology, D.M. , E.C., V.K.K., and P.T.; software, V.K.K., A.B.L.R., and P.T.; validation, T.A.N., S.P., and S.J.; formal analysis, P.T., and T.A.N.; investigation, A.B.L.R., S.P., and D.M.; resources, A.B.L.R., and T.A.N.; data curation, E.C.; writing—original draft preparation, V.K.K., A.B.L.R.,and S.P.; writing—review and editing, V.K.K, D.M., and S.P., visualization, D.M. and E.C.; supervision, D.M.; project administration, D.M.; funding acquisition, D.M. All authors have read and agreed to the published version of the manuscript.

Funding: This research was supported by the Basic Science Research Program through the National Research Foundation of Korea (NRF) funded by the Ministry of Education (No. 2020R1A6A 1A03046811). This work was supported by the National Foundation of Korea (NRF) grant funded by the Korea government (Ministry of Science and ICT (MIST)) (No. 2021R1A2C209494311).

Institutional Review Board Statement: Not applicable.

Informed Consent Statement: Not applicable.

Data Availability Statement: Not applicable.

Conflicts of Interest: The authors declare no conflict of interest.

References

1. Hu, X.; Wang, L.; Wong, K.K.; Tao, M.; Zhang, Y.; Zheng, Z. Edge and central cloud computing: A perfect pairing for high energy efficiency and low-latency. *IEEE Trans. Wirel. Commun.* **2019**, *19*, 1070–1083. [CrossRef]
2. Chen, Y.; Zhang, N.; Zhang, Y.; Chen, X.; Wu, W.; Shen, X.S. TOFFEE: Task offloading and frequency scaling for energy efficiency of mobile devices in mobile edge computing. *IEEE Trans. Cloud Comput.* **2019**, *9*, 1634–1644. [CrossRef]
3. Kang, M.; Li, X.; Shi, Z.; Rahman, T. Collaboration-oriented Boundary Energy Efficiency Detection for Continuous Objects in Edge Networks. *Procedia Comput. Sci.* **2022**, *202*, 128–134. [CrossRef]
4. Zhang, J.; Zheng, R.; Zhao, X.; Zhu, J.; Xu, J.; Wu, Q. A computational resources scheduling algorithm in edge cloud computing: From the energy efficiency of users' perspective. *J. Supercomput.* **2022**, *78*, 9355–9376. [CrossRef]
5. Guo, M.; Li, Q.; Peng, Z.; Liu, X.; Cui, D. Energy harvesting computation offloading game towards minimizing delay for mobile edge computing. *Comput. Netw.* **2022**, *204*, 108678. [CrossRef]
6. Chen, Y.; Zhang, N.; Zhang, Y.; Chen, X.; Wu, W.; Shen, X. Energy efficient dynamic offloading in mobile edge computing for internet of things. *IEEE Trans. Cloud Comput.* **2019**, *9*, 1050–1060. [CrossRef]
7. Liu, P.; An, K.; Lei, J.; Zheng, G.; Sun, Y.; Liu, W. SCMA-Based Multiaccess Edge Computing in IoT Systems: An Energy-Efficiency and Latency Tradeoff. *IEEE Internet Things J.* **2021**, *9*, 4849–4862. [CrossRef]
8. Zhou, Z.; Li, Y.; Li, F.; Cheng, H. An Intelligence Energy Consumption Model based on BP Neural Network in Mobile Edge Computing. *J. Parallel Distrib. Comput.* **2022**, *167*, 211–220. [CrossRef]
9. Li, C.; Zhang, Y.; Gao, X.; Luo, Y. Energy-latency tradeoffs for edge caching and dynamic service migration based on DQN in mobile edge computing. *J. Parallel Distrib. Comput.* **2022**, *166*, 15–31. [CrossRef]
10. Ren, D.; Li, X.; Zhou, Z. Energy-efficient sensory data gathering in IoT networks with mobile edge computing. *Peer-Netw. Appl.* **2021**, *14*, 3959–3970. [CrossRef]
11. Khan, I.; Tao, X.; Rahman, G.S.; Rehman, W.U.; Salam, T. Advanced energy-efficient computation offloading using deep reinforcement learning in MTC edge computing. *IEEE Access* **2020**, *8*, 82867–82875. [CrossRef]
12. Zeng, Q.; Du, Y.; Huang, K.; Leung, K.K. Energy-efficient resource management for federated edge learning with CPU-GPU heterogeneous computing. *IEEE Trans. Wirel. Commun.* **2021**, *20*, 7947–7962. [CrossRef]
13. Zheng, X.; Shah, S.B.H.; Bashir, A.K.; Nawaz, R.; Rana, U. Distributed hierarchical deep optimization for federated learning in mobile edge computing. *Comput. Commun.* **2022**, *194*, 321–328. [CrossRef]
14. Hu, Y.; Huang, H.; Yu, N. Resource Optimization and Device Scheduling for Flexible Federated Edge Learning with Tradeoff Between Energy Consumption and Model Performance. *Mob. Netw. Appl.* **2022**, *27*, 2118–2137. [CrossRef]
15. Pérez, S.; Arroba, P.; Moya, J.M. Energy-conscious optimization of Edge Computing through Deep Reinforcement Learning and two-phase immersion cooling. *Future Gener. Comput. Syst.* **2021**, *125*, 891–907. [CrossRef]
16. Ali, Z.; Khaf, S.; Abbas, Z.H.; Abbas, G.; Muhammad, F.; Kim, S. A deep learning approach for mobility-aware and energy-efficient resource allocation in MEC. *IEEE Access* **2020**, *8*, 179530–179546. [CrossRef]
17. Ale, L.; Zhang, N.; Fang, X.; Chen, X.; Wu, S.; Li, L. Delay-aware and energy-efficient computation offloading in mobile-edge computing using deep reinforcement learning. *IEEE Trans. Cogn. Commun. Netw.* **2021**, *7*, 881–892. [CrossRef]
18. Irtija, N.; Anagnostopoulos, I.; Zervakis, G.; Tsiropoulou, E.E.; Amrouch, H.; Henkel, J. Energy efficient edge computing enabled by satisfaction games and approximate computing. *IEEE Trans. Green Commun. Netw.* **2021**, *6*, 281–294. [CrossRef]
19. Lu, Y.; Liu, L.; Gu, J.; Panneerselvam, J.; Yuan, B. EA-DFPSO: An intelligent energy-efficient scheduling algorithm for mobile edge networks. *Digit. Commun. Netw.* **2022**, *8*, 237–246. [CrossRef]
20. Wen, J.; Yang, J.; Wang, T.; Li, Y.; Lv, Z. Energy-efficient task allocation for reliable parallel computation of cluster-based wireless sensor network in edge computing. *Digit. Commun. Netw.* **2022**. [CrossRef]
21. Xie, Z.; Poovendran, P.; Premalatha, R. Retention based energy harvesting technique for efficient internet of things aided edge devices. *Sustain. Energy Technol. Assess.* **2021**, *47*, 101424. [CrossRef]
22. Dai, H.; Wu, J.; Wang, Y.; Xu, C. Towards scalable and efficient Deep-RL in edge computing: A game-based partition approach. *J. Parallel Distrib. Comput.* **2022**, *168*, 108–119. [CrossRef]
23. Zhang, Y.; Fu, J. Energy-efficient computation offloading strategy with tasks scheduling in edge computing. *Wirel. Netw.* **2021**, *27*, 609–620. [CrossRef]
24. Alsubhi, K.; Imtiaz, Z.; Raana, A.; Ashraf, M.U.; Hayat, B. MEACC: An energy-efficient framework for smart devices using cloud computing systems. *Front. Inf. Technol. Electron. Eng.* **2020**, *21*, 917–930. [CrossRef]
25. Li, C.; Cai, Q.; Luo, Y. Multi-edge collaborative offloading and energy threshold-based task migration in mobile edge computing environment. *Wirel. Netw.* **2021**, *27*, 4903–4928. [CrossRef]
26. Zhou, Z.; Shojafar, M.; Abawajy, J.; Yin, H.; Lu, H. ECMS: An Edge Intelligent Energy Efficient Model in Mobile Edge Computing. *IEEE Trans. Green Commun. Netw.* **2021**, *6*, 238–247. [CrossRef]
27. Liu, X.; Liu, J.; Wu, H. Energy-aware allocation for delay-sensitive multitask in mobile edge computing. *J. Supercomput.* **2022**, *78*, 16621–16646. [CrossRef]
28. Xie, W.; Li, B.; Xiong, Y.; Liu, W.; Ou, J.; Fan, D. Energy efficient collaborative computation for double-RIS assisted mobile edge networks. *Phys. Commun.* **2022**, *2*, 101774. [CrossRef]

Review

Review of Next-Generation Wireless Devices with Self-Energy Harvesting for Sustainability Improvement

James Deva Koresh Hezekiah [1], Karnam Chandrakumar Ramya [2], Sathya Bama Krishna Radhakrishnan [3], Vishnu Murthy Kumarasamy [4], Malathi Devendran [5], Avudaiammal Ramalingam [6] and Rajagopal Maheswar [7,*]

1   Department of Electronics and Communication Engineering, KPR Institute of Engineering and Technology, Coimbatore 641407, Tamil Nadu, India; jamesdevakoresh@gmail.com
2   Department of Electrical and Electronics Engineering, Sri Krishna College of Engineering and Technology, Coimbatore 641008, Tamil Nadu, India; ramyakc@skcet.ac.in
3   Department of Computer Science and Engineering, Sathyabama Institute of Science and Technology, Chennai 600119, Tamil Nadu, India; rsathyarajeswari@gmail.com
4   Department of Electrical and Electronics Engineering, Sri Krishna College of Technology, Coimbatore 641042, Tamil Nadu, India; drkvm25@gmail.com
5   Department of Electronics and Communication Engineering, Kongu Engineering College, Erode 638060, Tamil Nadu, India; malathid2001@gmail.com
6   Department of Electronics and Communication Engineering, St. Joseph's College of Engineering, Chennai 600119, Tamil Nadu, India; avudaiammalr@stjosephs.ac.in
7   Department of Electronics and Communication Engineering, Centre for IoT and AI (CITI), KPR Institute of Engineering and Technology, Coimbatore 641407, Tamil Nadu, India
*   Correspondence: maheshh3@rediffmail.com

Abstract: Wireless methodologies are the focal point of electronic devices, including telephones, computers, sensors, mobile phones, laptops, and wearables. However, wireless technology is not yet utilized extensively in underwater and deep-space communications applications, and it is also not applied in certain critical medical, military, and industrial applications due to its limited battery life. Self-energy-harvesting techniques overcome this issue by converting ambient energy from the surroundings into usable power for electronic devices; devices that use such techniques are next-generation wireless devices that can operate without relying on external power sources. This methodology improves the sustainability of the wireless device and ensures its prolonged operation. This article gives an in-depth analysis of the recent techniques that are implemented to design an efficient energy-harvesting wireless device. It also summarizes the most preferred energy sources and generator systems in the present trends. This review and its summary explore the common scope of researchers in narrowing their focus in designing new self-energy-harvesting wireless devices.

Keywords: eco-friendly devices; self-energy harvesting; sustainable devices; wireless power transfer; next-generation networks

Citation: Hezekiah, J.D.K.; Ramya, K.C.; Radhakrishnan, S.B.K.; Kumarasamy, V.M.; Devendran, M.; Ramalingam, A.; Maheswar, R. Review of Next-Generation Wireless Devices with Self-Energy Harvesting for Sustainability Improvement. *Energies* **2023**, *16*, 5174. https://doi.org/10.3390/en16135174

Academic Editor: Chunhua Liu

Received: 6 June 2023
Revised: 27 June 2023
Accepted: 3 July 2023
Published: 5 July 2023

1. Introduction

Electronic gadgets and tools that are used to communicate device-to-device and device-to-human through wireless communication are termed wireless devices. Radio waves and infrared signals are some of the widely preferred wireless communication methods due to their reliability and flexible installation. Wireless communication technologies are categorized based on their signal strength and communication distance. Smartphones, laptops, smartwatches, Wi-Fi routers, Bluetooth speakers, and wireless headphones are some of the familiar wireless devices used in day-to-day life. The wireless connectivity in such devices is established through internal or external adapters and receivers. Therefore, wireless devices have the ability to move around anywhere in the network area; furthermore, the constraints of physical cables and wires are eliminated. In recent years, wireless communication devices have been incorporated into various sectors for improving their mobility and productivity [1–5]. Figure 1 shows sectors that have been widely updated with wireless communication in recent days.

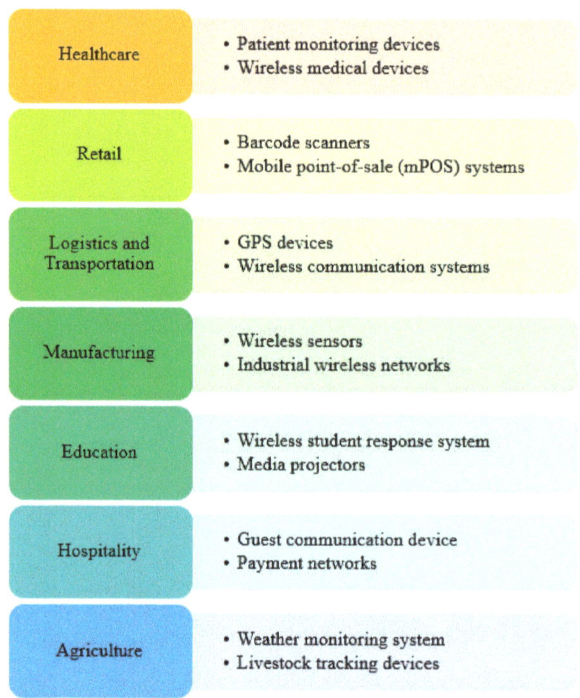

Figure 1. Wireless devices in various sectors.

1.1. Healthcare

Wearable sensors and remote monitoring devices are implemented in recent days to provide a continuous patient-monitoring environment. In some applications, such devices are programmed to create an alert signal upon the critical condition of a patient. The measured readings and signals are transferred to healthcare providers for treatment planning. Medical devices such as pacemakers, insulin pumps, and implantable devices are also equipped with wireless connectivity for improving healthcare system reliability.

1.2. Retail

Handheld barcode scanners have been developed with wireless connectivity for assisting retail workers in providing remote inventory management and price-checking processes. Mobile-POS (point of sale) is the system that is used in retail applications to make contactless payments. Hence, it reduces the purchasing time and improves the flexibility on the payment process.

1.3. Logistic and Transportation

Wireless devices that can access the Global Positioning System (GPS) are utilized in logistic applications for tracking the location of shipments and vehicles in real-time. Similarly, two-way radios and mobile devices are employed in transportation systems to make efficient and timely communication between transport personnel and logistic coordinators.

1.4. Manufacturing or Industrial Sector

Sensors that are equipped with wireless systems are efficient in transmitting the performance and environmental state of the connected equipment. At the same time, such systems are effective in planning process optimization and predictive maintenance. The industrial wireless networks are utilized to provide communication signals between the connected devices, machines, and control units for making an efficient automation system.

1.5. Education

Projectors are connected with computers/laptops/mobile phones for providing wireless media access to the user. This allows the student to view digital content in a perfect way without any peripheral connection. Handheld devices are utilized in certain cases, such as student response systems for marking attendance; other interaction activities such as quizzes are provided thorough additional mobile apps.

1.6. Hospitality

Smartphones and special devices that are equipped with special apps allow customized communication between a guest and the service provider for various requests. To increase customer convenience, direct and indirect payments are made with a wireless point-of-service device.

1.7. Agriculture

The wireless weather monitoring system is one of the most widely used applications of wireless devices; in this device, the sensors are connected to gather plenty of data from the environment to assist the farmer in planning irrigation, crop prediction, and other related activities. In certain applications, the sensors are utilized to monitor the health status of crops from a remote location. This enables farmers to notice several plant diseases at the beginning stage.

The performance of any wireless device may degrade with respect to the functionality factors presented in Table 1.

Table 1. Performance-degrading factors of wireless devices.

Functionality Factors	Description
Signal strength	Distance between the access points, and interference from other devices and obstacles are some of the issues that affect the signal strength of a wireless device.
Interference	Signals from devices that use the same frequency range, such as microwave ovens, cordless mics, and Bluetooth headphones, may interfere with each other and degrade signals' performance in terms of response speed and disconnection.
Frequency band	2.4 GHz and 5 GHz are the most preferred frequency bands for wireless device operation. Comparatively, 2.4 GHz devices receive interference more easily than 5 GHz devices, but 5 GHz devices can be active only for shorter distances of operation.
Channel congestion	Congestion happens in a crowded environment where multiple wireless devices are active. This degrades wireless device performance in terms of response speed and irregular operation.
Environmental factors	Weather conditions, building materials, geographical terrain, and reflective surfaces are some of the environmental factors that impact the signal strength and coverage of a wireless system.
Power source	Wireless devices such as smartphones and laptops are highly dependent upon the energy availability in the device's battery source. Battery level also has a small impact over the signal strength of a wireless connection.

2. Significance of Energy Harvesting for Wireless Devices

2.1. Sustainability

Most wireless devices such as smartphones and wearables work based on the power availability in their connected battery source. However, the energy stored in the battery may drain in an irregular manner based on the operational speed and performance of the connected device. Therefore, it is always expected that users will monitor the energy level of a battery to provide uninterrupted service. Recharging is one of the primary methods that allows the battery to restore its energy. In some cases, the battery is replaced with a new or recharged battery for the device's continuous operation. Self-energy harvesting is incorporated in very rare cases in recent days for increasing the sustainability of the battery-connected device [6].

2.2. Extended Battery Life

Ambient sources such as solar and kinetic energy are widely preferred in wireless devices for restoring the energy in battery modules. This minimizes the downtime and enhances the operation without a frequent recharge or replacement of a battery. Self-energy harvesting methodologies can also increase the operational time of wireless devices in some critical locations [7].

2.3. Mobility

Wireless devices are expected to be independent and more mobile than any other devices. Self-energy harvesting allows wireless devices to meet such expectations, and it allows devices to be operated in off-grid and remote locations for many hours without an external power source requirement. This allows wireless devices to act as outdoor sensors, which can be integrated with IoT technology [8].

2.4. Scalability and Flexibility

Wireless devices with self-energy harvesting can be deployed at any critical locations that cannot be facilitated with a power infrastructure. Therefore, the scalability of such devices can be improved to a certain extent, and the flexibility of such wireless devices allows the system to be incorporated with any other distributed applications [9].

2.5. Reliability and Redundancy

Self-energy-harvesting wireless devices are highly reliable, as they store energy for their own operation. Wire-connected devices operate based on the energy available in their connected terminal; this type of device really suffers during power-outage periods. In addition, the power fluctuation is comparatively minimal in battery-connected devices over the traditional AC circuit. Therefore, it enhances the life of wireless devices. Similarly, the redundancy of wireless devices is also very high, as they do not require any frequent or periodic maintenance [10].

2.6. Environmental Impact

Wireless devices are highly operated from renewable energy sources, and that reduces their carbon footprint. Therefore, wireless devices that are incorporated with a self-sustainable power source are widely acceptable in various sectors and nations [11].

2.7. Cost Efficiency

The operational cost of a wireless device is very low when it is implemented with a self-sustainable energy source as it does not require any replacement or recharging infrastructure [12]. Table 2 indicates the merits of self-energy-harvesting devices.

Table 2. Merits of self-energy-harvesting wireless devices.

Merits	Description
Hassle-free deployment	Absence of wire allows these devices to be plug-and-play.
Non-stop operation	Guarantees continuous operation, as the battery does not need to be replaced.
Compactness	Suitable for ultra-low-power industrial circuits.
Feasibility	Can be employed in places which are unsafe and hard to reach for maintenance.
Power backup	Harvested energy can be stored in a battery source if required.

3. Limitations of Self-Harvesting Wireless Devices

Self-harvesting wireless devices are designed to address the energy source constraint in regular wireless devices [13–15]. However, self-harvesting wireless devices also have certain limitations, as specified in Figure 2.

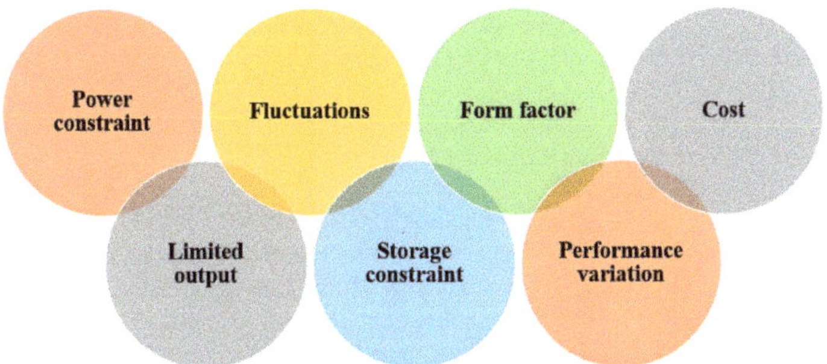

Figure 2. Performance-degrading factors of self-harvesting wireless devices.

3.1. Power Generation Constraints

Self-harvesting systems for wireless devices are highly reliant on environmental considerations. The amount of energy generated is limited when there is reduced solar, thermal, or kinetic energy in the surroundings. For example, solar-powered wireless devices may struggle with respect to solar energy intensity and availability in their surroundings. The performance may degrade on shaded and low-light conditions.

3.2. Limited Power Output

The power generated by a self-harvesting methodology cannot be the same as that of battery-powered energy sources. The generated energy can be sufficient for some low-power operations and low-energy devices; similarly, it cannot be utilized for high-power or high-performance applications.

3.3. Intermittent Power

Self-energy devices are open to energy fluctuations based upon the available energy source. The changes in irregular environmental conditions may lead generators to harvest irregular power outcomes. This kind of intermittent supply may damage the critical circuits of a wireless device, and in some cases, it may lead to interruption and data loss.

3.4. Energy Storage Constraint

Even when wireless devices are equipped with a self-harvesting energy supply, they may require an additional battery source for storage purposes. The saved energy in the battery may be required for some future operation. Rechargeable batteries and capacitors are some of the most used storage methodologies. However, such methodologies are heavily affected by degradation over time and self-discharge. Therefore, the use of batteries can affect the overall efficiency and reliability of the system.

3.5. Device Design and Form Factor

The integration of self-harvesting methodology into a wireless system may increase its size and weight. Therefore, it can limit the device design and form factor on miniaturization, portability, and wearable operation.

3.6. Cost

The installation and manufacturing cost of self-energy wireless devices are comparatively higher than battery-powered devices. The cost of energy-harvesting components, energy storage systems, and power regulatory models makes such systems costlier.

3.7. Performance Variability

The performance of self-harvesting wireless devices may vary based on the technology used for energy generation, user behaviour, and environmental conditions. Location, exposure to the energy source, orientation, and movement patterns are also some of the functionality factors that affect the performance of wireless devices. In some cases, these factors cause the reliability and predictability of wireless devices to be questionable. However, certain power optimization algorithms and methodologies have been developed in recent years to address such issues. The following section explores the present trend in such optimization models.

4. Energy-Harvesting Methodologies for Wireless Devices

Wireless devices are structured with different methodologies for converting ambient energy into useful energy. Solar, kinetic, and thermal energy utilization are some of the widely implemented energy-harvesting methods in wireless devices. Figure 3 indicates the energy-harvesting methods that are used in various applications.

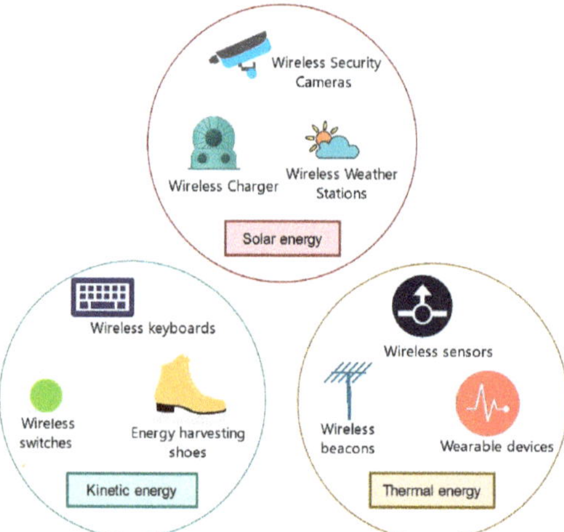

Figure 3. Applications of self-energy-harvesting techniques in wireless devices.

4.1. Solar Energy Harvesting

Solar energy harvesting is one of the most commonly used methods for energy harvesting; it utilizes photovoltaic cells to convert sunlight into electrical energy. The photovoltaic cell absorbs photons from sunlight to generate an electric current. Its use can be seen in devices from a small-size calculator to large-scale solar power plants.

4.2. Thermal Energy Harvesting

Thermal-energy-harvesting methods generate electricity through sensing units that observe temperature differences. Thermoelectric materials are utilized to observe heat gradients for generating electrical voltage. Therefore, these methodologies can be implemented in industrial applications where there is a huge temperature differential.

4.3. Kinetic Energy Harvesting

Kinetic energy harvesting can also be represented as a motion-based energy-harvesting method where the movement or motion of an object is used for generating electric power. This is a most common methodology used in wearable devices and self-powered sensor units. Electromagnetic induction and piezoelectric materials are used to convert mechanical motion into electric power.

4.4. Hybrid Energy-Harvesting Technology

Hybrid energy-harvesting methods are designed by integrating multiple energy-harvesting methods. This method is preferred for wireless devices due to its high reliability and better energy efficiency. Some of the important hybrid energy systems used for wireless devices are discussed below.

4.4.1. Solar-Energy-Based Hybrid Energy-Harvesting Methods

Arora et al., 2022 [16] developed a hybrid system based on a solar, thermal, and piezoelectric model for underwater WSN applications. A novel optimization method was also proposed in this work, and the theoretical calculation provides an energy outcome of 22.3 KJ per 24 h. Chen et al., 2022 [17] proposed a solar-panel- and wind-turbine-based hybrid energy system for mobile edge-computing systems where a dynamic offloading algorithm is utilized to regularize the generated power. Xiao et al., 2023 [18] made a hybrid model with solar PV and a thermoelectrical generator to convert the ambient light and heat energy in a room into useful power. The work uses a five-sided PV panel for this operation, and the generated power is used for IoT sensors for increasing its sustainability. Kim et al., 2022 [19] designed a hybrid system with a raindrop- and solar-energy-harvesting method. The work contains a triboelectric nanogenerator constructed with an inbuilt charge storage layer in the PV panel. The design was implemented in an invisible IoT security system and obtained a satisfactory outcome in terms of energy efficiency due to a better light-transmittance rate.

4.4.2. Thermal-Energy-Based Hybrid Energy-Harvesting Methods

Bakytbekov et al., 2022 [20] designed an RF- and thermal-based self-energy-harvesting dual-function triple-band antenna for IoT application. The experimental test indicates that this methodology provides an energy outcome of 13.6 µW with 250 mV. Kim et al., 2022 [21] framed a hybrid energy-harvesting model by integrating triboelectric and thermoelectric generators for wearable device applications. The work converts body motion and body heat into useful electric energy, and the experimental work took 240 s to store energy in a 3.3 mF capacitor of 3 V. Yang et al., 2022 [22] developed a hybrid energy-harvesting device that includes s triboelectric nanogenerator and thermoelectric generator. The work utilizes the inevitable heat produced from the Seebeck effect and that gives a betterment of 28 times over the traditional methods. Bakytbekov et al., 2023 [23] designed a multisource energy-harvesting system using an RF- and thermal-energy harvester. The experimental study gives a betterment of 10% over the traditional methods and produced 3680 µWh per day; further, it had the ability to transfer the data to the destination every 3.5 s.

4.4.3. Kinetic-Energy-Based Hybrid Energy-Harvesting Methods

Zhao et al., 2022 [24] structured a hybrid energy-harvesting model with piezoelectric and electromagnetic models for converting rotation and vibration energy into electric power. A triboelectric nanogenerator is utilized in this method for such energy conversion, and the work is implemented for wireless tire-pressure monitoring systems. Bai et al., 2023 [25] proposed a vibration- and rotation-energy-conversion model for wearable sensor electronics. The design is structured with electromagnetic and turboelectric generators and produces 300 mW from jogging and 800 mW from sprinting. The capability of the model can charge a wearable smartband rated with 400 mW power. Liu et al., 2023 [26] developed a marine-mammal-condition monitoring system with a triboelectric nanogenerator and micro-thermoelectric generator for making a self-powered device. The experimental work finds a betterment of 4.93% than the single energy source method of charging a battery. Cheng et al., 2023 [27] structured a hybrid energy-harvesting method that consists of piezoelectric, electromagnetic, and magnetostrictive generators. The experimental outcome gives a maximum output of 2.674 mW in Bluetooth wireless communication of humidity sensor data.

Hybrid energy systems are widely preferred in wireless devices for improving the reliability and sustainability of the connected systems. The self-hybrid energy systems are good for improving the battery life of the connected system by enhancing the energy-harvesting space to certain extent. Similarly, the hybrid model improves the system flexibility by adapting it for various applications. However, there are certain limitations in self-energy-harvesting systems, which are described in the following section.

5. Literature Review of Energy Optimization Models in Wireless Devices

Energy optimization is a methodology utilized to improve the power efficiency of energy harvesting by minimizing energy wastage. Usually, energy auditing will be implemented in every self-energy-harvesting system for analyzing its energy utilization pattern, and from that, a customized energy optimization algorithm will be developed. Table 3 represents some of the energy optimization methods utilized in wireless sensor networks and IoT applications.

Table 3. Literature review of energy optimization methods.

Methodology	Year	Energy Source	Generator Type	Application	Outcome
Multistage Dickson charge pump circuit [28]	2023	Electromagnetic wave	Multi-band RF antenna	IoT/WSN	78.3% of power conversion efficiency
Relay selection method [29]	2023	Electromagnetic wave	RF antenna	WBAN	7.8% of pocket reception rate improvement
Power control method [30]	2022	Electromagnetic wave	RF antenna	IoT sensor with NodeMCU	Reduced power requirement from 225 mW to 264 μW
Adaptive power transfer algorithm [31]	2022	Electromagnetic wave	RF antenna	IoT	49.5 mW outcome of wireless power transfer
Distributed resource management algorithm [32]	2021	Electromagnetic wave	RF antenna	B5G	Eight times reduced pocket loss compared to greedy algorithm
10-stage cross connected rectifier optimization [33]	2021	Electromagnetic wave	RF antenna	IoT	42.4% of peak end-to-end efficiency
Self-designed data and energy integrated network [34]	2021	Electromagnetic wave	RF antenna	IoT	Minimizes the sampling data to improve the sleeping time

Table 3. *Cont.*

Methodology	Year	Energy Source	Generator Type	Application	Outcome
Simultaneous wireless information and power transfer algorithm [35]	2021	Electromagnetic wave	RF antenna	5G/B5G IoT	7.81×10^{-11} ESA is achieved
Intelligent dynamic energy flow control algorithm [36]	2021	Electromagnetic wave	RF antenna	WSN	0.16 µV output per second
Rectenna array [37]	2020	Electromagnetic wave	RF antenna	IoT	67% of high energy conversion efficiency over Vivaldi rectenna
Hybrid spectrum access mode [38]	2020	Electromagnetic wave	RF antenna	Industrial IoT	Achieved larger transmission of data with less power
Broadband rectifier and a novel matching network [39]	2020	Electromagnetic wave	RF antenna	LTE	42% efficiency improvement
Supercapacitor with hybrid optimization [40]	2022	Hybrid solar, wind, and kinetic energy	Photovoltaic cell, wind turbine, and electromagnetic generator	Railway wireless sensors	2660 mW power generated from 5.5 m/s wind
Game theory and perturbed Lyapunov optimization theory [41]	2021	Hybrid vibration and kinetic energy	Piezoelectric and electromagnetic transducers	IoT	Better energy efficiency than naïve and greedy offloading
Parametric model optimization strategy [42]	2020	Hybrid vibration and kinetic energy	Piezoelectric and electromagnetic generators	IoT	25.45 mW power generated on 0.5G vibration
Rational adaptive mechanical design [43]	2022	Hybrid wind and kinetic energy	Triboelectric and electromagnetic generators	Wireless environment monitoring	60 times better output than the traditional model
Rotational tapered rollers [44]	2021	Hybrid wind and kinetic energy	Triboelectric and electromagnetic generators	IoT	63.8 mW output
Customized boxlike structure [45]	2020	Hybrid wind and kinetic energy	Triboelectric and electromagnetic nanogenerators	5G IoT	18.66 mW power output at 15 m/s wind speed
Magnetic flux intensity control [46]	2023	Magnetic field	Magneto-mechano-electric generator	WSN	5.5 mW power generated from 100e magnetic field
Energy per operation optimization [47]	2020	Solar	Photovoltaic cell	Wearable IoT	2.4 times better outcome than manual optimization
Boosted by boost converter [48]	2020	Solar	Photovoltaic cell	WSN	5.88 V generated at full sunlight
Efficient energy and radio resource management framework [49]	2020	Solar	Photovoltaic cell	UAV	Generated 13,000 J for 20 s

Table 3. *Cont.*

Methodology	Year	Energy Source	Generator Type	Application	Outcome
Prediction-based adaptive duty cycle MAC protocol [50]	2023	Solar	Photovoltaic cell	WSN	76.4% improvement on total energy consumption
Smart connector for energy balance [29]	2023	Thermal	Two thermoelectric generators	Bluetooth smart grid	4.9% energy improvement on sleep mode
Tapered nonlinear vibration energy harvester with MPPT [51]	2021	Vibration	Piezoelectric device	IoT	2660 $\mu W/cm^3 g^2$ of power density is obtained
Vibration enhancement mechanism [52]	2021	Vibration	Piezoelectric stack	IoT/WSN	2.622 W power output at 8.5 ms^{-1} wind speed
Cantilever and impact method [53]	2021	Vibration	Two piezoelectric devices	Zigbee wireless sensor	1.5 μW of maximum power output
Polymer film thickness alteration [54]	2022	Wind	Microwind generator	Wireless sensor	60 μW of maximum power output
In situ carbon dispersion method [55]	2022	Wind	High-power triboelectric nanogenerators	Wireless control	75.2 W/m^2 power density

Apart from optimized energy-generation techniques, certain optimization algorithms have been implemented in recent years to enhance the efficiency of energy utilization. A hybrid whale optimization algorithm–moth flame optimization was proposed to select the optimal cluster head for data transmission. It helps a wireless device to save its normalized network energy [56]. A slow-movement particle swarm optimization algorithm was designed to improve the scheduling process on a mobile edge device application. The experimental outcome indicates better computational and energy efficiency over the conventional particle swarm optimization [57]. A fuzzy-constraints-based cluster optimization methodology was developed to optimize the performances of cluster heads on data transmission in wireless ad hoc networks. The simulated outcome indicates a better network lifetime over the traditional LEACH and MPO methods [58]. An ad hoc on-demand multipath distance vector routing protocol was structured to enhance the energy efficiency in mobile ad hoc networks by enhancing the routing process. The performance of the proposed method gives better energy efficiency along with minimal data loss [59]. A machine-learning-based intelligent opportunistic routing protocol was proposed for WSN healthcare applications. The simulation result presents an acceptable outcome of energy consumption rate over the traditional MDOR and EEOR approaches [60].

6. Discussion on Emerging Methodologies

The literature section indicates that wireless devices are widely incorporated with RF antennas for extracting energy from electromagnetic waves. Following that, hybrid technologies are widely preferred for energy harvesting. In hybrid methods, wind with kinetic energy and vibration with kinetic energy are the most utilized methods. Electromagnetic generators are employed to convert kinetic energy into electrical energy, and piezoelectric sensors are utilized to convert vibration energy. Triboelectric nanogenerators and wind turbines are employed for harvesting energy from the wind. Individual solar-energy harvesting was found to be the third most preferred option; however, this was not found to be the focus of recent research studies in the way that electromagnetic wave energy generation has been studied.

The review work was performed to observe the research trends in energy harvesting in wireless devices, and therefore, the literature study was conducted between 2020 and 2023. Figure 4 represents the ratio of energy sources that were proposed for self-energy harvesting in wireless devices. Electromagnetic wave seems to be the topmost energy source, occupying 41.38% of space in the total self-sustainable models. Hybrid energy sources occupy 20.69% and solar energy takes 13.79% in the energy utilization space. Vibration and wind energy cover 10.34% and 6.9%, respectively. Magnetic field and thermal energy sources occupy each 3.45% of energy source space.

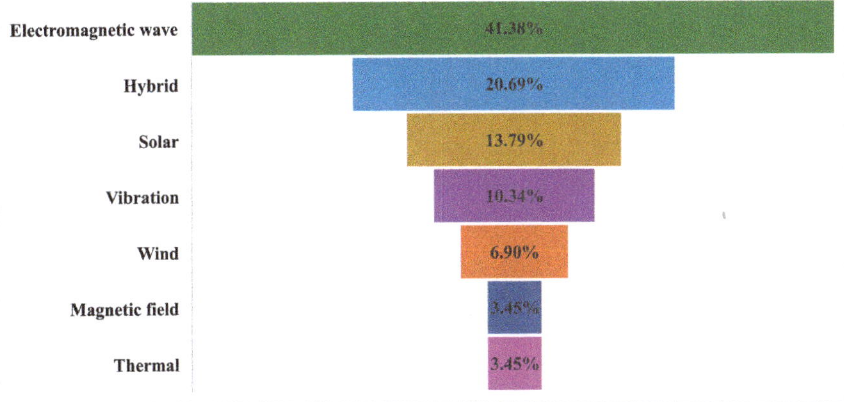

Figure 4. Distribution of energy source utilization by percentage.

Figure 5 represents the split-up proportion of the energy-harvesting generators utilized for wireless networks. It indicates the RF antenna as the most employed generator system in wireless devices. Wireless signals such as Wi-Fi, mobile, and other mobile communication have some low-level energy, and the RF antenna attracts such energy and stores it in a battery storage with the help of rectifier unit. In most of the wireless applications, the devices are structured with an RF antenna for signal transmission. The same RF antenna is utilized for energy harvesting in most of the systems [61]. Therefore, it does not require any additional energy-harvesting modules. Hence, it is widely used in wireless devices. Similar to RF antennae, an optical nanoantenna called a rectenna is also implemented in a few applications. However, its load power and energy conversion efficiency are comparatively poorer than other methods [62]. These kinds of self-harvesting methodologies may assist the Agriculture 4.0 methodologies, which are implemented with sensors and remote sensing units [63].

The electromagnetic generator seems to be the second most utilized power generator system in wireless devices. It creates electrical energy from flowing water and wind. However, the electromagnetic generator system cannot be placed in closed-environment wireless devices. It is majorly employed in open-place wireless communication systems. Similarly, the photovoltaic cell is also utilized in open-place wireless communication systems for harvesting energy from the solar energy source. However, the photovoltaic cell has the ability to generate only in the daytime. Hence, the solar energy source is utilized as one of the energy sources in hybrid systems. Piezoelectric sensors and triboelectric generators are also the most common models that are used in hybrid systems. The piezoelectric sensor generates electrical energy from vibrations, and the triboelectric generator generates electrical energy from wind energy. The wind turbine is also employed in a very few wireless devices for generating electrical energy. Table 4 represents the future directions and challenges of designing a self-energy-harvesting method for wireless applications.

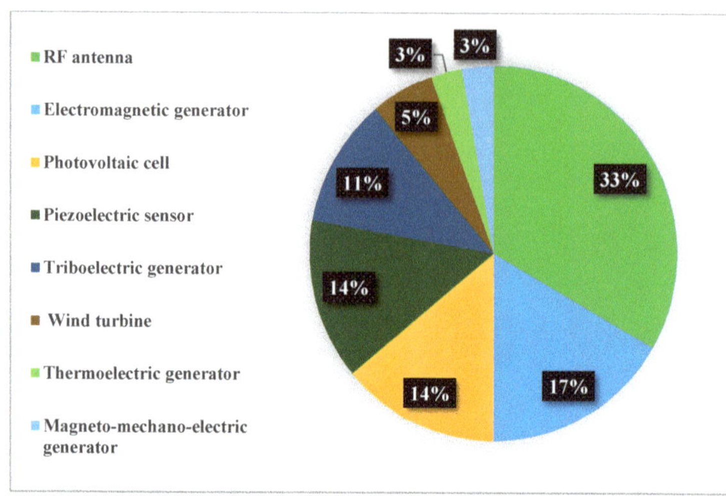

Figure 5. Distribution chart of preferred energy-harvesting generators.

Table 4. Future directions and challenges of self-energy-harvesting methods.

Methodology	Future Directions	Challenges
Electromagnetic wave	Minimizing the size of the RF antenna while maintaining its performance	General properties of the RF antenna material
Solar	Intelligent solar panel direction estimations	Space requirement and climate constraint
Vibration	Increasing the lifespan of the sensors	Cannot be suitable for several applications
Wind	Effienct windflow direction estimation	Climate constraint
Magnetic field	Improving the power density observation from the magnetic field	Not suitable for living area
Thermal	Increasing the energy conversion efficiency	Requires constant heat source

The performance of wireless energy-harvesting systems is measured in terms of power generated by them, and the following list indicates some of the other parameters that are used for estimation:

- Power conversion efficiency is a metric used for comparing a model with an existing technique [28].
- The performance of the self-harvesting wireless devices is optimized with an automatic algorithm, and in such cases, its performance is measured with the power requirement for some specific operation. The same operation is enforced in other algorithms, also for proving its efficiency [30,42,46].
- The energy optimization algorithms are also included in some wireless devices, and that reduces the pocket loss of signal transmission [29,32].
- Sleeping time analysis is found to be one of the efficient parameters, and that represents the amount of saved data transmission [29,34].
- Amount of power spent for sending a specific amount of data gives a better view of the wireless device in terms of its sustainability [38,43].

7. Conclusions

Next-generation wireless devices are expected to be incorporated with a self-energy-harvesting technique by many users. Therefore, different kinds of self-energy-harvesting methods have been developed in recent years. This paper explored the requirements of self-energy-harvesting systems in wireless devices, and it also indicated the methodologies that are widely employed in such energy-harvesting applications. A brief literature study was conducted with recent-years research outcomes from 2020 to 2023 to represent the recent trends in self-energy-harvesting techniques. The review summary indicated that electromagnetic-based power generation systems are widely employed in many applications, as they require very minimal peripheral modules for operation. The review also found that the research on solar-based power generation is not preferred by researchers in recent years for making energy-harvesting wireless devices, as it requires a huge amount of space. Similarly, thermal methodologies are not preferred, as these require a constant heat source; furthermore, wind-based methodologies were not preferred due to climatic considerations. The analysis explored the possible scope in future for a hybrid self-energy-harvesting system which may contain an RF antenna as a primary module for generating power from wireless signals. Other techniques such as photovoltaic, triboelectric, piezo-electric, and thermoelectric sensors can be equipped in such hybrid models based on the available energy source in the place where the wireless device is to be installed.

Author Contributions: Conceptualization, J.D.K.H., K.C.R., S.B.K.R., V.M.K., M.D., A.R. and R.M.; literature review, J.D.K.H., K.C.R., S.B.K.R., V.M.K., M.D., A.R. and R.M.; investigation, J.D.K.H., K.C.R., S.B.K.R., V.M.K., M.D., A.R. and R.M.; writing—original draft preparation, J.D.K.H., K.C.R., S.B.K.R., V.M.K., M.D., A.R. and R.M.; writing—review and editing, J.D.K.H., K.C.R., S.B.K.R., V.M.K., M.D., A.R. and R.M.; supervision, J.D.K.H., K.C.R., S.B.K.R., V.M.K., M.D., A.R. and R.M. All authors have read and agreed to the published version of the manuscript.

Funding: This research received no external funding.

Data Availability Statement: Not applicable.

Conflicts of Interest: The authors declare no conflict of interest.

References

1. He, C.; Chen, Y.-Y.; Phang, C.-R.; Stevenson, C.; Chen, I.-P.; Jung, T.-P.; Ko, L.-W. Diversity and Suitability of the State-of-the-Art Wearable and Wireless EEG Systems Review. *IEEE J. Biomed. Health Inform.* **2023**, 1–14. [CrossRef] [PubMed]
2. Rahmani, H.; Shetty, D.; Wagih, M.; Ghasempour, Y.; Palazzi, V.; Carvalho, N.B.; Correia, R.; Costanzo, A.; Vital, D.; Alimenti, F.; et al. Next-generation IoT devices: Sustainable eco-friendly manufacturing, energy harvesting, and wireless connectivity. *IEEE J. Microw.* **2023**, *3*, 237–255. [CrossRef]
3. Majid, M.; Habib, S.; Javed, A.R.; Rizwan, M.; Srivastava, G.; Gadekallu, T.R.; Lin, J.C.W. Applications of wireless sensor networks and internet of things frameworks in the industry revolution 4.0: A systematic literature review. *Sensors* **2022**, *22*, 2087. [CrossRef] [PubMed]
4. Yang, Z.; Chen, M.; Wong, K.K.; Poor, H.V.; Cui, S. Federated learning for 6G: Applications, challenges, and opportunities. *Engineering* **2022**, *8*, 33–41. [CrossRef]
5. Ananthi, J.V.; Jose, P.S.H. Performance Analysis of Clustered Routing Protocol for Wearable Sensor Devices in an IoT-Based WBAN Environment. In *Intelligent Technologies for Sensors: Applications, Design, and Optimization for a Smart World*; Apple Academic Press: Palm Bay, FL, USA, 2023; p. 253.
6. Tirth, V.; Alghtani, A.H.; Algahtani, A. Artificial intelligence enabled energy aware clustering technique for sustainable wireless communication systems. *Sustain. Energy Technol. Assess.* **2023**, *56*, 103028. [CrossRef]
7. Ouyang, Q.; Chen, J. The Future of Lithium-Ion Battery Charging Technologies. In *Advanced Model-Based Charging Control for Lithium-Ion Batteries*; Springer Nature: Singapore, 2023; pp. 175–176.
8. Cornet, B.; Fang, H.; Ngo, H.; Boyer, E.W.; Wang, H. An overview of wireless body area networks for mobile health applications. *IEEE Netw.* **2022**, *36*, 76–82. [CrossRef]
9. Mertes, J.; Lindenschmitt, D.; Amirrezai, M.; Tashakor, N.; Glatt, M.; Schellenberger, C.; Shah, S.M.; Karnoub, A.; Hobelsberger, C.; Yi, L.; et al. Evaluation of 5G-capable framework for highly mobile, scalable human-machine interfaces in cyber-physical production systems. *J. Manuf. Syst.* **2022**, *64*, 578–593. [CrossRef]
10. Li, L.; Liu, Y.; You, I.; Song, F. A Smart Retransmission Mechanism for Ultra-Reliable Applications in Industrial Wireless Networks. *IEEE Trans. Ind. Inform.* **2022**, *19*, 1988–1996. [CrossRef]

11. Vishnuram, P.; Nastasi, B. Wireless Chargers for Electric Vehicle: A Systematic Review on Converter Topologies, Environmental Assessment, and Review Policy. *Energies* **2023**, *16*, 1731. [CrossRef]

12. Wu, Y.; Song, Y.; Wang, T.; Qian, L.; Quek, T.Q.S. Non-orthogonal multiple access assisted federated learning via wireless power transfer: A cost-efficient approach. *IEEE Trans. Commun.* **2022**, *70*, 2853–2869. [CrossRef]

13. Xia, L.; Ma, S.; Tao, P.; Pei, W.; Liu, Y.; Tao, L.; Wu, Y. A Wind-Solar Hybrid Energy Harvesting Approach Based on Wind-Induced Vibration Structure Applied in Smart Agriculture. *Micromachines* **2022**, *14*, 58. [CrossRef] [PubMed]

14. Sayed, D.M.; Allam, N.K. All-solid-state, self-powered supercapacitors: State-of-the-art and future perspectives. *J. Energy Storage* **2022**, *56*, 105882. [CrossRef]

15. Hassan, M.; Abbas, G.; Li, N.; Afzal, A.; Haider, Z.; Ahmed, S.; Xu, X.; Pan, C.; Peng, Z. Significance of flexible substrates for wearable and implantable devices: Recent advances and perspectives. *Adv. Mater. Technol.* **2022**, *7*, 2100773. [CrossRef]

16. Verma, G.; Arora, S.; Nijhawan, G. A Solar, Thermal, and Piezoelectric Based Hybrid Energy Harvesting for IoT and Underwater WSN Applications. *Int. J. Sens. Wirel. Commun. Control* **2022**, *12*, 651–660. [CrossRef]

17. Chen, Y.; Zhao, F.; Lu, Y.; Chen, X. Dynamic task offloading for mobile edge computing with hybrid energy supply. *Tsinghua Sci. Technol.* **2022**, *28*, 421–432. [CrossRef]

18. Xiao, H.; Qi, N.; Yin, Y.; Yu, S.; Sun, X.; Xuan, G.; Liu, J.; Xiao, S.; Li, Y.; Li, Y. Investigation of Self-Powered IoT Sensor Nodes for Harvesting Hybrid Indoor Ambient Light and Heat Energy. *Sensors* **2023**, *23*, 3796. [CrossRef]

19. Kim, B.; Song, J.Y.; Kim, M.C.; Lin, Z.H.; Choi, D.; Park, S.M. All-aerosol-sprayed high-performance transparent triboelectric nanogenerator with embedded charge-storage layer for self-powered invisible security IoT system and raindrop-solar hybrid energy harvester. *Nano Energy* **2022**, *104*, 107878. [CrossRef]

20. Bakytbekov, A.; Nguyen, T.Q.; Zhang, G.; Strano, M.S.; Salama, K.N.; Shamim, A. Dual-function triple-band heatsink antenna for ambient RF and thermal energy harvesting. *IEEE Open J. Antennas Propag.* **2022**, *3*, 263–273. [CrossRef]

21. Kim, W.-G.; Kim, D.; Lee, H.M.; Choi, Y.-K. Wearable fabric-based hybrid energy harvester from body motion and body heat. *Nano Energy* **2022**, *100*, 107485. [CrossRef]

22. Yang, O.; Zhang, C.; Zhang, B.; He, L.; Yuan, W.; Liu, Y.; Li, X.; Zhou, L.; Zhao, Z.; Wang, J.; et al. Hybrid Energy-Harvesting System by a Coupling of Triboelectric and Thermoelectric Generator. *Energy Technol.* **2022**, *10*, 2101102. [CrossRef]

23. Bakytbekov, A.; Nguyen, T.Q.; Zhang, G.; Strano, M.S.; Salama, K.N.; Shamim, A. Synergistic multi-source ambient RF and thermal energy harvester for green IoT applications. *Energy Rep.* **2023**, *9*, 1875–1885. [CrossRef]

24. Zhao, L.C.; Zou, H.-X.; Zhao, Y.-J.; Wu, Z.-Y.; Liu, F.-R.; Wei, K.-X.; Zhang, W.-M. Hybrid energy harvesting for self-powered rotor condition monitoring using maximal utilization strategy in structural space and operation process. *Appl. Energy* **2022**, *314*, 118983. [CrossRef]

25. Bai, S.; Cui, J.; Zheng, Y.; Li, G.; Liu, T.; Liu, Y.; Hao, C.; Xue, C. Electromagnetic-triboelectric energy harvester based on vibration-to-rotation conversion for human motion energy exploitation. *Appl. Energy* **2023**, *329*, 120292. [CrossRef]

26. Liu, C.; Qu, G.; Shan, B.; Aranda, R.; Chen, N.; Li, H.; Zhou, Z.; Yu, T.; Wang, C.; Mi, J.; et al. Underwater hybrid energy harvesting based on TENG-MTEG for self-powered marine mammal condition monitoring system. *Mater. Today Sustain.* **2023**, *21*, 100301. [CrossRef]

27. Cheng, M.; Wu, J.; Liang, X.; Mao, R.; Huang, H.; Ju, D.; Hu, Z.; Guo, J.; Liu, M. Hybrid multi-mode magneto-mechano-electric generator with enhanced magnetic field energy harvesting performance. *Sens. Actuators A Phys.* **2023**, *352*, 114194. [CrossRef]

28. Li, J.; Gong, W. Optimized High-efficiency Multi-band RF Energy Harvester. In Proceedings of the 2023 IEEE Wireless Communications and Networking Conference (WCNC), Glasgow, UK, 26–29 March 2023; pp. 1–6.

29. Liu, Y.; Riba, J.-R.; Moreno-Eguilaz, M. Energy Balance of Wireless Sensor Nodes Based on Bluetooth Low Energy and Thermoelectric Energy Harvesting. *Sensors* **2023**, *23*, 1480. [CrossRef]

30. Raghav, K.S.; Bansal, D. Power controlled system for self-sustained RF energy harvesting sensors. *Analog. Integr. Circuits Signal Process.* **2022**, *113*, 73–79. [CrossRef]

31. Thangarajan, A.S.; Nguyen, T.D.; Liu, M.; Michiels, S.; Yang, F.; Man, K.L.; Ma, J.; Joosen, W.; Hughes, D. Static: Low Frequency Energy Harvesting and Power Transfer for the Internet of Things. *Front. Signal Process.* **2022**, *1*, 15. [CrossRef]

32. Shi, Z.; Xie, X.; Lu, H.; Yang, H.; Cai, J.; Ding, Z. Deep Reinforcement Learning-Based Multidimensional Resource Management for Energy Harvesting Cognitive NOMA Communications. *IEEE Trans. Commun.* **2021**, *70*, 3110–3125. [CrossRef]

33. Noghabaei, S.M.; Radin, R.L.; Savaria, Y.; Sawan, M. A high-sensitivity wide input-power-range ultra-low-power RF energy harvester for IoT applications. *IEEE Trans. Circuits Syst. I Regul. Pap.* **2021**, *69*, 440–451. [CrossRef]

34. Wang, Y.; Yang, K.; Wan, W.; Zhang, Y.; Liu, Q. Energy-efficient data and energy integrated management strategy for iot devices based on rf energy harvesting. *IEEE Internet Things J.* **2021**, *8*, 13640–13651. [CrossRef]

35. Amjad, M.; Chughtai, O.; Naeem, M.; Ejaz, W. SWIPT-assisted energy efficiency optimization in 5G/B5G cooperative IoT network. *Energies* **2021**, *14*, 2515. [CrossRef]

36. Verma, G.; Sharma, V. A novel RF energy harvester for event-based environmental monitoring in Wireless Sensor Networks. *IEEE Internet Things J.* **2021**, *9*, 3189–3203. [CrossRef]

37. Song, C.; Lu, P.; Shen, S. Highly efficient omnidirectional integrated multiband wireless energy harvesters for compact sensor nodes of Internet-of-Things. *IEEE Trans. Ind. Electron.* **2020**, *68*, 8128–8140. [CrossRef]

38. Liu, X.; Hu, S.; Li, M.; Lai, B. Energy-efficient resource allocation for cognitive industrial Internet of Things with wireless energy harvesting. *IEEE Trans. Ind. Inform.* **2020**, *17*, 5668–5677. [CrossRef]

39. Wang, M.; Yang, L.; Shi, Y. A dual-port microstrip rectenna for wireless energy harvest at LTE band. *AEU-Int. J. Electron. Commun.* **2020**, *126*, 153451. [CrossRef]

40. Tairab, A.M.; Wang, H.; Hao, D.; Azam, A.; Ahmed, A.; Zhang, Z. A hybrid multimodal energy harvester for self-powered wireless sensors in the railway. *Energy Sustain. Dev.* **2020**, *68*, 150–169. [CrossRef]

41. Xia, S.; Yao, Z.; Li, Y.; Mao, S. Online distributed offloading and computing resource management with energy harvesting for heterogeneous MEC-enabled IoT. *IEEE Trans. Wirel. Commun.* **2021**, *20*, 6743–6757. [CrossRef]

42. Jung, I.; Choi, J.; Park, H.-J.; Lee, T.-G.; Nahm, S.; Song, H.-C.; Kim, S.; Kang, C.-Y. Design principles for coupled piezoelectric and electromagnetic hybrid energy harvesters for autonomous sensor systems. *Nano Energy* **2020**, *75*, 104921. [CrossRef]

43. Lee, D.; Cho, S.; Jang, S.; Ra, Y.; Jang, Y.; Yun, Y.; Choi, D. Toward effective irregular wind energy harvesting: Self-adaptive mechanical design strategy of triboelectric-electromagnetic hybrid wind energy harvester for wireless environmental monitoring and green hydrogen production. *Nano Energy* **2022**, *102*, 107638. [CrossRef]

44. Fang, Y.; Tang, T.; Li, Y.; Hou, C.; Wen, F.; Yang, Z.; Chen, T.; Sun, L.; Liu, H.; Lee, C. A high-performance triboelectric-electromagnetic hybrid wind energy harvester based on rotational tapered rollers aiming at outdoor IoT applications. *iScience* **2021**, *24*, 102300. [CrossRef] [PubMed]

45. Fan, X.; He, J.; Mu, J.; Qian, J.; Zhang, N.; Yang, C.; Hou, X.; Geng, W.; Wang, X.; Chou, X. Triboelectric-electromagnetic hybrid nanogenerator driven by wind for self-powered wireless transmission in Internet of Things and self-powered wind speed sensor. *Nano Energy* **2020**, *68*, 104319. [CrossRef]

46. Patil, D.R.; Lee, S.; Thakre, A.; Kumar, A.; Song, H.; Jeong, D.-Y.; Ryu, J. Boosting the energy harvesting performance of cantilever structured magneto-mechano-electric generator by controlling magnetic flux intensity on magnet proof mass. *J. Mater.* **2023**, *in press*. [CrossRef]

47. Park, J.; Bhat, G.; Nk, A.; Geyik, C.S.; Ogras, U.Y.; Lee, H.G. Energy per operation optimization for energy-harvesting wearable IoT devices. *Sensors* **2020**, *20*, 764. [CrossRef] [PubMed]

48. Antony, S.M.; Indu, S.; Pandey, R. An efficient solar energy harvesting system for wireless sensor network nodes. *J. Inf. Optim. Sci.* **2020**, *41*, 39–50. [CrossRef]

49. Zhang, J.; Lou, M.; Xiang, L.; Hu, L. Power cognition: Enabling intelligent energy harvesting and resource allocation for solar-powered UAVs. *Future Gener. Comput. Syst.* **2020**, *110*, 658–664. [CrossRef]

50. Sarang, S.; Stojanović, G.M.; Drieberg, M.; Stankovski, S.; Bingi, K.; Jeoti, V. Machine Learning Prediction Based Adaptive Duty Cycle MAC Protocol for Solar Energy Harvesting Wireless Sensor Networks. *IEEE Access* **2023**, *11*, 17536–17554. [CrossRef]

51. Paul, K.; Amann, A.; Roy, S. Tapered nonlinear vibration energy harvester for powering Internet of Things. *Appl. Energy* **2021**, *283*, 116267. [CrossRef]

52. Sheeraz, M.A.; Malik, M.S.; Rehman, K.; Elahi, H.; Butt, Z.; Ahmad, I.; Eugeni, M.; Gaudenzi, P. Numerical assessment and parametric optimization of a piezoelectric wind energy harvester for IoT-based applications. *Energies* **2021**, *14*, 2498. [CrossRef]

53. Kim, J.H.; Cho, J.Y.; Jhun, J.P.; Song, G.J.; Eom, J.H.; Jeong, S.; Hwang, W.; Woo, M.S.; Sung, T.H. Development of a hybrid type smart pen piezoelectric energy harvester for an IoT platform. *Energy* **2021**, *222*, 119845. [CrossRef]

54. Le Scornec, J.; Guiffard, B.; Seveno, R.; Le Cam, V.; Ginestar, S. Self-powered communicating wireless sensor with flexible aero-piezoelectric energy harvester. *Renew. Energy* **2022**, *184*, 551–563. [CrossRef]

55. Zhang, Z.; Zhang, Q.; Zhou, Z.; Wang, J.; Kuang, H.; Shen, Q.; Yang, H. High-power triboelectric nanogenerators by using in-situ carbon dispersion method for energy harvesting and self-powered wireless control. *Nano Energy* **2022**, *101*, 107561. [CrossRef]

56. Maddikunta, P.K.R.; Gadekallu, T.R.; Kaluri, R.; Srivastava, G.; Parizi, R.M.; Khan, M.S. Green communication in IoT networks using a hybrid optimization algorithm. *Comput. Commun.* **2020**, *159*, 97–107. [CrossRef]

57. Zhang, Y.; Liu, Y.; Zhou, J.; Sun, J.; Li, K. Slow-movement particle swarm optimization algorithms for scheduling security-critical tasks in resource-limited mobile edge computing. *Future Gener. Comput. Syst.* **2020**, *112*, 148–161. [CrossRef]

58. Mohammed, A.S.; Asha, S.B.P.N.; Venkatachalam, K. FCO—Fuzzy constraints applied cluster optimization technique for wireless adhoc networks. *Comput. Commun.* **2020**, *154*, 501–508. [CrossRef]

59. Bhardwaj, A.; El-Ocla, H. Multipath routing protocol using genetic algorithm in mobile ad hoc networks. *IEEE Access* **2020**, *8*, 177534–177548. [CrossRef]

60. Pham, Q.-V.; Mirjalili, S.; Kumar, N.; Alazab, M.; Hwang, W.-J. Whale optimization algorithm with applications to resource allocation in wireless networks. *IEEE Trans. Veh. Technol.* **2020**, *69*, 4285–4297. [CrossRef]

61. Sherazi, H.H.R.; Zorbas, D.; O'flynn, B. A comprehensive survey on RF energy harvesting: Applications and performance determinants. *Sensors* **2022**, *22*, 2990. [CrossRef]

62. Citroni, R.; Di Paolo, F.; Livreri, P. Evaluation of an optical energy harvester for SHM application. *AEU-Int. J. Electron. Commun.* **2019**, *111*, 152918. [CrossRef]

63. Barrile, V.; Simonetti, S.; Citroni, R.; Fotia, A.; Bilotta, G. Experimenting Agriculture 4.0 with Sensors: A Data Fusion Approach between Remote Sensing, UAVs and Self-Driving Tractors. *Sensors* **2022**, *22*, 7910. [CrossRef]

Article

Energy Saving Optimization Technique-Based Routing Protocol in Mobile Ad-Hoc Network with IoT Environment

Vinoth Kumar Krishnamoorthy [1], **Ivan Izonin** [2], **Sugumaran Subramanian** [3], **Shishir Kumar Shandilya** [4], **Sivasankaran Velayutham** [5,*], **Thillai Rani Munichamy** [6,*] **and Myroslav Havryliuk** [2]

[1]   Department of Electrical and Electronics Engineering, New Horizon College of Engineering, Bengaluru 560103, India
[2]   Department of Artificial Intelligence, Lviv Polytechnic National University, 79013 Lviv, Ukraine
[3]   Department of Electronics and Communication Engineering, Sreenivasa Institute of Technology and Management Studies, Chittoor 517127, India
[4]   School of Computing Science and Engineering, VIT Bhopal University, Bhopal 466114, India
[5]   School of Electrical and Electronics Engineering, VIT Bhopal University, Bhopal 466114, India
[6]   Department of Electronics and Communication Engineering, Sri Krishna College of Technology, Coimbatore 641042, India
*   Correspondence: pvs.sankaran@gmail.com (S.V.); thillairani.m@skct.edu.in (T.R.M.)

Abstract: The Mobile Ad-hoc Network is a self-configuring decentralized network, where the network topology is dynamically modifiable. The IoT (Internet of Things) based Wireless Sensor Network contains more sensors and shares information over the Internet to a cloud server. However, the IoT-based wireless sensor network channel has moderate security is poor compared to MANET and packet loss is increased due to attackers. In IoT, all the sensors forward the detected data frequently to the internet gateway, so the energy saving in the network is low compared to MANET. In this work, the smart environment of IoT, Wireless Sensor Networks (WSN) and MANET make a great heterogeneous network in IT Technology; the combination of this heterogeneous network has new challenging issues. In this heterogeneous network, MANET provides a trusted route between the sensor to gateway nodes into the IoT environment using Energy Saving Optimization Techniques [MANET-ESO in IoT]. It saves energy for each node and reduces the economic level. The results of the ns-3 simulation show that the proposed method provides better results in Alive node counts, residual Energy, throughput, packet delivery ratio and routing overhead.

Keywords: heterogeneous network; IoT; MANET; internet; cloud

Citation: Krishnamoorthy, V.K.; Izonin, I.; Subramanian, S.; Shandilya, S.K.; Velayutham, S.; Munichamy, T.R.; Havryliuk, M. Energy Saving Optimization Technique-Based Routing Protocol in Mobile Ad-Hoc Network with IoT Environment. *Energies* **2023**, *16*, 1385. https://doi.org/10.3390/en16031385

Academic Editor: Oscar Barambones

Received: 22 December 2022
Revised: 11 January 2023
Accepted: 23 January 2023
Published: 30 January 2023

1. Introduction

The nodes of the Mobile Ad-hoc Network (MANET) are the group of self-organized mobile nodes that are deployed randomly without predefined models. These mobile nodes follow adhoc modelling for data communication in the absence of a router or any centralized devices. MANETs are widely applied for observing terrestrial abnormalities such as earthquakes, forest disasters and other natural disasters. Generally, MANETs are deployed with a vast amount of adhoc nodes to cover a wide range of geographical areas to detect notable natural objects. In this case, the Internet of Things (IoT) is considered a decentralized MANET environment. IoT base is required to take MANET observations into the internet domain, or any processing node (cloud system) deployed outside the local network domain. In the same manner, WSN is a type of network that contains resource-limited sensor nodes with a homogeneous type of internal components. In this work, WSN has mostly energy-limited sensor nodes than MANET nodes. In addition, sensor nodes deployed under the WSN environment need a gateway node to reach the outer network. The combination of MANET and WSN under an IoT environment creates heterogeneous network characteristics. Under this environment, energy optimization and routing protocol optimization are challenging tasks. This work takes this challenging research problem to be solved.

The MANET topology changes dynamically between multiple nodes, Jamali et al. [1] and Devi et al. [2]. The routing protocols are providing a route from source to destination. At the same time, various types of routing attacks in the network are reducing the performance of the network and minimizing the rate of transferred packets from sender to receiver, Kumar Debnath et al. [3]. This kind of problem can be solved using the trust route method. Initially, available routes are identified using the route discovery process. Finally, the best and trusted route is selected from the available routes with the support of optimization techniques. WSN nodes are spatially dispersed and linked to the wireless medium, Hu, Z et al. [4]. The sensors collect data from the physical environment and transmit the data to a centralised location, Khokhlachova et al. [5]. The nodes of the IoT environment can collect, control and monitor environmental data such as weather, forecast, object detection, gas leakage, etc. In an IoT environment, the collected information is stored and processed in the cloud system.

The MANET carries more data than the WSN, but the WSN consists of more sensors than the MANET, Ray, S et al. [6] and Jayalakshmi et al. [7]. The number of WSN sensor nodes is static and sensor nodes consume less energy than a MANET node. In contrast, node mobility is higher in MANET than in WSN. The MANET provides better power utilization and increases network lifetime through various energy-saving routing algorithms. The MANET is a special network, it utilizes less amount of energy with good scalability. In a MANET, the energy-saving mode increases the lifetime of each node's energy level and controls energy utilization by the appropriate selection of the energy-saving algorithm.

The MANET connects the nodes to the Internet through the gateway node. The gateway node acts as a connector to connect the MANET node to the Internet. The source MANET node is connected directly to the gateway node, or connected to the gateway node with the support of the adjacent node. The gateway node connects the different networks by MANET routing protocol and DNS server.

In this paper, the destination node gathers all of the sensor nodes' details, like protocol address, position and mobility rate (meters/second) in MANET. Here, IoT communication is established by WSN, MANET and the Internet. Consequently, this kind of network is represented as a heterogeneous network. The source sensor nodes detect the physical information and transfer the data to destinations through a multi-hop fashion. The clustering method is used to form the MANET. The cluster node reduces power wastage while transmitting and receiving the data. Finally, the data is forwarded to the gateway node to reach the internet server point. The gateway node is connected to a cloud network and the cloud system collects the data and stores the details.

The following conditions are considered in the research:

1. Cluster Network: More than one sensor node is connected as a cluster in the MANET. Each cluster is connected through a selected cluster head.
2. Routing Protocol: MANET considers an optimized routing protocol to determine the shortest route between sensors to the gateway node.
3. Internet and IoT: Data collected from various nodes are stored in the internet cloud server.

The IoT is a recent technology widely applied in this electronic world. Generally, the open wireless medium has poor security and utilizes more energy to sense the data and transmit the data to the nearest gateway. The recent works developed for wireless data communication use secure internet protocol functions and transport layer functions. In this network, neighbouring sensor nodes are formed into a small group by one MANET node is called a cluster network. In the cluster, MANET nodes act as cluster heads (CH) to forward the information between sensors. This cluster formation is reducing the bandwidth utilization rate and increasing the sensor node's lifespan. In addition, MANET is creating a trusted route and shortest route between sensors in the Internet Cloud.

Generally, clusters are created in MANET to easily organise the topology. In this manner, recent techniques use various energy-efficient clustering approaches for improving the performance of routing protocol. On the scope, clustering algorithms such as Low Energy Adaptive Clustered Hierarchy (LEACH) is a well-known algorithm for choosing the cluster heads based on optimal energy constraints. LEACH is a cluster formation protocol in a tree-based (hierarchical) network environment. Notably, the clusters of mobile nodes or sensor nodes are formed under the base station (centralized control centre). Clusters of nodes under each base station provide hierarchical network architecture. The IoT environment of cluster nodes can communicate with other cluster nodes through a centralized base station (control unit). In this case, cluster heads (nodes with the highest residual energy) can reach the base station for communicating the data [8].

In the same manner, real-time challenges and flat architectures in making MANET clusters are significantly noted by various research works. Unlike hierarchical IoT structures, flat network clusters can be formed for the distributed scenario. Particularly, flat networks deploy multiple gateway nodes for collecting the field node's data through cluster heads. The chosen gateway nodes are deployed in each cluster. In this method, cluster heads and other cluster nodes that are nearer to gateway nodes of neighbour clusters can communicate with each other. This kind of mechanism reduces energy overhead happens in each cluster head. Thus, the effective formation of clusters creates a crucial impact on energy-efficient data communication in IoT environments (hierarchical and flat networks) [9].

In this article, the MANET-ESO performance in the IoT environment is experimented with using the modules such as clustering principles, MANET-based on-demand routing process, IoT data collection process and trusted route establishment policies.

The simulation environment is created by MANET mobile nodes and Wireless sensor nodes with Internet (cloud network). The Sensor nodes are divided into small clusters and grouped by Mobile Ad-hoc nodes. Upon the validation of energy levels, nodes are assigned as CH to control sensor nodes. The routes are discovered from the sensor source to the gateway node with the help of the MANET cluster and on-demand routing protocol. The best route is evaluated by the total energy of the route, time is taken from source to destination and minimum hop count. In this method, IoT communication is stabilized over a long period of time using node energy level. In this case, the trust validation algorithm improves the packet delivery ratio, throughput, and reduces the routing overhead of the selected route among clusters.

The rest of this paper follows as: Section 2 expresses the literature review of problem formulation. Section 3 offers suggestions for network topology measurement tools. Section 4 describes the suggested approach for heterogeneity network utilising MANET-ESO in IoT Algorithm. The simulation and algorithm results are discussed with various parameters in Section 5. Section 6 is concluding the article with future work.

2. Literature Survey

To enhance the performance of MANET and IoT environments, the authors suggest a new routing protocol. The existing techniques are compared in this section. The Dynamic Critical Node Identification (DCNI) method was suggested by the authors Jayalakshmi et al. [7] and Hou, Y.T. et al. [10], who also identified the MANET-IoT network critical nodes. In this section, the currently used methods are contrasted. The DCNI approach controls the dynamically changing topology to identify fusion nodes in one topology. The outcome of DCNI supports the identification of the crucial nodes and results with enhanced quality and time complexity. However, only simple DOS attacks were used to evaluate the DCNI technique, Alam et al. [11]. The author Niu Z et al. [12] and Narayandas et al. [13] discusses the operation of MANET systems in IoT. This is a basic MANET with an IoT technique that already exists. This technology uses programmers to run the MANET, where some systems have more end devices.

The Dynamic Range Clustering (DRC) method with a learning-based routing scheme implements the load balancing strategies that are optimising the performance of MANETs, Aroulanandam V et al. [14] (LR). The management of network data and energy is improved by this strategy. The cluster has increased network stability with distinctive neighbours for learning-based routing and enabled non-congested data transformation at various network traffic rates. Because of the dynamic topology, the LR method's delay time is getting longer.

A new multi-objective scheme in a heterogeneous network is introduced as a multi-characteristics model by Amiri et al. [15]. This scheme achieves better throughput, energy usage, packet delivery ratio, fewer dead nodes throughout more rounds, and increased network lifetime. Using various network metrics, the cluster creates the network path and transfers the data from the source node to the neighbour node. However, large routing networks have little fault tolerance and significant delay times in this case.

To avoid the pitfall of becoming stuck in local optima, the author devised the Energy Efficient Cluster Head Selection Using Improved Sparrow Search Algorithm (EECHS–ISSA–DE), Kathiroli et al. [16] method. Based on residual energy, this algorithm chooses CH. Although the CH is not distributed evenly, stability is good. This algorithm's results for handling vast amounts of data take less time, and the crossover and CH selection procedures are enhanced. Nonetheless, the breadth and connectivity of this approach are crucial. In WSN, the Butterfly and Ant approach is suggested by Maheswari et al. [17]. In WSN, route generation and the best CH selection are challenging tasks. The path is chosen based on the nodes' energy, their distance from one another and the base station and the node's degree. This methodology's simulation results are taken into account for the local base station and global network.

To choose the best energy-saving nodes, the author Rajpoot et al. [18] proposed a modified Bellman-Ford algorithm. This algorithm was tested in both static and dynamic networks. This scheme found that the static network provided accuracy between 34 and 48 percent and the dynamic network provided energy savings between 35 and 42 percent. However, we found that the amount of energy saved depends on the network's size. In addition, routing protocols play a major role in energy-saving plans. In comparison, Adhoc On-Demand Routing Protocol (AODV) and Energy Quantized (EQ-AODV) routing protocols, the Energy Efficient AODV (EE-AODV) in Mobile Ad-hoc network has increased the energy level of the nodes. Since the ties in this protocol are less stable, the scientists intend to use a genetic algorithm to strengthen the links in their subsequent work.

Li Y et al. [19] proposed a novel EED algorithmic approach. The authors used two parameters such as hop count and residual node energy to estimate an optimal load distribution-based routing mechanism. This work expected the outcomes like optimal energy usage, maximum packet delivery ratio and maximum network lifetime. However, this method implemented a load-balancing routing mechanism to the MANET without considering IoT.

In the description of Er-rouidi M et al. [20], MANET and IoT networks with energy optimization rules are suggested to increase network efficiency. Additionally, the improved cluster method for a heterogeneous network uses the sensor node energy level to determine the route. Due to insufficient load balancing, the tiny size of a node in the network consumes more battery power than the huge size of the network.

Similarly, Vu Q.K. et al. [21] found energy efficiency and safe weight of the cluster approach when MANET and WSN are combined. In this work, all routes' costs are represented by the cluster-weighted values from the sources to the receiver. In MANET, the dynamic genetic algorithm is utilised to determine the dynamic topology based on the many node characteristics of power, position, availability, and latency, Bruzgiene R et al. [22] Chatterjee B et al. [23]. This approach is one of the path-finding optimization techniques. However, in a large-scale network, the network's accuracy is lower. The aforementioned existing algorithm has improved network performance using a variety of techniques such as residual energy calculation and clustering procedures. It has also optimized packet loss, hop count, distance, increased throughput, and packet delivery ratio. It has also formed a heterogeneous network with low cost and high performance. However, in this approach, every author focuses on a distinct characteristic and omitting any one of the crucial ones has decreased the service's quality.

The existing works are formed by heterogeneous networks with different algorithms, like cluster creation and genetic algorithms, to improve the performance of the network. However, the existing work has not fulfilled the expectations. Some clustering algorithms do not distribute clusters evenly, consequently, CH energy is drained soon. The genetic algorithm saves the network energy but has less fault tolerance and a high time delay.

In this proposed work, the energy utilisation of network nodes is improved using clustering, S. Sugumaran et al. [24] and a trust route selection method. The nodes in the network are divided into small segments using cluster techniques. The cluster nodes are formed by mobile ad-hoc nodes and sensors. The sensors are linked to neighbouring mobile ad-hoc nodes. The node with more sensors is elected as CH. Otherwise, the mobile ad-hoc node is acting as an adjacent node between the sensor (source) and destination. The CH is re-elected by the residual energy of the node. The route discovery process Kanthe A.M et al. [25] is selecting the different routes from source to destination. However, the best and trusted route is evaluated by the total energy utilisation of the nodes, and the distance between the source and destination, S. Sugumaran et al. [26]. This type of route selection provides a better result in terms of saving the total energy of the route, Lorincz et al. [27] under both malicious (DoS) and non-malicious conditions.

3. The Measurement Tools of Network Topology

The clustering techniques manage the large size of the network effect and control traffic and improve the energy level of each node. Faulty nodes can be easily identified in a cluster and automatically replaced by a suitable algorithm. In the cluster network, Venkatasubramanian et al. [28], the graphical method G (N, E) is represented. Here, N is the number of nodes in the cluster, and E is the total energy utilized to transmit the packet from node i to j and i,j $\in$ N. CH energy is evaluated by the residual energy of the node E_{Ri}. The residual energy E_{Ri} is calculated from the total energy of the node to the utilized energy of the node and distance d_{ij} is equal to the velocity of sending the packet to the time required to send all the packets from i to j. The Energy required to transmit one packet E_p and a Total number of packets P_n.

$$E_{CH} = \frac{E_{Ri}}{\sum_{j=1}^{n}\left(\left(d_{ij} + (E_p \times P_n)\right)\right)} \tag{1}$$

The data packet travels from one sensor node to another sensor node and is significantly observed in the network. Particularly, the data throughput of the network is evaluated from the amount of data transferred between the source and destination within a second.

The time, S. Sugumaran et al. [29] is considered depending on the state of the node. At some point, the node acts as inactive, sleeping and idle. The total time is regarded as,

$$T_{Total} = T_{Active} + T_{Sleep} + T_{idle} \tag{2}$$

In active mode, nodes start to transfer and receive the packet with less delay. But few nodes enter sleep mode because of a less level of energy and other nodes act in idle mode. It takes a bit more time to complete the packet transmission and receive, S. Sugumaran et al. [29]. The active time is evaluated by the total bandwidth of the sender-to-receiver channel and the total number of nodes within a network

$$T_{Active} = \frac{\sum_{1}^{n} BW_{ij}}{N_{ij}} \tag{3}$$

Distance is very important in the routing algorithm; the shorter distance brings the data to the destination in less time. In this work, the distance is calculated as the sum of the distance between the adjacent node from i to j, and the total transfer and receive time of the packet. The bandwidth, S M, B et al. [30] is the capability of the route to take the data

$$D_{ij} = \sum_{i=1}^{j} d_{ij} \text{per Seconds} \left(T_{Total} \text{ (Sec)} \times \frac{1 \text{ min}}{60 \text{ Seconds}} \right) \tag{4}$$

The total residual energy, S M, B et al. [30] of the network is taken into account by the energy of the cluster network and the remaining energy of the MANET route.

$$E_T = \sum_{r=1}^{n} (E_{CH} + E_{MANET}) \tag{5}$$

where r is the route between the sender and destination, and E_{CH} is CH residual energy. The MANET discovered the route's residual energy determined by the sum of the energy level between the adjacent links. The energy level of each link is calculated from the residual energy of the packet sent and received. The hop count of the source sensor node to the destination node is evaluated on all the discovered routes in MANET. The hop count is evaluated between the adjacent sequential nodes from the sensor to the destination. Sequential count details are collected from the route discovery process. Different types of routing algorithms provide more than one number of routes to reach the gateway node discovered by any one of the routing algorithms. However, the trusted route (R_t) offers long life, less time delay, reduced packet loss and improved performance of the system, Poluboyina et al. [31]. In this work, MANET optimized energy-saving route algorithm for IoT is evaluated using residual energy level of the node and distance D_{ij}, hop count.

4. Proposed Work of MANET-ESO in IoT

The MANET is using the proposed MANET-ESO in an IoT environment, and it increases the rate of trust and successive communication between sensor nodes. The IoT is consisting of WSN techniques with internet service. The mutual internet world of the electronics domain is growing with intelligent network applications. The IoT sensors are successively gathering physical incidents to digital data in different time intervals. Nevertheless, IoT data transmission delay and power utilization are increased due to the uncertainty of sensor nodes. However, the proposed model increases throughput and minimizes energy factors. This MANET Algorithm has improved network performance through trust and the best route optimization method. In Figure 1, different sensor nodes are connected to the adjacent mobile nodes. Additionally, mobile nodes are forming the cluster using a high density of sensors in that network.

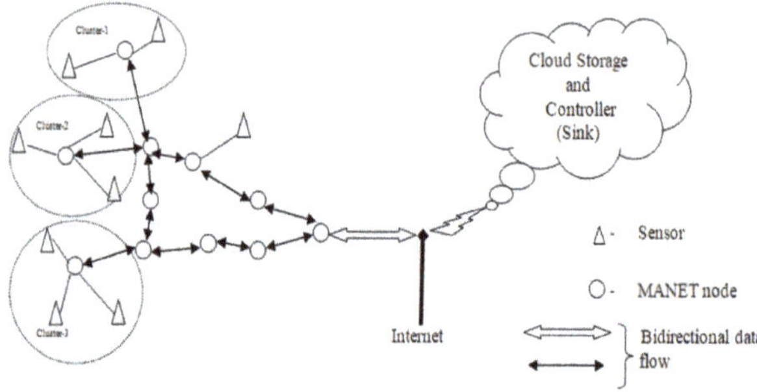

Figure 1. Heterogeneous Network Constructed by Sensor, MANET and Cloud storage through the Internet.

The CH controls, monitors and collects data from sensors. The CH is dynamically changing in the network in terms of the distance from the sensor to the mobile node and the energy level of the mobile node. In the MANET, the information is linked to the gateway node through the CH node, adjacent nodes and gateway nodes using the internet cloud network. The route created in MANET optimizes the energy and enables trusted route paths. Table 1 shows the abbreviations used in the manuscript.

Table 1. Abbreviations and Details.

Abbreviations	Details
MANET-ESO	Mobile Adhoc Network-Energy Saving Optimization
IoT	Internet of Things
WSN	Wireless Sensor Network
IP	Internet Protocol
DNS	Domain Name services
DoS	Denial of Service Attack
CH	Cluster Head
E_{Ri}	Residual Energy of the node
E_p	Energy required to transmit one packet
P_n	Total number of packets
T_{Total}	Total Time
T_{Active}	Active node Time
T_{Sleep}	Sleepy node Time
T_{idle}	Idle node Time
D_{ij}	Distance between adjacent node i to j
E_{CH}	CH residual energy
E_{MANET}	MANET discovered route residual energy
R_t	Trusted Route

In this case, available routes and the best routes are created sequentially using the proposed model. In addition, the MANET transmits and stores data or receives commands using the Internet service. Consequently, a heterogeneous network is created to transfer or receive the data through the internet between MANET and the cloud systems (via gateways). The gateway nodes have created the link between the MANET and the cloud routing algorithm.

In this case, the overall performance of the network is higher than a simple IoT network with optimal performance metrics (throughput, energy, delay, packet delivery ratio and routing overhead) while using MANET-ESO (Algorithm 1).

Algorithm 1: Algorithm for MANET-ESO in IoT

Initialized: Simulation area considered with n—number of sensor nodes
    CH—Cluster Head
    m—number of mobile nodes
    GW—Gate Way nodes
    E_{CH}—residual energy of CH
    r = number of discovered routes
    E_T = Total residual energy of the node
    E_{th} = Energy Threshold Level
    E_{min} = Route total Energy threshold level
    D_{ij} = distance between sensor nodes
    H_p = destination sequential count
If m is connected with more than one number of n, //cluster head selection
    Then, m = CH
else,
    m- act as an adjacent node in the network

Algorithm 1: *Cont.*

```
End if
If m is connected to Internet service,    // gateway node identification
    Then, m = GW
else,
    m–act as an adjacent node in the network
End if
if E_CH>= E_th,                // CH is reelected by Energy level of the node
    Then, retained as a CH
else,
    CH is re-elected
End if
ri=1                           // discovered routes are being checked one by one
While r_i ≤ r
    Energy = E_T
    Distance = Dij
    Hop Count = HP
    If E_T ≤ Emin then
            If D_ij ≤ Threshold-D, then
                    If H_p < Threshold–H, Then
                            r_i = Trusted        //All the conditions are satisfied then act as a
                                                 // trusted route
                    else
                            r_i = r_i++
Repeat the Trust condition
End while;
```

---

5. Simulation and Result of MANET-ESO in IoT Techniques

The simulation of MANET-ESO using IoT Techniques has been simulated by ns-3 on Ubuntu 20 platform.

The experimental configuration considers a MANET, cloud and wireless sensor network with the internet. Since the network has three separate network protocols and it is heterogeneous, the simulation area is estimated to be 1500×1500 m^2. All mobile nodes in the network are considered with the same memory size and random mobility model. A clustered network is made up of multiple sensors connected to CH. Therefore, CH in the MANET node is connected with more sensor nodes. All simulator parameters taken into consideration are shown in Table 2.

This situation is accomplished with internet Gateway nodes. The Gateway node is considered the right side centre of the network area, sensor nodes are considered the left side of the network area and MANET nodes are placed between the sensor to Gateway node in different positions. Every node in the network is dynamically adjusted to a different angle and orientation. In a simulation environment, the traffic generation models, mobility models, topology management models and traffic scheduling models give a realistic network platform. As illustrated in Table 2, a Poisson traffic model is configured under random packet generation mode. The packets generated in the traffic models are measured from 500 bits to 1500 bits in size based on the type of nodes in the IoT environment (mobile adhoc or sensor nodes). Generally, sensor nodes generate minimal-size packets compared to mobile adhoc nodes. Similarly, the mobility model has been configured with random waypoint procedures that allow mobile adhoc nodes to move independently at variable velocities (meters/seconds). In addition, the channel allocation models used in the 802.11 environments are working based on frequency-time allotment principles. The universal frequency band for IoT is configured in an ns-3 simulation environment as illustrated in Table 2. The topology followed in this IoT environment is a hybrid model that is configured with a star topology and mesh topology. Star topology is configured in the ns-3 framework to manage higher-level internet cloud points with other geographical nodes. At the lower level, mobile adhoc nodes and sensor nodes are adopted with mesh topology as they are communicating directly for signal propagation. This simulation considers this dual-topology model as optimal for constructing the entire IoT infrastructure. Anyhow, the sensor nodes are mobile nodes are using random packet distribution schemes based on their initial energy and residual energy rates.

Table 2. Simulation Parameters.

Parameters	Value
Field Area	1500×1500 m^2
Number of sensors nodes	50
Number of Mobile nodes	100
Percentage of cluster head	10%
Radio Propagation model	Free space
Mini. And Max Position of 5G BS	[0, 200]
Antenna Model	Omni directional Antenna
Initial Energy of the sensor node	0.5 Jules
Packet Size	500–1500 bits
MAC Packet Scheduling Model	802.11
Mobility model	Random way
Traffic generation Model	Random Poisson Traffic Model
Initial Energy of the mobile adhoc node	1.25 Joules
Wireless Channel Allocation Models	Frequency and Time Division Model
Node's mobility	10, 15, 20, 25 and 30 m/s (Variable)
IoT Frequency Band	2.4–5 GHz
Proposed Method	MANET-ESO
Adhoc Routing Protocol	AODV
Simulation Tool	Network Simulator (NS-3)

The implementation part of the ns-3 simulation testbed contains the modules of the proposed protocol, existing techniques (EECHS-ISSA DE and Butterfly and Ant) and network characteristics. In the simulated network environment (Table 2), procedure 2 has been implemented among the nodes in the IoT platform. In a similar manner, EECHS-ISSA DE and Butterfly and Ant are implemented separately for evaluating the test cases. The protocols and traffic models used in the experiment are implemented as random Poisson event procedures, Adhoc On-Demand Routing Protocol (AODV), distributed LEACH protocol and energy models. On the basis of this experimental test bed, the performance metrics such as the number of active nodes, throughput rate, residual energy, packet delivery rate and average routing delay are calculated. The mentioned metrics are calculated for iteratively varying network parameters such as the number of nodes, test rounds and suspicious nodes. The prescribed ns-3 simulation model has been implemented using C++ modules (Initial network deployment). In addition, python codes are used for importing the library functions of routing protocols and traffic models.

5.1. Results and Analysis in Terms of Node Alive Node vs. Number of Rounds

When using IoT approaches, the node alive time of MANET-ESO is compared with existing algorithms such as EECHS-ISSA DE [16] and Butterfly and Ant [17] method. Figure 2 illustrates the testing of the proposed methodology and the existing algorithm using 50 sensor nodes, 100 MANET nodes and one Gateway node in the ns-3 simulator. This shows the performance of the number alive node is better than the existing algorithm in a different number of rounds. The WSN constructed with MANET nodes using 10% of clusters was tested for 1800 rounds without dead nodes and produced better results. However, compared to other existing algorithms, the MANET-ESO protocol achieved high alive nodes up to 4500 rounds.

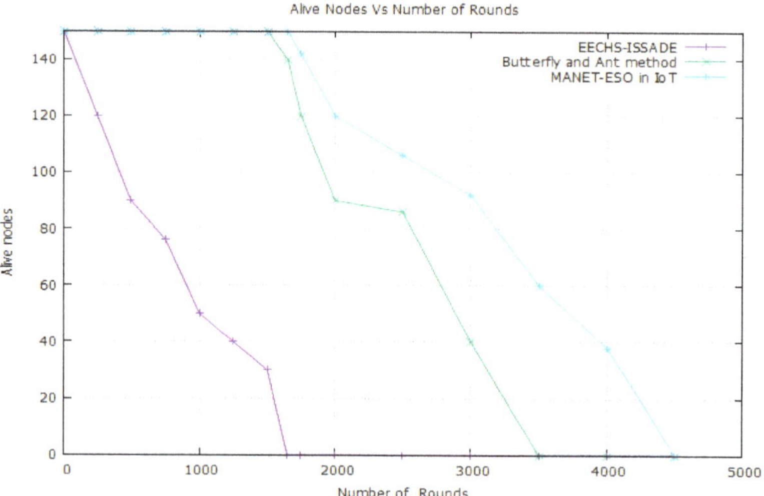

Figure 2. Live nodes vs. number of rounds.

In the EECHS-ISSA DE method, all the nodes are connected via the clustering technique, and the cluster head selection process requires more energy to discover the route. Hence live nodes are limited after 1800 rounds.

In the Butterfly and Ant method, the link to the BS station is discovered by two different algorithms. Firstly, it forms clusters and discovered routes. Both are utilizing less energy compared with EECHS-ISSA DE methods. However, the performance is not better than the proposed work. In the proposed work, the energy utilization of the network is reduced by the clustering algorithm. The cluster head is controlling all nodes in the cluster. So apart from cluster head remaining node energy utilization is less and increasing the alive in the network. Figure 2 shows the proposed algorithm performance, as the alive node is an average of 67.7% higher than EECHS-ISSA DE and an average of 10.7% higher than the Butterfly and Ant Method.

5.2. Results and Analysis in Terms of Residual Energy vs. Number of Rounds

This section discusses the residual energy of the node after 'n' rounds. The results are contrasted with the existing methodology. Figure 3 shows the results of residual energy vs. the number of rounds; the results of the proposed methods are compared to those of the EECHS-ISSA DE and Butterfly and Ant method. Both the cluster energy method and MANET trusted route algorithm compute and identify the best route to improve the residual energy of the route. The overall energy utilization is based on the number of adjacent nodes in the route. The performance of MANET-ESO in IoT is compared to other existing work and observed that EECHS-ISSA DE and Butterfly-Ant methods are wasting the energy level. The EECHS-ISSA DE reduces the energy level over the course of 1000 rounds. The energy balancing approach is being used in Butterfly and Ant method, which has maintained an energy level for up to 8200 rounds. Finally, when compared with the proposed method, the trusted route energy optimization method is choosing the best route based on the total energy utilization of the node, minimum distance and minimum hop count. So, the selected route is using less energy to complete the task and improved the performance of the network compared to other methods. The residual energy of nodes is increased by an average of 75% in this proposed algorithm compared to EECHS-ISSA DE and a 7.1% increase compared to Butterfly and Ant method.

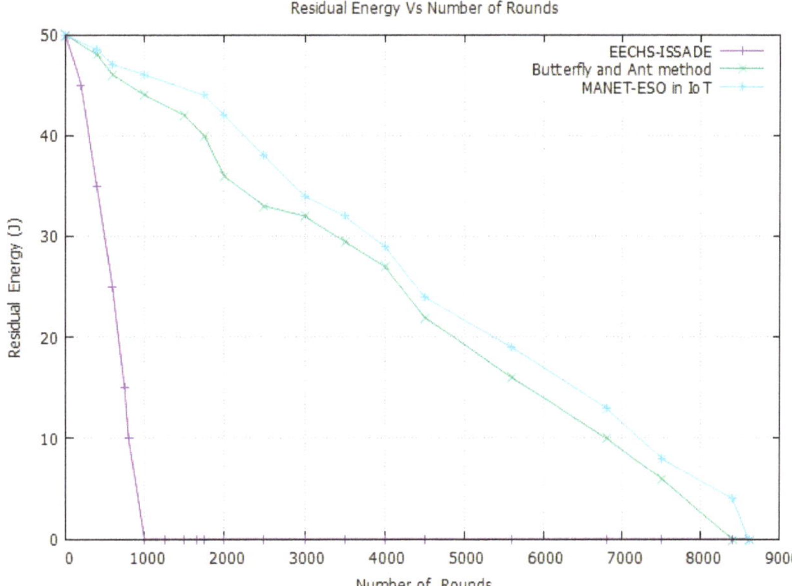

Figure 3. Residual Energy vs. number of rounds.

5.3. Results and Analysis in Terms of Throughput vs. Number of Rounds

The throughput is calculated by the total time taken to transport all packets from the source to the destination. This algorithm identifies the shortest distance by total time duration to transfer packets and less hop count of the discovered routes. It is carrying more packets within a minimum time. So, it increases the throughput of the network compared to other existing work. In EECHS-ISSA DE and Butterfly and Ant methods, throughputs are reduced due to node energy loss. As a result, nodes become dead and the throughput of the network is minimized. However, compared with existing works, the proposed method has outperformed up to 6800 rounds in the network. The Proposed algorithm has a throughput average of 75% higher than EECHS-ISSA DE and an average of 12.2% higher than the Butterfly and Ant Method (Figure 4).

5.4. Performance Evaluation of Packet Delivery Ratio (PDR) and Routing Overhead

This session compares the proposed work performance to EECHS-ISSA DE and Butterfly and Ant algorithms in terms of packet delivery ratio and routing overhead. Figure 5 illustrates the evaluation of the packet delivery ratio with respect to 50, 100, and 150 nodes. Comparing this proposed method to existing ones, the packet delivery ratio is high. The clustering method and trust routing algorithm in the network reduced the packet transmission delay and transferred the packet to a short distance. As a result, all packets reach the destination node with minimum loss than other existing methods. Under 50 and 100 nodes scenarios, the packet delivery ratio of the proposed method is increased by approximately 8% compared to EECHS-ISSA DE and approximately 4% increase compared to the butterfly ant method. The PDR is increased by 8% to 9% for 50 nodes when compared to EECHS-ISSA DE and Butterfly and Ant Method. Figure 6 shows the routing overhead of the proposed work in comparison to the existing algorithms.

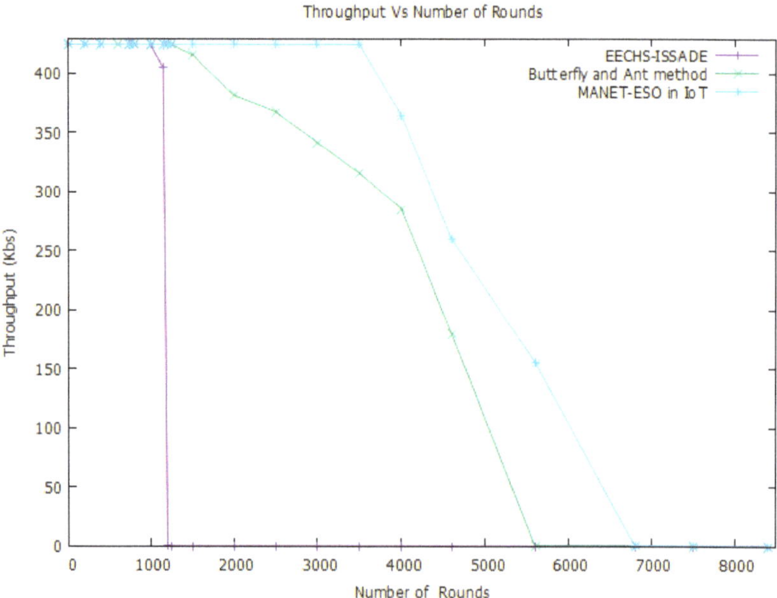

Figure 4. Throughput vs. number of rounds.

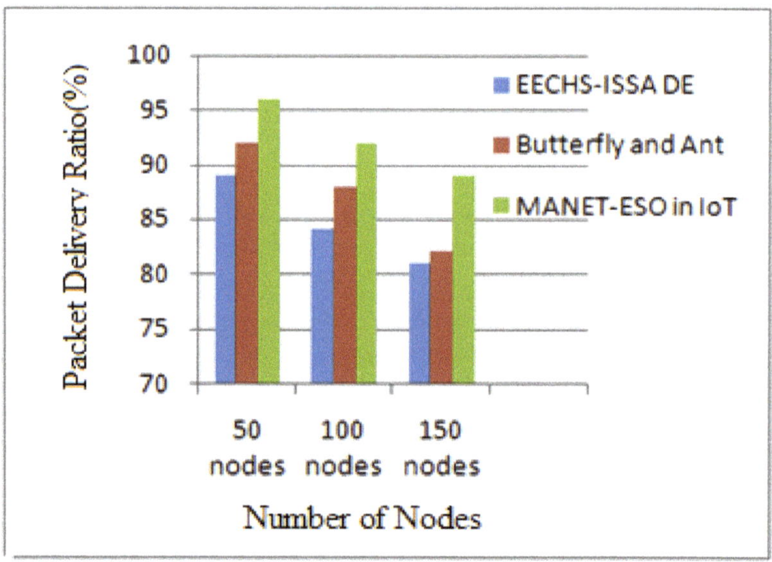

Figure 5. Packet Delivery Ratio vs. Number of Nodes.

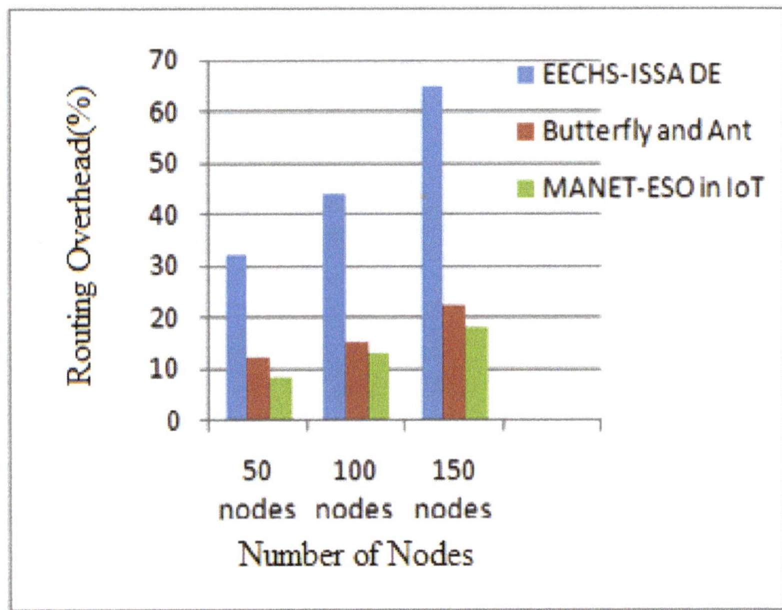

Figure 6. Routing Overhead vs. Number of Nodes.

The trusted routing method increases the route's life span and decreases routing overhead compared to other existing algorithms. In the simulation scenarios compared with 50, 100 and 150 nodes, the output result is 71% better compared to EECHS-ISSA DE and 20% better compared to the Butterfly and Ant method.

In this Figure 7, the packet delivery ratio of the MANET-ESO in the IoT algorithm is compared with the EECHS-ISSA DE, Butterfly and Ant algorithms with respect to the percentage of malicious nodes. The number of malicious nodes in the network increased by 10 and we observed that PDR improved.

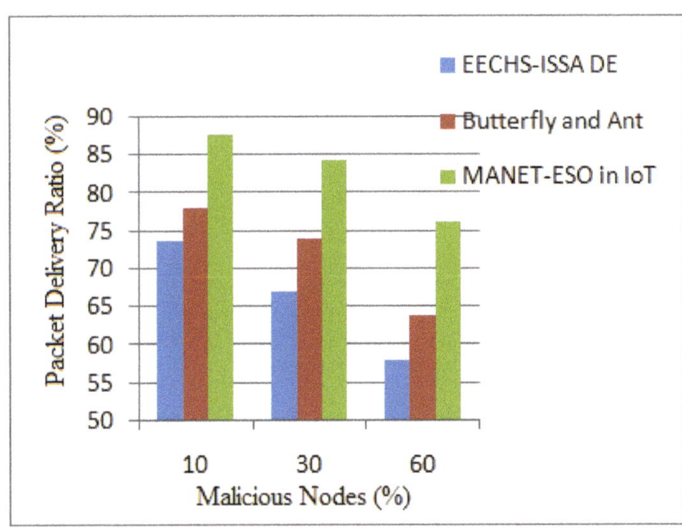

Figure 7. Packet Delivery Ratio vs. Malicious Nodes (%).

Again, the number of malicious nodes is increased to 30% and 60% respectively. Now the proposed algorithm's PDR ratio is reduced due to frequent route failures occurring from the source to the destination.

However, compared with the existing methods, the result of PDR is increased by 24% compared to EECHS-ISSA DE, and increased by 14% compared to the Butterfly and Ant method. Figure 8 shows the performance of the proposed algorithm for routing overhead as the number of malicious nodes increase in the network. The routing overhead is increased due to frequent route failure in the network. Because of the increased malicious activity, the best route selection process requires more time to complete all packet transmission.

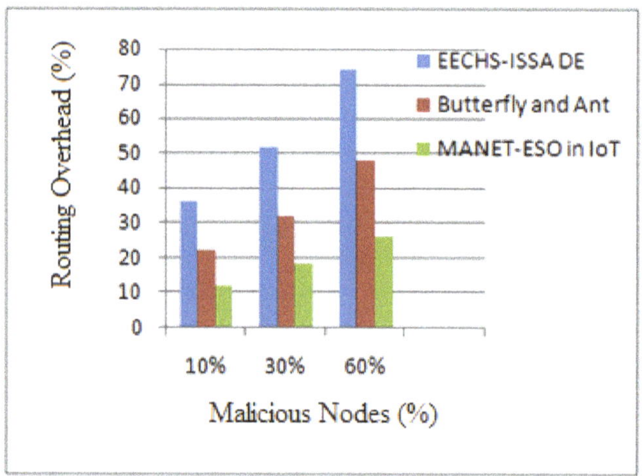

Figure 8. Routing Overhead vs. Malicious Nodes (%).

6. Conclusions

In the IoT, the proposed MANET-ESO algorithm has improved capabilities and reduces the energy utilization of nodes compared to the EECHS-ISSA DE algorithm, Butterfly and Ant method. The simulation environment is created by a heterogeneous network, such as a combination of a WSN, MANET and the internet cloud network. The sensor nodes are connected to MANET using a clustering algorithm. The CH is connected to a variety of sensors and its stability is excellent. The clustering algorithm improved the residual energy of the nodes in the network compared to the existing algorithms. Hence, the number of live nodes is more after the completion of more rounds in the network compared to the existing algorithm. The trusted route algorithm chooses the best route from discovered routes based on the total energy of the node, the distance of the route and the number of hop counts of the route. Under both malicious and benign conditions, the proposed algorithm produced a better packet delivery ratio, optimal throughput and minimal routing overhead. Thus the overall results are better than the existing methods. In the Future, various phishing and spoofing attacks will be used to test the network security. In this way, the level of security will be improved using one of the heuristic methods.

Author Contributions: Conceptualization, V.K.K., I.I. and M.H.; methodology, V.K.K., S.S. and S.V.; software, V.K.K.; validation, I.I. and S.K.S.; formal analysis, S.K.S.; investigation, V.K.K., S.S. and S.V.; resources, I.I.; data curation, T.R.M.; writing—original draft preparation, V.K.K., S.S. and S.V.; writing—review and editing, V.K.K., I.I. and S.K.S.; visualization, T.R.M.; supervision, I.I. and M.H.; project administration, S.K.S.; funding acquisition, I.I. All authors have read and agreed to the published version of the manuscript.

Funding: The National Research Foundation of Ukraine funded this research under project number 2021.01/0103.

Data Availability Statement: Not applicable.

Conflicts of Interest: The authors declare no conflict of interest.

References

1. Jamali, S.; Rezaei, L.; Gudakahriz, S.J. An Energy-Efficient Routing Protocol for MANETs: A Particle Swarm Optimization Approach. *J. Appl. Res. Technol.* **2013**, *11*, 803–812. [CrossRef]
2. Devi, M.; Gill, N.S. Mobile ad hoc networks and routing protocols in IoT enabled. *J. Eng. Appl. Sci.* **2019**, *14*, 802–811.
3. Kumar Debnath, S.; Saha, M.; Islam, M.; Sarker, P.K.; Pramanik, I. Evaluation of Multicast and Unicast Routing Protocols Performance for Group Communication with QoS Constraints in 802.11 Mobile Ad-Hoc Networks. *IJCNIS* **2021**, *13*, 1–15. [CrossRef]
4. Hu, Z.; Odarchenko, R.; Gnatyuk, S.; Zaliskyi, M.; Chaplits, A.; Bondar, S.; Borovik, V. Statistical Techniques for Detecting Cyberattacks on Computer Networks Based on an Analysis of Abnormal Traffic Behavior. *IJCNIS* **2021**, *12*, 1–13. [CrossRef]
5. Khokhlachova, Y.; Hu, Z.; Sydorenko, V.; Opirskyy, I. Method for Optimization of Information Security Systems Behavior under Conditions of Influences. *IJISA* **2017**, *9*, 46–58. [CrossRef]
6. Ray, S.; Mishra, K.N.; Dutta, S. Sensitive Data Identification and Security Assurance in Cloud and IoT Based Networks. *IJCNIS* **2022**, *14*, 11–27. [CrossRef]
7. Jayalakshmi, D.S.; Hemanand, D.; Kumar, G.M.; Rani, M.M. An Efficient Route Failure Detection Mechanism with Energy Efficient Routing (EER) Protocol in MANET. *IJCNIS* **2021**, *13*, 16–28. [CrossRef]
8. Jesudurai, S.A.; Senthilkumar, A. An improved energy efficient cluster head selection protocol using the double cluster heads and data fusion methods for IoT applications. *Cogn. Syst. Res.* **2019**, *57*, 101–106. [CrossRef]
9. Rahman, T.; Ullah, I.; Rehman, A.U.; Naqvi, R.A. Notice of violation of IEEE publication principles: Clustering schemes in MANETs: Performance evaluation, open challenges, and proposed solutions. *IEEE Access* **2020**, *8*, 25135–25158.
10. Hou, Y.T.; Shi, Y.; Sherali, H.D. Rate Allocation and Network Lifetime Problems for Wireless Sensor Networks. *IEEE/ACM Trans. Netw.* **2008**, *16*, 321–334. [CrossRef]
11. Alam, T.; Rababah, B. Convergence of MANET in Communication among Smart Devices in IoT. *IJWMT* **2019**, *9*, 1–10. [CrossRef]
12. Niu, Z.; Li, Q.; Ma, C.; Li, H.; Shan, H.; Yang, F. Identification of Critical Nodes for Enhanced Network Defense in MANET-IoT Networks. *IEEE Access* **2020**, *8*, 183571–183582. [CrossRef]
13. Narayandas, V.; Archana, M.; Raman, D. The Role of MANET in Collaborating IoT End Devices: A New Era of Smart Communication. *Int. J. Interact. Mob. Technol.* **2021**, *15*, 80–92. [CrossRef]
14. Aroulanandam, V.; Latchoumi, T.; Balamurugan, K.; Yookesh, T. Improving the Energy Efficiency in Mobile Ad-Hoc Network Using Learning-Based Routing. *RIA* **2020**, *34*, 337–343. [CrossRef]
15. Amiri, I.S.; Prakash, J.; Balasaraswathi, M.; Sivasankaran, V.; Sundararajan, T.V.P.; Hindia, M.N.; Tilwari, V.; Dimyati, K.; Henry, O. DABPR: A Large-Scale Internet of Things-Based Data Aggregation Back Pressure Routing for Disaster Management. *Wirel. Netw.* **2020**, *26*, 2353–2374. [CrossRef]
16. Kathiroli, P.; Selvadurai, K. Energy Efficient Cluster Head Selection Using Improved Sparrow Search Algorithm in Wireless Sensor Networks. *J. King Saud Univ.-Comput. Inf. Sci.* **2021**, *34*, 8564–8575. [CrossRef]
17. Maheshwari, P.; Sharma, A.K.; Verma, K. Energy Efficient Cluster Based Routing Protocol for WSN Using Butterfly Optimization Algorithm and Ant Colony Optimization. *Ad Hoc Netw.* **2021**, *110*, 102317. [CrossRef]
18. Rajpoot, P.; Dwivedi, P. Multiple Parameter Based Energy Balanced and Optimized Clustering for WSN to Enhance the Lifetime Using MADM Approaches. *Wirel. Pers. Commun.* **2019**, *106*, 829–877. [CrossRef]
19. Li, Y.; Xiong, W.; Sullivan, N.; Chen, G.; Hadynski, G.; Banner, C.; Xu, Y.; Tian, X.; Shen, D. Energy Efficient Routing Algorithm for Wireless MANET. In Proceedings of the 2019 IEEE Aerospace Conference, Big Sky, MT, USA, 2–9 March 2019; pp. 1–9.
20. Er-rouidi, M.; Moudni, H.; Mouncif, H.; Merbouha, A. A Balanced Energy Consumption in Mobile Ad Hoc Network. *Procedia Comput. Sci.* **2019**, *151*, 1182–1187. [CrossRef]
21. Vu, Q.K.; Le, N.A. An Energy-Efficient Routing Protocol for MANET in Internet of Things Environment. *Int. J. Onl. Eng.* **2021**, *17*, 88–99. [CrossRef]
22. Bruzgiene, R.; Narbutaite, L.; Adomkus, T. MANET Network in Internet of Things System. In *Ad Hoc Networks*; Ortiz, J.H., de la Cruz, A.P., Eds.; InTech: London, UK, 2017; ISBN 978-953-51-3109-0.
23. Chatterjee, B.; Saha, H.N. Parameter Training in MANET Using Artificial Neural Network. *IJCNIS* **2019**, *11*, 1–8. [CrossRef]
24. Sugumaran, S.; Venkatesan, P. Attacks Reduction in MANET with Cluster Supported Trusted Routing Protocol. *J. Adv. Res. Dyn. Control. Syst.* **2017**, *6*, 2023–2032.
25. Kanthe, A.M.; Simunic, D.; Prasad, R. Comparison of AODV and DSR on-demand routing protocols in mobile ad hoc networks. In Proceedings of the 2012 1st International Conference on Emerging Technology Trends in Electronics, Communication & Networking, Surat, India, 19–21 December 2012; pp. 1–5.
26. Sugumaran, S.; Venkatesan, P. Optimized Trust Path for control the Packet dropping and collusion attack using Ant Colony in MANET. *Int. J. Eng. Adv. Technol.* **2019**, *8*, 4833–4841. [CrossRef]

27. Lorincz, J.; Ukic, N.; Begusic, D. Throughput comparison of AODV-UU and DSR-UU protocol implementations in multi-hop static environments. In Proceedings of the 2007 9th International Conference on Telecommunications, Zagreb, Croatia, 13–15 June 2007; pp. 195–202.

28. Venkatasubramanian, S.; Suhasini, A.; Vennila, C. Cluster Head Selection and Optimal Multipath Detection Using Coral Reef Optimization in MANET Environment. *IJCNIS* **2022**, *14*, 88–99. [CrossRef]

29. Sugumaran, S.; Venkatesan, P.; Chitra, M.G.; Sivasakthiselvan, S.; Jayarajan, P. Best Optimized Route in MANET using Token Economy Management System. In Proceedings of the 2022 IEEE 4th International Conference on Advances in Electronics, Computers and Communications (ICAECC), Bengaluru, India, 10–11 January 2022. [CrossRef]

30. Benakappa, S.M.; Kiran, M. Energy Aware Stable Multipath Disjoint Routing Based on Accumulated Trust Value in MANETs. *IJCNIS* **2022**, *14*, 14–26. [CrossRef]

31. Poluboyina, L.; Reddy, V.S.; Prasad, A.M. Evaluation of QoS Support of AODV and Its Multicast Extension for Multimedia over MANETs. *IJCNIS* **2020**, *12*, 13–19. [CrossRef]

Article

MHSEER: A Meta-Heuristic Secure and Energy-Efficient Routing Protocol for Wireless Sensor Network-Based Industrial IoT

Anshika Sharma [1], Himanshi Babbar [1], Shalli Rani [1,*], Dipak Kumar Sah [2], Sountharrajan Sehar [3] and Gabriele Gianini [4,*]

[1] Chitkara University Institute of Engineering and Technology, Chitkara University, Rajpura 140401, Punjab, India; anshika.sharma@chitkara.edu.in (A.S.); himanshi.babbar@chitkara.edu.in (H.B.)
[2] Department of Computer Engineering and Application, GLA University, Mathura 281406, Uttar Pradesh, India; dipak.sah@gla.ac.in
[3] Department of Computer Science and Engineering, Amrita School of Computing, Chennai 601103, Amrita Vishwa Vidyapeetham, India; s_sountharrajan@ch.amrita.edu
[4] Dipartimento di Informatica, Universita degli Studi di Milano, 20133 Milano, Italy
* Correspondence: shalli.rani@chitkara.edu.in (S.R.); gabriele.gianini@unimi.it (G.G.)

Citation: Sharma, A.; Babbar, H.; Rani, S.; Sah, D.K.; Sehar, S.; Gianini, G. MHSEER: A Meta-Heuristic Secure and Energy-Efficient Routing Protocol for Wireless Sensor Network-Based Industrial IoT. *Energies* 2023, *16*, 4198. https://doi.org/10.3390/en16104198

Academic Editors: R. Maheswar, M. Kathirvelu and K. Mohanasundaram

Received: 22 April 2023
Revised: 14 May 2023
Accepted: 15 May 2023
Published: 19 May 2023

Abstract: Several industries use wireless sensor networks (WSN) for various tasks such as monitoring, data transmission, and data gathering. They find applications in the industrial internet of things (IIoT). WSNs are utilized to track and monitor changes in the environment. Since they include multiple small sensor nodes (SN), they are severely constrained, so resource management geared toward energy efficiency is crucial in this kind of network. Minimizing the power to interpret, transmit, and store data between various sensors poses important challenges. Experts have considered various ways to address these issues that unavoidably affect the network's performance: reducing energy usage while maintaining system throughput remains the primary research issue. Another important concern relates to network security. Specifically, intrusion detection and avoidance are major concerns. In this work, we introduce the meta-heuristic-based secure and energy-efficient routing (MHSEER) protocol for WSN-IIoT. The protocol learns the forwarding decisions using the number of hops, connection integrity characteristics, and accumulated remaining energy. To make the method more secure, the protocol also employs counter-encryption mode (CEM) to encrypt the data. A meta-heuristics study designed to achieve reliable learning is used in the suggested protocol. The protocol consists of two stages. The first stage uses a heuristics method to improve the option for dependable data routing. Security based on a computationally simple and random CEM is accomplished in the second stage. The proposed MHSEER protocol has been compared to the secure trust routing protocol for low power (Sectrust-RPL), heuristic-based energy-efficient routing (HBEER), secure and energy-aware heuristic-based routing (SEHR), and secure energy-aware meta-heuristic routing (SEAMHR) in terms of packet drop ratio, throughput, network delay, energy usage, and faulty pathways. The proposed protocol increases throughput to 95.81% and decreases the packet drop ratio, packet delay, energy consumption, and faulty pathways to 5.12%, 0.10 ms, 0.0102 mJ, and 6.51%, respectively.

Keywords: wireless sensor networks; industrial internet of things; sensor nodes; energy efficiency; meta-heuristic

1. Introduction

The industrial internet of things (IIoT) paradigm relies heavily on wireless sensor networks (WSNs) [1], which are wireless networks without infrastructure facilities that are deployed using a wide variety of wireless sensors to assess operational, physical, or environmental conditions. In WSN, sensor nodes (SNs) with an inbuilt CPU manage the system and are connected among them and to the base station (BS) [2] (which in turn can be linked to the internet). The nodes perform sensing, data processing, and transmission, whereas data collection, analysis, and delivery to the end user for decision-making are the BS's responsibilities [3]. Due to their versatility, WSNs have found application in several domains such as IoT, tracking and detection systems, conditions monitoring (relative humidity, temperature, and air density), patient evaluation, and agribusiness [4,5].

The key characteristic of WSN is that not only the storage capacity, memory, and CPU processor capabilities of the nodes are limited but their power consumption is considerably constrained [6,7]. Energy-efficient routing (EER) algorithms are therefore essential to reduce power consumption and extend the useful life of the network [8]. Furthermore, the network must be operated manually, which presents certain difficulties [9].

When designing protocols and hardware architectures, researchers should prioritize the effective utilization of the energy storage of SNs [10]. The approaches that have received the most attention include cluster formation and different data transmission communication methods. SNs are organized into a number of subsets, i.e., clusters, to reduce power usage for lengthy transmission [11]. Removing associated data that could reduce the total volume of data exchanged with the BS is the responsibility of a cluster head (CH) [2,12]. The CH transfers the combined information to the BS.

Since they are frequently placed in hazardous or distant locations and are therefore challenging to physically safeguard, sensor nodes are more susceptible to security attacks. They might also lack the ability to process information and the memory necessary to execute effective security regulations, which would make them more vulnerable to attacks. The probability of illegal access and compromised network security and integrity increases greatly as a result of such restrictions [13]. Most proposed solutions aim to increase resource efficiency and the timely distribution of data but neglect to address the reliability of sensor data, leaving a gap for hackers. For industrial operations, inaccurate or manipulated sensor data can have major repercussions, including malfunctioning equipment, delayed production, and safety risks. As a result, it is crucial to use secure protocols, encrypted data, and authentication methods to guarantee the integrity and validity of sensor data [14].

1.1. Objective

WSN-IIoT has the distinct benefit of enabling the continuous surveillance and oversight of industrial operations, which boosts productivity, safety, and effectiveness [15]. With the use of WSNs, preventative upkeep may be carried out on manufacturing equipment and processes, resulting in less downtime. WSNs can also aid in waste minimization and energy optimization, which can save money and assist the environment. WSN-IIoT may, in general, automate and revolutionize manufacturing operations by facilitating decision-making based on data. In order to balance the power usage and avoid unauthorized access, data manipulation, and exposure by malicious nodes on the network, an energy-efficient routing (EER) protocol solution should be created for IoT-based WSNs [16,17]. The purpose of this study is to propose a WSN routing technique that is secure and energy-efficient. The primary goals of the proposed MHSEER protocol are energy efficiency, stability, and reliability, in data transfer despite limited resources. To obtain the best possible data routing, some heuristics are utilized. To learn the routing decision, the MHSEER protocol uses integrated minimal resources, hop count, and connection integrity measures [18]. The suggested IIoT protocol has advantages such as improved security, decreased power consumption, and increased reliability. A secure routing protocol can shield critical data from cyberattacks and stop unauthorized users from connecting to the network. It helps IIoT devices use less energy, which extends their lifespans and minimizes their operating expenses, and by lowering the likelihood of network congestion and packet loss, the protocol can increase the reliability of IIoT systems. The MHSEER protocol is reliable thanks to encoding and the detection and segregation of hostile nodes; furthermore, it offers a method for dynamic routing [13]. The network monitoring between the BS and its surroundings and the energy-efficiency signals together provide effective load balancing and lessen the issue of network segregation. We will show that, compared to existing state-of-the-art solutions, the MHSEER protocol improves the performance of low-power SNs in terms of different performance metrics [19]. The suggested Protocol in IIoT also has several drawbacks, such as complexity and compatibility. A safe routing protocol's implementation might be difficult and expensive because it calls for specialized knowledge and skills. Certain protocols might not work with some IIoT devices, which could restrict their operation or necessitate extra hardware modifications.

1.2. Contributions

The paper's contributions are the following.

- An architecture of WSN-IIoT has been developed for the MHSEER protocol that helps minimize the delay and energy consumption of the network with maximum stability [14].
- To solve the above-mentioned issues, an MHSEER approach has been proposed, which enhances the choice for reliable data routing using a heuristic function and uses the encoding and decoding of data package based on counter encryption mode (CEM) [9].
- The proposed protocol compares the different parameters of a routing protocol such as throughput, network delay, packet drop rate, faulty pathways, and energy usage with the above-mentioned existing approaches. MHSEER increases the throughput and decreases metrics such as the packet drop rate and energy usage.

1.3. Structure

The structure of this article is as follows: Section 2 outlines the state-of-the-art in energy-efficient routing protocols. Section 3 introduces the proposed architecture of WSN-IIoT for the secured and energy-efficiency protocol. A detailed discussion of the MHSEER protocol and the stages designed is presented in Section 4. Section 5 shows the experimental findings about MHSEER and compares them to the competing approaches. Finally, Section 6 summarizes the findings of the paper.

2. Related Work

Various works aim at providing routing protocols that are secure and energy-efficient. In 2019, Hamzah et al. [2] discussed a fuzzy approach for selecting cluster heads (CHs) based on five features: density, residual energy, suitability, and distance from the BS. Utilizing FL-EEC/D (fuzzy logic-based energy-efficient clustering) depending on minimal segregation among CHs, WSNs are created. SNs are evaluated for energy efficiency using the Gini index; clustering techniques normalize resource allocations between WSNs. The results show that the energy usage of the SN stabilized and the energy efficiency with respect to the lifetime of a network improved. In regard to the first cluster dead and half clusters dead, the outcomes demonstrate an average growth.

In 2019, Liu et al. [10] suggest a revised routing protocol to increase WSN resource efficiency. Residual network energy and network average energy are taken into account by the IEE-LEACH protocol in this study. In order to further increase the network's energy consumption, the proposed protocol also adopts a threshold for choosing CHs amongst some of the SNs and makes use of single-hop, multi-hop, and composite connections. The simulation findings demonstrate that this method greatly decreases WSN power use if compared with a number of current routing strategies.

In 2020, Hayajneh et al. [5] communicated with the OSI layers (physical, data link, network, topology, and application) to examine cyber threats in WSNs. The number of significant attacks is calculated, and security precautions are set up to detect attacks. A security technique is created to fix the flaws and identify other problems that require more investigation. This method was only used with a small number of SNs, which negatively impacted the performance performance of the network and did not increase network safety when there were many SNs present. However, the complexity, consumption, and transmit time would all significantly increase, making this potentially inappropriate for WSNs.

In 2020, Binu et al. [14] created an innovative African Buffalo-based two-tier data dissemination technique (AB-TTDD) to check the energy-drained unit early, before information transfer. A brand new temporary energy mapping algorithm (TEMA) was also created to sustain the pathway by producing the reference node rather than the energy-drained component. This innovative method has significantly decreased both power usage and the packet flow ratio. The current study demonstrates that routing maintenance and optimization in WSN may decrease energy usage. However, the proposed methodology's processing takes longer.

In 2020, Haseeb et al. [6] suggested SEHR for WSNs to detect and prevent data manipulation while achieving greater performance. The method provided reliable and insightful learning through the use of heuristic evaluation, which was taken from artificial intelligence (AI). This technique uses a heuristic approach to spot and guard against data breaches. The current metaheuristic method provides less accurate data categorization. The key generation portion of this approach needs to be enhanced because the counter block allows the key value to be identified. By taking into account asynchronous operations among the sensor nodes, energy consumption and routing efficiency can be further enhanced.

In 2022, Gurram et al. [9] discussed a SEAMHR technique in order to choose the best route to the target while preserving data integrity. The learned path is used to identify the most ideal neighbourhood, which serves as a redirector to send the information to the intended receiver, and mutation elephant herding optimization (MEHO) is used to enhance the meta-heuristic function. The SEAMHR method uses the AEDL approach and CTR-AEDL mode encrypted using private keys to encrypt and decrypt data. The MATLAB tool is used to conduct the experiment.However, the suggested protocol will not be expanded to take into account mobility requirements and multi-hop network interactions.

In 2022, Seyfollahi et al. [16] provided a composite energy-aware strategy for data forwarding in IoT, considering the need to extend IoT technologies and the NP-hardness of power management in diverse and dispersed IoT networks. By integrating the support vector machine (SVM), a popular machine learning (ML) approach, and the meta-heuristic heat transfer optimizer (HTOA) approach, this paper tried to implement an optimal method for routing and transmitting data. The results demonstrate that merging ML and HTOA has produced the best possible energy-conscious routing for the IoT.However, it should be taken into account that HTOA, like another optimizer, could become trapped in local optima.

In 2022, Behera et al. [4] briefly discussed low-energy adaptive clustering hierarchy (LEACH)-based and bioinspired protocols, their advantages and disadvantages, their underlying presuppositions, and the selection criteria for the CH to comprehend routing protocols with various structures, innovative tactics, and improved efficiency in the WSN environment. The scalability, durability, and packet-delivery rates of different protocols are contrasted and considered as performance aspects. Additionally, the exploration of developing cryptographic techniques for verified encryption in WSNs that support privacy and network safety is an option.

In 2023, ref. [20] the WSN is responsible for gathering and arranging sensed data before sending it to the base station (BS). As the battery power of sensor nodes is limited, it is crucial to employ effective techniques for data collection and transmission to ensure the extended operation of the sensor network. In this study, the researchers utilized the particle swarm optimization (PSO) method to establish the cluster in WSN. Additionally, they proposed an energy-efficient routing protocol (E-FEERP) based on fuzzy logic. The E-FEERP algorithm considers various factors such as battery energy, the average distance between sensor nodes (SN) and the BS, node density, and communication quality to transmit data from the cluster head (CH) to the BS optimally.

Other works study the problem of clustering and energy efficiency from a game theory perspective [21,22]. Indeed, game theory is routinely used as a framework for the modeling and analysis of performance and security networked systems [23–29]. The work by Gemeda et al. [30] addresses the issue of energy efficiency by proposing a protocol, called GREET, based on non-coalitional game theory: the protocol is not power-aware, but it improves over LEACH in terms of network lifetime.

The current research increases throughput rates and decreases the packet loss rates, energy usage, defective routes, and packet delay of the WSN as compared to the existing energy-efficient protocol (Table 1), as indicated in the prior research.

Table 1. Comparative analysis of existing literature.

Ref. No.	Year/ Author Name	Objective	Software Used	Parameter	Future Scope
[1]	2019/ Airehrour et al.	By integrating the SecTrust system into the RPL protocol, a simulated exercise was conducted to demonstrate the SecTrust system's effectiveness at fending off Rank and Sybil assaults.	PhD Research Lab of Auckland University of Technology.	Throughput, packet drop rate.	To increase the network's integration of trustworthy nodes that have repaid their battery life by extending the SecTrust-RPL.
[2]	2019/ Hamzah et al.	Utilize the gain ratio to assess how effectively the clustering methods can balance the energy distribution among WSN sensor nodes. A fuzzy logic-based CH election method, a k-means-based clustering method, and LEACH are contrasted with the suggested technique FL-EEC/D.	.NET	SN's residual power, the distance to the BS, the density of the SN, the compacting of the SN, and location appropriateness.	The Gini index is a reasonable assessment tool for assessing the routing protocols' energy effectiveness in WSNs for the metric of energy distribution balance.
[6]	2020/ Haseeb et al.	The method provided reliable and insightful learning through the use of heuristic evaluation, which was taken from AI. This technique uses a heuristic approach to spot and guard against data breaches	MATLAB	Throughput, the ratio of packet drops, significant delay, consumed energy, erroneous routes, overhead on networks, and computational cost.	To make the system smarter and fault-tolerant by employing certain lightweight machine learning-based approaches to enhance the SEHR technique.
[8]	2019/ Kuhlani et al.	Developed an accessible virtual structure that serves as an intermediary architecture between the sink and the nodes while sharing metadata and query messages in order to lessen the mobile sink's frequently current location to all nodes.	MATLAB	Average rate of delivery, average energy utilization, the lifespan of a network, and absolute delay.	The suggested approach can be further explained to understand the flow better.
[31]	2019/ Alami et al.	To reduce the energy consumption of WSNs, hierarchical techniques that utilize clustering hierarchy are proposed. Data collection and transmission to a base station could be carried out using the nodes with the highest residual energy.	MATLAB	Stable timeframe, HNA, the lifespan of a network, network traffic, and throughput .	The suggested technique can be expanded to manage a system of mobile sinks and will analyze the network lifespan optimization.

3. Proposed Architecture of WSN-IIoT for Energy Routing Protocol

The clustering process divides the network into various clusters in the WSN environment. Each cluster contains a CH node that transmits data collected from its SN network to the BS, as shown in Figure 1. In hierarchical protocols, choosing the CH node is a crucial decision that adds to the system overhead. The network will experience significant overhead if the CH node fails. The objective is to suggest an energy efficiency technique that minimizes system overhead and maximizes stability. The proposed protocol allows all nodes in a cluster to be CH nodes; however, the clusters are not required to contain a CH node [10]. The CH node is elected using a learned machine. In fact, consuming less energy, which is currently being spent on frequent selections of CH nodes, will hopefully lengthen network longevity. The nodes cooperate to send data to the BS; as a result, the nodes closest to the BS will use more resources than most of the other nodes. As a result, it is likely that the clusters near the BS change to a non-connected condition. As a result, the detected data are not sent to the BS. The clusters next to the BS have smaller sizes to address this issue. Hence, energy for transmitted data will be saved [11].

The proposed approach does not require the SNs to recognize the CH node. An SN uses its cluster-ID and some basic details about its distant neighbors in one hop to deliver data to the BS. The routing protocol employs a distributed and localized strategy to use a learning system to identify the best CH nodes for each cluster. Each node in this scenario can choose the optimum path for data transfer or act as a CH node [2].

An energy-efficient routing protocol is created. An algorithm is initialized right away to preserve the route to identify erroneous data proclamation from the nearby nodes and predicts link failure using its fitness function, as shown in Figure 2. As a result, the algorithm helps extend the node lifetime by reconstructing the path and optimizing the data drop ratio. Hence, the information is securely transmitted without being interrupted.

So, an algorithm is implemented to track the announcement of misleading data. The inclusion of the proposed approach in the routing protocol helps to detect fraudulent data announcements. The protocol also foresees energy-drained nodes early on and raises an alarm as a result. This alarm helps to start the maintenance of the route of the protocol that is elaborated in Figure 2 [14]. Some SNs are included in the homogeneous structure of the energy-efficient routing protocol, and each node is aware of its own position. The fitness function evaluates the node's full energy; therefore, it has an early awareness of connection failure based on the energy of the node. As a result, the algorithm is initiated when the monitoring node issues an alarm in the event that a connection failure occurs within a few moments. An energy-efficient routing protocol is an effective, expandable, and adaptable routing protocol. In wireless architecture, the data can be transmitted to the source and destination nodes via a variety of routes. There are some basic functions such as route preservation, maximizing the lifespan of a network, and minimizing packet drops for secure data transmission [32].

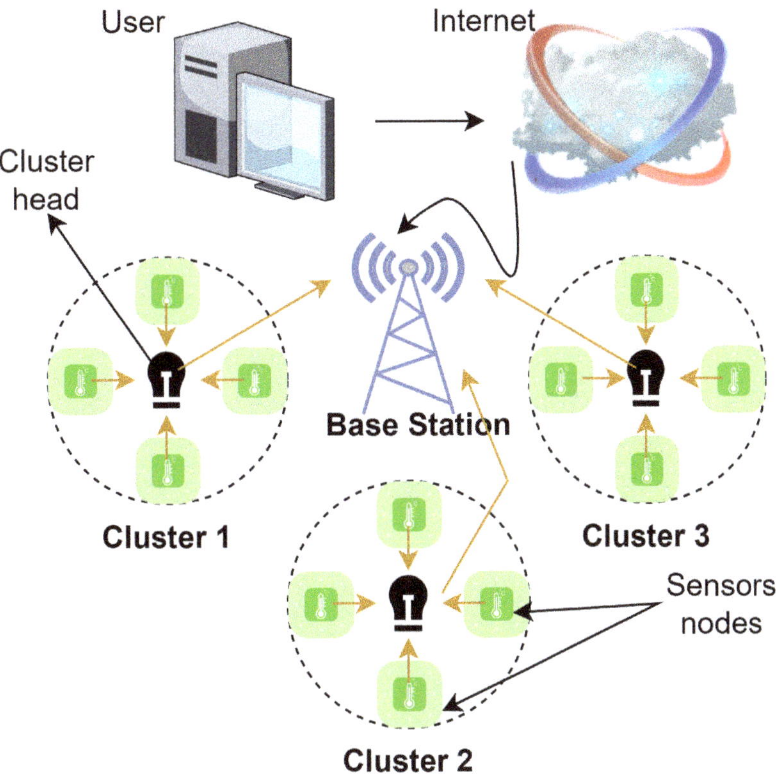

Figure 1. Framework of WSN in IIoT.

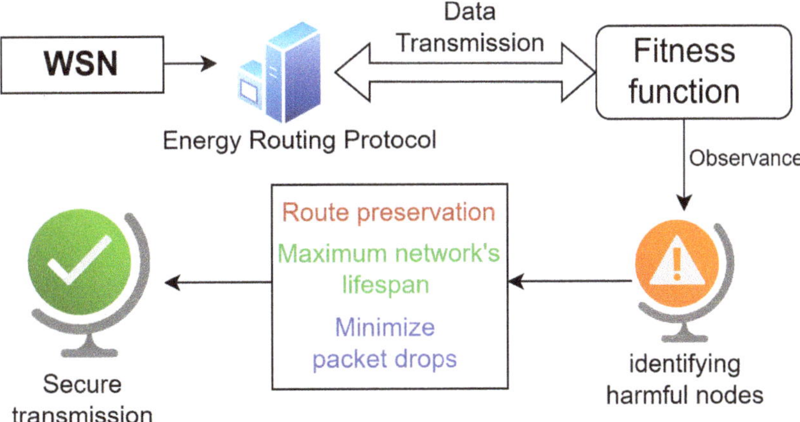

Figure 2. Architecture of WSN for energy-efficient routing protocol.

4. Methodology

This section describes the phases of the proposed MHSEER protocol in detail. The protocol uses link stability metrics, hop counts, and accumulated residual energies to make routing decisions. To begin with, the suggested MHSEER protocol uses a meta-heuristic study built to produce reliable and wise learning [9]. The protocol is a metaheuristic technique that enhances reliable data routing. There are two primary stages to this project. The first stage of the proposed protocol enhances the choice for reliable data routing using a heuristic algorithm. The beam heuristic reduces the memory requirements of the nodes and provides efficient next-hop selection by utilizing a number of characteristics and link reliability [6]. In the second stage, a protocol that is secure, legitimate, and based on a computationally simple and random CEM is achieved [14]. The proposed protocol (Figure 3) simultaneously uses the encryption and decryption of data packets while utilizing less computational power from the nodes. Furthermore, alarm and traffic assessment techniques reduce the likelihood of link failure and uneven energy use in the network area [33].

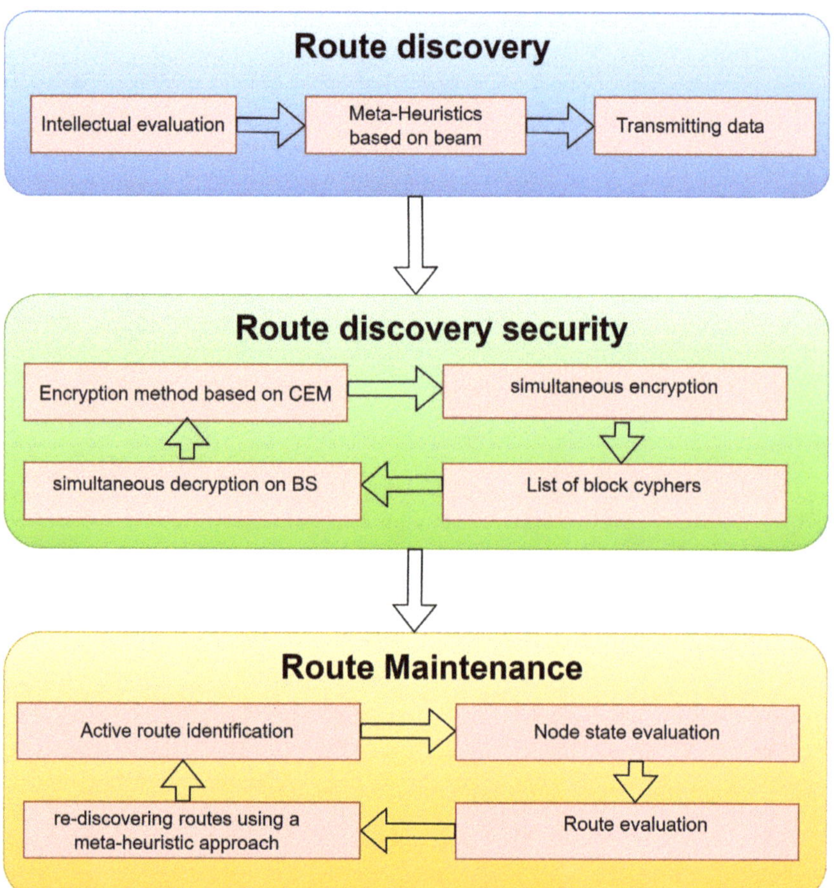

Figure 3. Design of the proposed method.

4.1. Route Discovery

Suppose that the SNs are structured as graphs G, where $G(D, L)$ represents the graph, L represents the optimal links between the two closely linked nodes n_1 and n_2, and D is the density of the SNs. The MHSEER protocol uses the heuristic method in the first stage to estimate the relative weight for locating the best node as the next hop. The decision to route in the direction of the target node is guided by the heuristic function [34]. The source node validates its route entrance against the BS in a local database to start the discovery process. If it is discovered that the route meets the criteria for a number of hops to the BS

(h), the degree of integrity of the link (dl_i), and the accumulated energy (e_i), the source node is selected as the next hop, and the packages are transferred directly to it [35]. Additionally, suppose that there is no viable route available that satisfies the routing criteria in the local database of the source node. In that case, each of the neighboring nodes must participate in the process of choosing the next hop. While accessing the affected links, the link integrity degree uses the hash function of cryptography to ascertain if the transmitted number of packets has been altered. Suppose node i endangers a probe packet with p bits and sends it to node j [36]. To identify its message code, M_c, node i first delivers the probe packet p_1 to the hash function. Second, the received M_c is added as $p_1 + M_c$ to the real probe packet and sent to node j. Node j recalculates M_c of the probe packet received by p_1 after receiving it along with M_c. One is used to denote the high threshold value, and zero is used to denote the low threshold value. The link's high threshold value indicates that it is very trustworthy and has a low fault rate. Residual energy aggregation e_i is estimated in two stages [9]. The residual level e_l is first continuously watched by each node, and then the residual energy's rate e_r is calculated using Equation (1).

$$\sum me_n - ce_n \tag{1}$$

where n is the neighbour, me_n is the maximal energy, and ce_n is the neighbour's energy consumption.

The node with the most accumulated residual energy is consequently assigned the highest precedence. Finally, the node examines the equivalent value that represents the hop count to the BS in its local table. The closer the node is to the BS, the less communication is required to transfer data, as indicated by a lower value. The heuristic function h_f is then determined by adding all of the measured values of the accumulated residual energy, the integrity of the link, and the number of hops in a stack, as shown in the Equation (2).

$$h_f = a * (e_l + e_r) + b * \frac{1}{h} + c * dl_i \tag{2}$$

where $a, b,$ and c are the weighting coefficients and provide the heuristic function with a meaningful effect. Following the computation of h_f, every neighbor node communicates its knowledge to the SN to forecast the best course of action to take to reach the BS. The neighbor node with the highest h_f indicates a superior rank for choosing the next hop. The proposed method uses the meta-heuristic protocol, which maximizes the suitable next-hop in a subset, to analyze the linked graph. In order to save memory and preserve the untrusted node while finding the next hop, the protocol creates node databases [37]. The beam meta-heuristics provide the method for choosing the best node as the next hop based on each node's greatest value. If we assume that, using beam heuristics, $h_{f0}, h_{f1}, \ldots, h_{fn}$, which have the highest weight value, then the weighted total $S(f_n)$ can be denoted as provided in the Equation (3)

$$S(f_n) = \sum h_{fi} \tag{3}$$

where $i = 0,1 \ldots ,n$.

4.2. Route Discovery Security

The proposed protocol concentrated on data protection for the selected beam based on heuristics routing in the second stage. To guarantee connectivity solutions with the least propagation delay, it also provides a route maintenance plan [6]. The proposed protocol uses a CEM that allows nodes to simultaneously encrypt data packets with randomization [38]. Three elements are required to start the CEM encoding method: a data package (D_i), a private key (K), and counter bits (C). To create the distinctive pattern of the counter block

B_i shown in Equation (4), the node N_i and counter bits are combined together and then sent through an encoding function E_f using key K.

$$B_i = E_f(K(N_i + C))$$

(4)

An encoder enables the encoding and decoding of the packet header, and a key generator is used to carry out the learning process. The network's parameters were optimized through a number of experiments, enabling the system to encode and decode data packets and B_i very quickly. The activation function used in this experiment is given by Equation (5).

$$f(x) = \frac{1}{1 + e^{-x}}$$

(5)

where $f(x)$ is the sigmoid function. The originating nodes n_i and E_f are used to create the new counter block B_{i+1} for the data packets D_{i+1} after being concatenated with N_i and C. The ciphertext sequences $c_i, \ldots, c_{n-1}, c_n$ nodes for $n_i, \ldots, n_{n-1}, n_n$ are consequently obtained in the BS, indicating that the decoding function D_f with the same key is used to obtain authentic data packets D_i where $i = 1, \ldots, n$ applying Equation (6).

$$D_i = [D_f(K(N_i + B_i))] \oplus c_i$$

(6)

4.3. Route Maintenance

Some other routes are found with the help of route maintenance if the energy rates of the selected next-hop nodes drop to a particular threshold value [39]. As a result, data degradation and re-transmission are prevented. When an energy-inefficient node is found during the routing stage, it stops transmitting data and sends an error message to the origin node. The next subsequent hop is chosen to continue the data routing after the originating node uses the heuristic function to identify the node with the highest weight. The application of an energy threshold for the maintenance of the route significantly decreases the time of network and route intrusions [40]. Additionally, the quantity of packets received that the next-hops of the BS's neighbors broadcast to them determines the congestion ratio. The proposed method is used to determine the traffic rate (Tf_r) among the neighboring hops (i) or the BS given by Equation (7).

$$Tf_r = \frac{b_l - P_l}{B_m}$$

(7)

where b_l, P_l, and B_m are the bandwidth of the link, the transmitted packets on the link, and the maximal bandwidth from i to the BS, respectively.

5. Results and Discussion

This section describes the experimental findings and configuration of the proposed method over the Sectrust-RPL [1], SEHR [6], HBEER [6], and SEAMHR [9] approaches. The various network techniques are validated, tested, and verified using the MATLAB simulator tool to assess the results of the experiments. In terms of throughput, packet delay, defective routes, and energy consumption, the results are significantly better than the previous research's results. In addition, the performance of the linked methodologies and the proposed protocol is provided by taking into account variables such as end-to-end delay, fault pathways, throughput, packet drop ratio, and energy consumption. Table 2 displays the WSN parameter settings used in the simulation.

Table 2. Simulation parameters.

Parameter	Value
Area of simulation	$300 * 300$ m
SNs	250
Infected nodes	50
Size of packets	64 bits
Level of energy	4 Joules
Position of BS	$200, 600$
Beamwidth	4
Control messages	40 bits
Range of transmission	40 m
Type of traffic	CBR

Table 3 represents the comparative analysis of the proposed approach with the state of art approaches, and Figure 4 shows the graphical representation of the same.

Table 3. Comparative analysis of proposed approach with state-of-art approaches.

Parameters	Sectrust-RPL	HBEER	SEHR	SEAMHR	Proposed
Throughput (%)	74.94	79.75	86.16	94.64	95.81
Packet drop ratio (%)	25.66	19.71	13.32	7.34	5.12
End-to-end delay (ms)	0.178	0.162	0.136	0.114	0.10
Energy consumption (mJ)	0.0326	0.0252	0.0190	0.0154	0.0102
Faulty routes (%)	17.07	13.31	10.46	7.83	6.51

5.1. Throughput Analysis

Table 4 depicts the analysis of the proposed throughput with the four existing methods. Figure 5 represents the throughput with respect to the different number of nodes. Additionally, by utilizing the CEM mechanism for information security, the proposed protocol lessens the possibility of rogue nodes impairing the operation of data transportation between the SNs and the BS. Furthermore, improving the delivery of packets and network access results from traffic monitoring between neighboring nodes and the BS, while the throughput is calculated by Equation (8).

$$Throughput = \frac{Recieved \quad packets}{Total \quad time} \tag{8}$$

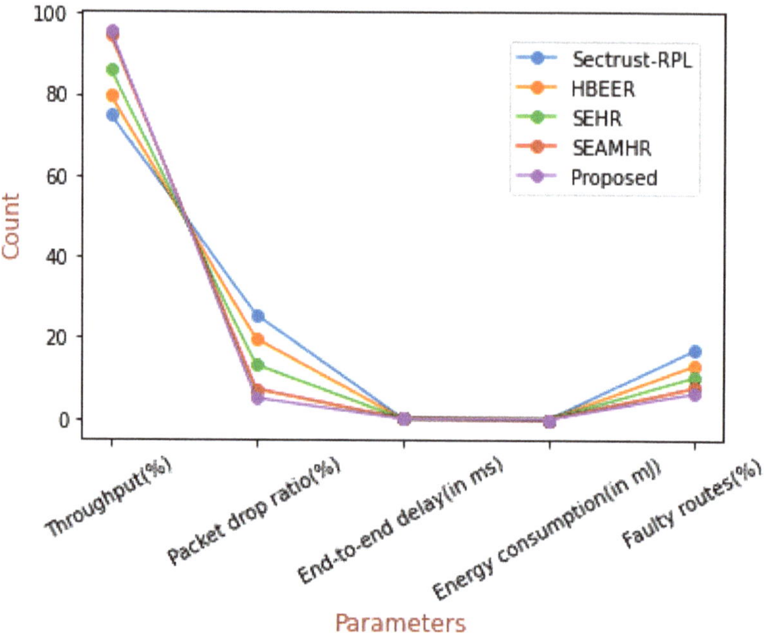

Figure 4. Comparative analysis of state-of-art approaches in terms of different parameters.

Table 4. Comparative analysis of proposed throughput (%) with state-of-art approaches.

No. of Nodes	Sectrust-RPL	HBEER	SEHR	SEAMHR	Proposed
50	78.8	85	89.8	95.2	95.8
100	77.4	84.4	89.2	94.9	96.2
150	76.2	82.3	86.3	94.76	95.9
200	74.94	79.75	86.16	94.64	95.12
250	70	76.2	85	92.8	93.25

Figure 5 shows the 250 nodes with respect to the throughput. In the case of 50 nodes, the proposed approach acquires the maximum throughput, i.e., 95.8%, as compared to the other existing approaches; in the case of 100 nodes, the proposed approach acquires 96.2% throughput in comparison to the existing approaches; in the case of 250 nodes, the proposed approach acquires 93.25% as compared to the existing approaches.

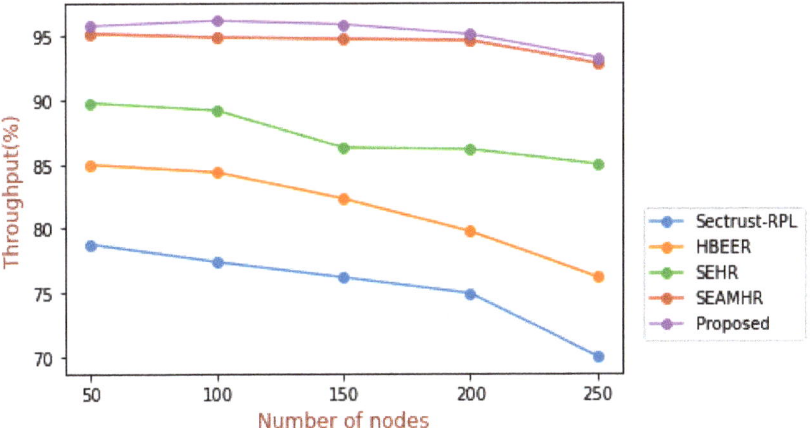

Figure 5. Throughput.

5.2. Packet Drop Ratio Analysis

Table 5 shows the comparative analysis of the packet drop ratio of the four existing methods with the proposed approach. Figure 6 represents the packet drop rate with respect to the different numbers of nodes. In contrast to existing methods, the MHSEER protocol uses a beam heuristic approach to construct a residual energy-based weighted function, the number of hops to the BS, and the connection integrity criteria. Such an approach allows for the choice of the most reliable and energy-efficient nodes for packet transmission, and the packet drop ratio can be calculated by Equation (9).

$$Packet\ drop\ ratio = \frac{Received\ packets}{Total\ packets} \tag{9}$$

Table 5. Comparative analysis of proposed packet drop ratio (%) with state-of-art approaches.

No. of Nodes	Sectrust-RPL	HBEER	SEHR	SEAMHR	Proposed
50	22.51	16.22	10.11	5	4.8
100	23.63	16.32	12	6.21	5.10
150	24.94	17.41	13.24	7.12	6.56
200	25.66	19.71	13.32	7.34	6.85
250	30	21.27	15	9.54	8.24

Figure 6 depicts the minimum packet drop ratio in the proposed approach. As shown, if there are 50 nodes, it is 4.8% as compared to the existing approaches. If there are 250 nodes, it is 30% for Sectrust-RPL; 21.27% for HBEER; 15% for SEHR; 9.54% for SEAMHR; and 8.24% for the proposed approach, which is the minimum that is required.

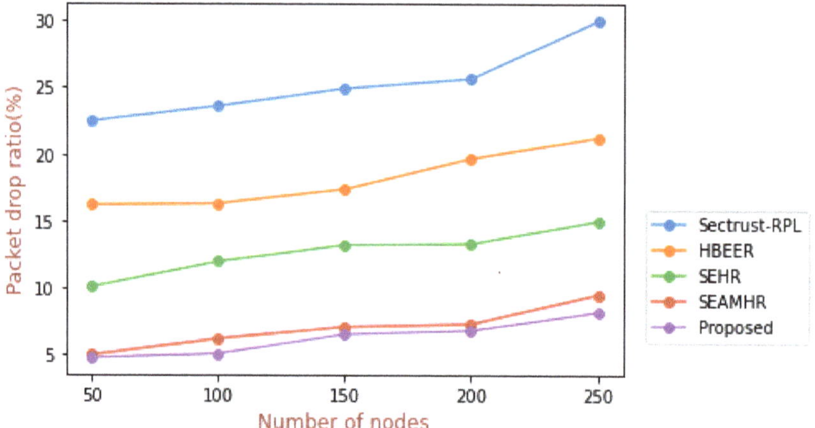

Figure 6. Packet drop ratio.

5.3. End-to-End Delay Analysis

Table 6 depicts the comparative analysis of the packet delay of the four existing methods with the proposed approach. Figure 7 represents the end-to-end delay with respect to the different number of nodes. The proposed protocol reduces the likelihood of path re-finding with the lowest number of re-transference, in contrast to existing solutions, by selecting the security-aware and reliable path for the routing. Such a routing technique eventually reduces congestion issues and effectively utilizes the bandwidth of wireless channels to send data packets with the least amount of network delay.

Table 6. Comparative analysis of proposed end-to-end delay (ms) with state-of-art approaches.

No. of Nodes	Sectrust-RPL	HBEER	SEHR	SEAMHR	Proposed
50	0.152	0.141	0.120	0.0912	0.015
100	0.163	0.149	0.122	0.101	0.018
150	0.171	0.154	0.132	0.110	0.100
200	0.178	0.162	0.136	0.114	0.100
250	0.192	0.173	0.141	0.121	0.104

Figure 7 depicts the minimum end-to-end delay in the proposed approach. In case there are 100 nodes, the proposed approach acquires 0.015 ms, i.e., minimum delay, as compared to the existing approaches. In the case of 250 nodes, it acquires 0.192 ms for Sectrust-RPL, 0.173 ms for HBEER, 0.141 ms for SEHR, and 0.121 ms for SEAMHR; 0.104 ms has been acquired for the proposed approach.

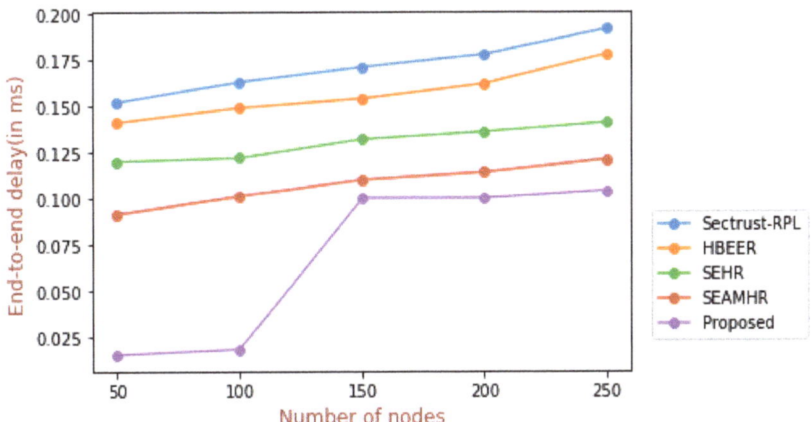

Figure 7. End-to-end delay.

5.4. Energy Consumption Analysis

Table 7 compares the energy usage of the four existing methods with the proposed approach.

Table 7. Comparative analysis of proposed energy consumption (mJ) with state-of-art approaches.

No. of Nodes	Sectrust-RPL	HBEER	SEHR	SEAMHR	Proposed
50	0.0276	0.0211	0.013	0.0131	0.0122
100	0.0301	0.0212	0.0156	0.0143	0.0130
150	0.0313	0.0223	0.0179	0.0150	0.0141
200	0.0326	0.0252	0.0190	0.0154	0.0142
250	0.0355	0.0271	0.0223	0.0162	0.0155

Figure 8 represents the energy consumption with respect to the different numbers of nodes. The proposed method shows a minimal consumption of energy of 0.0102 mJ, while the existing Sectrust-RPL, HBEER, SEHR, and SEAMHR approaches show a consumption of 0.0326 mJ, 0.0252 mJ, 0.0190 mJ, and 0.0154 mJ, respectively. Because of the smart and risk-tolerant routing approach used by the proposed protocol, the routing pathways were stable for a considerable amount of time.

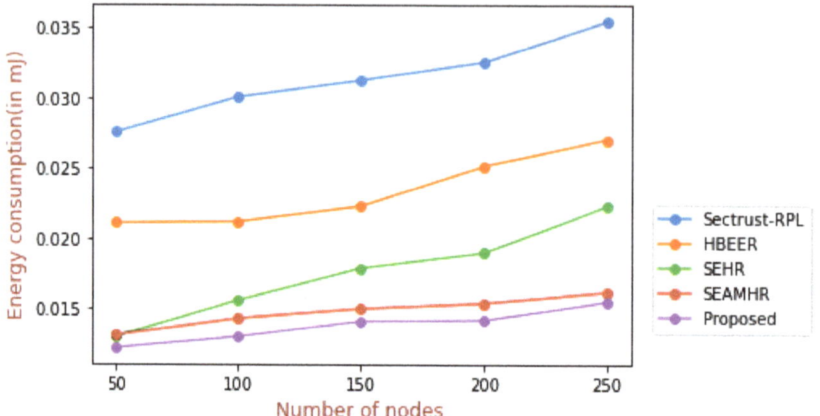

Figure 8. Energy consumption.

5.5. Faulty Routes Analysis

Table 8 shows the comparative analysis of the faulty pathways of the four existing methods with the proposed approach.

Table 8. Comparative analysis of proposed faulty routes (%) with state-of-art approaches.

No. of Nodes	Sectrust-RPL	HBEER	SEHR	SEAMHR	Proposed
50	13.57	10.78	7.81	5.56	4.45
100	14.68	11.23	8.43	6.13	5.50
150	16.25	12.45	9.98	7.24	5.95
200	17.07	13.31	10.46	7.83	6.54
250	22.34	15.11	12.12	9.35	8.45

Figure 9 represents faulty pathways as a function of the number of nodes. The proposed approach shows a minimal faulty pathway of 6.51%, while the existing approaches Sectrust-RPL, HBEER, SEHR, and SEAMHR show 17.07%, 13.31%, 10.46%, and 7.83%, respectively. This is a result of the improvement of the decision to route the heuristic function pathway in the proposed protocol. Furthermore, the heuristic function generates smart conclusions that take into account a variety of factors, such as connection integrity, that promote the consistency of the data.

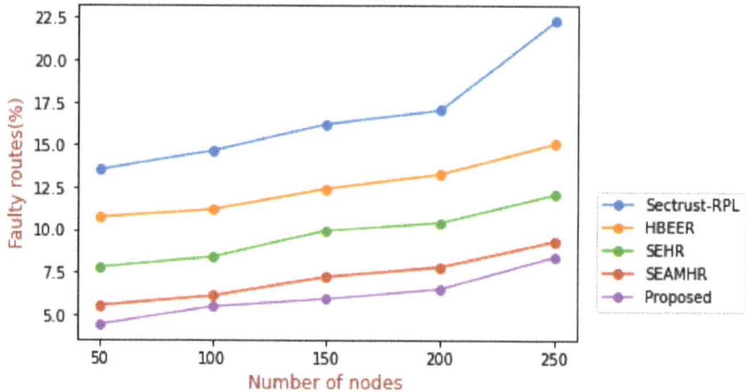

Figure 9. Faulty routes.

5.6. Findings and Implications for the Research

The major objective of this study is to suggest a WSN routing technique that is secure and energy-efficient. The proposed MHSEER protocol's primary goals are energy efficiency, stability, and reliable data transfer capability because of its scarce resources. The finding of the paper demonstrates that the throughput rate of the protocol is 95.81%, and the packet drop ratio, packet delay, energy consumption, and faulty pathways are 5.12%, 0.10 ms, 0.0102 mJ, and 6.51%, respectively. As a result, the suggested method reduces energy consumption, erroneous routes, and latency while also greatly extending the network lifetime.

6. Conclusions

To optimize the routing strategy with wise judgments against malicious nodes, this paper provides a safe EER protocol for WSNs. To achieve dependable communication for WSN, the proposed protocol focuses on important elements, including energy usage, secure data transfer, packet delay, and route maintenance. Using all parameters that have an impact on the energy effectiveness of the WSN protocol is recommended to obtain the highest results from energy-efficient routing protocols. To achieve efficient next-hop decisions and decrease the use of node memory, the heuristic function uses a variety of factors and network integrities. Using different parameters, performance evaluations with existing approaches are shown. The approach offers improved network performance metrics such as throughput and reduced packet drop ratios, energy consumption, end-to-end delays, and defective routes. The proposed protocol increases throughput to 95.81% and decreases the packet drop ratio, packet delay, energy consumption, and faulty pathways to 5.12%, 0.10 ms, 0.0102 mJ, and 6.51%, respectively, in comparison to the existing energy-efficient routing protocol. However, if the proposed approach can be implemented using realistic datasets, it can produce far better outcomes. In future work, the proposed energy-efficient protocol will be implemented using real-world datasets for better results.

Author Contributions: Conceptualization, H.B. and A.S.; methodology, S.R. and A.S.; validation, A.S., D.K.S. and H.B.; formal analysis, S.S., S.R., and H.B.; investigation, S.R. and G.G.; resources, S.R.; data curation, G.G., and A.S.; and writing—original draft preparation, A.S. and H.B. All authors have read and agreed to the published version of the manuscript.

Funding: This research received external funding.

Data Availability Statement: Not applicable.

Acknowledgments: The work was partially supported by the MUSA (Multilayered Urban Sustainability Action) project, funded by the European Union—NextGenerationEU, under the National Recovery and Resilience Plan (NRRP) Mission 4 Component 2 Investment Line 1.5: strengthening of research structures and creation of R&D "innovation ecosystems", set up of "territorial leaders in R&D" (CUP G43C22001370007, Code ECS00000037). The work was also partially supported by the SERICS project (PE00000014) under the NRRP MUR program funded by the EU—NextGenerationEU.

Conflicts of Interest: The authors declare no conflict of interest.

References

1. Airehrour, D.; Gutierrez, J.A.; Ray, S.K. SecTrust-RPL: A secure trust-aware RPL routing protocol for Internet of Things. *Future Gener. Comput. Syst.* **2019**, *93*, 860–876. [CrossRef]
2. Hamzah, A.; Shurman, M.; Al-Jarrah, O.; Taqieddin, E. Energy-efficient fuzzy-logic-based clustering technique for hierarchical routing protocols in wireless sensor networks. *Sensors* **2019**, *19*, 561. [CrossRef]
3. Sharma, S.; Kaur, A. Survey on wireless sensor network, Its Applications and Issues. In *Journal of Physics: Conference Series*; IOP Publishing: Bristol, UK, 2021; Volume 1969, p. 012042.
4. Behera, T.M.; Samal, U.C.; Mohapatra, S.K.; Khan, M.S.; Appasani, B.; Bizon, N.; Thounthong, P. Energy-Efficient Routing Protocols for Wireless Sensor Networks: Architectures, Strategies, and Performance. *Electronics* **2022**, *11*, 2282. [CrossRef]
5. Hayajneh, A.A.; Bhuiyan, M.Z.A.; McAndrew, I. A novel security protocol for wireless sensor networks with cooperative communication. *Computers* **2020**, *9*, 4. [CrossRef]

6. Haseeb, K.; Almustafa, K.M.; Jan, Z.; Saba, T.; Tariq, U. Secure and energy-aware heuristic routing protocol for wireless sensor network. *IEEE Access* **2020**, *8*, 163962–163974. [CrossRef]

7. Kumar, A.; Sharma, I. Enhancing Cybersecurity Policies with Blockchain Technology: A Survey. In Proceedings of the 2022 5th International Conference on Contemporary Computing and Informatics (IC3I), Greater Noida, India, 14–16 December 2022; pp. 1050–1054. [CrossRef]

8. Kuhlani, H.; Wang, X.; Hawbani, A.; Busaileh, O. Heuristic data dissemination for mobile sink networks. *Wirel. Netw.* **2020**, *26*, 479–493. [CrossRef]

9. Gurram, G.V.; Shariff, N.C.; Biradar, R.L. A Secure Energy Aware Meta-Heuristic Routing Protocol (SEAMHR) for sustainable IoT-Wireless Sensor Network (WSN). *Theor. Comput. Sci.* **2022**, *930*, 63–76. [CrossRef]

10. Liu, Y.; Wu, Q.; Zhao, T.; Tie, Y.; Bai, F.; Jin, M. An improved energy-efficient routing protocol for wireless sensor networks. *Sensors* **2019**, *19*, 4579. [CrossRef]

11. Qureshi, K.N.; Bashir, M.U.; Lloret, J.; Leon, A. Optimized cluster-based dynamic energy-aware routing protocol for wireless sensor networks in agriculture precision. *J. Sens.* **2020**, *2020*, 9040395. [CrossRef]

12. Sharma, S.; Guleria, K. Pneumonia Detection from Chest X-ray Images using Transfer Learning. In Proceedings of the 2022 10th International Conference on Reliability, Infocom Technologies and Optimization (Trends and Future Directions) (ICRITO), Noida, India, 13–14 October 2022; IEEE: Piscataway, NJ, USA, 2022; pp. 1–6.

13. Akin, E.; Korkmaz, T. Comparison of routing algorithms with static and dynamic link cost in software defined networking (SDN). *IEEE Access* **2019**, *7*, 148629–148644. [CrossRef]

14. Binu, G.; Shajimohan, B. A novel heuristic based energy efficient routing strategy in wireless sensor network. *Peer- Netw. Appl.* **2020**, *13*, 1853–1871. [CrossRef]

15. Gowri, S.; Pappa, C.K.; Tamilvizhi, T.; Nelson, L.; Surendran, R. Intelligent Analysis on Frameworks for Mobile App Development. In Proceedings of the 2023 5th International Conference on Smart Systems and Inventive Technology (ICSSIT), Tirunelveli, India, 23–25 January 2023; IEEE: Piscataway, NJ, USA, 2023; pp. 1506–1512.

16. Seyfollahi, A.; Taami, T.; Ghaffari, A. Towards developing a machine learning-metaheuristic-enhanced energy-sensitive routing framework for the internet of things. *Microprocess. Microsyst.* **2023**, *96*, 104747. [CrossRef]

17. Yun, W.K.; Yoo, S.J. Q-learning-based data-aggregation-aware energy-efficient routing protocol for wireless sensor networks. *IEEE Access* **2021**, *9*, 10737–10750. [CrossRef]

18. Daanoune, I.; Baghdad, A.; Ballouk, A. An enhanced energy-efficient routing protocol for wireless sensor network. *Int. J. Electr. Comput. Eng. (2088–8708)* **2020**, *10*, 5462–5469. [CrossRef]

19. Liao, R.F.; Wen, H.; Wu, J.; Pan, F.; Xu, A.; Jiang, Y.; Xie, F.; Cao, M. Deep-learning-based physical layer authentication for industrial wireless sensor networks. *Sensors* **2019**, *19*, 2440. [CrossRef]

20. Narayan, V.; Daniel, A.; Chaturvedi, P. E-FEERP: Enhanced Fuzzy-based Energy Efficient Routing Protocol for Wireless Sensor Network. *Wirel. Pers. Commun.* **2023**, 1–28. [CrossRef]

21. Gemeda, K.A.; Gianini, G.; Libsie, M. The effect of node selfishness on the performance of WSN cluster-based routing algorithms. In Proceedings of the AFRICON 2015, Addis Ababa, Ethiopia, 14–17 September 2015; IEEE: Piscataway, NJ, USA, 2015; pp. 1–5.

22. Gemeda, K.A.; Gianini, G.; Libsie, M. Collaborative packets forwarding to extend lifetime of multi-authority wireless sensor networks. In Proceedings of the 2017 International Conference on I-SMAC (IoT in Social, Mobile, Analytics and Cloud)(I-SMAC), Palladam, India, 10–11 February 2017; IEEE: Piscataway, NJ, USA, 2017; pp. 513–519.

23. Manshaei, M.H.; Zhu, Q.; Alpcan, T.; Bacşar, T.; Hubaux, J.P. Game theory meets network security and privacy. *ACM Comput. Surv. (CSUR)* **2013**, *45*, 25. [CrossRef]

24. Gianini, G.; Damiani, E.; Mayer, T.R.; Coquil, D.; Kosch, H.; Brunie, L. Many-player inspection games in networked environments. In Proceedings of the 2013 7th IEEE International Conference on Digital Ecosystems and Technologies (DEST), Menlo Park, CA, USA, 24–26 July 2013; IEEE: Piscataway, NJ, USA, 2013; pp. 1–6.

25. Lena Cota, G.; Mokhtar, S.B.; Lawall, J.; Muller, G.; Gianini, G.; Damiani, E.; Brunie, L. A framework for the design configuration of accountable selfish-resilient Peer-to-Peer systems. In Proceedings of the 2015 IEEE 34th Symposium on Reliable Distributed Systems (SRDS), Montreal, QC, Canada, 28 September–1 October 2015; IEEE: Piscataway, NJ, USA, 2015; pp. 276–285.

26. Lena Cota, G.; Mokhtar, S.B.; Gianini, G.; Damiani, E.; Lawall, J.; Muller, G.; Brunie, L. Analysing Selfishness Flooding with SEINE. In Proceedings of the 2017 47th Annual IEEE/IFIP International Conference on Dependable Systems and Networks (DSN), Denver, CO, USA, 26–29 June 2017; IEEE: Piscataway, NJ, USA, 2017; pp. 603–614.

27. Lena Cota, G.; Mokhtar, S.B.; Gianini, G.; Damiani, E.; Lawall, J.; Muller, G.; Brunie, L. RACOON++: A semi-automatic framework for the selfishness-aware design of cooperative systems. *IEEE Trans. Dependable Secur. Comput.* **2017**, *16*, 635–650. [CrossRef]

28. Gianini, G.; Mio, C.; Fossi, L.G.; Egyed-Zsigmon, E. A Watermark Inspection Game for IoT Settings. In Proceedings of the 2019 IEEE World Congress on Services (SERVICES), Milan, Italy, 8–13 July 2019; IEEE: Piscataway, NJ, USA, 2019; Volume 2642, pp. 29–34.

29. Gianini, G.; Viola, F.; Lena-Cota, G.; Lin, J. Hybrid Inspector-Inspectee-Agent Games in Mobile Cloud Computing. In Proceedings of the 16th ACM Symposium on QoS and Security for Wireless and Mobile Networks, Alicante, Spain, 16–20 November 2020; pp. 95–100.

30. Gemeda, K.A.; Gianini, G.; Libsie, M. An evolutionary cluster-game approach for Wireless Sensor Networks in non-collaborative settings. *Pervasive Mob. Comput.* **2017**, *42*, 209–225. [CrossRef]

31. El Alami, H.; Najid, A. ECH: An enhanced clustering hierarchy approach to maximize lifetime of wireless sensor networks. *IEEE Access* **2019**, *7*, 107142–107153. [CrossRef]
32. Yin, Y.; Li, Y.; Gao, H.; Liang, T.; Pan, Q. FGC: GCN based federated learning approach for trust industrial service recommendation. *IEEE Trans. Ind. Inform.* **2022**, *19*, 3240–3250. [CrossRef]
33. Gao, H.; Huang, W.; Liu, T.; Yin, Y.; Li, Y. Ppo2: Location privacy-oriented task offloading to edge computing using reinforcement learning for intelligent autonomous transport systems. *IEEE Trans. Intell. Transp. Syst.* **2022**. [CrossRef]
34. Babbar, H.; Rani, S. Software-defined networking framework securing internet of things. In *Integration of WSN and IoT for Smart Cities*; Springer: Cham, Switzerland, 2020; pp. 1–14.
35. Babbar, H.; Rani, S.; Islam, S.M.; Iyer, S. QoS based Security Architecture for Software-Defined Wireless Sensor Networking. In Proceedings of the 2021 6th International Conference on Innovative Technology in Intelligent System and Industrial Applications (CITISIA), Sydney, Australia, 24–26 November 2021; IEEE: Piscataway, NJ, USA, 2021; pp. 1–5.
36. Khalaf, O.I.; Abdulsahib, G.M. Energy efficient routing and reliable data transmission protocol in WSN. *Int. J. Adv. Soft Comput. Appl.* **2020**, *12*, 45–53.
37. Lilhore, U.K.; Khalaf, O.I.; Simaiya, S.; Tavera Romero, C.A.; Abdulsahib, G.M.; Kumar, D. A depth-controlled and energy-efficient routing protocol for underwater wireless sensor networks. *Int. J. Distrib. Sens. Netw.* **2022**, *18*, 15501329221117118. [CrossRef]
38. Elsmany, E.F.A.; Omar, M.A.; Wan, T.C.; Altahir, A.A. EESRA: Energy efficient scalable routing algorithm for wireless sensor networks. *IEEE Access* **2019**, *7*, 96974–96983. [CrossRef]
39. Maheshwari, P.; Sharma, A.K.; Verma, K. Energy efficient cluster based routing protocol for WSN using butterfly optimization algorithm and ant colony optimization. *Ad. Hoc. Netw.* **2021**, *110*, 102317. [CrossRef]
40. Xu, C.; Xiong, Z.; Zhao, G.; Yu, S. An energy-efficient region source routing protocol for lifetime maximization in WSN. *IEEE Access* **2019**, *7*, 135277–135289. [CrossRef]

Article

Multi-Channel Assessment Policies for Energy-Efficient Data Transmission in Wireless Underground Sensor Networks

Rajasoundaran Soundararajan [1], Prince Mary Stanislaus [2], Senthil Ganesh Ramasamy [3], Dharmesh Dhabliya [4], Vivek Deshpande [5], Sountharrajan Sehar [6] and Durga Prasad Bavirisetti [7,*]

[1] Department of Networking and Communications, SRM Institute of Science and Technology, Chennai 603203, India
[2] Department of Computer Science and Engineering, Sathyabama Institute of Science and Technology, Chennai 600119, India
[3] Department of Electronics and Communication Engineering, Sri Krishna College of Engineering and Technology, Coimbatore 641008, India
[4] Department of Information and Technology, Vishwakarma Institute of Information Technology, Pune 411037, India
[5] Department of Computer Engineering, Vishwakarma Institute of Information Technology, Pune 411037, India
[6] Department of Computer Science and Engineering, Amrita School of Computing, Amrita Vishwa Vidyapeetham, Chennai 601103, India
[7] Department of Computer Science, Norwegian University of Science and Technology (NTNU), 7491 Trondheim, Norway
* Correspondence: durga.bavirisetti@ntnu.no

Citation: Soundararajan, R.; Stanislaus, P.M.; Ramasamy, S.G.; Dhabliya, D.; Deshpande, V.; Sehar, S.; Bavirisetti, D.P. Multi-Channel Assessment Policies for Energy-Efficient Data Transmission in Wireless Underground Sensor Networks. *Energies* **2023**, *16*, 2285. https://doi.org/10.3390/en16052285

Academic Editor: Abu-Siada Ahmed

Received: 30 January 2023
Revised: 22 February 2023
Accepted: 23 February 2023
Published: 27 February 2023

Abstract: Wireless Underground Sensor Networks (WUGSNs) transmit data collected from underground objects such as water substances, oil substances, soil contents, and others. In addition, the underground sensor nodes transmit the data to the surface nodes regarding underground irregularities, earthquake, landslides, military border surveillance, and other issues. The channel difficulties of WUGSNs create uncertain communication barriers. Recent research works have proposed different types of channel assessment techniques and security approaches. Moreover, the existing techniques are inadequate to learn the real-time channel attributes in order to build reactive data transmission models. The proposed system implements Deep Learning-based Multi-Channel Learning and Protection Model (DMCAP) using the optimal set of channel attribute classification techniques. The proposed model uses Multi-Channel Ensemble Model, Ensemble Multi-Layer Perceptron (EMLP) Classifiers, Nonlinear Channel Regression models and Nonlinear Entropy Analysis Model, and Ensemble Nonlinear Support Vector Machine (ENLSVM) for evaluating the channel conditions. Additionally, Variable Generative Adversarial Network (VGAN) engine makes the intrusion detection routines under distributed environment. According to the proposed principles, WUGSN channels are classified based on the characteristics such as underground acoustic channels, underground to surface channels and surface to ground station channels. On the classified channel behaviors, EMLP and ENLSVM are operated to extract the Signal to Noise Interference Ratio (SNIR) and channel entropy distortions of multiple channels. Furthermore, the nonlinear regression model was trained for understanding and predicting the link (channel behaviors). The proposed DMCAP has extreme difficulty finding the differences of impacts due to channel issues and malicious attacks. In this regard, the VGAN-Intrusion Detection System (VGAN-IDS) model was configured in the sensor nodes to monitor the channel instabilities against malicious nodes. Thus, the proposed system deeply analyzes multi-channel attribute qualities to improve throughput in uncertain WUGSN. The testbed was created for classified channel parameters (acoustic and air) with uncertain network parameters; the uncertainties of testbed are considered as link failures, noise distortions, interference, node failures, and number of retransmissions. Consequently, the experimental results show that DMCAP attains 10% to 15% of better performance than existing systems through better throughput, minimum retransmission rate, minimum delay, and minimum energy consumption rate. The existing techniques such as Support Vector Machine (SVM) and Random Forest (RF)-based Classification (SMC), Optimal Energy-Efficient Transmission (OETN), and channel-aware multi-path routing principles using Reinforcement Learning model (CRLR) are identified as suitable for the proposed experiments.

Keywords: wireless underground sensor networks; channel; distortion; machine learning; reactive communication; energy efficiency; quality assessment

1. Introduction

WUGSNs are widely used in the field of object surveillance under the ground surface. WUGSNs need multiple underground sensor nodes deployed sparsely or densely under the surface level. The underground sensor nodes detect the environmental objects and resources (oil substances, water substances, soil materials, etc.). Extensively, these networks help to observe the real-time object movements in defense sectors [1]. WUGSN nodes are operated with a limited set of resources (memory, processor, energy, and lifetime) amongst real-time issues such as vulnerable medium, channel distortions, and uncertain environmental conditions. Amid these network issues, learning and predicting the wireless channel parameters are major problems.

Particularly, a WUGSN channel maintains three types of links for each data transmission. The channel creates the association between underground sensor node and surface sensor node. Next, the channel provides the link between surface sensor node and ground base station. In addition, a third type of link makes the data path between underground sensor nodes [2]. Each type of wireless channel carries network data under unique data traffic parameters and channel quality metrics. The channel qualities and traffic parameters are configured based on signal strength (amplitude), transmission energy, receiving energy, data transfer rate, transmission range, noise rate, interference rate, attenuation rate, traffic type, connection establishment rate, antenna type, medium access control policies, influence rate of legitimate sensor nodes, influence rate of malicious sensor nodes, and other uncertain events.

The channel quality parameters are uncommon for different types of WUGSN mediums established for underground transmission, underground to surface transmission, and surface to base station transmission [3–5]. In this case, the channel parameter analysis model needs suitable channel quality management principles, adaptive learning functions, and reactive backing systems [6,7]. The recent development in computing arena initiates the variants of Machine Learning (ML) and Deep Learning (DL) frameworks to make intelligent decisions against critical problems. Recently introduced channel quality assessment techniques use ML and DL approaches for predicting the channel behaviors. The artificial decision-making systems provide channel metric collection, data preprocessing, classification, decision making, and report generation phases.

The existing solutions contribute better understanding practices against uncertain channel measurements. Rehan et al. [8] proposed ML-based channel quality and stability evaluation procedures for Wireless Sensor Networks (WSNs). This novel channel assessment model achieved baseline channel prediction principles for evaluating channel rank points, signal strength indicator, and connection quality metrics. In this work, the common aggregated channel quality indicator value showed the stability of wireless channels. Similarly, Aldossari et al. [9] analyzed the channel modelling principles for wireless communication. The channel modelling is the process of computing channel measurement quantities using statistical data analysis, stochastic process, and ML principles. This work modelled the wireless channel qualities using signal fading rate, bandwidth rate, Doppler spread, and block error rate.

The ML-based channel measurement functions were computed for observing upper bound and lower bound characteristics of wireless channels. In the same manner, many research works have evolved to track and analyze the wireless channel parameters [10,11]. Moreover, the existing wireless channel models were limited against nonlinear event analysis policies and multi-channel attribute analysis policies. Firstly, the existing channel configuration models and parameter estimation models utilized default attributes of static wireless sensor networks. Secondly, the classification of channel distortions due to environmental issues and malicious events were not identified separately. Finally, the observation of various medium qualities (underground substances and open-air medium) was not attained through optimal ML and DL approaches. These are noted as crucial research problems for feasible data communication.

On the prospect, the proposed DMCAP creates the suitable signaling models, nonlinear channel assessment models, multi-channel interference models, and channel distortion analysis models against uncertain WUGSN channels. Particularly, the proposed research work is motivated to design and implement a reactive nonlinear channel learning models and IDS engines for supporting feasible channel quality estimation. This research work backs higher throughput attainment through the proposed DMCAP model with EMLP, ENLSVM, VGAN-IDS, and nonlinear entropy analysis procedures.

The proposed DMCAP model is significant against existing techniques in terms of differentiated channel assessment procedures. Notably, most of the existing techniques were implemented for analyzing the channel qualities using linear models for WUGSN. The existing techniques were not implemented accurately for evaluating multi-channel quality assessment factors such as nonlinear noise quantity, entropy rate, signal loss rate, and multi-channel interference rate (underground, underground to surface, and surface to surface). According to the motivation, this proposed model classifies the channel under three categories such as underground wireless channels, underground–surface channels and surface wireless channels. The proposed channel models and signal models are created by analyzing wireless channel attributes and signal attributes, respectively.

Particularly, the proposed DMCAP model analyzes the channel quality metrics and nonlinearity issues using EMLP and ENLSVM. In this case, EMLP is used to determine the classified entropy levels for various wireless channels. The development of EMLP (back propagation procedures) is initialized over the configured model of interference and energy optimization. EMLP is a useful procedure for classifying the channel interference on multiple attribute (multi-modal) validation schemes. In this manner, ensemble MLP units are created to learn and classify multi-channel entropy attributes in WUGSN environment.

In the next case, ENLSVM ensures multi-level classifier units for extracting SNIR from multiple channel quality metrics. ENLSVM consists of multi-channel SNIR distribution procedures, SNIR classification, and likelihood analysis procedures under uncertain network conditions. In the final case, VGAN-IDS analyzes the channel distortion rate initiated by malicious sensor nodes. On this basis, the proposed model observes and classifies the channel quality metrics of WUGSN through reactive channel learning and adaptation procedures. Thus, the proposed DMCAP model achieves optimal data transmission pattern and energy saving solution. The proposed DMCAP model has the motivation to optimize overall WUGSN communication quality depends upon real-time multi-channel uncertainties. The technical contributions of proposed model are listed below.

- Development of multi-channel signaling and channel models;
- Development of multi-channel ensemble model and channel attribute classification model;
- Implementing channel entropy classification procedures (EMLP);
- Configuring SNIR distribution and nonlinear regression procedures (ENLSVM);
- Channel distortion analysis against malicious events;
- Supporting optimal wireless channel utilization and data communication solutions through proposed channel behavior learning techniques.

Unlike other existing techniques, the testbed of the proposed model was configured for both surface level channel parameters and underground channel parameters. According to real-time assumptions, the proposed DMCAP model was enriched with a dual channel propagation model such as acoustic (underground) and ground (surface) features. These are the notable features of the proposed DMCAP model compared with existing techniques. On these conditions, the proposed DMCAP model significantly classifies real-time multi-channel attributes based on SNIR, entropy, malicious events, and other nonlinear distortions.

The contributions of the proposed DMCAP model has the benefits such as optimal energy utilization rate, optimal link establishment rate, minimal routing delay, and maximum secure throughput rate compared with existing techniques. Compared with other existing techniques, the proposed model was specially developed for WUGSN. Furthermore, the proposed model was modelled to predict multi-channel quality metrics for reducing the impacts of channel uncertainties in WUGSN. The results provided in Section 4 illustrate the benefits of proposed model against existing techniques practically.

On the basis of research motivation, the manuscript has notable research works that contribute to the channel evaluation and attribute assessment model in Section 2. Section 3 of this manuscript explains the technical contributions and system design of proposed DMCAP model. Section 4 provides the experiment details and performance evaluation. This section shows the implementation details, network configuration parameters, channel configurations, and results. Finally, Section 5 summarizes the crucial contributions of the research work with appropriate future philosophies.

2. Related Works

In wireless networks, channel models perform a major role in implementing a reliable data communication system. Flawless data transmission highly depends on the quality of wireless channels and the rate of distortions. On this basis, any intelligent (ML and DL) communication models should properly learn and predict the active channel conditions. The wireless channel modelling procedures and the data communication models are closely related to each other. There are different types of wireless networks found as wireless sensor networks, mobile ad-hoc networks, wireless personnel area networks, and wireless local area networks.

Among these networks, wireless sensor networks are deployed for collecting the ecological data from various objects on the surface, sea, and underground areas. Compared with other sensor networks, WUGSNs are highly dominated by channel distortions and underground obstacles.

Similarly, WUGSNs consist of both underground links and surface links. On this environment, the nature of each channel is configured uniquely with crucial parameters (antenna type, interference, noise, energy, etc.). Understanding the real-time channel distortions and modelling the channels according to the need aids better reactive communication systems. The proposed DMCAP model implements a realistic WUGSN channel model and channel quality prediction model with the help of multi-channel quality assessment policies. The scope of the proposed research work was initiated from various related research works. This section describes the recently developed channel assessment and channel quality estimation policies.

Bogena et al. [12] proposed a novel signal attenuation assessment model against soil contents for hybrid WUGSNs. In this contribution, the low-cost soil surface network was created to estimate the wireless signal transmission. This work evaluated various signaling possibilities against different types of soil thickness. This work stated that the WUGSN can communicate with other nodes via 5 cm soil surface (thickness). This work enabled radio communication channel using ZigBee network protocol to build soil-net environment. In this environment, ZigBee communication channels were created to share the information between underground sensor nodes to evaluate the ability of signal attenuation levels. Similarly, the channels were enabled to connect underground soil nodes and surface nodes. This work stated that the attenuation rate created by soil thickness levels affected the channel ability. However, this work missed the versatile configurations of various uncertainties such as noise, link failures, node failures, and dropped packets.

Sharma et al. [13] analyzed the technical benefits and limitations of Internet of Things (IoT) environment and WSNs. This work provided the details such as network heterogeneity, energy optimization, scalability, routing delay, network security, channel flexibility, and data throughput. This work classified different types WSNs and the characteristics in terms of mobility, energy resources, deployment models, and architectures. Yan et al. [14]

proposed game theory approaches for clustering the sensor nodes in order to reduce the energy consumptions. Game theory is the technique considered for various decision-making systems. In this case, each sensor node of the WSN was treated as a player node on the field. On this deployment model, the sensor nodes were clustered on the basis of their current states (active or passive). This work provided a solution for energy-based clustering solutions in WSNs. The energy-efficient game theory approach and clustering, mechanisms introduced in this work were configure to identify the migration status of active node in to sleep state and vice versa. In the same manner, this protocol introduced penalty principles that were working against greedy nodes or selfish nodes available in sensor networks. Notably, these penalty procedures were applied to control the energy violations created by communicating nodes in the network. However, this technique was not developed with well-defined intelligent approaches, uncertain channel models and ML procedures.

Among the solutions developed against various problems of WSNs, the accurate detection and prediction of channel events play a crucial part. O'Mahony et al. [15] proposed a method of analyzing the channel characteristics of WSNs using Support Vector Machine (SVM) and Random Forest (RF)-based Classification model (SMC). SVM and RF are the ML approaches used in the work to understand the nature of real-time wireless channel qualities. This work developed an experimental base for analyzing the channel noise rates, jamming problems, data transfer difficulties and other signaling properties. This work contributed for wireless channel quality assessment practices. This mechanism provided a proper data point collection principles to observe the channel irregularities and uncertainties of wireless sensor networks. At the same time, the suggested channels with optimal conditions were identified as suitable for data communication. On the other side, the supervised models need nonlinear data analysis support systems.

In addition, the effort of this mechanism did not identify multi-path channel disturbances, underground uncertainties and unique properties of multiple channels. Singh et al. [16] proposed Optimal Energy-Efficient Transmission (OETN) with naked mole-rate principles. The need for channel parameter estimation and the quality of service model are the important features for WUGSN. The provided model in this work integrated naked mole-rat algorithm and cross layer multi-channel assessment policies to improve wireless channel stability. Since the energy-efficient channel stability model was defined properly, this work achieved the reliable data communication. Notably, the mole-rat algorithm was applied with magnetic induction procedures. This work stated that the generic electromagnetic signaling mechanisms were seriously affected by channel uncertainties. In this regard, this work found the magnetic induction technique for the underground sensor network. The influence of the above work found the solution for ensuring better throughput and minimal energy consumption in WUGSN. In contrast, uncertain induction rates were not evaluated for multiple channels of WUGSN.

Di et al. [17] proposed channel-aware multi-path routing principles using Reinforcement Learning model (CRLR) for underwater sensor networks. Compared with previous works, CRLR delivered a complex channel analysis and learning practices towards underwater channel estimations. In this regard, the CRLR model was proposed to improve wireless routing protocol functions and ensure optimal data communication possibilities under single-path and multi-path routing strategies. In addition, this work found Reinforcement Learning model to optimize the channel energy utilization factors. At the same time, this work confirmed real-time rewards for channel events to improve the channel throughput with minimal energy consumption. However, the CRLR was not developed to meet uncertain channel conditions and vulnerable channels.

Similarly, other recent research works proposed various wireless communication strategies, ML applications, data distribution principles, and security measures for WSNs [18–20]. Cortés et al. [21] proposed wireless channel observation techniques against signal jamming attacks with the help of collaborative node mechanisms. The work analyzed the possibilities of jamming problems initiated from other malicious nodes to legitimate sensor nodes. Particularly, the cooperative signal detection model was enabled for maintaining feasible data communication channels for industrial sensor networks. However, this system was limited in terms of heterogeneity features and uncertain network conditions. The existing systems described above found various real-time problems on handling the wireless channels and underground channels. Furthermore, the underground substances, rock displacements, and oil contents were noted as crucial obstacles. Further, the works suggested the applications of suitable ML techniques for channel stability predictions.

Tam et al. [22] proposed multi-objective teaching–learning scheme for handling the problems of sensor networks such as coverage probability and lifetime enhancement. This scheme implemented optimized evolutionary algorithms, genetic algorithm, and multi-objective policies. In the implementation, sensor spacing solutions, sensor dominated solutions, and optimal node quantity solutions were attained through network attribute learning models. This work determined the possibilities of better network lifetime and node coverage. At the same time, this scheme stated the importance of a continuous network optimization problem to determine lifetime and coverage abilities. Though solutions are expected to be improved with well-trained ML and DL approaches. Singh et al. [23] debated wireless sensor underground infrastructures and underground monitoring problems. Particularly, this work discussed soil monitoring methods and other environmental observations using WSNs.

In the same manner, the involvement of the work led to the development of WSN infrastructure using magnetic induction principles that are suitable for underground applications. In addition, the effort of the work was continued with the future scopes of WSN-based underground applications. Yet, this article was incomplete in terms of research innovations.

In the same manner, Sun et al. [24] delivered the potentials of border surveillance using WSNs. As WSNs are manufactured using resource-limited components, the significance of lifetime and energy factors is inevitable. In order to improve the lifetime, energy optimization, and transmission coverage quality, Ehlali et al. [25] and Pal et al. [26] developed different types of coverage analysis models and lifespan improvement strategies for WSNs. However, most of the existing systems were not taking WUGSN characteristics and channel uncertainties under research constraints seriously; this is needed and this problem is estimated to be resolved. Additionally, the available channel assessment techniques and WSN communication protocols were not ensuring feasible data throughput under dynamic channel conditions [27,28]. The literature analysis provides various technical details with the following limitations.

- Multi-channel characteristics are not considered for improving data transmission quality;
- The channels of WUGSNs are not assumed with realistic conditions (acoustic and air-based channel parameters);
- The reasons for data loss are not classified under malicious behaviors and channel behaviors;
- Channel models and attributes are not properly analyzed through multi-classifier units and nonlinear functions. Since WUGSN has heterogeneous channel behaviors (distortions), these are necessary for future channel assessment plans.

The proposed model was implemented with appropriate nonlinear data analysis models, ensemble channel attribute classification models, and malicious event analysis models to improve the entire network communication quality.

Consequently, this proposed model attempts to reduce the number of retransmissions, routing delay, and energy wastages by effectively assessing multiple channel properties of WUGSN. Table 1 shows the comparison of previous works (limitations) and the proposed idea to implement DMCAP. Practically, WSNs, WUGSNs, and other wireless network channels are more vulnerable to uncertain channel qualities. Furthermore, WUGSN channel properties are uncommon for underground mediums and surface mediums with crucial real-time distortions. In this field, the deep multi-channel assessment principles are required to maintain the reliable communication against channel problems and malicious events (intruders). The proposed DMCAP model technically approaches all these issues and ensures the solutions against the mentioned problems.

Table 1. Previous works and motivation of proposed work.

Previous Techniques	Motivations of Proposed Work
Homogeneous wireless channel assessment techniques are proposed.	Multi-channel (heterogeneous) assessment techniques are required.
Linear and moderate channel models are created for implementing the networks.	WUGSNs are expected to be considered with more realistic network parameters.
Uncertain conditions are not produced in the network model effectively.	More uncertain conditions must be imposed on underground and open medium of WUGSN.
Energy optimization is not taken crucially for WUGSN with differentiated channel qualities.	Energy optimization must be taken crucially for WUGSN with differentiated channel qualities.

3. DMCAP System

WUGSN has the collection of UG_N underground sensor nodes that are deployed beneath the surface. Each sensor node UG_i has maximum 5 m of circular transmission range through underground obstacles. The underground obstacles between the sensor nodes can be observed as soil substance, rocks, colloidal surfaces, oil substances, and others. Sensor nodes transmit the data to surface nodes or neighbor nodes through electromagnetic signals or acoustic signals. G_s surface sensor nodes receive the underground data streams continuously from different sensors. This proposed system designs the G_s surface nodes as mobile ad-hoc sensor nodes that can move around the surveillance area gradually. In the next level, BG_s number of ground base stations receive the signals from surface nodes. This hybrid WUGSN has different types of channel environments between UG_N, G_s and BG_s points. According to this network design, each channel link needs significant traffic parameters, signal models, and noise models for different links. UG_N nodes forward the data to neighbor nodes and surface nodes (red nodes). At the same time, surface sensor nodes can detect the neighbor sensor node or nearest base station to deliver the data.

3.1. Signaling and Channeling Models

The proposed system formulates complex signaling parameters and channel distortion factors as provided below. The multi-channel signal models and noise models are provided in the following equations.

Signal Attenuation factor:

$$S\alpha^i = \omega\sqrt{\frac{me}{2}}\left(\sqrt{1+\left(\frac{e_\prime}{e}\right)^2}-1\right) \tag{1}$$

Equations (1) and (2) state ω as signal wavelength. m and e are electromagnetic permeability and real permittivity factor respectively. $e_\prime$ indicates imaginary portion of dielectric permittivity factor.

Signal Phase shifting factor:

$$S\beta^i = \omega\sqrt{\frac{me}{2}}\left(\sqrt{1+\left(\frac{e_\prime}{e}\right)^2}+1\right) \tag{2}$$

signal reflection factor:

$$S(R)^i = \frac{1-\sqrt{e_\prime}}{1+\sqrt{e_{\prime\prime}}} \tag{3}$$

attenuation due to reflection:

$$S(RA)^i = 10 \log \frac{2 \cdot S(R)^i}{1 + S(R)^i} \tag{4}$$

Equations (3) and (4) illustrate reflection models. Similarly, the underground particles such as soil contents, water contents and other particles create signal attenuation between all sensor nodes. The underground sensor signals are affected due to reflection, absorption, refraction, and scattering of electromagnetic signals. The receiving power at different links are derived as provided in Equations (5)–(7).

Receiving power at UG_N link,

$$P(R)^{Ui-Ui} = P^{tr} + G^{tr} + G^{rx} - Loss^{UG} \tag{5}$$

- P^{tr}-Sensor Node's Transmission Power;
- G^{tr}-Transmisison Gain;
- G^{rx}-Receiving Gain;
- $Loss^{UG}$-Underground Signal Loss.

Receiving power at G_s and UG_N link:

$$P(R)^{Ui-Si} = P^{tr} + G^{tr} + G^{rx} - \left[Loss^{UG} + Loss^0 + Loss^A\right] \tag{6}$$

- $Loss^0$-Free space loss;
- $Loss^A$-Loss due to adversarial events.

Receiving power at G_s and BG_s link:

$$P(R)^{Si-Bi} = P^{tr} + G^{tr} + G^{rx} - \left[Loss^0 + Loss^A\right] \tag{7}$$

Over these signaling models, Signal to Noise Interference Ratio (SNIR) is determined for uncertain and lossy conditions under different channels and links. In the determination phase, set SNIR threshold range between $\vartheta1$ and $\vartheta2$. In this work, SNIR is adjusted for different channels (Equation (8)). The SNIR is tuned based on novel training algorithm.

$$SNIR(Ui - Ui, \ Ui - Si, \ Si - Bi) \propto C^Q \cdot N^C \cdot T^F(dt) \tag{8}$$

- $SNIR(Ui - Ui, \ Ui - Si, \ Si - Bi) < \vartheta1$, Signal Dropped;
- $SNIR (Ui - Ui, \ Ui - Si, \ Si - Bi) \geq \vartheta2$, Signal recieved at the sink;
- $SNIR (Ui - Ui, \ Ui - Si, \ Si - Bi) \in R(\vartheta1, \vartheta2)$, Adversarial block;
- C^Q-Channel Type;
- N^C-Node Type;
- T^F-Learning Factor at regular interval;
- $\vartheta1$-Lower bound;
- $\vartheta2$-Upper Bound.

As mentioned in Equation (8), three types of channels (links) are created to transfer the information from set of underground sensor nodes to base station via surface sensor nodes. These channels are configured with separate frequencies, bandwidths, energy limits, and other communication needs. Particularly, the link $Ui - Ui$ is a static under the ground level. At the same time, the links such as $Ui - Si$ and $Si - Bi$ are dynamic surface links in the WUGSN [29,30].

Since the surface nodes are moving from one location to another location randomly, these links are separately maintained with the help of unique mobility patterns. In addition, the communication parameters and interference rates of the links are changed against $Ui - Ui$ link attributes. As illustrated in Figure 1, the brown color nodes are UG_i, the green color node is G_i and the base station is BG_i. Figure 1 shows the channels (links) between the sensor nodes and base station. As multiple channels are needed for this heterogeneous WUGSN, the implementation of multi-channel SNIR model with learning factors is a crucial task. The time bounded streams of signaling and interference factors are collected and packed as individual tuple as provided below. The WUGSN channel parameters and distortion cases are determined continuously through complex mathematical functions. Furthermore, these details are modelled as $T(S)\alpha^i$ and $T(P)\alpha^i$. These are channel monitoring tuples used as the sequence of inputs to the proposed ML techniques.

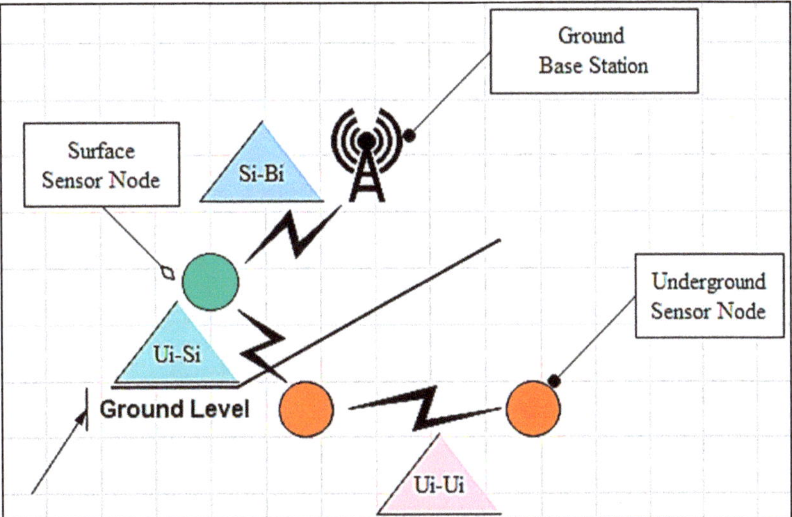

Figure 1. Links and nodes.

Signal attenuation tuple:

$$T(S)\alpha^i = (S\alpha^i,\ S\beta^i, S(R)^i, S(RA)^i)\cdot\frac{d\tau}{dt}\forall\, L^N \tag{9}$$

Signal energy tuple:

$$T(P)\alpha^i = \left(P(R)^{Ui-Ui},\ P(R)^{Ui-Si}, P(R)^{Si-Bi}\right)\cdot\frac{d\tau}{dt}\forall\, L^N \tag{10}$$

Equations (9) and (10) illustrate the details of $T(S)\alpha^i$ and $T(P)\alpha^i$. In Equations (9) and (10), τ states signaling interval; L^N denotes total number of wireless links in the network. These channel quality management tuples are provided in the ML network layers to compute learning factor and adaptive channel quality weight factor [31–33].

3.2. Multi-Channel Ensemble Model and Channel Attribute Classification Model

As WUGSN needs multi-channel data transmission model, signaling parameters, and channel distortions are dynamically determined at each signaling intervals. The statistical channel analysis models and other conventional models analyze the channel qualities with nominal assumptions. In this case, the requirement of multi-channel ensemble classification model is more crucial.

Algorithm 1 shows the procedure of multi-channel interference and energy model based on channel quality tuples, $T(S)\alpha^i$ and $T(P)\alpha^i$. Algorithm 1 initiates various classification procedures for analyzing the real-time quality of WUGSN channels. Most importantly, the multi-channel quality evaluation procedures are initialized in each sensor node deployed at either underground locations or surface points.

Algorithm 1: Multi-Channel Interference and Energy Modelling Procedure

---

Input: Channel Parameters and Identifiers
Output: Ensemble Classifier Activation
1: Get channel parameters as tuple, $T(S)\alpha^i$ and $T(P)\alpha^i$
2: Get channel identifiers, $Ch^i \forall S(Ui - Ui,\ Ui - Si,\ Si - Bi)$
3: Initiate traffic analyzer for all active nodes
4: Compute ensemble DL engine function, $D^E(F)$ and Build ensemble classifier units
        a. Information Entropy Classifier
        b. SNIR Classifier
        c. Attack Classifier
5: Install $D^E(F)$ for all active nodes
6: Initiate $D^E(F)$ for all active communication under svc mode.
7: Call classifiers of $D^E(F)$
8: Redo

End

The subsets of Algorithm 1 lead to enable channel entropy classification practices, SNIR classification practices, attenuation/energy classification practices, and malicious event classifications on each wireless channel. In this deep channel quality analysis and learning model, lightweight ML and DL techniques are utilized for ensuring reliable data transmission conditions [34]. Figure 2 shows the basic signaling model. According to the basic model, the transmitted signal quality, receiving signal quality, channel distortions, and other adversarial factors are evaluated under uncertain mediums. In the first phase of channel quality assessment process, information entropy is determined and evaluated for multiple wireless channels. The information entropy model plays a major role in finding the actual liveliness of each wireless channel of WUGSN. In this regard, the proposed system implements conditional joint entropy functions with Ensemble Multi-Layer Perceptron (EMLP) classifier units in each sensor node. The proposed EMLP procedures are reactive against live channel parameters [35,36].

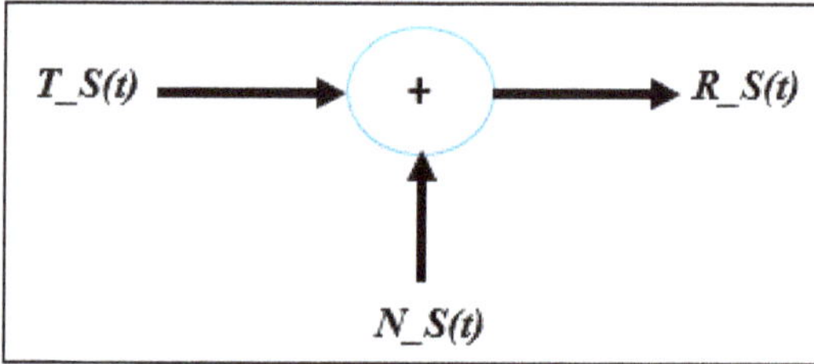

Figure 2. Basic signal model.

The sensor nodes participating in underground communication and surface communication lead to significant information entropy. The entropy model is defined with joint conditional distribution function as provided in Equation (11).

$$H(t_{n_{ai}}|t_i) = E_{t_{n_{ai}}|r_{n_{ai}}}[-\log p(t_{n_{ai}}|r_{n_{ai}}) = -\sum_{t_i\,\epsilon\, t_{n_{ai}}} p(t_{n_{ai}}|r_{n_{ai}})\log p(t_{n_{ai}}|r_{n_{ai}}) \qquad (11)$$

Let the Equation (11) as E_N^i be an entropy pair for active channel at time τ. The information entropy varies continuously for each data transmission. The proposed system implements EMLP network for classifying the continuous streaming of entropy determinations. The training phase of EMLP observes the determinations from Equation (11) and the relationship model for entropy function is provided in Equations (12) and (13). Equation (13) illustrates mutual channel entropy value in the network [37].

$$H(t_{n_{ai}}|r\_n_{ai}) = H(t_{n_{ai}}|r\_n_{ai}) - H(r_{n_{ai}}) \tag{12}$$

$$I(t_{n_{ai}}; r_{n_{ai}}) = \sum_{t_{n_{ai}}, r_{n_{ai}}} p(t_{n_{ai}}, r\_n_{ai}) \cdot \log \frac{p(t_{n_{ai}}, r_{n_{ai}})}{p(t_{n_{ai}}) \cdot p(r_{n_{ai}})} \tag{13}$$

Similarly, the channel data transfer capacity at τ is defined with $I(t_{n_{ai}}; r_{n_{ai}})$ as mentioned in Equation (14).

$$C(I) = \max(I(t_{n_{ai}}; r_{n_{ai}}, \tau, c)) \tag{14}$$

Equations (11)–(14) are denoting the channel entropy variances and determinations of EMLP (Algorithm 2). In this case, let the sensor nodes in an active communication channel are t_n_{ai} and r_n_{ai}. In this circumstance, t_n_{ai} denotes the active transmitter and r_n_{ai} denotes the active receiver. $E_{t_{n_{ai}}|r_{n_{ai}}}$ denotes joint entropy function related to each active communication pair. $H(t_{n_{ai}}|t_i)$ and $I(t_{n_{ai}}|t_i)$ are denoting joint conditional functions of entropy model and information model. According to Equation (11), $H(t_{n_{ai}}|t_i)$ ensures the data dissemination through other nodes (multi-path) based on logarithmic entropy distribution.

Algorithm 2: EMLP for Entropy Learning and Classification

Input: $H(t_{n_{ai}}|r\_n_{ai})$, $I(t_{n_{ai}}; r_{n_{ai}})$, $C(I)$
Output: Classified results and Entropy Knowledge Base Creation
1: Get all channel entropy determinations
2: Train the EMLP using training data, $TE(x, y, \tau)$
3: Determine the entropy bias rate, $\varnothing(Ui - Ui)$, $\varnothing(Ui - Si)$ and $\varnothing(Si - Bi)$
4: Set minimal bias function as steepest decent for all data samples
5: EMLP backpropagation unit is modelled as below
Do for all nodes:
$$\frac{d\omega_{x,y}}{d\tau} = -\frac{\partial\varnothing}{\partial\omega_{x,y}}, \frac{d\delta_{x,y}}{d\tau} = -\frac{\partial\varnothing}{\partial\delta_{x,y}}$$
6: Update the knowledge base, $D(E(x_i, y_i, \tau))$
7: Redo for all nodes in WUGSN.

End

In this case, data shall be fragmented and distributed according to divide and conquer approach since multiple paths are active to destination. The equation is remodeled as indicated to state the relationship between transmitter and forwarding nodes (sink) of multiple paths under logarithmic scale. Equation (12) indicates $H(t_{n_{ai}}|r\_n_{ai})$ as a training function of distributed entropy model with neglected local entropy of each sink (forwarding node) since this proposed model uses EMLP to strictly consider channel entropy conditions for each session. In the same manner, Equation (13) $I(t_{n_{ai}}; r_{n_{ai}})$ represents cumulative channel information distribution probability between t_n_{ai} and r_n_{ai} for each active session in WUGSN. Under this scenario, r_n_{ai} can be considered as any forwarding node (sink) or destination node in multi-path channel model.

In addition, the crucial determination over channel data transfer capacity is defined as $C(I)$ under real-time conditions of WUGSN. Equation (14) has the maximum $I(t_{n_{ai}}; r_{n_{ai}})$ at τ as the real-time data transfer capacity for a particular channel, c. In this manner, the training phase of EMLP for multi-channel entropy model can be expressed for effective WUGSN communication model. This helps to learn the real-time entropy variations for each channel $(Ui - Ui, Ui - Si, Si - Bi)$.

3.3. SNIR Distribution and Analysis Using Multi-Channel Nonlinear Regression Model

The generic nonlinear channel regression model is determined as shown in Equation (15). Generally, SNIR determined at the channel is nonlinear in nature. Similarly, the production of multi-channel SNIR falls completely under nonlinear functions. In this nonlinear production of timeline data of SNIR, the prediction and determination can be modelled with the help of multi-channel nonlinear regression model. As provided in Equation (15), the nonlinear function of channel quality management tuples shall be determined.

$$y(n) = \theta^0 + (T(S)\alpha^i + T(P)\alpha^i)\cdot\theta^1 + \theta^1\cdot\theta^0 + (T(S)\alpha^i + T(P)\alpha^i)^2 + \ldots \tag{15}$$

$$y(n) = \frac{\theta^0}{1 + \theta^{1(T(S)\alpha^i + T(P)\alpha^i) - \theta^2}} \tag{16}$$

- θ^i-Nonlinear quantity factor.

From the nonlinear model determination, the nonlinear ability of each channel impacts the channel can be calculated. Let the multi-channel nonlinear regression model, $m(y(n))$ be expressed as

$$y(n) \propto C(I) \tag{17}$$

$$m(y(n)) = N\cdot y(n)\cdot\frac{dCi}{d\tau} \tag{18}$$

Equation (18) denotes Ci as channel identifier and N as number of channels. The channel capacity and attenuation/energy nonlinearity determinations are closely coupled with SNIR production rate. As the rates of previous channel qualities and quantities vary, SNIR leads to distortions. At this moment, the learning and classification practices of nonlinear SNIR sequence make significant effects in WUGSN communication. This is modelled with the help of Ensemble Nonlinear SVM (ENLSVM). Generally, SNIR determines the upper bounds in information transferring for a wireless channel. The optimal determination of SNIR for each wireless channel shows the quality of wireless links against various signal distortions [38].

In WUGSN, SNIR is defined as provided Equation (19).

$$\text{SNIR}\left(l^i\right) = \frac{P^{sl}}{I^l + \text{Noise}^l} \pm \varphi \tag{19}$$

Equation (19) illustrates the SNIR determination at the active link, l^i. In this equation, P^{sl} represents required signal quality, I^l and Noise^l denote interference and noise impacts in the channel, respectively. Equation (20) shows the multi-channel SNIR determination function for all active channels. The distribution of multi-channel $\text{SNIR}\left(l^i\right)$ is determined with the Gaussian distribution model with φ as provided in Equation (21).

$$S\left(\text{SNIR}\left(l^i\right)\right) = \text{SNIR}\left(l^i\right)\cdot ds\cdot d\tau \ \forall \ S(L) \tag{20}$$

- l^i-a link at time τ;
- l^i can be either $Ui - Ui$ or $Ui - Si$ or $Si - Bi$;
- φ-nonlinear event variance, where $0 \leq \varphi \leq 1$.

In this case, m is the mean and φ^2 is the variance in distribution model. The interference likelihood function is formulated for given distribution model in Equation (21).

$$D(V) = \frac{1}{(2\pi\varphi^2)^{1/2}} \exp\left\{-\frac{1}{2\varphi^2}\left(\text{SNIR}\left(l^i\right) - m\right)^2\right\} \tag{21}$$

$$l\left(\text{SNIR}\left(l^i\right)/m, \varphi^2\right) = \prod_{n=1}^{N} G\left(\frac{\text{SNIR}(l^n)}{m}, \text{SNIR}\left(l^i\right)^2\right) \tag{22}$$

$$1\left(\mathrm{SNIR}\left(l^{i}\right)\right) = \sum_{i=1}^{k} \pi_{i} G(\mathrm{SNIR}\left(l^{i}\right)/m_{i}, \varphi_{i}^{2}) \tag{23}$$

- k—Total number of interference classes;
- i—Interference data class for current channel;
- π_{i}—Mixing coefficient varies from 0 to 1;
- $\mathrm{SNIR}\left(l^{i}\right)$—SNIR for channel 'i'.

The proposed DMCAP effectively uses both nonlinear regression model with ENLSVM principles to classify SNIR data for multiple channels. Moreover, the effective data distribution function helps to handle the SNIR data space optimally for ENLSVM evaluation procedures. Algorithm 3 describes the multi-channel SNIR evaluation procedures.

Algorithm 3: ENLSVM

Input: $S\left(\mathrm{SNIR}\left(l^{i}\right)\right)$, $D(V), 1\left(\mathrm{SNIR}\left(l^{i}\right)\right)$
Output: Classified data distributions
1: Get training samples for k rounds
2: Set threshold for $\mathrm{SNIR}\left(l^{ith}\right), D\left(V^{th}\right), 1\left(\mathrm{SNIR}\left(l^{ith}\right)\right)$ -> $\{Ui - Ui \text{ or } Ui - Si \text{ or } Si - Bi\}$
3: Do training for i = (1,k)
$\mathrm{SNIR}\left(l^{i}\right)$ for k1 samples/SNIR Classifier
$D(V)$ for k2 samples/Distribution Classifier
$1\left(\mathrm{SNIR}\left(l^{i}\right)\right)$ for k3 samples/Likelihood Classifier
4: Configure φ as nonlinear component for all channels
5: Do testing for i = (1,k), Ensemble Test

$$e = \max_{\varphi, threshold} \sum \varphi \cdot [\mathrm{SNIR}, D(V), 1(\mathrm{SNIR})]$$

6: Do recurrent training and testing

End

3.4. Channel Quality Distortions Due to Malicious Events

Evaluation of the changes and distortions happens due to abnormal environmental properties, and the distortions initiated due to malicious attacks are the major risks of WUGSN. In this concern, the WUGSN channels are vulnerable to jamming attacks, wormhole attacks, packet dropping attacks, Distributed Denial of Service (DDoS) attacks, and other malicious injections. On the field, the channel security system needs to classify the attacks and other distortions regularly.

The proposed system configures both channel assessment policies and security practices to monitor the wireless communication parameters [39]. Algorithm 4 presents VGAN engine to be active in sensor nodes to monitor the wireless transmissions. The VGAN-enabled IDS works against different types of attacks based on attack knowledge installed in the node's local storage [40]. VGAN-IDS initiates event generator functions and event discriminator functions to monitor the adversarial activities involved in the channels.

Algorithm 4: VGAN-IDS
Input: Encoded Data Sequences, Channel Quality Metrics **Output**: Vulnerability Logs and Attack Classifications 1: Initiate VGAN associated IDS in sensor node 2: Initiate Attack Dataset, Channel attribute dataset 3: Call VGAN (Sample generator, Data discriminator) 4: Set passive capturing mode (low energy) or active capturing mode (optimal energy) 5: Set M = 1 for sensor monitor (VGAN-IDS) 6: Set M = 0 for forwarding nodes 7: Extract the channel data packets and evaluate using VGAN-IDS 8: Classify the data and suspicious events 9: Share alert reports 10: Redo for all sessions
End

VGAN-IDS performs actively over multi-channel neighbor interactions to classify channel distortions into malicious issues and environmental issues separately. Thus, the proposed DMCAP model monitors environmental distortions and malicious events independently to estimate the quality of wireless channels under uncertain conditions. The VGAN-IDS model and channel assessment policies proposed in the article safeguard the multi-channel data transmission against channel distortions and suspicious events. The proposed system was implemented as shown in Section 4. Section 4 provides an immense impact on the development of proposed DMCAP model and performance evaluations. Figure 3 illustrates the overall system model implemented inside each sensor node supports for reliable network communication. The sensor node has hardware internal components and software internal components. As shown in Figure 3, the software module of each WUGSN sensor node contains proposed DMCAP procedures, VGAN-IDS engine, reactive channel estimator, and DMCAP/IDS activator procedures on demand. In addition, Figure 4 illustrates the overall DMCAP system design and the integral phases used for managing reactive data transmission. The proposed internal learning system and channel assessment procedures was initiated for active channels and nodes [41,42].

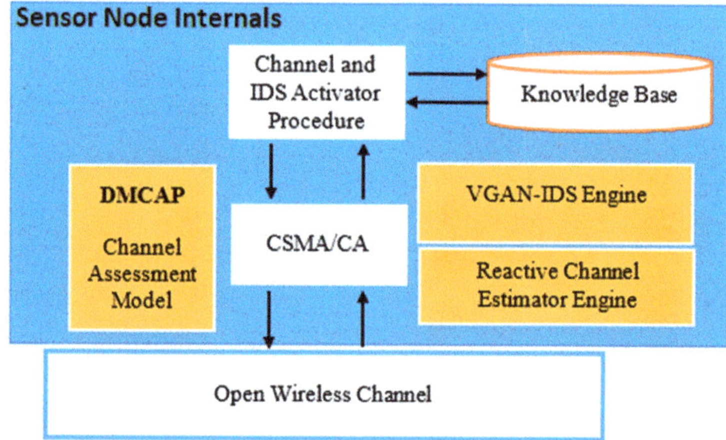

Figure 3. Proposed DMCAP system in sensor node.

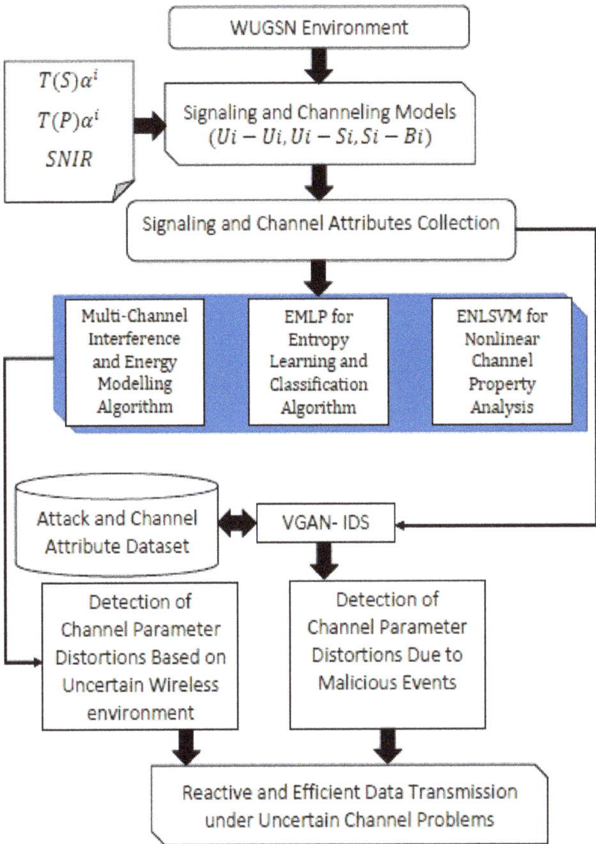

Figure 4. System design of proposed DMCAP phases.

4. Experiments and Results

The experimental circumstance provides the WUGSN design using Network Simulator 3.0 (NS 3.0). In this experiment, wireless sensor network patches are installed and WUGSN parameters are configured as illustrated in Tables 2 and 3. The NS 3.0 sensor network package provides the configuration features related to noise, interference, amplitude, and underground channel constraints. Additionally, the proposed network model consists of electromagnetic signal characteristics and acoustic characteristics. Consequently, the proposed techniques are developed using C++ and Python 3.8 languages in the modelled network base.

Table 2. WUGSN-surface channel configuration parameters.

Test Bed Features	Features
Tool	NS-3.0
MAC	CSMA/CA
Channel Assessment Model	DMCAP
Routing Protocol	AOMDV
Number of Sensor Nodes	40, 80, 100
Attack Dataset	KDD'99
Transmitter Power (W)	0.56
Receiver Power (W)	0.31
Channel Throughput (Kbps)	Variable
Coverage (meters)	50 (maximum)
Channel	Surface (Air)
Mobility	Surface-Random Way Point
Propagation	Two Ray Ground

Table 3. WUGSN configuration parameters.

Test Bed Features	Features
Tool	NS-3.0
MAC	CSMA/CA
Channel Assessment Model	DMCAP
Routing Protocol	AOMDV
Number of Sensor Nodes	40, 80, 100
Attack Dataset	KDD'99
Transmitter Power (W)	0.78
Receiver Power (W)	0.58
Channel Throughput (Kbps)	Variable
Coverage (meters)	25 (maximum)
Channel	Underground
Mobility	Underground-Random Way
Propagation	Two Ray Acoustic

The configured WUGSN has the proposed techniques under Medium Access Control (MAC) policies as inbuilt patches. The network works under 3.5 GHz band around 500 m^2 area. The built-up network area contains the sensor nodes with underground channel characteristics and open surface characteristics. Table 2 shows the WUGSN characteristics of surface sensor nodes and Table 3 illustrates the configurations of underground channel parameters.

The dual-type channel quality configurations ensure close real-time WUGSN channel assumptions [43]. As provided in Figure 3, the knowledge base maintains channel parameter attributes and malicious event attributes (Knowledge Discovery in Databases (KDD'99)). Figure 5 shows the variations in major channel quality parameters such as channel interference rate, channel data transfer rate, and overall channel uncertainty rate. Moreover, Figure 5 denotes the distortion rates at probability scale measurement. The observed measurements present the reduction in data transfer rate as uncertainty and interference rates are increasing over time. The experiment confirms that underground and surface level channel distortions severely affect sensor node's data communication efforts [44]. The test bed of WUGSN initially observes the wireless channel quality metrics and channel distortions as provided in Figures 5 and 6. Figure 6 relates the quantity of data retransmission rate and channel noise production rate. Generally, noise is defined as unwanted signals that disturb the original data communication.

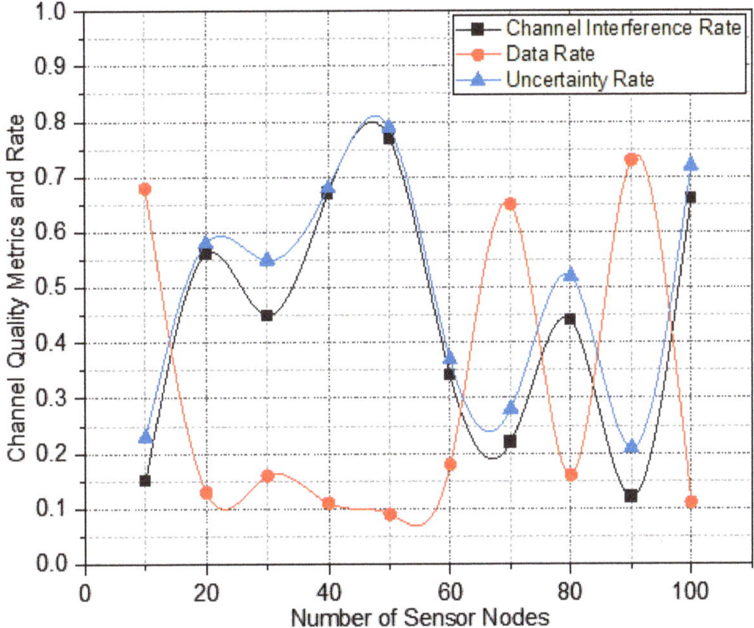

Figure 5. Channel quality measurements.

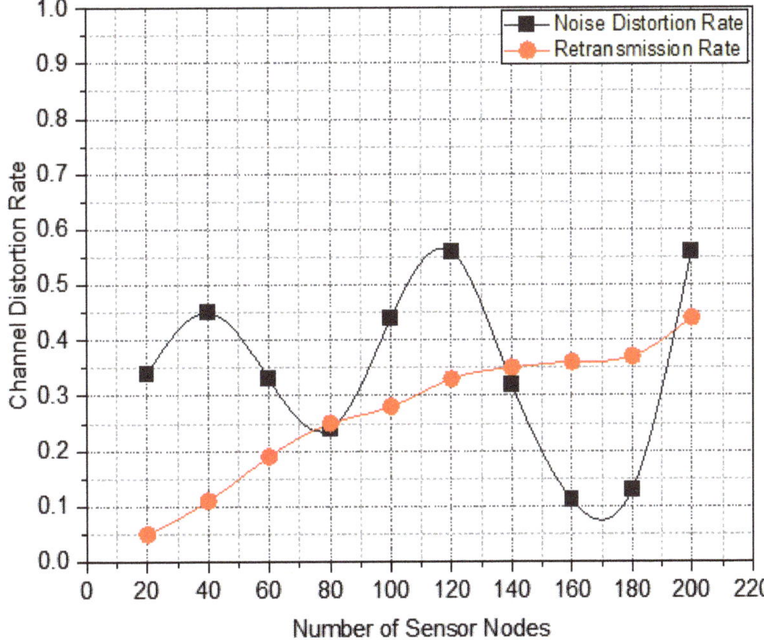

Figure 6. Channel distortion rate.

$$\text{Channel Interference Rate} = \frac{S^I}{S^O} * 100 \tag{24}$$

- S^I-Signal from communicating parties;
- S^O-Signals from other nodes.

$$\text{Channel Data Rate} = \frac{TH}{SD} \tag{25}$$

- TH-Data Transfer Rate between nodes in bits per second;
- SD-Session Duration in seconds.

$$\text{Channel Uncertainty Rate} = \frac{\sum (n^d \cdot l^d) d\tau}{\text{Nactive}} \tag{26}$$

- n^d-Number of node failures in a session;
- l^d-Number of link failures in a session;
- Nactive-Number of active nodes in a session.

$$\text{Channel Noise Distortion Rate} = \frac{\text{Noise (M)} - \text{Noise }(\tau)}{\text{Nactive}} * 100 \tag{27}$$

- Noise (M)-Mean Noise Rate;
- Noise (τ)-Noise rate at time τ.

$$\text{Channel Retransmission Rate} = \frac{NL}{\text{Number of links in a channel}} * 100 \tag{28}$$

- NL-Total Number of Retransmissions between each link.

Equations (24)–(27) describe the network uncertain conditions enabled in the simulation environment. The uncertain channel conditions can be imposed in the WUGSN testbed as channel interference rate, noise distortion rate, retransmission rate, and overall channel uncertainty rate as denoted in the equations. The detailed illustrations of imposed uncertain conditions are provided in Figures 5 and 6. The changes in channel uncertainties are depicted over the changing number of sensor nodes in the network.

Interference is the signal generated from other sources. Both types of signals crucially interrupt the channel quality measures [45,46]. According to the real-time channel distortions, the proposed network model was configured and experimented for evaluating the system performance. As illustrated in Figures 5 and 6, the total number of sensor nodes in WUGSN varies from 20 to 200. The total number of sensor nodes contains both underground sensor nodes and surface nodes. Figure 6 describes the gradual hike of data retransmission rate against uncertain noise production rate in the wireless channels. In this case, the data retransmission rate increases from 5% (0.05) to 45% (0.45) rapidly against the noise rate (0.35 to 0.55). The network channel characteristics experimented in this section show the competent assumptions of the proposed network model [47].

As illustrated in Figure 6, the rapid hike of sensor nodes' data retransmission rate impacts energy consumption rate and network lifetime. Accordingly, the need for efficient channel assessment and learning system is mandatory to activate channel-aware data transmission principles for uncertain WUGSNs. The proposed DMCAP model and the existing models compete to provide reactive data transmission procedures against channel distortions. The models are evaluated using various learning-support performance metrics. In this regard, the proposed and existing techniques are evaluated using system precision, link quality prediction, routing delay, and secure channel throughput against channel uncertainty rate and other factors.

Figure 7 clarifies precision rate of different channel assessment techniques. Generally, system precision rate is determined as provided in Equation (29).

$$\text{Prec(System)} = \frac{C\_t}{T\_t} * 100 \tag{29}$$

- $C_t \rightarrow$ Correctly predicted True channel distortion events;
- $T\_t \rightarrow$ Total predicted events as true channel distortions.

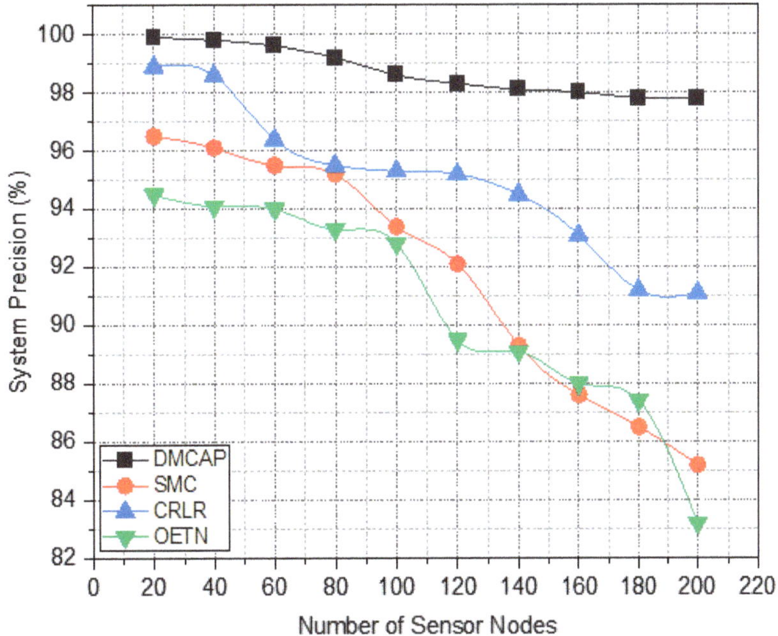

Figure 7. System precision rate.

The system precision rate justifies the actual perfectness of the proposed technique (Equation (29)). In this case, the DMCAP outperforms the existing techniques (SMC, CRLR, and OETN). The maximum precision rate of DMCAP is measured as 99.9% where other techniques fall between 94% and 98.5%. Similarly, the minimal system precision rate is recorded as 97.8% for the DMCAP model. At the same time, the existing techniques secure the precision rate from 80% to 91%.

In the existing techniques, SMC is identifying optimal channels based on signal phase parameters (in-phase and quadrature phase) through software-enabled radio signaling mechanisms. The samples collected from in-phase and quadrature phase (I and Q) are applied in to SVM and RF functions. The supervised learning approaches such as SVM and RF are conventional for analyzing the phase samples of signals deeply. In the developed testbed, this proposed article assumes more complex uncertain conditions for channel optimization. Under this case, the existing SMC is not qualified to obtain better precision and prediction rate (link quality) compared with the proposed model. It attains 85% of precision rate and 82% of link quality prediction rate. This is lower than the proposed DMCAP and CRLR (91% of precision and 87% of prediction). The reason behind the better performance of CRLR is related to effective training procedures of reinforcement network over channel parameters. At the same time, OETN attains minimal growth in terms of system precision and link quality prediction. Since the OETN procedures and following baseline node evaluation procedures to optimize the channel utilization and energy utilization, the result is not significant under uncertain underground conditions.

As provided in Figures 8 and 9, the testbed was adapted to evaluate the performance of all techniques against changing uncertainty rate (Equation (26)). Link quality predictions can be observed against channel distortions and malicious events. Comparably, the proposed DMCAP shows 8% of better precision rate than other techniques.

In this experiment, the proposed techniques observe the major channel distortion issues using multi-channel assessment principles and channel quality tuple analysis procedures. Additionally, the proposed DMCAP ensures multi-level ML and DL evaluation schemes for predicting noise, SNIR, entropy, and other significant distortions [48,49]. The absence of deeply trained evaluation procedures affects the existing techniques against uncertain WUGSN channels.

Figures 8 and 9 show the measurements of link quality prediction rate against environmental distortions and malicious event distortions. Predicting the link quality to transmit the data without loss is a primary goal of this proposed system. The goal can be achieved when the sources of data loss are detected properly. The liveliness of channels is affected due to either environmental disturbances or malicious events. Finding and treating the issues ensure the better link prediction rate. Considering the practical issues of link quality management, the proposed DMCAP system initiates dual case channel quality monitoring practices. In this manner, DMCAP implements both environmental property assessment models and malicious event assessment models. The experiments show the benefit of resilience proposed DMCAP system as provided in Figures 8 and 9. Figure 8 depicts link quality prediction rate against environmental issues.

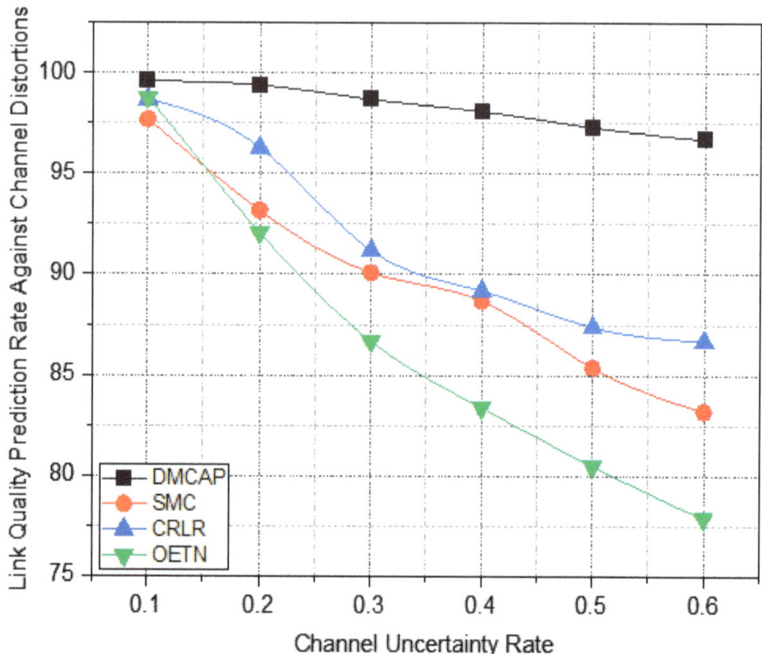

Figure 8. Link quality prediction rate against channel distortion.

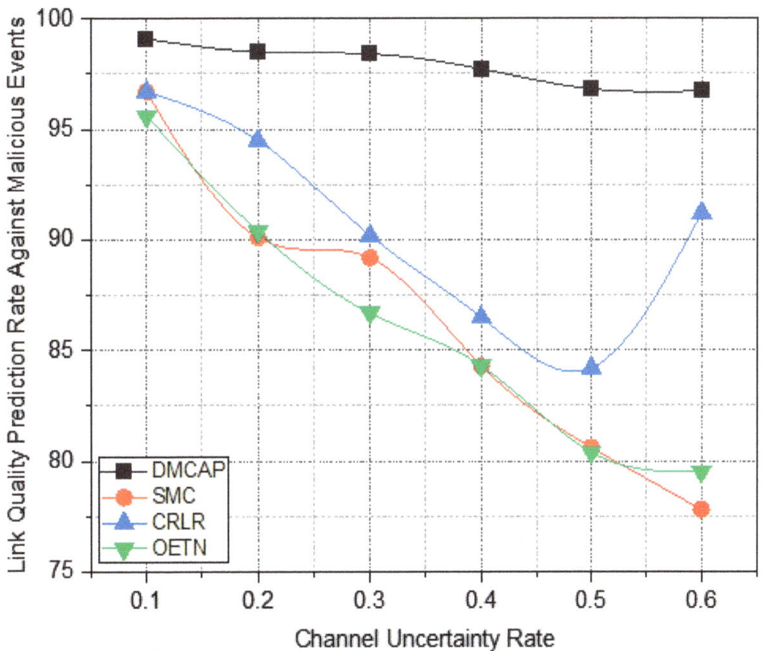

Figure 9. Link quality prediction rate against malicious events.

Compared with other techniques, the performance of DMCAP is commendable (96% of prediction rate against maximum uncertainty rate). On the other hand, CRLR produces optimal prediction and security benefits against OETN and SMC. As CRLR has channel-aware RL engines in each node, the impact shows on better results (92%). At the same time, the existing techniques such as SMC and OETN are not efficient against malicious events under uncertain wireless networks. SMC and OETN are the good procedure for analyzing the channel metrics under limited channel assumptions with minimal rate of dynamic network conditions. However, the development of multi-channel assessment and reactive route selection are primary goals for next generation networks. On the scope, the proposed DMCAP is performing uniquely compared with other techniques.

DMCAP maintains the link quality prediction rate between 99.8% and 96.6% against raising uncertainty rate. The uncertainty rate of each link denotes the overall signal interruptions produced as noise, interference, link failures, node failures, and intruder activities on the channel. Likewise, DMCAP holds 99.7% to 97% of link quality prediction rate against malicious events (Figure 9).

Correspondingly, the proposed DMCAP achieves a better link prediction rate at the optimal time complexity (milliseconds) as illustrated in Table 4. As shown in Table 4, the time complexity of DMCAP for initiating reactive transmission falls between 328 and 489 milliseconds.

Table 4. Time complexity.

Number of Active Links	DMCAP (ms)	SMC (ms)	CRLR (ms)	OETN (ms)
1000	328	444	399	456
1200	356	558	446	567
1400	389	720	478	787
1600	435	780	498	810
1800	468	897	525	940
2000	489	1116	578	1208

The measurements are huge for existing techniques compared with the proposed model. The time complexity denoted in Table 4 is calculated in milliseconds (time taken to complete the operations). Each channel has multiple links between source node and destination node. In this experiment, the number of channel links are varied from 1000 to 2000. As proposed, DMCAP uses suitable nonlinear regression models and multi-channel ensemble classifiers, channel assessment operations taken in each sensor node are evenly balanced. Consequently, the time taken to complete channel assessment procedures is minimized (328 ms to 489 ms) compared with other works.

In this evaluation, SMC and OETN struggle to analyze the real-time channel data using conventional approaches. Hence these existing techniques are consuming more time to obtain classified attributes of multiple channels (1116 ms and 1208 ms, respectively). At the same time, the precision rate of these systems are not optimal. On the other side, the CRLR is consuming procedure operational time between 399 ms and 578 ms which is comparably closer to the proposed method. In this case, the proposed method utilizes the efficiency of multi-channel classifiers (ensemble) over reinforcement procedures to reduce the time complexity around the network.

Figure 10 and Table 5 show the average routing delay (ms) and reactive link establishment rate (%) against channel uncertainties, respectively. As provided in Figure 10, the routing delay for each channel shall be reduced with proposed active channel assessment and reactive channel handling policies. Initiating the reactive data transmission in the network is impossible when the channel assessment system is not reactive and inefficient against channel interruptions. At this moment, the existing techniques produce 280 ms to 220 ms of routing delay for each channel.

Table 5. Reactive link establishment rate (%).

$e(\tau)$	DMCAP	SMC	CRLR	OETN
20	97.1	79.1	86.3	78.4
40	97.5	79.7	86.9	78.5
60	98.2	80.6	87.8	78.9
80	98.9	81.7	88.2	79.2
100	99.4	82.4	90.2	80.1
120	99.7	83.2	90.8	82.3

Significantly, the routing delay is directly proportional to the reactive link establishment rate during node or link failures on the channel. As indicated in Figure 10, the routing delay of the proposed model is minimal compared with other techniques for various iterations (10 iterative experiments). This can be related to Table 5 results of the proposed model. The results provided in Table 5 show the successful link establishment rate during failures.

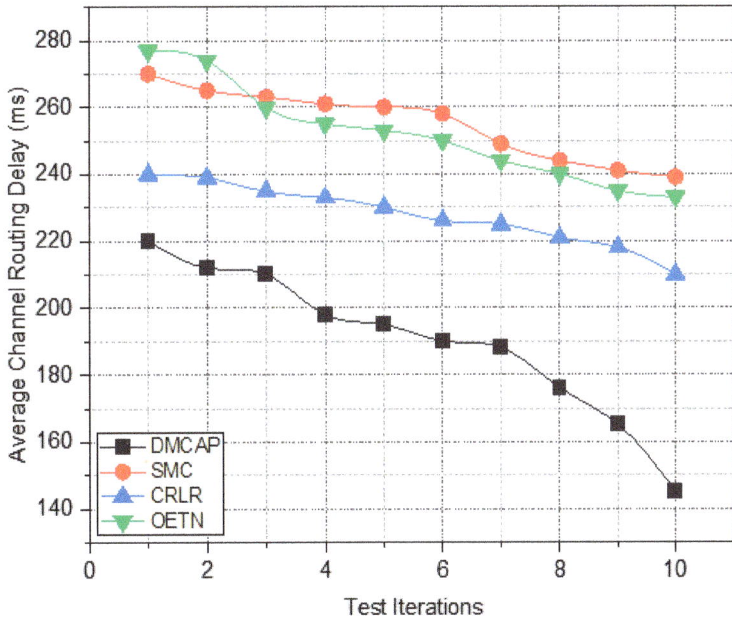

Figure 10. Average routing delay.

Against the performance of existing techniques, the proposed model shows limited routing delay as it is predicting the channel uncertainties actively. On the same way, Table 5 provides the ability of successful link establishment rate among uncertain channel problems and link disabilities. As the number of epochs increases, the proposed DMCAP model updates reactive channel management quality by obtaining the channel assessment attributes. This mechanism works better than other systems and DMCAP obtains 99.7% of reactive link establishment rate. At the same time, CRLR is the only existing technique producing average routing delay (210 ms as minimum) and better link establishment rate (90.8%) compared with other existing techniques.

The reinforcement learning system of CRLR is the reason for optimal performance. In contrast, SMC and OETN are evenly generating more routing delay as they are not deeply understanding the channel behaviors through crucial learning and classification principles. Thus, they are attaining maximum routing delays of 230 ms and 240 ms, respectively. Similarly, the performance is not optimal for link recreation phase as provided in Table 5.

Figure 11 and Table 6 illustrate closely related performance metrics such as secure throughput achievement rate and retransmission reduction rate for each channel. Nevertheless, secure throughput rate shows the amount of successful data transmission against malicious interruptions. The secure throughput of each channel is ensured and obtained with VGAN-IDS engine installed in each sensor node. In this experiment, the proposed DMCAP achieves secure throughput from 17.2 Kilobits Per Seconds (Kbps) to 23.9 Kbps when the number of training epochs are increasing gradually. Moreover, the throughput of CRLR obtains better states compared with OETN and SMC. OETN and SMC maintain only nominal security against malicious event (15.6 Kbps and 13.1 Kbps). This experiment reveals that the proposed technique securely manages the data transmission under VGAN-IDS initiatives and alert systems. Due to the immense experiment-based observations, the number of retransmissions initiated at each link is crucially reduced in WUGSN (DMCAP).

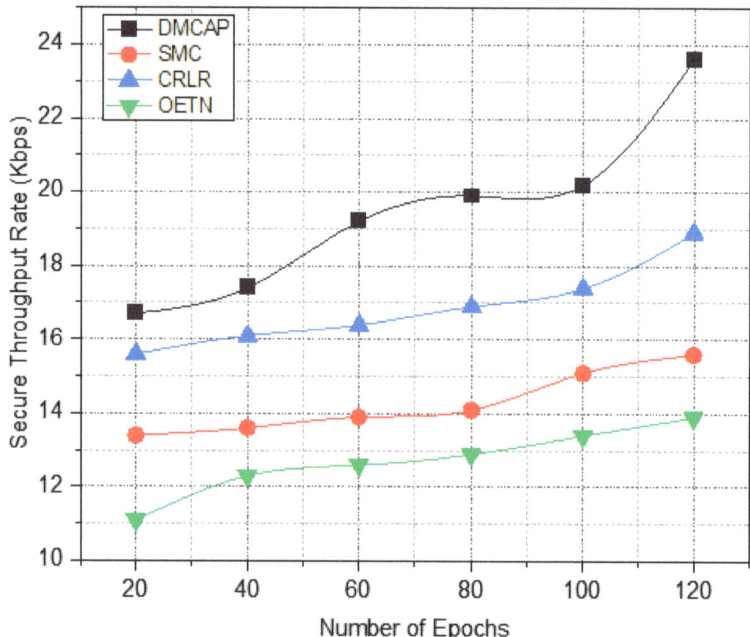

Figure 11. Secure throughput rate.

Table 6. Retransmission reduction rate (%).

e(τ)	DMCAP
20	90.1
40	96.7
60	97.3
80	97.9
100	98.9
120	99.5

Table 6 shows the observed results for the DMCAP system's retransmission reduction rate. The rate of retransmission is reduced as the number of epochs increases. The retransmission reduction rate is defined as the number of retransmissions required at each sensor node against environmental interrupts and malicious interrupts on the link. As illustrated in Table 6, the proposed system reduces the retransmission rate (%) from 90.1% to 99.5% successfully. This indicates that the energy and lifetime of each sensor node on the link is saved with the benefit of proposed DMCAP procedures.

As per the theoretical and experimental clarifications, the proposed system was implemented as distributed multi-channel assessment and activation protocol in each sensor node. The proposed DMCAP is a light-weight protocol installed in both underground and surface sensor nodes for ensuring reliable data communication under uncertain WUGSN conditions. This protocol is fast and reactive in the uncertain channel environment. Particularly, DMCAP uses ensemble multi-channel attribute assessment procedures (ENLSVM) and VGAN-IDS engines as sensor internal procedures. Thus, the proposed DMCAP overcomes data communication problems that happen due to nonlinear productions and channel uncertainty issues optimally compared with other techniques (SMC, CRLR, and OETN). Notably, the proposed model effectively detects and predicts the channel attributes for managing the quality of wireless communication. The contributions and advantages of the proposed DMCAP model was illustrated through various testbed experiments against channel uncertainties.

To achieve significant benefits from proposed channel assessment policies, the testbed was configured with dual-link characteristics to meet the conditions of underground channel assumptions and surface level channel assumptions. The configuration of classified channel configuration parameters helps to observe the crucial and real-time performance of proposed DMCAP model (Tables 2 and 3). However, these dual-link configuration properties are not used in existing techniques. On the realistic testbed, experiments are taken to illustrate the real-time uncertainty conditions of WUGSN using the measured quantities of noise distortion rate, retransmission rate, channel interference rate, data rate, and other uncertainties (link and node failures).

Figures 5 and 6 infer the dynamic nature of the WUGSN setup. On the simulation platform, the performance metrics such as link quality prediction rate, system precision rate, average routing delay, and secure throughput rate are measured for the proposed model and the existing models (SMC, CRLR, and OETN). In addition, the reactive link establishment rate and retransmission reduction rate are considered as crucial factors for ensuring the stability and efficiency of proposed DMCAP model against exiting techniques.

$$\text{Reactive Retransmission Rate} = \frac{\text{DMNR\_t}}{\text{UTt}} * 100 \tag{30}$$

- $DMNR_t$–Number of Retransmission taken at one iteration by DMCAP;
- UTt–Number of network or link failures happens at one iteration.

$$\text{Retransmission Reduction Rate} = \frac{\text{DMNR\_t}}{\text{NRTt}} * 100 \tag{31}$$

- $NRTt$–Total number of retransmissions taken without proposed DMCAP.

Equations (30) and (31) illustrate the importance of the growth shown in Tables 5 and 6 for the proposed DMCAP model. As discussed, the proposed model uses more unique multi-channel modelling schemes for differentiating the characteristics of each wireless channel (air medium or acoustic medium). In addition, the combination of channel distortion analysis and malicious turbulences over the channel are taken seriously to predict the link stability to initiate multi-path transmission in WUGSN. The detailed channel analysis and attribute assessment schemes improve the data throughput over uncertain channels.

Consequently, Table 7 illustrates the energy optimization rate achieved through the reduction in retransmission rate. Furthermore, this article found the coverage problems during wireless communication in the WUGSN and near ground sensor networks. Likewise, a few notable works are considered under this technical scope [50,51]. As the uncertain channel qualities and coverage irregularities create crucial network problems, these are expected to be considered as major parts under any research cases [52].

Table 7. Energy optimization rate.

$e(\tau)$	DMCAP
20	0.21
40	0.26
60	0.29
80	0.32
100	0.36
120	0.39

5. Conclusions

Generally, WSNs and WUGSN are massively applied at defense, industrial, medical, and environmental monitoring conditions. The existence and the deployment condition of WUGSNs create more channel interruptions. WUGSNs maintains both deep underground medium and air medium to lead multi-hop wireless links. According to the link nature, the channel configuration parameters change from one link to another link in the same channel. Correspondingly, the impact of underground channel distortions and surface channel distortions create major problems for WUGSN communication. The channel is disturbed due to underground noise, interference, and malicious interactions. The rate of channel interruptions is not common for surface transmission with widely varying uncertainty rates [53–55].

Against the significant problems, the proposed DMCAP model was developed as a distributed sensor agent. The DMCAP model contained multi-channel signaling models, EMLP, ENLSVM, and VGAN-IDS procedures for providing reactive data transmission against critical channel distortions. The procedures developed inside the sensor node initiated channel attribute evaluation functions, reactive channel activation functions, malicious event monitoring functions, and reactive channel estimator functions. The proposed novel functions and complex data analysis models guarantee the channel protection and reliable communication [56–58]. Consequently, the proposed DMCAP commits the reduction in the retransmission rate, time complexity, and routing delay as illustrated in Section 4 against the existing systems such as CRLR, OETN, and SMC. Consequently, the DMCAP saves overall network energy and lifetime under uncertain channel conditions [59]. However, the proposed DMCAP procedures are lacking lightweight encryption mechanisms to enable data confidentiality and channel masking facilities. Additionally, this proposed model was not evaluated for multiple sensor nodes applied in WUGSNs. These are considered as the limitations of this proposed article. In future, the secure DMCAP is estimated to be designed and implemented for WUGSNs.

Author Contributions: R.S. and S.S.; Conceptualization, R.S. and D.P.B.; methodology, software, validation, P.M.S. and S.G.R.; formal analysis, D.D. and V.D.; investigation, resources, data curation, writing—original draft preparation, R.S. and D.P.B.; writing—review and editing, visualization, supervision, project administration, funding acquisition. All authors have read and agreed to the published version of the manuscript.

Funding: This research received no external funding.

Institutional Review Board Statement: Not applicable.

Informed Consent Statement: Not applicable.

Data Availability Statement: Not applicable.

Conflicts of Interest: The authors declare no conflict of interest.

References

1. Akyildiz, I.F.; Stuntebeck, E.P. Wireless underground sensor networks: Research challenges. *Ad Hoc Netw.* **2006**, *4*, 669–686. [CrossRef]
2. Banaseka, F.K.; Katsriku, F.; Abdulai, J.D.; Adu-Manu, K.S.; Engmann, F.N.A. Signal Propagation Models in Soil Medium for the Study of Wireless Underground Sensor Networks: A Review of Current Trends. *Wirel. Commun. Mob. Comput.* **2021**, *2021*, 8836426. [CrossRef]
3. Trang, H.T.H.; Dung, L.T.; Hwang, S.O. Connectivity analysis of underground sensors in wireless underground sensor networks. *Ad Hoc Netw.* **2018**, *71*, 104–116. [CrossRef]
4. Banaseka, F.K.; Franklin, H.; Katsriku, F.A.; Abdulai, J.-D.; Ekpezu, A.; Wiafe, I. Soil Medium Electromagnetic Scattering Model for the Study of Wireless Underground Sensor Networks. *Wirel. Commun. Mob. Comput.* **2021**, *2021*, 8842508. [CrossRef]
5. Liu, G. Data Collection in MI-Assisted Wireless Powered Underground Sensor Networks: Directions, Recent Advances, and Challenges. *IEEE Commun. Mag.* **2021**, *59*, 132–138. [CrossRef]
6. Abdorahimi, D.; Sadeghioon, A.M. Comparison of Radio Frequency Path Loss Models in Soil for Wireless Underground Sensor Networks. *J. Sens. Actuator Netw.* **2019**, *8*, 35. [CrossRef]
7. Malik, P.S.; Abouhawwash, M.; Almutairi, A.; Singh, R.P.; Singh, Y. Comparative analysis of magnetic induction based communication techniques for wireless underground sensor networks. *PeerJ Comput. Sci.* **2022**, *8*, e789. [CrossRef]
8. Rehan, W.; Fischer, S.; Rehan, M. Machine-learning based channel quality and stability estimation for stream-based multi-channel wireless sensor networks. *Sensors* **2016**, *16*, 1476. [CrossRef]

9.   Aldossari, S.M.; Chen, K.-C. Machine Learning for Wireless Communication Channel Modeling: An Overview. *Wirel. Pers. Commun.* **2019**, *106*, 41–70. [CrossRef]

10. Ozdemir, O.; Niu, R.; Varshney, P.K. Channel Aware Target Localization With Quantized Data in Wireless Sensor Networks. *IEEE Trans. Signal Process.* **2008**, *57*, 1190–1202. [CrossRef]

11. Trevlakis, S.E.; Boulogeorgos, A.-A.A.; Chatzidiamantis, N.D.; Karagiannidis, G.K. Channel Modeling for In-Body Optical Wireless Communications. *Telecom* **2022**, *3*, 136–149. [CrossRef]

12. Bogena, H.R.; Huisman, J.A.; Meier, H.; Rosenbaum, U.; Weuthen, A. Hybrid wireless underground sensor networks: Quantification of signal attenuation in soil. *Vadose Zone J.* **2009**, *8*, 755–761. [CrossRef]

13. Sharma, S.; Verma, V.K. An Integrated Exploration on Internet of Things and Wireless Sensor Networks. *Wirel. Pers. Commun.* **2022**, *124*, 2735–2770. [CrossRef]

14. Yan, X.; Huang, C.; Gan, J.; Wu, X. Game Theory-Based Energy-Efficient Clustering Algorithm for Wireless Sensor Networks. *Sensors* **2022**, *22*, 478. [CrossRef]

15. O'Mahony, G.D.; Harris, P.J.; Murphy, C.C. Investigating Supervised Machine Learning Techniques for Channel Identification in Wireless Sensor Networks. In Proceedings of the 2020 31st Irish Signals and Systems Conference (ISSC), Letterkenny, Ireland, 11–12 June 2020; IEEE: New York, NY, USA, 2020; pp. 1–6.

16. Singh, P.; Singh, R.P.; Singh, Y. An optimal energy-throughput efficient cross-layer solution using naked mole rat algorithm for wireless underground sensor networks. *Mater. Today Proc.* **2021**, *48*, 1076–1083. [CrossRef]

17. Di Valerio, V.; Presti, F.L.; Petrioli, C.; Picari, L.; Spaccini, D.; Basagni, S. CARMA: Channel-aware reinforcement learning-based multi-path adaptive routing for underwater wireless sensor networks. *IEEE J. Sel. Areas Commun.* **2019**, *37*, 2634–2647. [CrossRef]

18. Osamy, W.; Khedr, A.M.; Salim, A.; AlAli, A.I.; El-Sawy, A.A. Recent Studies Utilizing Artificial Intelligence Techniques for Solving Data Collection, Aggregation and Dissemination Challenges in Wireless Sensor Networks: A Review. *Electronics* **2022**, *11*, 313. [CrossRef]

19. Li, N.; Nguyen, H.; Rostami, J.; Zhang, W.; Bui, X.N.; Pradhan, B. Predicting rock displacement in underground mines using im-proved machine learning-based models. *Measurement* **2022**, *188*, 110552. [CrossRef]

20. Osanaiye, O.A.; Alfa, A.S.; Hancke, G.P. Denial of Service Defence for Resource Availability in Wireless Sensor Networks. *IEEE Access* **2018**, *6*, 6975–7004. [CrossRef]

21. Cortés-Leal, A.; Del-Valle-Soto, C.; Cardenas, C.; Valdivia, L.J.; Del Puerto-Flores, J.A. Performance Metric Analysis for a Jamming Detection Mechanism under Collaborative and Cooperative Schemes in Industrial Wireless Sensor Networks. *Sensors* **2021**, *22*, 178. [CrossRef]

22. Tam, N.T.; Hoang, V.D.; Binh, H.T.T.; Vinh, L.T. Multi-objective teaching–learning evolutionary algorithm for enhancing sensor network coverage and lifetime. *Eng. Appl. Artif. Intell.* **2021**, *108*, 104554. [CrossRef]

23. Singh, P.; Singh, R.P.; Singh, Y.; Chohan, J.S.; Sharma, S.; Sadeghzadeh, M.; Issakhov, A. Magnetic Induction Technology-Based Wireless Sensor Network for Underground Infrastructure, Monitoring Soil Conditions, and Environmental Observation Applications: Challenges and Future Aspects. *J. Sens.* **2022**, *2022*, 9332917. [CrossRef]

24. Sun, Z.; Wang, P.; Vuran, M.C.; Al-Rodhaan, M.A.; Al-Dhelaan, A.M.; Akyildiz, I.F. BorderSense: Border patrol through advanced wireless sensor networks. *Ad Hoc Netw.* **2011**, *9*, 468–477. [CrossRef]

25. Ehlali, S.; Sayah, A. Towards Improved Lifespan for Wireless Sensor Networks: A Review of Energy Harvesting Technologies and Strategies. *Eur. J. Electr. Eng. Comput. Sci.* **2022**, *6*, 32–38. [CrossRef]

26. Mini; Pal, A. Coverage sensitivity analysis of a wireless sensor network with different sensing range models considering boundary effects. *Mater. Today Proc.* **2021**, *49*, 3640–3645. [CrossRef]

27. Sharma, P.; Singh, R.P.; Mohammed, M.A.; Shah, R.; Nedoma, J. A Survey on Holes Problem in Wireless Underground Sensor Networks. *IEEE Access* **2021**, *10*, 7852–7880. [CrossRef]

28. Levintal, E.; Ganot, Y.; Taylor, G.; Freer-Smith, P.; Suvocarev, K.; Dahlke, H.E. An underground, wireless, open-source, low-cost system for monitoring oxygen, temperature, and soil moisture. *Soil* **2022**, *8*, 85–97. [CrossRef]

29. Himanshu; Khanna, R.; Kumar, A. Artificial intelligence applications for target node positions in wireless sensor networks using single mobile anchor node. *Comput. Ind. Eng.* **2022**, *167*, 107998. [CrossRef]

30. Lou, Y.; Ahmed, N.; Lou, Y.; Ahmed, N. *MI Wireless Sensor Networks. Underwater Communications and Networks*; Academic Press: Cambridge, MA, USA, 2022; pp. 353–363.

31. Lin, K.; Hao, T. Experimental link quality analysis for LoRa-based wireless underground sensor networks. *IEEE Internet Things J.* **2020**, *8*, 6565–6577. [CrossRef]

32. Vuran, M.C.; Akyildiz, I.F. Channel model and analysis for wireless underground sensor networks in soil medium. *Phys. Commun.* **2010**, *3*, 245–254. [CrossRef]

33. Lin, K.; Hao, T. Link Quality Analysis of Wireless Sensor Networks for Underground Infrastructure Monitoring: A Non-Backfilled Scenario. *IEEE Sens. J.* **2020**, *21*, 7006–7014. [CrossRef]

34. Salam, A.; Vuran, M.C. *EM-Based Wireless Underground Sensor Networks*; Academic Press: Cambridge, MA, USA, 2018; pp. 247–285. [CrossRef]

35. Han, M.; Zhang, Z.; Yang, J.; Zheng, J.; Han, W. An Attenuation Model of Node Signals in Wireless Underground Sensor Net-works. *Remote Sens.* **2021**, *13*, 4642. [CrossRef]

36. Leevy, J.L.; Hancock, J.; Khoshgoftaar, T.M.; Peterson, J.M. IoT information theft prediction using ensemble feature selection. *J. Big Data* **2022**, *9*, 6. [CrossRef]

37. Safaei, M.; Driss, M.; Boulila, W.; Sundararajan, E.A.; Safaei, M. Global outliers detection in wireless sensor networks: A novel approach integrating time-series analysis, entropy, and random forest-based classification. *Softw. Pract. Exp.* **2022**, *52*, 277–295. [CrossRef]

38. Wen, J.; Dargie, W. Characterization of Link Quality Fluctuation in Mobile Wireless Sensor Networks. *ACM Trans. Cyber-Phys. Syst.* **2021**, *5*, 1–24. [CrossRef]

39. Zhao, D.; Zhou, Z.; Wang, S.; Liu, B.; Gaaloul, W. Reinforcement learning–enabled efficient data gathering in underground wireless sensor networks. *Pers. Ubiquitous Comput.* **2020**, *1*, 1–18. [CrossRef]

40. Cohen, G.; Giryes, R. Generative adversarial networks. *arXiv* **2022**, arXiv:2203.00667.

41. Wu, S.; Austin, A.C.M.; Ivoghlian, A.; Bisht, A.; Wang, K.I.-K. Long range wide area network for agricultural wireless underground sensor networks. *J. Ambient. Intell. Humaniz. Comput.* **2020**, *1*, 1–17. [CrossRef]

42. Engmann, F.; Adu-Manu, K.S.; Abdulai, J.-D.; Katsriku, F.A. Network Performance Metrics for Energy Efficient Scheduling in Wireless Sensor Networks (WSNs). *Wirel. Commun. Mob. Comput.* **2021**, *2021*, 9635958. [CrossRef]

43. Engmann, F.; Adu-Manu, K.S.; Abdulai, J.-D.; Katsriku, F.A. Applications of prediction approaches in wireless sensor networks. In *Wireless Sensor Networks-Design, Deployment and Dpplications*; 2021; Volume 1, pp. 1–16. Available online: https://www.walmart.com/ip/Wireless-Sensor-Networks-Design-Deployment-and-Applications-Hardcover-9781838809096/785806007 (accessed on 20 February 2023). [CrossRef]

44. Sahu, S.; Silakari, S. Analysis of energy, coverage, and fault issues and their impacts on applications of wireless sensor net-works: A concise survey. *Network* **2021**, *10*, 13.

45. Donmez, B.; Mitra, R.; Miramirkhani, F. Channel Modeling and Characterization for VLC-Based Medical Body Sensor Networks: Trends and Challenges. *IEEE Access* **2021**, *9*, 153401–153419. [CrossRef]

46. Adel, A.; Norsheila, F. Probabilistic routing protocol for a hybrid wireless underground sensor networks. *Wirel. Commun. Mob. Comput.* **2011**, *13*, 142–156. [CrossRef]

47. Akkaş, M.A.; Sokullu, R. Wireless Underground Sensor Networks: Channel Modeling and Operation Analysis in the Terahertz Band. *Int. J. Antennas Propag.* **2015**, *2015*, 780235. [CrossRef]

48. Ali, F.; Habib, U.; Muhammad, F.; Khan, Y.; Armghan, A.; Alenezi, F.; Abbas, Z.H.; Ali, A.; Qamar, M.S. Alleviation of nonlinear channel effects in long-haul and high-capacity optical transmission networks. *Int. J. Commun. Syst.* **2021**, *35*, e5050. [CrossRef]

49. Nourani, V.; Kheiri, A.; Behfar, N. Multi-station artificial intelligence based ensemble modeling of suspended sediment load. *Water Supply* **2021**, *22*, 707–733. [CrossRef]

50. García, L.; Parra-Boronat, L.; Jimenez, J.M.; Lloret, J.; Abouaissa, A.; Lorenz, P. Internet of Underground Things ESP8266 WiFi Coverage Study. In Proceedings of the INNOV 2019, The Eighth International Conference on Communications, Computation, Networks and Technologies, Valencia, Spain, 29 November 2019; IARIA XPS Press: Lisbon, Portugal, 2019; pp. 1–6.

51. Botella-Campos, M.; Parra, L.; Sendra, S.; Lloret, J. WLAN IEEE 802.11 b/g/n Coverage Study for Rural Areas. In Proceedings of the 2020 International Conference on Control, Automation and Diagnosis (ICCAD), Paris, France, 7–9 October 2020; IEEE: New York, NY, USA, 2020; pp. 1–6.

52. Botella-Campos, M.; Jimenez, J.M.; Sendra, S.; Lloret, J. Near-Ground Wireless Coverage Design in Rural Environments. In Proceedings of the ALLSENSORS 2020, The Fifth International Conference on Advances in Sensors, Actuators, Metering and Sensing, Valencia, Spain, 21–25 November 2020; IARIA XPS Press: Lisbon, Portugal, 2020; pp. 14–19.

53. Mao, J.; Zhao, Y.; Xia, Y.; Yang, Z.; Xu, C.; Liu, W.; Huang, D. Revisiting Link Quality Metrics and Models for Multichannel Low-Power Lossy Networks. *Sensors* **2023**, *23*, 1303. [CrossRef]

54. Wang, Z.; Wei, S.; Zou, L.; Liao, F.; Lang, W.; Li, Y. Deep-Learning-Based Carrier Frequency Offset Estimation and Its Cross-Evaluation in Multiple-Channel Models. *Information* **2023**, *14*, 98. [CrossRef]

55. Weedage, L.; Stegehuis, C.; Bayhan, S. Impact of Multi-connectivity on Channel Capacity and Outage Probability in Wireless Networks. *IEEE Trans. Veh. Technol.* **2023**, 1–14. [CrossRef]

56. Kim, W.; Ahn, Y.; Kim, J.; Shim, B. Towards deep learning-aided wireless channel estimation and channel state information feedback for 6G. *J. Commun. Netw.* **2023**, 1–15. [CrossRef]

57. Hu, C.-H.; Chen, Z.; Larsson, E.G.; Larsson, E.G. Scheduling and Aggregation Design for Asynchronous Federated Learning over Wireless Networks. *IEEE J. Selected Areas Commun.* **2023**, 1. [CrossRef]

58. Sui, J.-Y.; Liao, S.-Y.; Li, B.; Zhang, H.-F. High sensitivity multitasking non-reciprocity sensor using the photonic spin Hall effect. *Opt. Lett.* **2022**, *47*, 6065. [CrossRef]

59. Wan, B.-F.; Zhou, Z.-W.; Xu, Y.; Zhang, H.-F. A Theoretical Proposal for a Refractive Index and Angle Sensor Based on One-Dimensional Photonic Crystals. *IEEE Sens. J.* **2020**, *21*, 331–338. [CrossRef]

Article

IEDA-HGEO: Improved Energy Efficient with Clustering-Based Data Aggregation and Transmission Protocol for Underwater Wireless Sensor Networks

Shubham Joshi [1], T.P Anithaashri [2], Ravi Rastogi [3], Gaurav Choudhary [4,*] and Nicola Dragoni [4]

[1] Department of Computer Engineering, SVKM'S NMIMS Mukesh Patel School of Technology Management and Engineering, Shirpur 425405, Maharashtra, India

[2] Institute of Computer Science and Engineering, Saveetha School of Engineering, Saveetha Institute of Medical and Technical Sciences, Chennai 602105, Tamilnadu, India

[3] Department of C.S.E., Koneru Lakshmaiah Education Foundation, Vaddeswaram 522302, Andhra Pradesh, India

[4] DTU Compute, Department of Applied Mathematics and Computer Science, Technical University of Denmark (DTU), 2800 Kongens Lyngby, Denmark

*   Correspondence: gauch@dtu.dk

Abstract: With the emerging technology in underwater wireless sensor networks (UWSN), many researchers are undergoing this field since it cannot maintain the batteries and recharge them manually. Network duration should be taken into account because they can easily be recharged by a non-conventional resource like solar energy. When coming to the data collection process, clustering is an effective method to construct vitality effective UWSNs. The clustering properties of UWSNs differ from those of terrestrial wireless sensor networks (TWSNs) due to the sparse deployment of nodes as well as the dynamic nature of the channel. This paper proposes improved efficient data aggregation in a Hexagonal grid with energy optimization (IEDA-HGEO) protocol for effective data transmission with an optimal clustering process. It is further compared with ERP^{2}R n energy-efficient routing protocol and EGRC (Energy-efficiency Grid Routing based on 3D Cubes). The three techniques mentioned above are specifically examined for their applicability to underwater communication, and their performance is compared in terms of energy consumption, efficiency, throughput, packet delivery ratio, and delay. The proposed method achieved the following metrics: delay 41%, energy consumption 48%, efficiency 95%, throughput 95%, and PDR 92%.

Keywords: UWSN; clustering; multi-hop; energy consumption

Citation: Joshi, S.; Anithaashri, T.; Rastogi, R.; Choudhary, G.; Dragoni, N. IEDA-HGEO: Improved Energy Efficient with Clustering-Based Data Aggregation and Transmission Protocol for Underwater Wireless Sensor Networks. *Energies* **2023**, *16*, 353. https://doi.org/10.3390/en16010353

Academic Editor: Oscar Barambones

Received: 23 November 2022
Revised: 22 December 2022
Accepted: 23 December 2022
Published: 28 December 2022

1. Introduction

UWSNs have become a key piece of method for underwater monitoring and exploration, including scientific, commercial, and military applications, over the past ten years [1,2]. UWSNs have various advantages over their remote sensing competitors that can deliver localized and more accurate data collecting. They can also use a wider range of sensors, such as chemical, temperature, light, and motion sensors, among others. Traditional underwater instrumentation equipment is being replaced by UWSN technology. In the past, large sensor nodes with data-storage capabilities have been physically placed in the target space below the water. For the duration of the operation, each node runs autonomously to collect readings in accordance with a predetermined program [3]. SNs are picked up at the conclusion of the operation, and the information gathered is recovered and processed. To relay real-time data to an offshore or even on-shore control station for immediate analysis, UWSN technology gives underwater sensor nodes networking capabilities. The underwater sensor network deployment can be controlled interactively by sending control signals from the control station to underwater SNs via a communication channel. In comparison to conventional instrumentation methods, UWSNs provide significant benefits [4].

Because UWSN-SN are powered by batteries, which are challenging to replace or recharge in aquatic environments [5], energy conservation is a serious challenge. A basic research problem is the creation of routing protocols that are reliable, scalable, and energy-efficient in these networks. Due to the fact that the majority of data forwarding protocols currently in use were created for stationary networks, they cannot be used directly with ground-based sensor networks [6].

Figure 1 denoted the basic clustering formation in the UWSN network, whereas two types of sinks, namely onshore and surface sinks, are there to transfer the collected information to the satellite. Moreover surface node, cluster head and underwater sink node have interconnection with each other to collect the data [7]. An autonomous underwater vehicle is used here for moving purposes from one cluster to another cluster or contact with the surface node in case of any emergency during the transmission of packets [8].

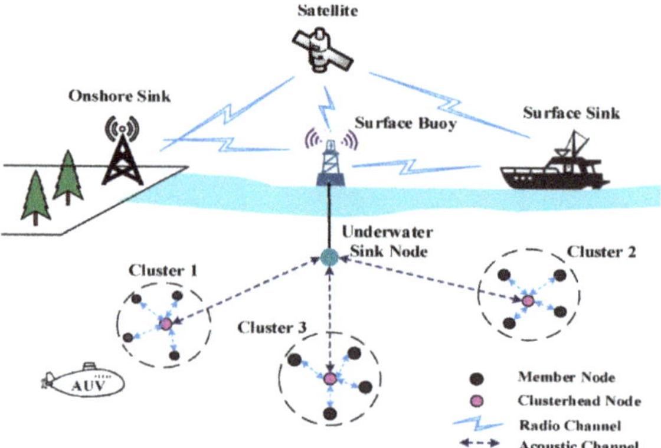

Figure 1. Basic clustering process in underwater wireless sensor networks.

The contribution of this research is as follows:

1. To take into account the energy consumption and travel time of AUVs, we propose a data-gathering protocol based on AUV path planning.
2. To propose improved efficient data aggregation in IEDA-HGEO protocol for effective data transmission with optimal clustering process.
3. It is further compared with ERP^{2}R n energy-efficient routing protocol and EGRC.

The paper is organized as follows: Section 1 gives a brief explanation of UWSN and its applications, difficulties during the transmission process, existing energy-efficient protocols, and the clustering process in the network. Section 2 explained some surveys about existing protocols in the clustering process of UWSN. Section 3 explains the proposed methodology for efficient clustering and data transmission process. Section 4 gives the detailed structure of results by comparing them with existing techniques. Finally, Section 5 ends with a conclusion and future work.

2. Literature Review

A new distributed energy-aware routing protocol is DUCS [9]. It is specifically made for long-term, non-time-critical aquatic monitoring applications utilizing UWSNs without GPS support and random node mobility. A suggested Underwater Positioning Scheme (UPS) in [10] has ordinary nodes listening exclusively to the signals from beacon nodes, and after receiving four beacon messages, the time difference is converted into a range distance. The authors of [11] employ Cayley–Menger to use a mobile beacon node to find the coordinates of sensor nodes. The combined radio and acoustic signals, which are immune to multipath fading, are used to calculate the separation between nodes. In [12], scientists put out a number of mobile node-based strategies, like AUVs, to lessen the impact of unequal energy use. These techniques involve an AUV travelling around the network on a tour path and stopping at a designated location known as a tour point to obtain the gathered data from static nodes in a neighbourhood. For WSNs, Selvi et al. introduced the UCAPN method, which lengthens network lifespan [13]. To balance node energy consumption as well as extend network lifetime, UCAPN groups SNs into clusters of varying sizes. Information from non-cluster-head nodes is first transmitted directly to the closest cluster-head, and then it is transmitted to the sink node. With the help of a control method wherein BS regulates the number of CHs and CHs regulate cluster members, Gulnaz Ahmed et al. [14] presented a MOCHs selection for WSNs. This technique provides robust clustering while addressing the issue of backward transmission. By including energy-collecting techniques, this protocol can be made much better [15]. According to the node density, Khan et al. established a protocol in [16] that can adapt to three different types of networks. Underwater sensors are allocated cubic areas in [17] and placed regionally inside such a framework. To avoid void nodes, another work in [18] is introduced. Every second hop of the packet in this work is examined to see if the node's status is void or not. A void-aware pressure routing (VAPR), which addresses the void problem in this class of greedy routings, is proposed in [19]. [20] makes a proposal for Clustered Vector-Based Forwarding (CVBF) routing protocol. CVBF is made for areas of seawater that are both sparse and dense. According to the authors, CVBF enhances the data delivery ratio as well as decreases end-to-end delay. However, in these protocols, cluster reconstruction for switching CHs by AUVs is repeated until their missions are accomplished to balance the energy consumption among sensor nodes. AUVs can, therefore, run out of energy more quickly before completing their missions as a result of the foregoing constraints' significant energy consumption. What is more, the CHN determination conspires, and the information transmission process is further developed in the PE-Filter convention. Nonetheless, the conventions in [21,22] are intended for TWSNs, and they ought to be altered for UWSNs. Wang et al. took on an energy-proficient lattice directing in light of 3D solid shapes (EGRCs) for UWSNs, where the organization is separated into heaps of little blocks, and each 3D square is viewed as a group [23]. In addition, the EGRC convention streamlines the CHN choice and further develops the quest cycle for the following bounce hub. Be that as it may, the EGRC does not present the detail of the information combination instrument as the overt information repetitiveness might exist and ought to be diminished. In [24], a submerged bunching convention based on the fluffy c means and the moth-fire advancement (FCMMFO) was proposed to upgrade the presentation of UWSNs. Be that as it may, the multi-jump steering way has not been advanced in [25]. Ahmed et al. presented a submerged grouping convention as indicated by repetitive transmission control (RTC), which takes out the overt information repetitiveness at the CHN level and at the area head level [26].

3. Proposed Methodology

Improved efficient data aggregation in Hexagonal grids with energy optimization (IEDA-HGEO) in UWSNs is proposed. The features of our suggested protocol are as follows: The first is to cover as much ground as possible. To cover most nodes, we divided our WSN into hexagonal areas. The best cell shape for clustering in a network is hexagonal. One node in each cell serves as the cell's CH, and the cluster head is selected using a unique method that involves picking the node in smaller cells with the highest residual energy and closest location to the base station (BS) to handle data aggregation as well as transmission.

3.1. Deployment of Gateway Nodes

The Surface Gateways (SG) and Cluster head make up the underwater network (CH). SGs are static nodes fastened to surface-based buoys. They have electromagnetic and acoustic interfaces, respectively. Through an electromagnetic interface, SGs link the underwater network to the Internet. SG uses an acoustic interface to transmit and receive packets to the underwater network. One or more CHs may be connected to each SG. To relay packets from SGs to the active AUVs at the ocean floor and vice versa, CHs are positioned underwater at various depths is shown in Figure 2.

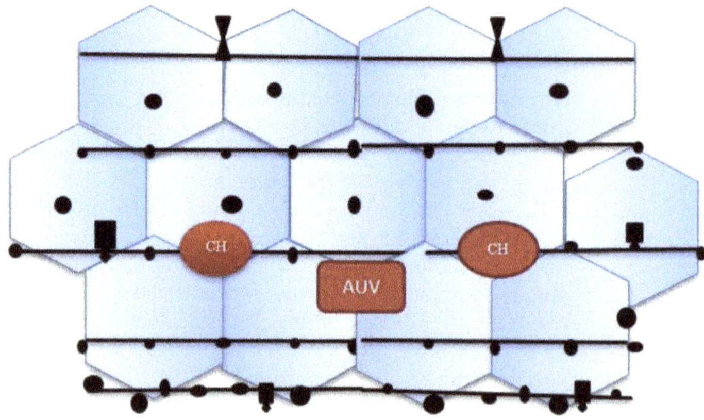

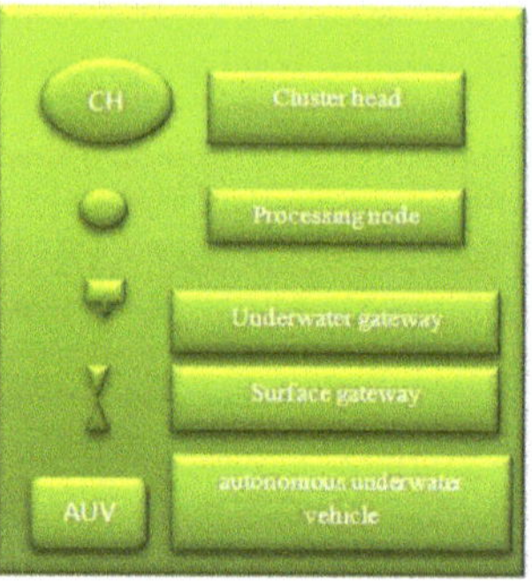

Figure 2. Node deployment in the network.

Due to the characteristics of an underwater communication network, the energy consumption scheme of UWSNs sets it apart from the energy-consuming strategy of WSNs. Equation (1) illustrates the numerical solution.

$$E(\text{distance, } fc) = \text{EnergyTh (distance, } fa) \tag{1}$$

I here stands for a node's frequency. It is equivalent to Equations (2) and (3)

$$fa = 10^{fa(fc)/10} \tag{2}$$

$$fa(fc) = \left[0.11 * \left((fc)^2 * \left(\frac{1}{(1+fc)}\right)^2\right) + 0.22\left((fc)^2 * \left(\frac{1}{(1+fc)}\right)^2\right) \cdots + n\left((fc)^n * \left(\frac{1}{(1+fc)}\right)^n\right) \tag{3}$$

Since the exposure zone of every sensing device is boundary-based, Equation (4) may be used to describe private network radial distance NR, node density ND, sector width $RingW$, hop limit $MaxH$, and the total number of participants Np.

$$Np = \sum_{i=0}^{n} NR * ND * MaxH * Ring * W \tag{4}$$

From inside to outside, the UWSN's connectivity is separated into various little ring sections or Ar1 to Arn. RingW/OT, where RingW is the network radius and OT is the ideal connection path length threshold for sensing devices, can be used to identify the maximum number of ring regions. Dlmax is a representation of the route's maximum delay. The frequency of *Yij* is equal to 5 whenever nodes *i* and *j* are connected. If not, *Yij*'s total value is 0. We view the search for multi-hop routes as a multi-objective optimization problem, with the sole objective of locating the best route at the most reasonable price. Following Equations (5)–(7) shows how an objective formula (Fobj) is obtained. The proposed system architecture is shown in Figure 3.

$$\text{Min Fobj} = \sum_{i=1}^{n} \sum_{j=1}^{n} Y_{ij*Dij} \tag{5}$$

$$Dij = Dtij * \text{MinPower} * Kij + Drj \tag{6}$$

$$\sum_{i=1}^{n} \sum_{j=1}^{n} Dij\ Yij < \text{Dlmax} \tag{7}$$

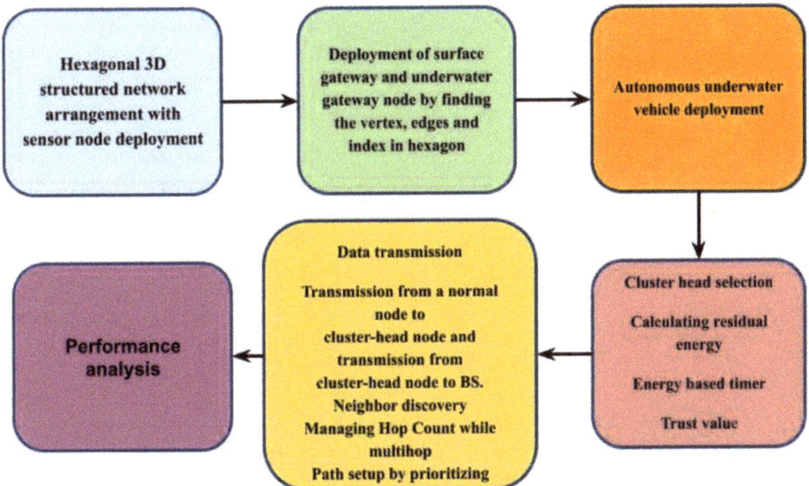

Figure 3. Proposed system architecture.

3.2. Cluster Head Selection

In this study, two processes of TCH and FCH selection are used to determine cluster heads. Using a tentative CH selection method based on EBT and Trust Value, the cluster head is chosen. To choose TCH, the node is given a timer, and trust values are calculated using the node's total Trust value. The TCH is determined by the node with the highest trust value and energy. Additionally, the planned head count, node degree, and competition range are used to determine the final cluster head is shown in Figure 4.

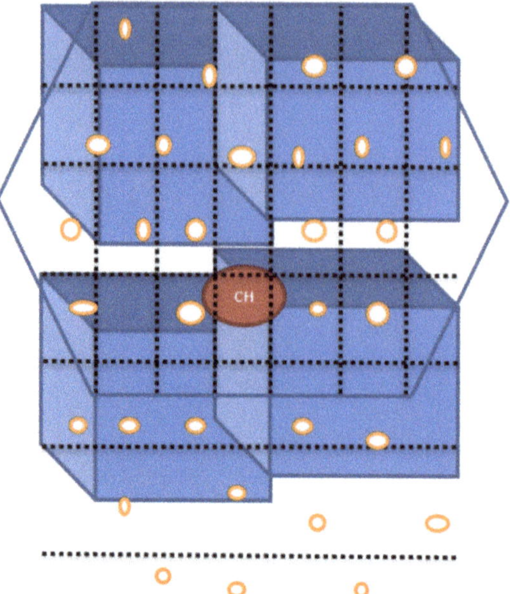

Figure 4. Clustering process at one coverage area.

3.2.1. TCH Selection Based on Energy-Based Timer (EBT)

According to the energy of each sensor node, a timer is allocated to the nodes. The nodes' allotted waiting times are determined by energy. High energy nodes are supported in this phase as the potential new cluster head. If not, CH is the same node with maximum transmission energy. The model description for this timer with an energy basis is as follows. If each node can determine the average energy value of its neighbours, then let us say that node I has k neighbours. $S_i = \{i_1, i_2, i_3, \ldots i_n \ldots i_k, \}$, where i_n is the nth neighbour node. The average energy of node I is given by following Equation (8):

$$Average\ Energy(i) = \begin{cases} \frac{1}{k} \sum_{n=1}^{k} Energy(i_n) & k > 0 \\ 0 & k = 0 \end{cases} \tag{8}$$

TCH is chosen using an energy-based timer from the SNs. The following equation is utilized to evaluate the energy-based waiting time value for any sensor node ID S_i by Equation (9).

$$Wait\ Time\ (s_i) = \frac{Avg\ Energy\ of\ s_i\ Neighbor\ node}{Energy\ of\ S_i} \tag{9}$$

According to the equation above, the waiting time gets shorter as node energy rises. Less waiting time will be allotted to the node with more energy. The potential Cluster Head is chosen to be this node. Other SNs leave CH selection when they receive this message before the start of their waiting time. Selected tentative CH broadcasts a tentative CH message within its broadcast range.

In a hierarchical fashion, Fn gathers data. Fn acts as a parent node and gathers information from its offspring. Distance (d) between parent and child nodes is equal to $2r_{dt} < d < 2R_f$ and rdt are communication and dth ranges of Fnby Equation (10).

$$S = \left\{ C \in S \mid 2r_{dt} < d_{F \to C} < 2R_f \right\} \tag{10}$$

The total amount of data gathered at Fnis if each C transmits a packet of 1 bits by Equation (11).

$$F_{\text{data}} = \sum_i^{|S|} \ell_i \tag{11}$$

Following data gathering, we compute Fn's energy consumption using Equation (12) as follows:

$$E_{F_n} = e_s + e_t \times R\left(\sum_i^{|S|} \ell_i + \ell\right)\phi + EDA + SNR \tag{12}$$

where R is the radius, φ is the data aggregation factor, es is sensing energy, et is electronic energy per bit during transmission, and EDA is data aggregation energy.

The volume of $S1$ is evaluated as follows by Equation (13),

$$\begin{aligned} V_{S1} &= \int_0^r \int_0^{\frac{\pi}{2}} \int_0^{2\pi} \varrho^2 \cos\phi\, d\theta d\varrho \\ &= \pi \int_0^r (R^2 - Z^2) dz \\ &= \tfrac{2}{3}\pi R^3 \end{aligned} \tag{13}$$

Similarly, the volume of $S2$ is evaluated as follows by Equations (14) and (15),

$$V_{S2} = \frac{2}{3}\pi r^3 \tag{14}$$

$$V_{ss} = V_{S2} - V_{S1} = \frac{2}{3}\pi\left(R^3 - r^3\right) \tag{15}$$

The number of nodes in the bounded spherical segment is given by the following equation if the network's node density is ψ by Equations (16) and (17).

$$N_{ss} = \frac{2}{3}\pi\psi\left(R^3 - r^3\right) \tag{16}$$

$$E_{F_n - all}^{rcv} = e_r + \frac{2}{3}\pi\psi\left(R^3 - r^3\right) \times \ell \tag{17}$$

3.2.2. TCH Selection Based on Trust Value

To identify node behaviour, node quality, and node services, Trust Value (TV) is utilized. Additionally, it is utilized for sensor node routing, reconfiguration, and data aggregation. It offers a method for calculating the reliability of SNs. In this study, the trust value is employed to gather information and keep track of various node activities. Trust value is utilized to locate potential CH along with Energy based Timer (EBT). To maximize the effectiveness of optimal CH selection, tentative CH selection uses the EBT and TV techniques. The trust value of nodes is determined using Equation (18) below.

$$Trust\ Value(TV)_{nodes} = \frac{N_{FD}}{N_{REC}} \tag{18}$$

where NFD stands for the quantity of packets forwarded, and NREC for the quantity of packets received. The node with the greatest trust value is chosen as the temporary cluster leader after the trust values of each individual node are determined. The final CH process is then carried out. The outcome of the TCH selection is finally returned by the EBT and the TV.

3.3. Data Transmission

All nodes handshake with one another and communicate their attributes with the predecessor node in the route to start the network's functioning and route discovery. By calculating the distance between the two nodes using their position attribute obtained during handshaking, the neighbour node is selected. To find the route, each node needs to be handshaking with other nodes. When a route is discovered, node N1 is taken for the source node, and the other nodes are taken into account when selecting the optimal neighbour node. Currently, N1 is sending "HELLO" messages to every other node and is receiving "REPLY" messages in return. Figure 5 depicts the proposed data transmission flowchart.

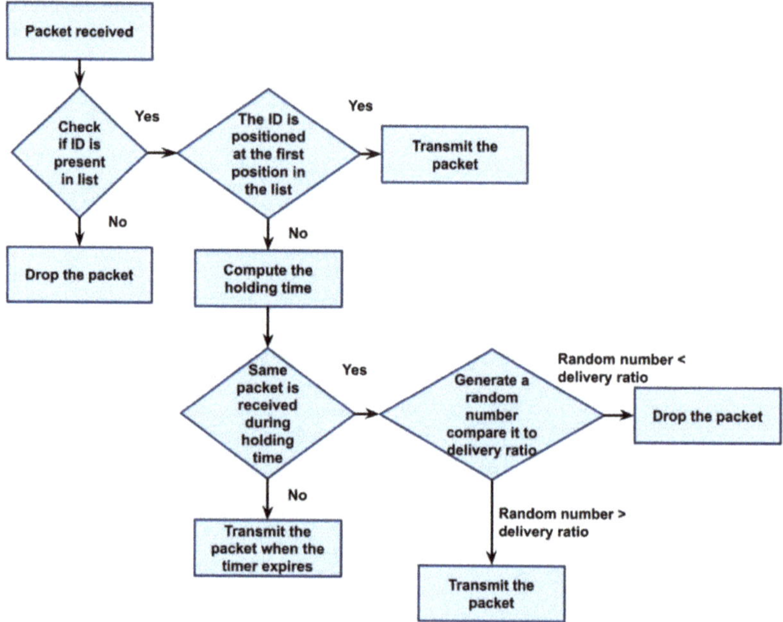

Figure 5. Proposed data transmission flowchart.

Now, node N1 sends a message to node N2, who then begins sending "HELLO" packets to all of the other nodes and receiving "REPLY" messages from them. It is possible to obtain multipath in a similar way between the source and destination nodes.

The forwarder determines the candidate node's distance from both the forwarder and the sink for each transmission. The priority rises with decreasing value. The highest priority node in each forwarder set is the first to be enlarged, followed by nodes whose distances are less than the node transmission radius (Cr), and finally, nodes in the Ci whose distances are less than the communication radius for all nodes already present in the cluster. The forwarder chooses the highest priority set as the next-hop forwarder set after performing the calculation. If the highest priority set is unable to transmit, the remaining forwarder set will only transmit the packets in order.

We introduce an EL to balance energy consumption; to do this, we divide the node's original energy into m equal parts. If the current energy of a node is greater than $i-1$ parts and less than or equal to I parts, the node's EL is i ($1 \leq i \leq m$). Each uw-sensor has 2 transmission options during the steady data transmission phase: sending the packets via MT or sending them straight to the uw-sink (DT). The Algorithm 1 represents the proposed IEDA_HEGO system.

Algorithm 1: Algorithm of IEDA-HGEO.

Require:
  $-$ Set_rn(SN, SN_x) $\rightarrow$ + USet $_rn(SN)$
Initially : $HC(rn_x) = 0$
  hop $_{-}$set$(rn_y) = 0$
Ensure:
$-SN_{Ry}$: elected as RN
1 : if USet $rn(SN)$ =null then
2 : Elect RN
(USet_rn(SN) is a set of SN responding to SPK packets to NN)
3 : USet_rn $(SN) =$ USet $rn(SN) +$ Set$_{-}rn(SN_y, SN_x)$
where $SN_y \in SN_{Ry}$, with a value of $Set_{-}rn(SN_y, SN_x)$ in RPKpacket
4 : $HC(rn_x) = HC(rn_x) + 1$
5 : Repeat Steps 1 to 5 to elect the next neighbour node for SN_y Dest_Node
6 : hop $-$ set$(rn_y) =$ hop set$(m_y) + 1$
Return
5 : end if
8: if no SPK control packets are returned, then
10: End if
11 : for everySPK is generated from SN_z do
15: compute Confidence_level using Equation (4)
16 : USet $rn(SN) =$ USet $rn(SN) +$ Set $rn(SN_z, SN_x)$
17 : if Confidence_level$(y) <$ Confidence_level(z)) then
18 : Remove SN_{Ry} from USet $rn(SN)$
19 : Add the SN_{Rz} as the next node for SN_x
20: end if
21 : drop this SPK control packet
22: end for
23: $HC(SN_x) = HC(SN_x) + 1$

---

4. Experimental Analysis

Settings for many parameters can be seen in Table 1. Unlike earlier studies, the network is deployed in three dimensions. The region in three dimensions is $1000 \times 2000 \times 250 \text{ m}^3$. Where BS is located at the ocean's surface, there are 1055 sensor nodes. Nodes are spread out at random between 1000 and 250 metres. The 260-m transmission radius of SN allows for packet delivery in three to four hops. There are four different packet sizes: 1500, 2000, 3000, and 4000 bytes. RNs, CHs, and CCOs aid in achieving network connectivity. The RPK and SPK control packet sizes are set to 13B and 9B, respectively. In IEDA-HGEO, a levelling parameter of 0.85 has been selected for link quality. For data packets, the transmission power is set at 2.8 W, while for control packets, it is 1.5 W. The SNs can produce random packets, and the chain made up of the RNs, CHs, and CCOs should be used to forward these packets to the BS. In the simulations, 4300 packets are created with various payload counts, and energy usage is noted. The MATLAB simulator (R2021a) was used to carry out the experiments. It is a shorter version of "matrix and laboratory" and was invented by MathWorks. For research investigations, structural engineering, and a wide range of scientific issues needing precise numerical estimates, Matplotlib offers a complete answer.

This section evaluates and compares the performance of the proposed improved efficient data aggregation in IEDA-HGEO with that of two existing widely used routing protocols: ERP2R and EGRC. These two algorithms were chosen because they are well-known in the literature and share the same objectives as the method under consideration. The following measures are utilized to assess the performance of the EE-DHS: Energy consumption, energy efficiency, throughput, network lifetime, and delay are all factors is represented in Table 2.

Table 1. Simulation parameters.

Parameters	Value
Network area	$1000 \times 2000 \times 250$ m^3
Transmission radius	260 m
Number of nodes	1055
Size of data packets	1500, 2000, 3000, and 4000 bytes
Transmission power for data packet	2.8 (W)
RPK and SPK control packets size	13 bytes and 9 bytes
Transmission power for a control packet	1.5(W)

Table 2. Comparison of various parameters for the proposed protocol.

Parameters	ERP^{2}R	EGRC	IEDA-HGEO [Proposed]
Energy consumption	55	51	48
Efficiency	88	92	95
Throughput	91	93	95
Packet delivery ratio	85	87	92
Delay	45	43	41

Energy consumption is the term used to describe how much energy each individual node uses to process information. The Equation Eelec* k (1) is used to describe energy consumption while data transmission through the packet to connected element 'j'. The K bit of the data packet is receiving sensing element I while using energy which is represented by Equation (19).

$$Tx(x,y) = Eene*M + Eamp * d2(x,y) * M \tag{19}$$

Weight between connected nodes I and j are denoted by dij.
When one bit of energy is transmitted, Equation (20) is

$$etx(d) = pd1 + ptd*dn \tag{20}$$

where pd1 is the power dissipated by sending 1 bit of data, ptd is the energy required to transfer the nodes across a long distance

Figure 6 shows the energy consumption of our suggested EE-DHS, which results in 30% more resource conservation. The current methods, ERP2R and EGRC, use 50% and 65% of the resources, respectively. Data PDR, as shown in Figure 6, is calculated by dividing the total number of data packets generated by the source by the number of data packets that were successfully delivered. It is common for some nodes to have low energy levels while other nodes have high energy levels while the system is operating. Nodes with very little remaining energy must cut back on energy use because their operational lifetime is almost up. To determine the delivery ratio of data packets, packet delivery ratio trace files are post-processed. Specifically, the relationship between sent as well as received packets. The average rate of successfully delivered messages via a communication connection is known as throughput. This information may travel over a logical or physical link or go through a specific network node. Typically, the throughput is expressed in bits per second, though it can also be expressed in data packets per second.

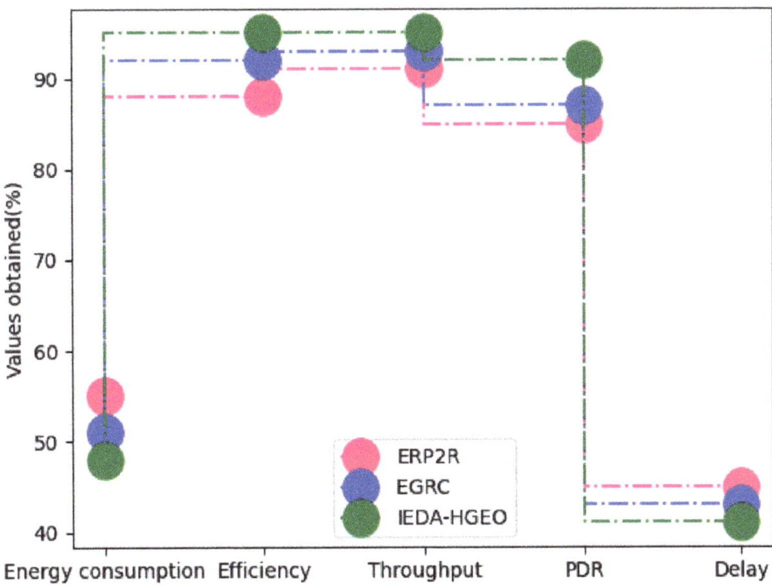

Figure 6. Parametric comparison between proposed and existing techniques.

Discussion: This work centres around reproduction analysis instead of real execution. In the real execution, loads of submerged sensor hubs and a boat on the ocean surface are required. The hubs are furnished with sensors to detect and procure data, a battery to give energy, a memory gadget to store information, a processor to accomplish controlling and handling capabilities, an acoustic modem to accomplish submerged remote acoustic correspondences, a power enhancer, the waterproof gadget, etc. As far as handling, the hubs ought to be fast, stable, and energy-saving. In memory, they need to have an enormous stockpiling limit and guarantee that no information is lost after the passing of hubs. With respect to the submerged remote correspondence innovation, we are attempting to accomplish low dormancy, low blunder rate, and significant distance interchanges.

5. Conclusions

In the future networking field of underwater acoustic sensor networks, network coverage and energy consumption are the key concerns. In this paper, we introduced the IEDA-HGEO algorithm, an improved energy-efficient data collection method for hexagonal grid structures in wireless sensor networks. Here, we use a grid structure with multi-hop routing as a data transmission mechanism and use energy, distance, and end-to-end delay as factors to find the next hop. With reference to energy consumption, throughput, PDR, and delay, our suggested method was thoroughly compared to two other well-known routing methods, the ERP2R and EGRC. Simulation results demonstrated that our technique performed better than previous algorithms as well as successfully extended network lifetime. Proposed method achieved the following metrics: delay 41%, energy consumption 48%, efficiency 95%, throughput 95%, and PDR 92%.

Author Contributions: Conceptualization, S.J., R.R., T.P.A., and G.C.; methodology, S.J., R.R., T.P.A., and G.C.; software, R.R. and G.C.; validation, N.D. and G.C.; analysis, S.J., R.R., T.P.A., and G.C.; investigation, G.C. and N.D.; resources, N.D.; data curation, S.J., R.R., T.P.A., and G.C.; writing—original draft preparation, S.J., R.R., T.P.A., and G.C.; writing—review and editing, G.C and N.D.; visualization, S.J., R.R., T.P.A., and G.C.; supervision, N.D.; project administration, N.D.; funding acquisition, G.C., and N.D. All authors have read and agreed to the published version of the manuscript.

Funding: This work has been supported by project TRANSACT funded under H2020-EU.2.1.1.—INDUSTRIAL LEADERSHIP—Leadership in enabling and industrial technologies—Information and Communication Technologies (grant agreement ID: 101007260).

Data Availability Statement: Data will be shared for review based on the editorial reviewer's request.

Conflicts of Interest: The authors declare no conflict of interest.

References

1. Lilhore, U.K.; Khalaf, O.I.; Simaiya, S.; Tavera Romero, C.A.; Abdulsahib, G.M.; Kumar, D. A depth-controlled and energy-efficient routing protocol for underwater wireless sensor networks. *Int. J. Distrib. Sens. Netw.* **2022**, *18*, 15501329221117118. [CrossRef]
2. Khan, Z.A.; Karim, O.A.; Abbas, S.; Javaid, N.; Zikria, Y.B.; Tariq, U. Q-learning based energy-efficient and void avoidance routing protocol for underwater acoustic sensor networks. *Comput. Netw.* **2021**, *197*, 108309. [CrossRef]
3. Yang, Y.; Wu, Y.; Yuan, H.; Khishe, M.; Mohammadi, M. Nodes clustering and multi-hop routing protocol optimization using hybrid chimp optimization and hunger games search algorithms for sustainable energy efficient underwater wireless sensor networks. *Sustain. Comput. Inform. Syst.* **2022**, *35*, 100731. [CrossRef]
4. Iqbal, S.; Hussain, I.; Sharif, Z.; Qureshi, K.H.; Jabeen, J. Reliable and energy-efficient routing scheme for underwater wireless sensor networks (UWSNs). *Int. J. Cloud Appl. Comput. (IJCAC)* **2021**, *11*, 42–58. [CrossRef]
5. Subramani, N.; Mohan, P.; Alotaibi, Y.; Alghamdi, S.; Khalaf, O.I. An efficient metaheuristic-based clustering with routing protocol for underwater wireless sensor networks. *Sensors* **2022**, *22*, 415. [CrossRef]
6. Liu, Z.; Jin, X.; Yang, Y.; Ma, K.; Guan, X. Energy-efficient guiding-network-based routing for underwater wireless sensor networks. *IEEE Internet Things J.* **2022**, *9*, 21702–21711. [CrossRef]
7. Nain, M.; Goyal, N. Energy efficient localization through node mobility and propagation delay prediction in underwater wireless sensor network. *Wirel. Pers. Commun.* **2022**, *122*, 2667–2685. [CrossRef]
8. Gola, K.K.; Gupta, B. Underwater acoustic sensor networks: An energy efficient and void avoidance routing based on grey wolf optimization algorithm. *Arab. J. Sci. Eng.* **2021**, *46*, 3939–3954. [CrossRef]
9. Chaaf, A.; Saleh Ali Muthanna, M.; Muthanna, A.; Alhelaly, S.; Elgendy, I.A.; Iliyasu, A.M.; El-Latif, A.; Ahmed, A. Energy-efficient relay-based void hole prevention and repair in clustered multi-AUV underwater wireless sensor network. *Secur. Commun. Netw.* **2021**, *2021*, 9969605. [CrossRef]
10. Menaka, D.; Gauni, S. An energy efficient dead reckoning localization for mobile Underwater Acoustic Sensor Networks. *Sustain. Comput. Inform. Syst.* **2022**, *36*, 100808. [CrossRef]
11. Mohan, P.; Subramani, N.; Alotaibi, Y.; Alghamdi, S.; Khalaf, O.I.; Ulaganathan, S. Improved metaheuristics-based clustering with multihop routing protocol for underwater wireless sensor networks. *Sensors* **2022**, *22*, 1618. [CrossRef]
12. Khan, Z.U.; Gang, Q.; Muhammad, A.; Muzzammil, M.; Khan, S.U.; Affendi, M.E.; Ali, G.; Ullah, I.; Khan, J. A Comprehensive Survey of Energy-Efficient MAC and Routing Protocols for Underwater Wireless Sensor Networks. *Electronics* **2022**, *11*, 3015. [CrossRef]
13. Wang, C.; Shen, X.; Wang, H.; Mei, H. Energy-efficient collection scheme based on compressive sensing in underwater wireless sensor networks for environment monitoring over fading channels. *Digit. Signal Process.* **2022**, *127*, 103530. [CrossRef]
14. Chenthil, T.R.; Jesu Jayarin, P. An energy-efficient distributed node clustering routing protocol with mobility pattern support for underwater wireless sensor networks. *Wirel. Netw.* **2022**, *28*, 3367–3390. [CrossRef]
15. Moussaoui, D.; Hadjila, M.; Irid, S.M.H.; Souiki, S. Clustered chain founded on ant colony optimization energy efficient routing scheme for under-water wireless sensor networks. *Int. J. Electr. Comput. Eng.* **2021**, *11*, 5197. [CrossRef]
16. Noorbakhsh, H.; Soltanaghaei, M. EEGBRP: An energy-efficient grid-based routing protocol for underwater wireless sensor networks. *Wirel. Netw.* **2022**, *28*, 3477–3491. [CrossRef]
17. Sivakumar, V.; Kanagachidambaresan, G.R.; Arif, M.; Jackson, C.; Arulkumaran, G. Energy-Efficient Markov-Based Lifetime Enhancement Approach for Underwater Acoustic Sensor Network. *J. Sens.* **2022**, *2022*, 3578002. [CrossRef]
18. Anuradha, D.; Subramani, N.; Khalaf, O.I.; Alotaibi, Y.; Alghamdi, S.; Rajagopal, M. Chaotic search-and-rescue-optimization-based multi-hop data transmission protocol for underwater wireless sensor networks. *Sensors* **2022**, *22*, 2867. [CrossRef]
19. Goutham, V.; Harigovindan, V.P. Full-duplex cooperative relaying with NOMA for the performance enhancement of underwater acoustic sensor networks. *Eng. Sci. Technol. Int. J.* **2021**, *24*, 1396–1407. [CrossRef]
20. Su, Y.; Li, L.; Fan, R.; Liu, Y.; Jin, Z. A Secure Transmission Scheme with Energy-Efficient Cooperative Jamming for Underwater Acoustic Sensor Networks. *IEEE Sens. J.* **2022**, *22*, 21287–21298. [CrossRef]
21. Sampathkumar, A.; Mulerikkal, J.; Sivaram, M. Glowworm swarm optimization for effectual load balancing and routing strategies in wireless sensor networks. *Wirel. Netw.* **2020**, *26*, 4227–4238. [CrossRef]
22. Murugan, S.; Sampathkumar, A.; Kanaga Suba Raja, S.; Ramesh, S.; Manikandan, R.; Gupta, D. Autonomous Vehicle Assisted by Heads up Display (HUD) with Augmented Reality Based on Machine Learning Techniques. In *Virtual and Augmented Reality for Automobile Industry: Innovation Vision and Applications*; Hassanien, A.E., Gupta, D., Khanna, A., Slowik, A., Eds.; Studies in Systems, Decision and Control; Springer: Cham, Switzerland, 2022; Volume 412. [CrossRef]
23. Reshma, G.; Al-Atroshi, C.; Nassa, V.K.; Geetha, B.; Sunitha, G.; Galety, M.G.; Neelakandan, S. Deep Learning-Based Skin Lesion Diagnosis Model Using Dermoscopic Images. *Intell. Autom. Soft Comput.* **2022**, *31*, 621–634.

24. Arun, A.; Bhukya, R.R.; Hardas, B.M.; Kumar, T.C.A.; Ashok, M. An Automated Word Embedding with Parameter Tuned Model for Web Crawling. *Intell. Autom. Soft Comput.* **2022**, *32*, 1617–1632.
25. Li, G.; Liu, F.; Sharma, A.; Khalaf, O.I.; Alotaibi, Y.; Alsufyani, A.; Alghamdi, S. Research on the natural language recognition method based on cluster analysis using neural network. *Math. Probl. Eng.* **2021**, *2021*, 9982305. [CrossRef]
26. Alsufyani, A.; Alotaibi, Y.; Almagrabi, A.O.; Alghamdi, S.A.; Alsufyani, N. Optimized intelligent data management framework for a cyber-physical system for computational applications. *Complex Intell. Syst.* **2021**, 1–13. [CrossRef]

Article

Improved Secure Encryption with Energy Optimization Using Random Permutation Pseudo Algorithm Based on Internet of Thing in Wireless Sensor Networks

S. Nagaraj [1], Atul B. Kathole [2], Leena Arya [3], Neha Tyagi [4], S. B. Goyal [5,*], Anand Singh Rajawat [6], Maria Simona Raboaca [7,*], Traian Candin Mihaltan [8], Chaman Verma [9] and George Suciu [10,*]

[1] Department of CSE, Mallareddy University, Hyderabad 500043, India
[2] Department of Information Technology, Pimpri Chinchwad College of Engineering, Pune 411044, India
[3] Department of CSE, Koneru Lakshmaiah Education Foundation, Vaddeswaram 522502, India
[4] Department of IT, G.L Bajaj Institute of Technology & Management, Knowledge Park 3, Greater Noida 201306, India
[5] Faculty of Information Technology, City University, Petaling Jaya 46100, Malaysia
[6] School of Computer Sciences & Engineering, Sandip University, Nashik 422213, India
[7] ICSI Energy Department, National Research and Development Institute for Cryogenics and Isotopic Technologies, 240050 Ramnicu Valcea, Romania
[8] Faculty of Building Services, Technical University of Cluj-Napoca, 40033 Cluj-Napoca, Romania
[9] Faculty of Informatics, University of Eötvös Loránd, 1053 Budapest, Hungary
[10] R&D Department Beia Consult International, 041386 Bucharest, Romania
* Correspondence: sb.goyal@city.edu.my (S.B.G.); simona.raboaca@icsi.ro (M.S.R.); george@beia.ro (G.S.)

Citation: Nagaraj, S.; Kathole, A.B.; Arya, L.; Tyagi, N.; Goyal, S.B.; Rajawat, A.S.; Raboaca, M.S.; Mihaltan, T.C.; Verma, C.; Suciu, G. Improved Secure Encryption with Energy Optimization Using Random Permutation Pseudo Algorithm Based on Internet of Thing in Wireless Sensor Networks. *Energies* **2023**, *16*, 8. https://doi.org/10.3390/en16010008

Academic Editors: R. Maheswar, M. Kathirvelu and K. Mohanasundaram

Received: 19 September 2022
Revised: 13 November 2022
Accepted: 15 December 2022
Published: 20 December 2022

Abstract: The use of wireless and Internet of Things (IoT) devices is growing rapidly. Because of this expansion, nowadays, mobile apps are integrated into low-cost, low-power platforms. Low-power, inexpensive sensor nodes are used to facilitate this integration. Given that they self-organize, these systems qualify as IoT-based wireless sensor networks. WSNs have gained tremendous popularity in recent years, but they are also subject to security breaches from multiple entities. WSNs pose various challenges, such as the possibility of numerous attacks, their innate power, and their unfeasibility for use in standard security solutions. In this paper, to overcome these issues, we propose the secure encryption random permutation pseudo algorithm (SERPPA) for achieving network security and energy consumption. SERPPA contains a major entity known as a cluster head responsible for backing up and monitoring the activities of the nodes in the network. The proposed work performance is compared with other work based on secure IoT devices. The calculation metrics taken for consideration are energy, overheads, computation cost, and time consumption. The obtained results show that the proposed SERPPA is very significant in comparison to the existing works, such as GKA (Group Key Agreement) and MPKE (Multipath Key Establishment), in terms of data transfer rate, energy consumption and throughput.

Keywords: wireless sensor network (WSN); IoT; security; attacks; cluster mechanism; time consumption

1. Introduction

The development of communication and network systems, design advancements, and the implementation of microprocessor devices has resulted in the creation of intelligent systems to monitor and manage complex operations. Wireless Sensor Networks (WSN) as well as IoT devices are examples of communication devices that use the internet infrastructure as well as protocols for managing the world of IoT-linked objects. Technology advancements allow for the progression of various technological and living processes. They have resulted in intelligent robot behaviors, intelligent cities, robotization, and autonomous vehicles [1–4]. Information security plays a vital role in data integrity and privacy, preventing catastrophic consequences, even a general calamity [5–9]. To avoid unexpected behaviors, extra security mechanisms included within devices and systems are utilized to enhance the security measures incorporated into internet protocols. Furthermore, many these devices, such as surveillance systems, cameras, and sensors, are operated in real-time to ensure the that stated security methods operate without disturbing the overall system functionality. They must be designed so that they are simple to implement in both hardware as well as software and that their use does not disrupt system behaviors and is efficient [10,11].

There are a variety of IoT applications that have become an intrinsic part of our lives. Many of them may be divided into common sectors such as smart home, smart grid, smart healthcare, intelligent transportation, etc. However, due to widespread adoption, various IoT issues have emerged, including a lack of processing capacity, memory resources, different hardware operating characteristics, enormous amounts of data transmission, heterogeneous data, and various kinds of networks [12]. The other significant IoT challenges that must be overcome, especially for low-resource devices and heterogeneous technologies, are data integrity, data confidentiality, and personal privacy [13]. One of the most efficient ways to protect data and communication confidentiality is cryptography. Cryptography can also be used to ensure information integrity and authentication services.

In the IoT era, it is vital to note that most IoT solutions have a "closed design", making it difficult or complex to add extra security features after the manufacturing process is completed. On the other hand, the suite of cryptographic methods that are executed is narrowing due to IoT devices' limited software and hardware resources. Hence, there is a need for a proper balance between the desired level of security as well as execution capabilities. Various cryptographic methods that provide roughly the same level of security may consume various amounts of power and resources, so the user must choose the most suitable for their needs, taking into account the restrictions of the IoT application as well as the deployed hardware [14]. The public-key cryptography methods consume considerably more power and resources than symmetric cryptography algorithms due to their long processing times [15]. The implementation of the symmetric method in IoT security solution design is a sensible choice. The work [16] describes a detailed examination as well as comparison of symmetric block-type methods in IoT devices, including LEA, Twofish, RC6, AES, SPECK128 ChaCha20-Poly1305 algorithms. Stream or sequential symmetric key ciphers are often faster than block-type ciphers. Generally, block ciphers require greater memory resources to encrypt/decrypt bigger chunks (blocks) of data, whereas sequential ciphers usually take one or a few bits at a time. They also utilize modest memory requirements and thus are ideal for implementation in constrained contexts. As a subclass of symmetric cryptography algorithms, stream cryptography algorithms are among the most widely used cryptography data protection approaches.

In work [17], the pseudo-random generator generates a series of bits instead of a random sequence of encryption bits. This concept is extracted from Shannon's one-time pad scheme. The plain-text encryption sequence produced by pseudo-random generator is employed, and its attributes define the security of the protected data. As a result, in stream cryptography methods, the primary goal is to create pseudo-random bit/symbol generators with good cryptographic properties. In the previous fifty years, many concepts have been executed with varying degrees of success.

The RC4 generator, created by Ron Rivest in 1987, is the most popular as well as the most utilized pseudo-random sequence generator. Reverse engineering of the RSA INC program revealed the algorithm description, and Rivest himself confirmed the accuracy of the algorithm description found [18].

The popularity of the RC4 algorithm can be attributed to its simplicity and ease of implementation in software as well as hardware. The cryptanalytic community has taken notice of its tremendous popularity and application. The findings of a rigorous and in-depth investigation of this algorithm revealed several flaws in the method. The empirically observed shortcomings and the actual results of the authors of the article are theoretically proven, which provides a detailed assessment of the weaknesses identified.

The implementation methods in security protocols also contributed to the algorithm's vulnerability; hence its usage in security protocols has been discouraged since 2015.

The idea of the RC4 algorithm and its elegance suggests the possibilities of its commercialization. Our goal is to describe a low-complexity, high-efficiency general method of pseudo-random generator that does not suffer from RC4 flaws and is suited for security solution implementation in computationally restricted microprocessor contexts WSN and IoT. Different mathematical methods are utilized to analyze the probability distribution of the output sequence, information leakage, and correlation properties between the state of the generator as well as output sequence periodicity to prove plausible cryptographic properties of the proposed pseudo-random generator.

WSN raises several research questions, e.g., if it is an effective security management system, has an efficient topology, an efficient WSN architecture, and is an optimized energy-efficient approach. Because of the sensitive data it deals with as well as monitors, security is a rapidly growing study topic in WSN. Effective solutions that ensure security as well as energy efficiency are required. Numerous attacks demand a robust security solution to protect WSN. WSN's energy resource, on the other hand, is often limited. As a result, the various security protocols are not immediately relevant to diverse security challenges. More resources are required for state-of-the-art security solutions. Planetary security and cryptography techniques are primarily used in wired and wireless networks. Still, the most pressing concern is how we might apply these protocols to WSN platforms. Determining the optimal level of application of these protocols for security insurances is a difficult task.

The key motivation for this research is that WSNs monitor various actions, the majority of which are quite sensitive. As a result, we require a comprehensive security mechanism for these operations. We employ a cryptographic approach in the study to safeguard sensitive applications for a variety of applications. However, robust security algorithms require a lot of resources, such as energy, bandwidth, and memory. It is difficult to use any robust cryptographic method in a resource-constrained WSN that limits the capabilities of the resources. The packet delivery and average delay are performed by trust management between sensor nodes with critical management in a dynamic environment. Furthermore, data privacy is effectively secured.

This paper is organized as follows: Section 1 presents the introduction, in Section 2 related works and drawbacks from existing work are discussed, Section 3 presents the proposed mechanism and workflow, Section 4 presents the results and discussion, and finally, Section 5 presents the conclusions.

2. Related Work

A complete survey of most qualified research work is provided in this section. The significant problem of these systems is mainly included in that literature. Those systems that have a considerable problem with the suggested scheme are primarily included in that literature. Several major methods for crucial management in WSNs are critically examined. Following a thorough examination, various research shortcomings in the available literature are identified. A novel efficient key agreement mechanism has been presented based on deficiencies identified in the literature review. In the following are some of the most noteworthy schemes and protocols. In WSNs, ref. [19] proposed an elliptic curve cryptography-based group key agreement approach. The proposed protocol offers implicit sensor node verification. The proposed protocol's main advantages are its low communication and computing costs. The fundamental problem with the proposed work is that every node is involved in intricate security processes. The whole calculation cost is massive, affecting the scheme's overall performance. In the WSN situation, ref. [20] presented a unique group key agreement protocol. The suggested technique manages session keys in a very efficient manner. Furthermore, the suggested approach allows for many passive and active attacks within the network while also ensuring forward as well as backward secrecy. Recovering key computation in the given case is complex. As a result, obtaining private key information is challenging for any hacker. However, the primary flaw in this study is that the clustering algorithm made extensive use of WSN resources. In reference [21] a technique based on round optimization and Group Key Agreement authentication (GKA) was proposed. The suggested protocol validates WSN group key agreement. In addition, the overall number of rounds in the system is kept to a minimum. The cost of processing reduces dramatically, and the complexity of network computing improves as well. The GKA protocol, on the other hand, adds network computing costs.

Similarly, ref. [22] suggested a safe key management approach for WSN. In the suggested method, elliptic curve cryptography was employed for key management, which ensures authentication and secure key management. Furthermore, the suggested technique protects against a variety of security assaults by employing a new technique based on discrete logarithmic issue, which provides increased security. In references [22,23], techniques for managing group session keys between sensor nodes as well as servers were presented. This approach is based on the well-known symmetric key encryption AES. These designs are very costly and security-efficient, according to the extensive reviews of the planned work. The primary difficulty with these systems is that the AES decryption is conducted at base station (BS). As a result, in a given WSN scenario, adequate sensor node authentication is difficult. Virtual and Lu [24–26] suggested techniques based on ECC cryptography as well as the Alike method for IoT platforms. Resistance cryptography is an energy-efficient approach for energy optimization that requires the least amount of energy while providing confidence between sensor nodes. ECC-based cryptography is also a fantastic key management method. The performance of RSA as well as ECC-based algorithms is evaluated in this study. The ECC-based algorithm outperforms the RSA-based approach, according to the output results. However, in the IoT domain, utilizing RSA for the same technique required a variety of network resources. In [27], a key agreement system was suggested. A proposed work session key is established for a defined time between sensor nodes as well as the gateway (GW). The session key is rebuilt for a specific session if there is a change or a fault in the network. Furthermore, the proposed technique provides implicit authentication of WSN nodes. The suggested approach is scalable and protects against network errors. The key challenges with this project, however, are the high communication and computing expenses. Multipath Key Establishment is a technique described by [28] for secure message communication (MPKE). The key value of the suggested research work is the secure and dependable transfer of information. Reed-Solomon codes have been utilized for secure session key agreement, with each round completed utilizing the protocol for Perfectly Secure Message Transmission (PSMT). The main agreement steps are not explicitly explained in this research effort, which is a limitation. As a result, resource-constrained WSNs are ineffective and energy-inefficient.

Some new and attractive research approaches for security and privacy protection have been offered in other relevant works. In this study [29], cryptographic techniques were applied, particularly ECC (Elliptic Curve Cryptography), which has a low processing complexity for cryptographic operations. Furthermore, these schemes consume extremely little energy and have very low communication costs [30–35]. The previous approach contributes about the use of IoT in the different fields to improve the performance [36–38]. This section examines and analyses a complete assessment of related research work [39–41] offered by writers for effective key management in WSN. According to critical analysis of the research work, asymmetric cryptography methods are very appropriate for crucial data communication security. Asymmetric cryptography techniques, on the other hand, can be employed to secure critical operations. Heavy activities are viable as well as necessary for robust security preservation. Symmetric key methods, on the other hand, have been frequently utilized. Symmetric key management techniques, on the other hand, are not very good at defending against a wide range of assaults. The primary problem with public-key cryptography methods is that we cannot use them directly in a WSN with limited resources. Rewards that are given out after state transitions are used to address the convergence problem. The transition mechanism considers different device energy levels and deep learning methods to make the changes in energy consumptions [42–44]. Various schemes are proposed for perfect secret sharing mechanism in CRT [45]. The suggested model seeks to maximize network lifetime and energy usage [46–49]. We can, however, use them professionally at a dissimilar phase, where complex processes are effectively decreased.

3. Contribution of the Proposed Work

1. The proposed protocol aims to enhance performance of communication and network lifetime. It can be achieved by minimizing energy consumption and communication delay.
2. The implementation of secure encryption random permutation pseudo algorithm (SERPPA) for enhancing energy efficient communication.
3. The proposed model contains work cluster member who is responsible for cluster head selection.
4. The cluster head has a significant role, and its activities monitor and backup the nodes' activities in the network.
5. In this work, several existing protocols are described with their advantages and disadvantages. The proposed model is different from the existing models, and the overall result will enhance the data traffic, energy consumption, and throughput rate through stable routing.

Figure 1 shows the proposed architecture diagram; multiple sensor networks are required for data flow among various sources into a single sink. Sensor nodes are subject to limited capacity, memory, and computing resources which need optimal resource utilization. In the network, all sensor nodes use the same variables, and there is a possibility of redundant sensed data. Along with the application specifications, steering methods are designed according to the node's limitations and characteristics. The routing restrictions commonly used are short-latency, adaptive redundancy, QoS performance constraints, severe impact, etc.

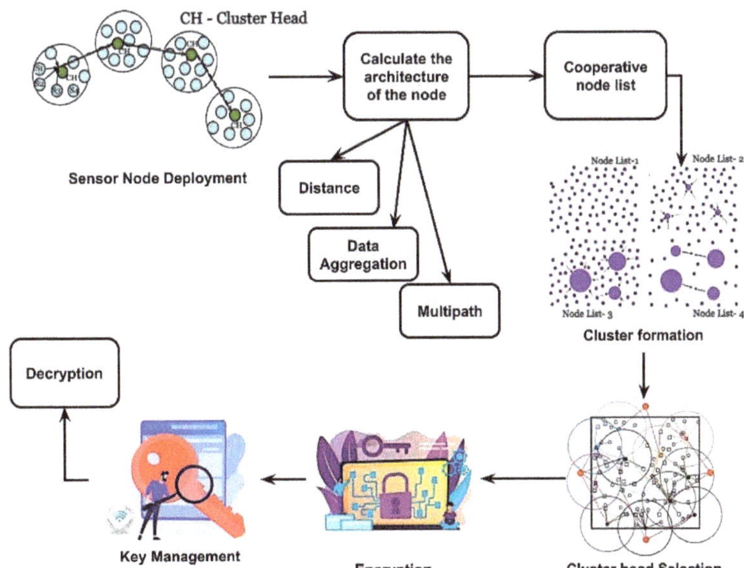

Figure 1. Proposed architecture diagram.

A WSN protocol mainly utilizes one-hop neighborhood information. Implementing two-hop neighborhood information maximizes the configuration complexity, which has a lack of routing decisions results and invalid evasion in WSN. The traffic balance around the network is essential for enhanced network life and transmission time along with tolerable reliability or energy efficiency. Two adaptive routing protocols are proposed to fulfill the implementation demands and challenges, e.g., energy efficiency, timely distribution service, latency, and reliability.

Based on the sensor node, nodes are indexed and connected, respectively. When the nodes are first deployed, they all have the same energy level. "Hello" messages are used to establish a connection between these nodes. The nodes' position, energy level, and distance between them are collected based on request as well as response. We request that all neighboring sensor nodes deliver "hello" messages. A "hello" message is received by each sensor node, which acknowledges it to sync to the other nodes. Each node's initial energy is found before packet transmission begins, and these cluster formations are conducted in various groups. Following the creation of a cluster model, the cluster supervisor's next working procedure begins.

Secure Encryption Random Permutation Pseudo Algorithm (SERPPA)

The secure encryption random permutation pseudo algorithm (SERPPA) translation cipher mechanism is extracted from the Advanced Encryption Standard (AES). AES is a symmetric-based encryption and block cipher algorithm. The proposed SERPPA can manage the message length of about 128,192,256,512 bits [31]. It ensures an enhanced security level with minimum energy consumption. SERPPA contains four AES operations: byte substitution, mixing columns, shifting rows, and round key. In SERPPA, the key management system is employed for both encryption and decryption on the sensor nodes. The byte substitution comprises key management processes. The sensor nodes collect the meaningful length of 512 bits, and the repeated characters are removed. The fixed key value is assigned and the default character is considered for further process. A special character is filled on the space between the characters. Next, the encryption process begins, and the decryption process is processed through the key management systems represented in Algorithm 1. The architecture of SERPPA is shown in Figure 2.

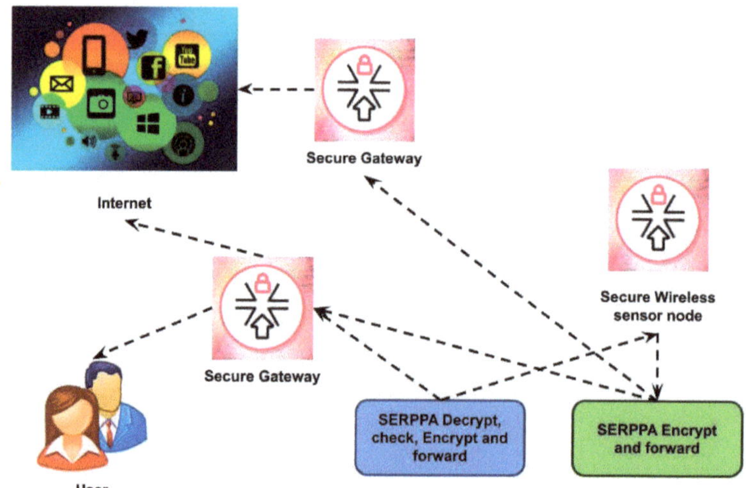

Figure 2. Architecture of SERPPA.

Algorithm 1: of SERPPA

Begin

    start the process of getting 512-bitlength inputs

    remove the presence of same character at multiple times

do

    assign the encryption vector values [5,15,22,8,9,11] for the 512 bits

where "x" is first character

ex: x = 5 of key length is assigned

then

    taken key [15,22,8,9,11]

assign to remaining characters of 512 bits

if

    presence of space between a letter

    *assign a special character (\, #, $, &, *)*

end

do

encryption

        apply k = [5,15,22,8,9,11] values to first set of 512 input message

  next

        apply k = [11,9,8,22,15,5] to next set of same 512 input message

    Continue . . .

    Stop once the character get over

end

end

end

Figure 3 shows the data transfer between the sensor nodes in the network. The proposed secure encryption random permutation pseudo algorithm (SERPPA) workflow operates with 512 bits unmeaningful message encryption with limited energy consumption. Initially, the proposed work implementation begins with sensor node deployment. The deployed sensor nodes form a cluster with the cluster head (CH). CH selection is based on the low energy adaptive clustering hierarchy (LEACH) method. The LEACH algorithm works according to the time division multiple access (TDMA) system with the medium access control (MAC) protocol. The proposed system's main function is to achieve minimized energy consumption and enhanced lifetime for the sensor cluster nodes in the networks.

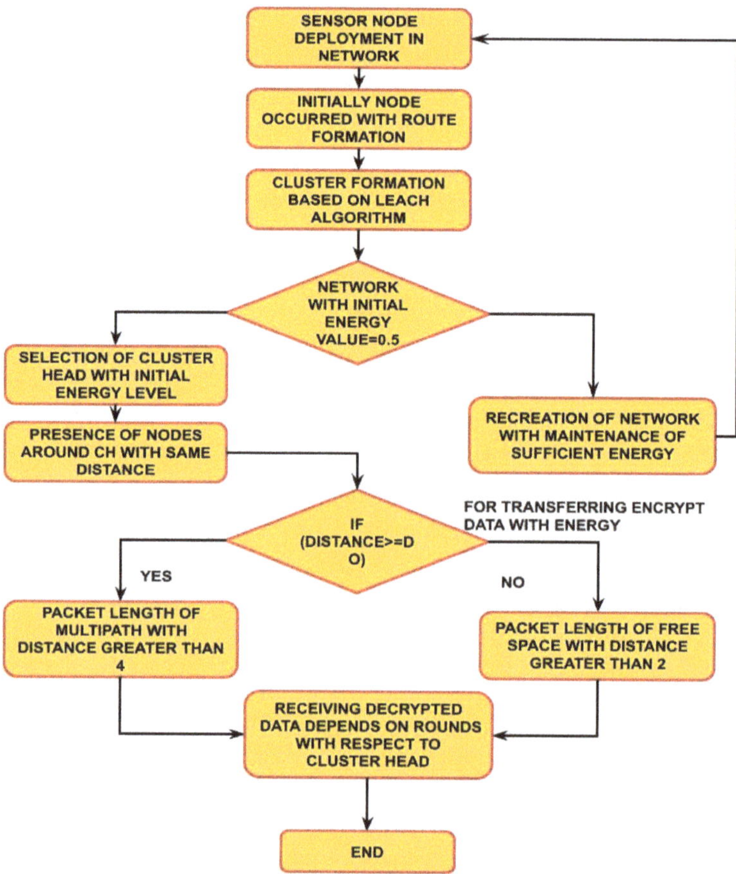

Figure 3. Transfer of encrypted data and receiver of decrypted data with energy consumption.

Additionally, CDMA minimizes the interference between the clusters. Initially, 0.5 is the energy level where the network formation begins. The encrypted data transmission is based on distance between nodes, and rounds of CH determine the reception of decrypted data is shown in Algorithm 2.

Algorithm 2: for Secure Encryption Random Permutation Pseudo Algorithm (SERPPA) for Energy Consumption

---

1. Begin
2. Step-1 creates the simulation network (1500 × 1500)
3. {
4. Set of nodes are deployed randomly; $S_{n(x,y)}$;
5. Check the node's availability state (idle, active)
6. {
7. Send the request " hello" to all neighboring sensor nodes [Hello $\rightarrow S_{n(x,y)}$];
8. If received the message and send the acknowledgment nodes states is active $\rightarrow S_{n(x,y)}$;
9. Else if
10. Message not received the sensor node, the node is the idle state [idle $\rightarrow S_{n(x,y)}$];
11. Collect the cooperative node list;
12. }
13. Ready to form the network [Ready $\rightarrow S_{(ex,y)}$];
14. }
15. Step-2 creates the cluster formation and CH election process-based energy factor.
16. Create the cluster according to with cooperative node list
17. $S_n(d) = \text{distance} \rightarrow \sqrt{\left(x_{i,x} - x_{j,y}\right)^2 + \left(y_{i,y} - y_{j,x}\right)^2}$ // $s_n[i][j]$, $(x_{i,x} - x_{j,y})$ x,i coordinates and, y,j coordinates $\left(y_{i,y} - y_{j,x}\right)$ calculate with x and y coordinate distance
18. Energy factor is patched up with the node cluster for transferring to BS
19. $S_n(e)\text{energy} = \left[\frac{I_{ex} - I_{er}}{R_t}\right]$ //node energy defines
     Sex $\rightarrow$ initialenergy; $I_{er} \rightarrow$ ResidualEnergyofanode$R_t \rightarrow$ Responsetime;
20. Packet transfer received Total ratio $\left[\frac{\text{packets received by all destination node)}}{\text{Total packets send by all source node}}\right]$;
21. EDA $= \frac{\text{hop}(i)-1}{\text{hop}_{max}}$ hop$_{max}$ is the maximum number of node hops in network, and hop(i) is the hop numbers from node i to Sink.
22. ETX $= \frac{1}{D_f * D_r}$ where Df represents the calculated likelihood that a packet will be received by a neighbour and Dr is the calculated likelihood that an acknowledgement packet will be successfully received.
23. Determine optimum distance of a node;
24. if (distance >= d0)
     nodeArch. Node(chNo). energy =
     *energy − (ETX + EDA) * packet length + Emp * packet length * (distance⁴)*
     else
     nodeArch. Node(chNo). energy =
     *energy − (ETX + EDA) * packet length + Efs * packet length * (distance²)*
     end
     nodeArch. Node(chNo). energy = nodeArch. Node(chNo). energy –ctrPacketLength * ERX * round (nodeArch.numNode/clusterModel. NumCluster);
25. end

a. ClusterModel. NodeArch = nodeArch;
26. end

---

4. Experimental Results

Network simulator NS-2 is taken for the proposed SERPPA execution. The experimental setup consists of a simulation environment in which the network is formed with 120 nodes between 1800 × 1800 m². The nodes in the deployed network are dynamic and independent of each other, and they use the random way mobility model. The link-layer protocol in this configuration adheres to the IEEE 802.11 Mac standard. Network traffic is generated using multicast at constant bit rates. WLAN heterogeneous traffic is used for execution, including IEEE 802 and 802.11b. A TCP or UDP network topology is used to set up the data connection. Two thousand bytes per packet are used for network transmission at 24 Mbps data rates. Table 1 defines the simulation parameters taken for execution.

Table 1. Simulation parameters and their values.

Simulation Parameter	Value
Simulator	Network Simulator-2
Number of nodes	100
Simulation time	200 s
Mac Protocol	IEEE 802.11
Simulation area	1800×1800 m^2
Mobility model	accidental waypoint model
Radio range	100 m
Data rate	24 Mbps
Antenna	Omnidirectional antenna
Traffic type	Multicast constant bit ratio
Packet size	512 bytes
Node speed	10–35 m/s

Energy consumption is one of the important factors which determines the efficiency of the algorithms. Figure 4 illustrates the performance in terms of energy consumption with a total number of nodes obtained by each algorithm. The algorithm taken for consideration is the herein proposed SERPPA with existing GKA and MPKE. The X-axis determines the total number of nodes that take part in observation, and the Y-axis defines the total energy consumed. The consumed energy is calculated in terms of joules. At regular intervals, the total number of nodes that take part is gradually increased by ten. Observation of each set of nodes with respect to the algorithms is plotted graphically for analysis. From the herein proposed SERPPA, energy consumption of 10 nodes is 230 joules, 20 nodes is 310 joules, 30 nodes is 500 joules, 40 nodes is 650 joules, 50 nodes is 710 joules, 60 nodes is 800 joules, 70 nodes is 980 joules, 80 nodes is 1200 joules, 90 nodes is 1500 joules, and 100 nodes is 1900 joules. Whereas GKA energy consumption of 10 nodes is 550 joules, 20 nodes is 720 joules, 30 nodes is 810 joules, 40 nodes is 930 joules, 50 nodes is 1100 joules, 60 nodes is 1250 joules, 70 nodes is 1650 joules, 80 nodes is 1800 joules, 90 nodes is 2200 joules, and 100 nodes is 2800 joules. Energy consumed by MPKE for 10 nodes is 650 joules, 20 nodes is 720 joules, 30 nodes is 900 joules, 40 nodes is 1200 joules, 50 nodes is 1630 joules, 60 nodes is 1850 joules, 70 nodes is 2100 joules, 80 nodes is 2500 joules, 90 nodes is 2900 joules, and 100 nodes is 3200 joules. The graphical representation clearly represents that the proposed SERPPA results are more efficient than those of GKA and MPKE.

Network overhead determines the additional information taken while transmitting the packets. More overhead leverages the network performance; hence overhead has a significant impact in the overall network performance. Figure 5 illustrates the performance of overhead with the total number of nodes obtained by each algorithm. The algorithm taken for consideration is proposed SERPPA with existing GKA and MPKE. The X-axis determines the total number of nodes that take part in observation, and the Y-axis defines the total overhead attained by each algorithm. The consumed overhead is calculated in terms of kb. At regular intervals, the total number of nodes that take part is gradually increased by ten. Observation of each set of nodes concerning the algorithms is plotted graphically for analysis. From the proposed SERPPA, the overhead of 10 nodes is 5 kb, 20 nodes is 8 kb, 30 nodes is 12 kb, 40 nodes is 19 kb, 50 nodes is 25 kb, 60 nodes is 30 kb, 70 nodes is 35 kb, 80 nodes is 50 kb, 90 nodes is 65 kb, and 100 nodes is 75 kb. Whereas GKA attained overhead for 10 nodes is 12 kb, 20 nodes is 18 kb, 30 nodes is 21 kb, 40 nodes is 30 kb, 50 nodes is 38 kb, 60 nodes is 50 kb, 70 nodes is 60 kb, 80 nodes is 65 kb, 90 nodes is 70 kb, and 100 nodes is 85 kb. The MPKE attained overhead for 10 nodes is 20 kb, 20 nodes is 28 kb, 30 nodes is 35 kb, 40 nodes is 40 kb, 50 nodes is 45 kb, 60 nodes is 60 kb,

70 nodes is 65 kb, 80 nodes is 70 kb, 90 nodes is 78 kb, and 100 nodes is 90 kb. The graphical representation clearly shows that proposed SERPPA overhead is more efficient than GKA and MPKE.

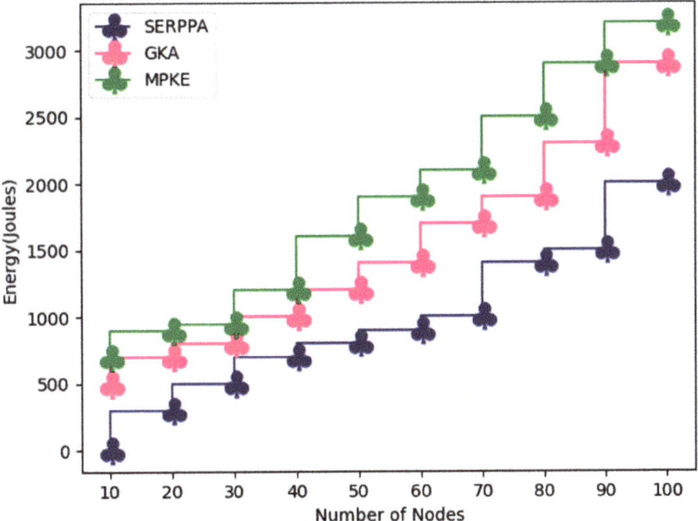

Figure 4. Energy vs. number of nodes.

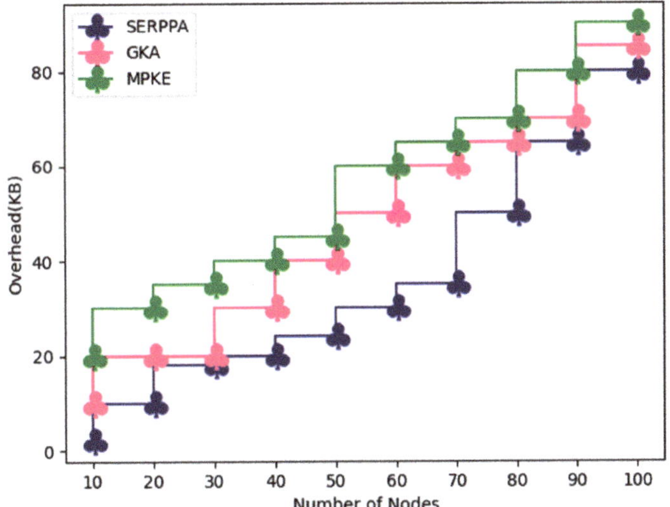

Figure 5. Overhead vs. number of nodes.

The computation cost is the communication cost taken for each transmission by a set of nodes. Figure 6 illustrates the performance of computation cost with a total number of nodes obtained by each algorithm. The algorithm taken for consideration is proposed SERPPA with existing GKA and MPKE. The X-axis determines the total number of nodes that take part in observation, and the Y-axis determines the total computation cost attained by each method. At regular intervals, the total number of nodes that take part is gradually increased by ten. The observation of each set of nodes concerning the algorithms is plotted graphically for analysis. From the proposed SERPPA, the computation cost of 10 nodes is 15, with 20 nodes is 21, with 30 nodes is 28, with 40 nodes is 40, with 50 nodes is 50, with 60 nodes is 65, with 70 nodes is 80, with 80 nodes is 95, with 90 nodes is 110, and 100 nodes is 140. The computation cost attained by GKA for 10 nodes is 35, with 20 nodes is 50, with 30 nodes is 65, with 40 nodes is 78, with 50 nodes is 90, with 60 nodes is 110, with 70 nodes is 125, with 80 nodes is 138, with 90 nodes is 150, and with 100 nodes is 165. The computation cost attained by MKPE for 10 nodes is 50, with 20 nodes is 65, with 30 nodes is 75, with 40 nodes is 80, with 50 nodes is 95, with 60 nodes is 120, with 70 nodes is 140, with 80 nodes is 165, with 90 nodes is 170, and with 100 nodes is 180. The graphical representation clearly shows that the proposed SERPPA results are very low compared to GKA and MPKE.

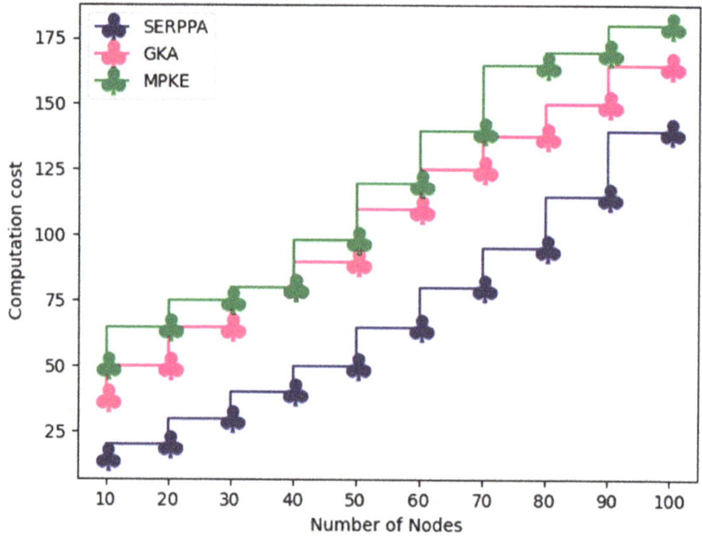

Figure 6. Computation cost vs. number of nodes.

Quick packet transmission without any data loss is a vital requirement. An algorithm with minimum time and successful transmission determines its efficiency. Figure 7 shows the performance of time consumed with the total number of nodes obtained by each algorithm. The algorithm taken for consideration is proposed SERPPA with existing GKA and MPKE. The X-axis determines the total number of nodes that take part in observation, and the Y-axis defines the total time consumed by each algorithm. The time consumption is calculated in terms of milliseconds (ms). At regular intervals, the total number of nodes that take part is gradually increased by ten. The observation of each set of nodes with respect to the algorithm is plotted graphically for analysis. From which the proposed SERPPA time consumed for 10 nodes is 3 ms, time consumed for 20 nodes is 5 ms, time consumed for 30 nodes is 8 ms, time consumed for 40 nodes is 11 ms, time consumed for 50 nodes is 15 ms, time consumed for 60 nodes is 19 ms, time consumed for 70 nodes is 22 ms, time consumed for 80 nodes is 25 ms, time consumed for 90 nodes is 30 ms and time consumed for 100 nodes is 35 ms. Whereas the time consumption of GKA for 10 nodes is 10 ms, time consumed for 20 nodes is 12 ms, time consumed for 30 nodes is 15 ms, time consumed for 40 nodes is 18 ms, time consumed for 50 nodes is 20 ms, time consumed for 60 nodes is 22 ms, time consumed for 70 nodes is 25 ms, time consumed for 80 nodes is 28 ms, time consumed for 90 nodes is 40 ms and time consumed for 100 nodes is 50 ms. The MKPE time consumption for 10 nodes is 12 ms, time consumed for 20 nodes is 15 ms, time consumed for 30 nodes is 20 ms, time consumed for 40 nodes is 25 ms, time consumed for 50 nodes is

28 ms, time consumed for 60 nodes is 30 ms, time consumed for 70 nodes is 35 ms, time consumed for 80 nodes is 40 ms, time consumed for 90 nodes is 55 ms and time consumed for 100 nodes is 60 ms. The graphical representation clearly shows that the time consumed by the proposed SERPPA is very low in comparison to GKA and MPKE.

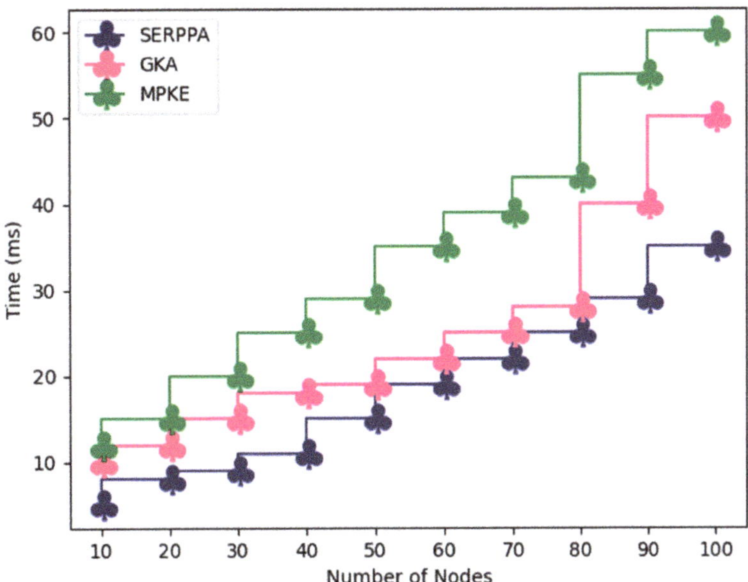

Figure 7. Time consumed vs. number of nodes.

5. Conclusions

In this research, we proposed the secure encryption random permutation pseudo algorithm (SERPPA) to secure the data transmitted through WSN. SERPPA is extracted from AES, which is an asymmetric-based encryption and block cipher algorithm. The secure encryption random permutation pseudo algorithm (SERPPA) is a cluster-based mechanism for enhancing energy-efficient communications. The proposed model contains work cluster member, which is responsible for cluster head selection. The cluster head contains the major role, such as monitoring and backing up the activities of the nodes in the network. The performance of the proposed work is determined by a comparison work carried out with GKA and MPKE. The evaluation metrics taken for consideration are energy, overheads, computation cost, and time consumption. Compared to previous methods, data traffic and energy consumption was decreased and throughput rate was increased through stable routing. The experimental results were carried in an NS-2 simulation environment. The proposed system observations are plotted and represented graphically with the result obtained by GKA and MPKE. In all evaluation metrics, the proposed SERPPA results are more efficient than the others. It shows the proposed SERPPA performance is far better than the existing algorithms. Some limitations of the proposed algorithm have not been evaluated across the various algorithms; in future research, it will be evaluated with more different approaches.

Author Contributions: Conceptualization, S.N.; supervision, A.B.K. and S.B.G.; writing—original draft, L.A. and N.T.; validation, A.S.R. and S.B.G.; writing—review and editing, A.S.R., S.B.G., M.S.R., T.C.M., C.V., G.S. All authors have read and agreed to the published version of the manuscript.

Funding: This paper was partially supported by UEFISCDI Romania and MCI through BEIA projects NGI-UAV-AGRO, SOLID-B5G, T4ME2, DISAVIT, AISTOR, MULTI-AI, ADRIATIC, Hydro3D, EREMI, MUSEION, iPREMAS, IPSUS, U-GARDEN, CREATE, STACK, ENTA and by European Union's Horizon 2020 research and innovation program under grant agreement No. 883522 (S4ALLCITIES). This work is supported by Ministry of Research, Innovation, Digitization from Romania by the National Plan of R & D, Project PN 19 11, Subprogram 1.1. Institutional performance-Projects to finance excellence in RDI, Contract No. 19PFE/30.12.2021 and a grant of the National Center for Hydrogen and Fuel Cells (CNHPC)—Installations and Special Objectives of National Interest (IOSIN). Ministry of Investments and European Projects: Human Capital Sectoral Operational Program 2014–2020, Contract no. 62461/03.06.2022, SMIS code 153735.

Data Availability Statement: Data will be shared for review based on the editorial reviewer's request.

Conflicts of Interest: The authors declare no conflict of interest.

References

1. Rehman, R.A.; Khan, B. IoT Elements, Layered Architectures and Security Issues: A Comprehensive Survey. *Sensors* **2018**, *18*, 2796.
2. Salah, K. *The Era of Internet of Things*, 2nd ed.; Springer: Cham, Switzerland, 2019.
3. Rayes, A.; Samer, S. *Internet of Things from Hype to Reality*, 2nd ed.; Springer: Cham, Switzerland, 2019.
4. Atlam, H.; Walters, R.J.; Wills, G.B. Internet of Things: State-of-the-art, Challenges, Applications, and Open Issues. *Int. J. Intell. Comput. Res.* **2018**, *9*, 928–938. [CrossRef]
5. Costa, D.G.; Figuerêdo, S.; Oliveira, G. Cryptography in Wireless Multimedia Sensor Networks: A Survey and Research Directions. *Cryptography* **2017**, *1*, 4. [CrossRef]
6. Kambourakis, G.; Marmol, F.G.; Wang, G. Security and Privacy in Wireless and Mobile Networks. *Future Internet* **2018**, *10*, 18. [CrossRef]
7. Ziegler, S. *Internet of Things Security and Data Protection*, 2nd ed.; Springer: Cham, Switzerland, 2019.
8. Cheruvu, S.; Kumar, A.; Smith, N.; Wheeler, D.M. *Demystifying Internet of Things Security: Successful IoT Device/Edge and Platform Security Deployment*; Apress: Berkeley, CA, USA, 2019.
9. Mahmood, Z. (Ed.) *Security, Privacy and Trust in the IoT Environment*; Springer: Cham, Switzerland, 2019.
10. Banday, M.T. *Cryptographic Security Solutions for the Internet of Things*; IGI Global: Hershey, PA, USA, 2019.
11. Biryukov, A.; Perrin, L. State of the Art in Lightweight Symmetric Cryptography. IACR Cryptology ePrint Archive. 2017. Available online: https://eprint.iacr.org/2017/511 (accessed on 28 October 2019).
12. Jing, Q.; Vasilakos, A.V.; Wan, J.; Lu, J.; Qiu, D. Security of the Internet of Things: Perspectives and challenges. *Wirel. Netw.* **2014**, *20*, 2481–2501. [CrossRef]
13. Frustaci, M.; Pace, P.; Aloi, G.; Fortino, G. Evaluating Critical Security Issues of the IoT World: Present and Future Challenges. *IEEE Internet Things J.* **2017**, *5*, 2483–2495. [CrossRef]
14. Hamad, F.; Smalov, L.; James, A. Energy-aware Security in M-Commerce and the Internet of Things. *IETE Tech. Rev.* **2009**, *26*, 357–362. [CrossRef]
15. Bilal, M.; Kang, S.G. An Authentication Protocol for Future Sensor Networks. *Sensors* **2017**, *17*, 979. [CrossRef]
16. Saraiva, D.A.F.; Leithardt, V.R.Q.; de Paula, D.; Sales Mendes, A.; González, G.V.; Crocker, P. PRISEC: Comparison of Symmetric Key Algorithms for IoT Devices. *Sensors* **2019**, *19*, 4312. [CrossRef]
17. Von zur Gathen, J. *CryptoSchool*; Springer: Berlin/Heidelberg, Germany, 2015.
18. Rivest, R.; Schuldt, J. Spritz—A Spongy RC4-Like Stream Cipher and Hash Function. Available online: https://en.wikipedia.org/wiki/RC4#cite_note-Rivest2014-14 (accessed on 27 October 2014).
19. Tang, H.; Zhu, L.; Zhang, Z. A Novel Authenticated Group Key Agreement Protocol Based on Elliptic Curve Diffie-Hellman. In Proceedings of the 2008 4th International Conference on Wireless Communications, Dalian, China, 12–17 October 2008; pp. 1–4.
20. Zhang, Z.; Jiang, C.; Deng, J. A novel group key agreement protocol for wireless sensor networks. In Proceedings of the 2010 International Conference on Measuring Technology and Mechatronics Automation, Changsha, China, 13–14 March 2010; pp. 230–233.
21. Pramod, N.D.B.; Sharnappa, G.R. Review on fault detection and recovery in WSN. *Int. J. Adv. Res. Comput. Sci. Softw. Eng.* **2015**, *5*.
22. Abi-Char, P.E.; Mhamed, A.; El-Hassan, B. A secure authenticated key agreement protocol based on elliptic curve cryptography. In Proceedings of the 3rd International Symposium on Information Assurance and Security, Manchester, UK, 29–31 August 2007; pp. 89–94.
23. Meingast, M.; Roosta, T.; Sastry, S. Security and privacy issues with health care information technology. In Proceedings of the 2006 International Conference of the IEEE Engineering in Medicine and Biology Society, New York, NY, USA, 30 August–3 September 2006; pp. 5453–5458.
24. Singelée, D.; Latré, B.; Braem, B.; Peeters, M.; De Soete, M.; De Cleyn, P.; Preneel, B.; Moerman, I.; Blondia, C. A secure low-delay protocol for wireless body area networks. *Adhoc. Sens. Wirel. Netw.* **2010**, *9*, 53–72.

25. Ertaul, L.; Lu, W. *ECC Based Treshold Cryptography for Secure Data Forwarding and Secure Key Exchange*; University of Waterloo: Waterloo, ON, Canada, 2005.

26. Ertaul, L.; Chudinov, P.; Morales, B. IoT security: Authenticated lightweight key exchange (ALIKE). In Proceedings of the International Conference on Wireless Networks (ICWN), Rome, Italy, 30 June–4 July 2019; pp. 45–50.

27. Eldefrawy, M.H.; Khan, M.K.; Alghathbar, K. A key agreement algorithm with rekeying for wireless sensor networks using public key cryptography. In Proceedings of the 2010 International Conference on Anti-Counterfeiting, Hoboken, NJ, USA, 22–24 March 2010; pp. 1–6.

28. Wu, J.; Stinson, D.R. Three improved algorithms for multipath key establishment in sensor networks using protocols for secure message transmission. *IEEE Trans. Dependable Secur. Comput.* **2010**, *8*, 929–937.

29. Gope, P.; Das, A.K.; Kumar, N.; Cheng, Y. Lightweight and Physically Secure Anonymous Mutual Authentication Protocol for Real-Time Data Access in Industrial Wireless Sensor Networks. *IEEE Trans. Ind. Inform.* **2019**, *15*, 4957–4968. [CrossRef]

30. He, D.; Ma, M.; Zeadally, S.; Kumar, N.; Liang, K. Certificateless Public Key Authenticated Encryption with Keyword Search for Industrial Internet of Things. *IEEE Trans. Ind. Inform.* **2018**, *14*, 3618–3627. [CrossRef]

31. Dua, A.; Kumar, N.; Das, A.K.; Susilo, W. Secure Message Communication Protocol Among Vehicles in Smart City. *IEEE Trans. Veh. Technol.* **2017**, *67*, 4359–4373. [CrossRef]

32. Tyagi, S.; Kumar, N. A systematic review on clustering and routing techniques based upon LEACH protocol for wireless sensor networks. *J. Netw. Comput. Appl.* **2013**, *36*, 623–645. [CrossRef]

33. He, D.; Zeadally, S.; Kumar, N.; Lee, J.-H. Anonymous Authentication for Wireless Body Area Networks with Provable Security. *IEEE Syst. J.* **2016**, *11*, 2590–2601. [CrossRef]

34. Yousefpoor, M.S.; Barati, H. Dynamic key management algorithms in wireless sensor networks: A survey. *Comput. Commun.* **2018**, *134*, 52–69. [CrossRef]

35. Yousefpoor, M.S.; Barati, H. DSKMS: A dynamic smart key management system based on fuzzy logic in wireless Complexity 9 sensor networks. *Wirel. Netw.* **2020**, *26*, 2515–2535. [CrossRef]

36. Sampathkumar, A.; Tesfayohani, M.; Shandilya, S.K.; Goyal, S.B.; Jamal, S.S.; Shukla, P.K.; Bedi, P.; Albeedan, M. Internet of Medical Things (IoMT) and Reflective Belief Design-Based Big Data Analytics with Convolution Neural Network-Metaheuristic Optimization Procedure (CNN-MOP). *Comput. Intell. Neurosci.* **2022**, *2022*, 2898061. [CrossRef]

37. Ramanan, M.; Singh, L.; Kumar, A.S.; Suresh, A.; Sampathkumar, A.; Jain, V.; Bacanin, N. Secure blockchain enabled Cyber-Physical health systems using ensemble convolution neural network classification. *Comput. Electr. Eng.* **2022**, *101*, 108058. [CrossRef]

38. Arumugam, S.; Shandilya, S.K.; Bacanin, N. Federated Learning-Based Privacy Preservation with Blockchain Assistance in IoT 5G Heterogeneous Networks. *J. Web Eng.* **2022**, *21*, 1323–1346. [CrossRef]

39. Bedi, P.; Goyal, S.B.; Rajawat, A.S.; Shaw, R.N.; Ghosh, A. Application of AI/IoT for Smart Renewable Energy Management in Smart Cities. In *AI and IoT for Smart City Applications*; Studies in Computational Intelligence; Piuri, V., Shaw, R.N., Ghosh, A., Islam, R., Eds.; Springer: Singapore, 2022; Volume 1002. [CrossRef]

40. Sharma, S.; Rani, M.; Goyal, S. Energy Efficient Data Dissemination with ATIM Window and Dynamic Sink in Wireless Sensor Networks. In Proceedings of the 2009 International Conference on Advances in Recent Technologies in Communication and Computing, Kerala, India, 27–28 October 2009; pp. 559–564. [CrossRef]

41. Sharma, S.; Goyal, S.B.; Qamar, S. Four-Layer Architecture Model for Energy Conservation in Wireless Sensor Networks. In Proceedings of the 2009 4th International Conference on Embedded and Multimedia Computing, Jeju, Republic of Korea, 10–12 December 2009; pp. 1–3. [CrossRef]

42. Kaliappan, V.K.; Lalpet Ranganathan, A.B.; Periasamy, S.; Thirumalai, P.; Nguyen, T.A.; Jeon, S.; Min, D.; Choi, E. Energy-Efficient Offloading Based on Efficient Cognitive Energy Management Scheme in Edge Computing Device with Energy Optimization. *Energies* **2022**, *15*, 8273. [CrossRef]

43. Ibrahim, B.; Rabelo, L.; Gutierrez-Franco, E.; Clavijo-Buritica, N. Machine Learning for Short-Term Load Forecasting in Smart Grids. *Energies* **2022**, *15*, 8079. [CrossRef]

44. Banuselvasaraswathy, B.; Sampathkumar, A.; Jayarajan, P.; Sheriff, N.; Ashwin, M.; Sivasankaran, V. A Review on Thermal and QoS Aware Routing Protocols for Health Care Applications in WBASN. In Proceedings of the 2020 International Conference on Communication and Signal Processing (ICCSP), Chennai, India, 28–30 July 2020; pp. 1472–1477. [CrossRef]

45. Subrahmanyam, R.; Rukma Rekha, N.; Subba Rao, Y.V. Multipartite Verifiable Secret Sharing Based on CRT. In *Computer Networks and Inventive Communication Technologies*; Lecture Notes on Data Engineering and Communications Technologies; Smys, S., Bestak, R., Palanisamy, R., Kotuliak, I., Eds.; Springer: Singapore, 2022; Volume 75. [CrossRef]

46. Subramani, N.; Mohan, P.; Alotaibi, Y.; Alghamdi, S.; Khalaf, O.I. An Efficient Metaheuristic-Based Clustering with Routing Protocol for Underwater Wireless Sensor Networks. *Sensors* **2022**, *22*, 415. [CrossRef]

47. Bacanin, N.; Arnaut, U.; Zivkovic, M.; Bezdan, T.; Rashid, T.A. Energy Efficient Clustering in Wireless Sensor Networks by Opposition-Based Initialization Bat Algorithm. In *Computer Networks and Inventive Communication Technologies*; Lecture Notes on Data Engineering and Communications Technologies; Smys, S., Bestak, R., Palanisamy, R., Kotuliak, I., Eds.; Springer: Singapore, 2022; Volume 75. [CrossRef]

48. Lakshmanna, K.; Subramani, N.; Alotaibi, Y.; Alghamdi, S.; Khalafand, O.I.; Nanda, A.K. Improved Metaheuristic-Driven Energy-Aware Cluster-Based Routing Scheme for IoT-Assisted Wireless Sensor Networks. *Sustainability* **2022**, *14*, 7712. [CrossRef]
49. Mehrotra, S.; Sharan, A. Comparative Analysis of K-Means Algorithm and Particle Swarm Optimization for Search Result Clustering. In *Smart Trends in Computing and Communications*; Springer: Singapore, 2020; pp. 109–114. [CrossRef]

Article

Energy-Efficient Network Protocols and Resilient Data Transmission Schemes for Wireless Sensor Networks—An Experimental Survey

Dharmesh Dhabliya [1], Rajasoundaran Soundararajan [2], Parthiban Selvarasu [3], Maruthi Shankar Balasubramaniam [4], Anand Singh Rajawat [5], S. B. Goyal [6,*], Maria Simona Raboaca [7,*], Traian Candin Mihaltan [8], Chaman Verma [9] and George Suciu [10,*]

[1] Department of IT, Vishwakarma Institute of Information Technology, Pune 411048, India
[2] School of Computing Science and Engineering, VIT Bhopal University, Sehore 466114, India
[3] Department of CSE, Saveetha School of Engineering, Saveetha Institute of Medical and Technical Sciences, Chennai 602105, India
[4] Department of ECE, Sri Krishna College of Engineering and Technology, Coimbatore 641008, India
[5] School of Computer Science and Engineering, Sandip University, Nashik 422213, India
[6] Faculty of Information Technology, City University, Petaling Jaya 46100, Malaysia
[7] ICSI Energy Department, National Research and Development Institute for Cryogenics and Isotopic Technologies, 240050 Ramnicu Valcea, Romania
[8] Faculty of Building Services, Technical University of Cluj-Napoca, 40033 Cluj-Napoca, Romania
[9] Department of Media and Educational Informatics, Faculty of Informatics, Eötvös Loránd University, 1053 Budapest, Hungary
[10] R&D Department, Beia Consult International, 041386 Bucharest, Romania
* Correspondence: sb.goyal@city.edu.my (S.B.G.); simona.raboaca@icsi.ro (M.S.R.); george@beia.ro (G.S.)

Citation: Dhabliya, D.; Soundararajan, R.; Selvarasu, P.; Balasubramaniam, M.S.; Rajawat, A.S.; Goyal, S.B.; Raboaca, M.S.; Mihaltan, T.C.; Verma, C.; Suciu, G. Energy-Efficient Network Protocols and Resilient Data Transmission Schemes for Wireless Sensor Networks—An Experimental Survey. *Energies* 2022, *15*, 8883. https://doi.org/10.3390/en15238883

Academic Editor: Jose A. Afonso

Received: 23 September 2022
Accepted: 31 October 2022
Published: 24 November 2022

Publisher's Note: MDPI stays neutral with regard to jurisdictional claims in published maps and institutional affiliations.

Abstract: Wireless sensor networks (WSNs) are considerably used for various environmental sensing applications. The architecture and internal specifications of WSNs have been chosen based on the requirements of particular applications. On this basis, WSNs consist of resource (energy and memory)-limited wireless sensor nodes. WSNs initiate data communication from source to destination via physical layer management principles, channel slot scheduling principles (time division multiple access), wireless medium access control (WMAC) protocols, wireless routing protocols and application protocols. In this environment, the development of WMAC principles, routing protocols and channel allotment schemes play crucial roles in network communication phases. Consequently, these layering functions consume more energy at each sensor node, which leads to minimal network lifetime. Even though the channel management schemes, medium control protocols and routing protocols are functionally suitable, the excessive energy consumption affects the overall network performance. In this situation, energy optimization algorithms are advised to minimize the resource wastage of WSNs during regular operations (medium control and routing process). Many research works struggle to identify the optimal energy-efficient load balancing strategies to improve WSN functions. With this in mind, the proposed article has conducted a detailed literature review and notable experimental comparisons on energy-efficient MAC protocols, channel scheduling policies and energy-efficient routing protocols. To an extent, the detailed analysis over these wireless network operations helps to understand the benefits and limitations of recent research works. In the experimental section of this article, eight existing techniques are evaluated under energy optimization strategies (WMAC, channel allocation, sleep/wake protocols, integrated routing and WMAC policies, balanced routing and cooperative routing). The proposed review and the classified technical observations collected from notable recent works have been recognized as crucial contributions. The results infer the suggestions for feasible WSN communication strategies with optimal channel management policies and routing policies. Notably, the simulation results show that cross-layer or multi-layer energy optimization policies perform better than homogeneous energy optimization models.

Keywords: wireless sensor network; energy optimization; MAC; routing; review and data communication

1. Introduction

WSNs consist of tiny sensor nodes made for data sensing, data computation, data transmission and data reception tasks. Unlike wired network scenarios, wireless sensor nodes transmit the sequence of environmental data from one location to another location via open multi-hop channels. In this case, each sensor node is intended to broadcast beacon messages to recognized neighbor sensor nodes in order to establish multi-hop channels. At the same time, each sensor node must hear the requests coming from other sensor nodes. Thus, the sensor nodes crucially spend significant amounts of energy in listening states. As the sensor nodes are vulnerable to resource limitations in real-time conditions, wireless network protocols and communication frameworks are expected with energy consideration functions.

Generally, WSNs manage various services in their protocol stack, such as mobility management, security management, power management, event management and quality management. The application-specific WSN architectures are deployed in order to provide various levels of network services. Wireless physical layer functions, WMAC functions and wireless routing protocols are the major considerations for achieving superior data transmission in WSNs. In this regard, configured physical layer parameters and nodes' effective attributes initiate network operations. To an extent, wireless channel allotment policies (scheduled or random) are decided by WMAC procedures. Similarly, identifying optimal routes for multi-hop data communication along the channel from source node to destination node must be configured with the help of suitable wireless routing protocols. According to the deployment strategies of WSNs, the routing protocols are chosen to enable a multi-path routing process or a uni-path routing process.

Along the execution of layering functions in the network, each sensor node receives irregular energy distribution, computation load and memory utilization. Under the conventional or standard network functions, the lifetime of the WSN is unacceptably undersized. The lifetime of the WSN or sensor nodes can be increased through proper energy utilization policies and load distribution policies. These policies are expected in MAC and route management tasks. Initiating crucial experimental analysis over various WSN-based energy optimization techniques and load optimization techniques provides a new motivation for future research works. In this concern, the individual energy optimization rules established for WMAC and routing function in each node ensure the entire network's lifetime and link availability. The importance of energy-sensitive communication protocols is seriously considered for research under WMAC and routing layer functions. As WMAC and routing jobs are more closely related to channel liveliness than other layers, the need for controlling energy wastage along the respective channel is a critical task. In the same manner, multi-path routing protocols are noted as better solutions than uni-path routing mechanisms against security threats. Apart from physical architectures, the sensor nodes require logical neighbor association rules to build usable network channels. Logical network configuration and successful data communication are confirmed through effective WMAC principles and resilient routing protocols, respectively. On the other hand, energy-efficient WMAC procedures and routing protocols are widely expected in each wireless sensor node to save the individual node's energy. With this in mind, this experimental survey has been initiated from the study of the WSN's characteristics, types, future-generation policies and real-time network problems. Accordingly, the major problems and solutions are described, as given in Table 1.

Table 1. WSN studies and categories of works.

WSN Studies	Respective Categories of Studies
[1–4]	Energy optimization, mobilty, applications, general overview and security.
[5–11]	Channel rate allocation (lifetime management), reliable data collection, energy optimization, channel eavesdropping, network coverage, localization and mobility prediction.
[12–16]	Reliability and energy considerations, energy problems in IoT, channel scheduling, medical sensors and performance optimization in IoT.

As given in Table 1, the general aspects and WSN characteristics are discussed. In addition, other research works [17–27] discuss various resource allocation issues, routing protocols, smart sensor characteristics and energy problems of WSNs. The baseline understandings of WSNs and their application inspires researchers to focus on suitable WMAC and wireless routing strategies in real time. In the same manner, the articles used for experimental survey are categorized under WMAC and wireless routing protocols. Particularly, the related research works are classified under energy optimization policies, energy balancing policies, channel scheduling policies and machine learning techniques [28–88].

The detailed literature discussions and relative observations mainly target the optimized WSN functions. Notably, energy-optimized solutions on the relative functions between WMAC and routing protocols of WSNs (channel allocation and route identification) are vastly explained in this article. Moreover, the experimental section of this article details a crucial set of energy optimization policies and load balancing policies in order to suggest better WSN strategies on the basis of WMAC and wireless routing protocols. As many research works propose resource-constrained communication protocols (channel control and route control), the need for classified results is mandatory to obtain crucial aspects through appropriate experimental conditions.

In the same manner, research problems are widely noted under energy-efficient channel management policies, medium utilization quality, energy-aware WMAC routines, node liveliness, node connectivity and optimized neighbor discovery processes. As the battery-powered wireless sensor nodes are vulnerable to unplanned energy depletions, the overall functions of the entire WSN cannot be expected as static and stable in real-time conditions. In this regard, the proposed article analyzes and compares recent works conducted regarding the issues mentioned. In addition, this article mainly finds the technical benefits, limitations and scientific facts of crucial research works accompanied by the energy-efficient WMAC principles, optimal medium utilization and energy sensitive wireless routing protocols of WSNs.

The study contributions of the article are listed, and details are given in the respective sections.

- Discussing the types of WSNs, configuration details, resource management, channel rate allocation and future requirements.
- Taking a comparative study and experiments on WMAC principles, channel allocation strategies and energy optimization issues.
- Discoursing the functions, limitations and properties of various wireless routing protocols.
- Energy optimization issues in wireless routing environment and protocol support.
- Experimenting a detailed energy optimization scenario between different cases of WMAC functions and wireless routing protocol functions.

Many literature survey works are proposed under energy optimization policies for enabling feasible network communication between wireless sensor nodes. At any rate, the implications of multi-layer energy conservation policies (WMAC energy solutions, WMAC channel management, routing problems and load balancing problems) provide an integrated problem analysis and solution-making platform for future researchers. Crucially, the existing study articles consider mostly uniform energy-efficient solutions on a particular layer. The energy optimization solutions discussed on single-layer functions limits the relative energy-based interpretations between WMAC (channel allocation) functions and routing protocol functions. On the whole, the novelty and contributions of the proposed literature survey give diversified technical details on multi-layer network functions and energy considerations. On this basis, the research findings are taken in order to solve the energy optimization issues regarding wireless channel allotment schemes, WMAC protocol functions and wireless routing protocols.

The proposed review article has been classified under different sections. Section 2 of the article consists of the technical discussions on WSN architectures, network problems, energy-sensitive WMAC strategies and energy-efficient routing protocols. Section 3 includes practical investigations and performance comparisons between notable literatures. Section 4 of this article provides a detailed conclusion on the review findings and future scopes.

2. Materials and Methods

2.1. Technical Discussions on Related Works

As discussed earlier, the importance of WMAC and wireless routing protocols is seriously considered in many recent research works. Predominantly, the research initiatives focused on medium control, channel allotment, routing efficiency and energy constraints to ensure fault-protective sensor nodes during independent data transmissions. Additionally, the guarantee to increase network stability and lifespan is the most essential quality for well-organized WSNs. Among the unexpected environmental conditions and network uncertainties, the assurance of successful data communication can be attained through a deep technical survey and a newly created solution. The following sections of this proposed review article discuss the details of WMAC concerns and routing concerns to achieve energy-controlled lifetime enhancement routines.

On the whole, the MAC protocol basically has two variants, such as the carrier sense multiple access with collision detection approach (CSMA/CD) and the carrier sense multiple access with collision avoidance approach (CSMA/CA), used on wired networks and wireless networks, respectively. Under the CSMA/CD approach, the network interface card (NIC) of any node supervises the availability of the other node's activity on the wired channel (collision) to start data transmission. The detection of collision at the NIC delays the data transmission of a particular node. On the other hand, the CSMA/CA approach monitors the wireless channel for a random duration to prevent collision during data transmission. In addition to MAC layer functions, logical link control (LLC) functions are operated in a data link layer for multiplexing, de-multiplexing and network layer interfacing services. In any event, the WMAC (CSMA/CA), LLC and wireless routing functions consume much more energy in WSNs.

In this case, routing protocols execute neighbor discovery functions (route requests and route replies), routing table organizations, link establishment functions and data routing functions continuously. Particularly, neighbor discovery jobs consume crucial amounts of energy in the idle listening state. From these discussions, this article finds energy optimization challenges through multi-layer network protocols to assure the lifetime of the WSN.

2.1.1. Wireless Sensor Networks and Challenges

WSNs are vastly used for establishing autonomous communication environments, Internet of Things (IoT) platforms and other distributed networks. As a collection of tiny sensor nodes is responsible for multi-path distributed communication, the effective utilization of medium, channel slot management and routing processes executed are expected to be optimized under limited energy-consumption practices. Generally, WSNs are categorized as static WSNs, mobile WSNs, deterministic WSNs, uncertain WSNs, single base station (BS) WSNs, multiple BS WSNs, direct-hop WSNs, multi-hop WSNs, homogeneous WSNs and heterogeneous WSNs (IoT), etc. The environment of various types of WSNs consists of crucial network properties such as energy optimization, network scalability, node responsiveness, communication reliability and node mobility.

A wireless sensor node deployed in the field has internal components such as sensor units, data processing units, data communication units and optional software modules (operating system or system software). The sensor nodes used in WSNs are identified as generic, special purpose, gateway and other higher models. In this regard, Cardei et al. [1] proposed an energy-efficient data communication strategy for organizing the sensing, processing and data transmission tasks in each sensor node. The novel contribution of this work was identifying and managing disjoint connected dominating sets around the WSN. At the same moment, the nodes that were not participating in the communication were considered as disjoint entities. The connected dominating sets managed a logical association between various disjoint nodes in the network. The logical construction of connected dominating sets provided the optimal energy wastage spent from inactive sensor nodes. On the other hand, this work had not provided a crucial cross-sectional energy optimization model at different layers.

Ekici et al. [2], Zhang et al. [3] and Yick et al. [4] discussed various types of WSNs, wireless data communication and applications. Among these works, Ekici et al. explained the scenarios and practical difficulties of mobile sensor networks where the nodes are allowed to move around the geographical region. As discussed, each type of WSN architecture is important for appropriate field applications. Notably, mobile WSNs have the most flexible yet unstable network architectures against the uncertainties of wireless medium and network failures.

In this situation, the responsibilities of channel allocation strategies, WMAC transactions and routing protocols are crucial compared to other types of standard WSNs. On behalf of these wireless network environments, Zhang et al. and Yick et al. discussed the common functions of WSNs and the applications. These works deliver the recent needs of wireless communication technology and sensor platforms. The growth of WSNs is placed around the fields of agriculture, health monitoring systems, home automation systems, industrial automation systems, military security systems, ocean monitoring systems, wild animal tracking systems, underground sensor systems and other object surveillance systems.

Each type of application-specific system needs a suitable set of wireless sensor nodes with various configurations. According to that, the processor model, memory model, signal transmitter, signal receiver, signal converters, sensor standards and the type of power source are selectively considered for managing overall network functions. On that note, this work provided the details of data collection techniques, coverage properties and other communication strategies. Figure 1 gives the internal components of a generic wireless sensor node.

Hou et al. [5] analyzed the problems of distributed sensor data collection and accumulation throughput rate. This work had taken the idea of implementing lexicography-based minimum–maximum rate allocation principles, linear rate evaluation programming models and data parametric analysis models to stabilize the overall network lifetime. As the irregular data allocation and data accumulation strategies severely affect each node's performance and lifetime, the entire WSN has been disturbed in its functionality and liveliness. The problem of flexible rate allocation and lifetime stability assurance are considered as the most important issues in WSNs. With this in mind, this work observed the close relationship between individual data rate of each sensor node and the node's lifetime issues during data transmission.

The determinations of this existing work had crucial results, yet the need for power stability and channel properties were not considered seriously through MAC and routing cautions. Similarly, Wei et al. [6] prescribed the problems and reliable technical supports for collecting data through underwater sensor networks (wireless channel). As compared to other communication bands such as microwave frequencies and radio frequencies, underwater sensor networks require acoustic signal transmission for data dissemination. The water medium is not free to transmit signals as easily as possible using generic signal bands. Data dissemination through a water medium has many problems relevant to data rate sustainability, packet drops, route breakages, channel reliability and data collisions.

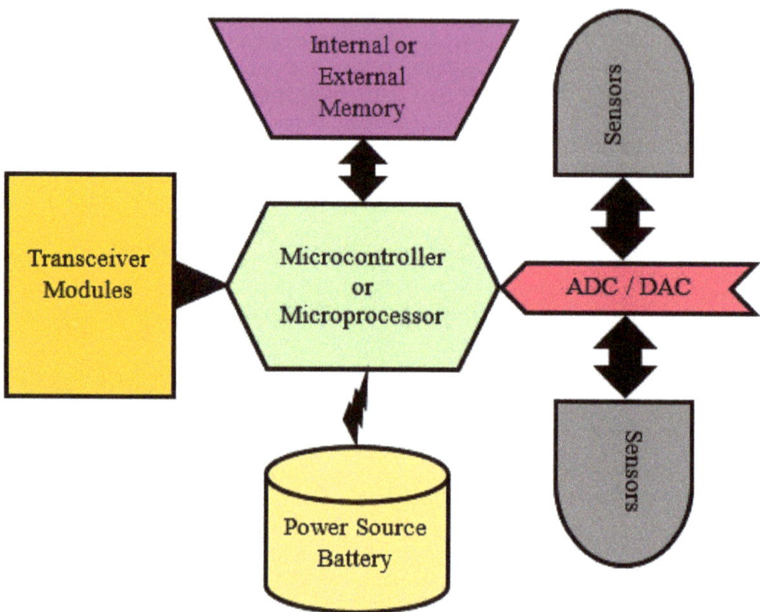

Figure 1. Internal components of sensor node.

The acoustic signals used for multi-hop communication through water become scattered, diverted and attenuated before reaching the destination node. In these situations, the energy spent by each sensor node is not useful. Thus, the need for underwater medium allocation and energy-controlling mechanisms are highly recommended against unequal channel distortions. Particularly, the applications of underwater sensor networks are useful in military-based underwater vehicles and submarine systems.

Optimizing the sensor node's resources ensures a long-time data communication through the wireless medium. Especially, the resources of sensor nodes are plunged by both the medium and the malicious activities injected around the network. Boubiche et al. [7] and Bashar et al. [8] expressed the importance of cyber threats and physical layer (channel) threats executed in WSNs. The cyber threats or attacks massively acquire energy resources through unauthorized activities. Particularly, active attacks such as denial of service (DoS) create a major problem for resource availability in WSNs. In contrast, passive attacks such as eavesdropping and wormhole attacks silently gather data dissemination efforts. With this in view, solutions have to be met to ensure stable energy-saving practices against cyber threats. At the same time, counter algorithms to stand against the attacks are expected with energy optimization techniques.

Cao et al. [9] and Kotiyal et al. [10] proposed innovative solutions on node coverage and connectivity problems. Among the above contributions, the earlier work developed a unique social spider optimization for improving the sensor node's coverage capability in heterogeneous sensor networks. Each sensor node can transmit the data based on the coverage probability, neighbor availability and neighbor discovery ability. However, the quality of coverage probability and neighbor discovery processes of each node was completely raised based on residual energy level. Generally, various types of technical contributions are motivated to improve the functionalities of the sensor node to attain quality communication. The disparity and the research problems with those works were consistent with inadequate energy optimization views. The second work focused on managing node localization procedures using a cuckoo search algorithm in WSNs. Likewise, a few other research works find solutions against node localization issues [11].

Energy-focused research is driven towards WSNs in order to extend network lifetime and node availability. Reliable energy considerations and energy-optimized computing frameworks are deeply analyzed with body sensor networks and IoT systems, respectively [12,13]. IoT architectures and software-defined networks depend highly on minimal energy consumptions under WSN pitches. Nweye et al. [14] provided the necessity of heat, ventilation and air conditioning (HVAC)-based communication schedules for saving the node's energy in wireless local area networks (WLANs). The technical diversity provided in the energy optimization field was mandatory for specific wireless communication technologies. The application of WSNs decides the required level of energy savings during data communication. With this in mind, Chandra et al. [15] proposed intelligent energy optimization techniques for cardio sensor networks.

In this work, Chandra et al. identified the use of sensor data compression possibilities against energy wastage. In the same way, many research works have been conducted to produce green computing application using WSNs [16,17]. The need for green computing technologies and energy optimization principles are mandatory for the innovative applications of WSNs such as IoT, smart city systems, intelligent farming systems, fault-tolerant energy models, edge computing systems and smart learning systems [18–22]. The well-defined energy optimization technique must focus on both WMAC policies and routing policies to save the node's energy. Specifically, the energy wastage of every sensor node happens during active listening, passive listening, data transmitting and data receiving activities.

Under these unavoidable circumstances, an efficient solution has a huge impact on each sensor node to save energy. In this manner, Nayak et al. [23] proposed machine learning (ML) techniques for enabling intelligent routing processes to reduce overhead and energy consumption rates. This work found the benefits and limitations under ML-based routing solutions. The benefits were noted, with this technique providing the optimal routing process with minimal latency. At the same time, the limitations were considered as how to develop lightweight ML algorithms in order to reduce the computation overhead and energy consumption rate. The detailed comparisons are illustrated in Table 2. The increasing demand of sensors and WSNs through military equipment, drone applications, IoT environments and secure sensor environments must be addressed with energy optimization models and load management models [24–27]. The detailed review about WSNs, deployment issues, MAC policies, channel assignment policies and routing schemes identifies the following research problems and future scopes:

- Energy-efficient WMAC principles are required for application-specific sensor nodes.
- Energy optimization and green computing models are expected for increasing the lifetime and availability of WSNs.
- Reactive energy-saving routing protocols and on-demand channel establishment strategies are estimated.
- Channel quality determination and reactive scheduling mechanisms are required to build green computing platforms for future WSN architectures.

Table 2. WSNs and recognized challenges.

Related Articles	Strategies	Problems Considered
Cardei et al. [1]	Energy optimization using disjoint and connected dominating sets	Energy optimization and live connectivity identification
Ekici et al. [2]	Understanding mobility issues and solutions	Connectivity, lifetime and mobility
Hou et al. [5]	Lexicographical order of channel rate Allocation and lifetime management (Linear programming and serial channel parametric analysis)	Optimal data rate allocation and network lifetime

Table 2. *Cont.*

Related Articles	Strategies	Problems Considered
Boubiche et al. [7]	Cyber security solutions and challenges using light weight methodologies	Network security and energy-efficient cyber security
Bashar et al. [8]	Physical layer interception probability model	Physical layer and channel-trapping attacks
Cao et al. [9]	Heterogeneous social spider energy optimization and coverage scheme	Energy optimization, network coverage and neighbor identification
Kotiyal et al. [10]	Optimized cuckoo search and locality management	Node localization error and neighbor identification
Nain et al. [11]	Propagation latency and delay prediction with energy optimization	Acoustic signaling and underwater energy wastage
Mishra et al. [13]	Nature-inspired algorithms (grey wolf Optimization, PSO and balanced clustering)	Energy optimization for IoT systems
Nweye et al. [14]	HVAC scheduling and energy plans (Wi-Fi)	Energy optimization for Wi-Fi systems
Chandra et al. [15]	Data compression and channel modelling systems	Energy optimization for body sensor networks (cardio health care systems)
Dhaya et al. [18]	Multi-modal resource allocation and load balancing systems	Energy optimization for agriculture IoT sensors
Humayun et al. [20]	Smart energy plans for fifth-generation IoT systems	Network lifetime and energy plans
Zhu et al. [21]	Artificial-IoT systems with energy-efficient scheduling frameworks	Energy optimization and channel timelines
Vashisht et al. [22]	A review on ML-based smart sensor platforms	Current and future challenges in WSNs
Nayak et al. [23]	A review on routing protocols, energy problems, lifetime and localization	Routing and neighbor monitoring challenges
Bhargava et al. [24]	Low-cost link establishment using cuckoo neural network System	Network lifetime and nonlinear network modelling
Haseeb et al. [25]	Multi-attribute learning and secure sensor modelling system	Wireless security, uncertainty and mobility models
Geetha et al. [26]	Green energy modelling, future load forecasting and energy balancing system	Load balancing, energy optimization, delay computing and distance management
Ren et al. [27]	Edge computing and energy modelling for smart city	Green energy and edge models

As mentioned above, the basic identifications and expectations of WSNs and their recognized challenges should be the motive for any researcher. This article has extended its deep dive into the current technologies of WMAC, wireless routing protocols and energy-saving limitations of wireless infrastructures around the world.

2.1.2. WMAC Strategies with Energy Optimization Techniques

Wireless communication techniques, medium access principles and channel modelling strategies are common in research aspects. At the same time, finding suitable techniques and proposing novel techniques with reliability are challenging tasks. Richert et al. [28] implemented the variances of MAC protocols such as carrier sense multiple access/collision avoidance (CSMA/CA), the sensor MAC (S-MAC) protocol, weak channel or signal detection policies, the timeout MAC (T-MAC) protocol and other variances. This experimental view of MAC policies extracts the functions of each MAC model including delay-sensitive MAC (DS-MAC), energy-sensitive time division multiple access (TDMA) policies and tree MAC policies. According to the details given in this work, MAC models can be expressed as shown in Table 3. In this case, collision models such as collision detection (CD), collision notification (CN) and weak signal detection (WSD) are experimented. Similarly, reduced TDMA (TDMA-R), schedule exchange protocol (SEP), adaptive neighbor election algorithm (AEA) and neighbor discovery protocol (NDP) are evaluated.

Table 3. Collision models and MAC models [28].

Collision Models	MAC Models	Channel Allocation
CSMA/CD	MAC	TDMA
CSMA/CA	WMAC	TDMA-R
CSMA/CN	S-MAC	SEP
CSMA/WSD	DS-MAC	AEA
	T-MAC	NDP

However, this experiment limits the assumptions in MAC policies only for energy wastage in WSNs. This assumption shall be extended into routing policies with the considerations of various technical glitches of wireless routing protocols. Jain et al. [29] discovered a novel energy-efficient network architecture with cluster head coordination principles and hot-spot analysis procedures. In this scheme, cluster heads are selected with the help of the Harris hawk optimization protocol and dynamic clustering protocol. In addition, this technique has been extended with the dynamic routing protocol to reduce the sensor node's energy depletion rate.

The implementation section of this work has provided performance metrics such as network energy depletion rate, lifespan of sensor nodes, packet transmission rate, network coverage factor, etc. At any rate, this contribution is limited to node selection procedures rather than channel organizing policies. Energy-efficient MAC models widely apply sleep and wakeup mechanisms to increase the lifespan of nodes in WSNs. The sleep and wakeup protocols work based on reactively initiated transmission schedules for multi-path channels. Chawra et al. [30] and Alzahrani et al. [31] proposed energy-efficient sleep and wakeup scheduling techniques in each sensor node to save energy. Particularly, the former work uses memetic techniques for organizing WMAC by determining the qualities of energy consumption rates of sensor node, coverage factor, neighbor connectivity rate and optimal wakeup duration for each channel. Accordingly, this technique ensures the benefit of energy control and liveliness of sensor nodes in WSNs.

The latter work analyzes the architectures of ad hoc sensor networks and establishes quorum-assistive sleep and wakeup protocols for organizing WMAC policies. In this case, the quorum properties of each sensor node are validated and slots are activated to minimize energy expenditure. However, both works are limited with WMAC perspectives. The efforts in producing sleep and wake models, duty-sensitive scheduling procedures and adaptive channel utilization with WMAC continue for various types of WSNs [32–34]. Moreover, the standpoint of energy-optimized WMAC establishment is a required aspect for all researchers.

Ranjan et al. [35] discussed impacts on energy diffusion rate and the network disturbances. According to this model, the network irregularities and traffic turbulences happened due to the excessive diffusion of energy in each node. At the time, the underprivileged deliberations of MAC management rules and irregular load distribution among sensor nodes were creating energy losses in each node. Hence, the entire WSN met communication problems. This work provided the experimental analysis cases for different types of WSNs such as cluster-based environments and random environments around different sizes of geographical regions. Under this testbed, this experiment revealed issues with the lifespan of each sensor node and the downtime of each node. This experiment helped to study the stability of the overall network during continuous energy drops.

In the same way, a few other research works tried energy-efficient MAC models in static WSNs [36]. Alablani et al. [37] recommended a novel energy-controlled MAC and routing protocol for underwater sensor networks. Generally, underwater sensor networks use acoustic signal propagation models and resource-constrained sensor nodes. Unlike other sensor networks, underwater sensor nodes are not feasible to be charged regularly through any modes (electricity or solar power). The hostile nature of underwater sensor networks requires more efficient energy-saving mechanisms for both MAC protocols and routing protocols. This scheme used network properties such as finite energy limits, a multi-hop communication session, narrow transmission possibilities, sleep mode executions and a uniform energy utilization factor. The observations of this work revealed the better utilization of channels and routing sessions to improve the quality of energy optimization principles.

In the same manner, Samal et al. [38] established MAC models for energy-sensitive body sensor networks. Compared to other types of WSNs, body sensor networks are extremely tiny and simple components. The diffusion of energy from each node severely affects the performance of each body sensor in its health monitoring functions. This leads to improper determinations of health recordings. With this in mind, this work proposes multi-channel scheduling techniques and sleep mode supports for biosensor units. Sakib et al. [39] implemented a new MAC policy for WSNs using quality of service (QoS) parameters (delay, throughput, jitter, bandwidth, packet loss rate, etc.) and data priority models. According to this method, multi-priority values are computed for each data packet for each session. The multi-hop data transmission was initiated based on priority values and priority-assistive MAC principles. On the other hand, the determinations of optimal QoS quantities for each node were taken for energy-controlled data communication sessions. Both techniques are better in terms of multi-property considerations to achieve energy-efficient MAC solutions.

In this concern, Darabkh et al. [40] proposed uncertainty-aware transmission scheduling models for clustered IoT systems through TDMA and spread spectrum-based MAC techniques. In this work, cluster heads were selected based on locality information, residual energy rate and balanced workload distribution models. In another way, this scheme contributed to observing the presence of uncertainties, interference, delay, power distortion and other channel problems. The focus on multiple channel properties supports allocating TDMA slots for wireless data transmission in IoT systems. In the same vein, Subramanyam et al. [41] and Gowda et al. [42] illustrated the possible ways of building on-demand duty cycle establishment principles and hybrid MAC policies, respectively, for energy-controlled wireless transmission. Most of the works executed under energy-efficient MAC policies were conventional in terms of limited hardware assumptions and resource considerations for various types of WSNs.

The contribution of Ajmi et al. [43] varied from other MAC policies under the attentions of logical modelling procedures. This scheme found an idea of configuring inter-cluster and intra-cluster MAC policies with cross-layered communication principles. Notably, the establishment of inter-cluster and intra-cluster MAC solutions provides independent handling of energy wastage in the IoT environment. The establishment of energy optimization techniques for MAC policies for wireless health monitoring systems are widely required around the world. In this case, mobility management protocols, static channel policies and energy harvesting models are created for autonomous health monitoring systems [44–46].

Udoh et al. [47] justified the relationship between MAC layer functions and radio signaling models with regards to energy consumption. Relating to other existing MAC solutions, this work validated malicious events related to energy wastage in each sensor node. Under this scenario, this work compared S-MAC and T-MAC policies based on duty slot allocation procedures. According to the establishment of active and idle slots, T-MAC and S-MAC principles managed the radio signal propagations. Against these signal propagation models, special-type attacks are generated such as denial of ideal/sleep attack and active channel attack. These attacks mainly target channel availability and energy-saving slots (sleep mode) of WSNs. Consequently, the residual energy in each node automatically reduces to the inactive state. This scheme provided security frameworks against attacks to minimize the impact of energy wastage. At any rate, the need for lightweight security models against channel attacks are ignored in this contribution.

In a similar fashion, Sadeq et al. [48] and Lakshmi et al. [49] created theoretical MAC models and heterogeneous MAC models for WSNs, respectively. These works found the maximization of packet delivery rate through energy optimization models. Sah et al. [50] intended an energy-efficient sensor management architecture for industrial applications. Conspicuously, industrial IoT systems are completely distributed and heterogeneous in nature. Hence, the provision of energy optimization procedures is complicated for different types of sensor nodes. Additionally, this technique implemented load balancing rules and MAC scheduling procedures to optimize the sensor node's energy in the distributed IoT platform to organize the industrial components. In this case, Yang et al. [51] concentrated on node clustering techniques, multi-hop routing principles and bearable energy solutions for underwater sensor networks. The contributions of the above works mainly found MAC-based and channel-based energy leakages due to various circumstances. At the same time, the future findings regarding green-MAC (G-MAC) computing models and reactive scheduling models are expected as follows:

- Adaptations of various WMAC principles are expected to save energy under tiny autonomous WSNs.
- Distributed and heterogeneous WMAC principles are highly anticipated for IoT-based WSNs.
- Multi-channel reactive scheduling models and control management principles are required for WSNs.
- Routing and medium coordination solutions are needed to minimize the energy with a cross-layered design.

Hence, the exhaustive literature search and contribution analysis provided future technical needs in order to improve the energy-efficient solutions for WMAC functions. To a certain extent, knowing the impacts of routing layer functions with energy-saving models is essential for confirming a better lifespan of WSNs.

2.1.3. Energy-Efficient Wireless Routing Strategies and Protocols

The significance of integrated WMAC principles and wireless routing protocols are identified around the research community arena. Most of the research works focused on the energy-saving plans of WSNs having determinations with WMAC policies only. However, route discovery, route management and data routing protocols heavily control the energy release rate in each sensor node. As a result, the inventions on individual energy-efficient WMAC protocols, energy-efficient routing protocols and hybrid interface energy management principles are inescapably required.

Zagrouba et al. [52] described various wireless routing solutions such as cluster-based routing protocols, hierarchical routing protocols, random (flat) routing protocols, node-centric routing protocols, data-centric routing protocols, static routing protocols, mobility-support routing protocols, time-sensitive routing protocols and other geography-aware routing protocols. In addition to these protocols, wireless networks and WSNs focus on application-specific routing protocols, medium-specific routing protocols and energy-efficient routing protocols. Each routing protocol has many variances, such as low-energy adaptive clustered hierarchy (LEACH), dynamic source routing (DSR), ad hoc on-demand distance vector routing (AODV), temporary ordered routing algorithm (TORA), the real-time routing protocol (RTRP), geographic and location-based routing protocol (GLRP), QoS-based routing protocol (QRP) and other protocols. Among these protocols, the strategy of data communication and route management policies are initiated in a proactive manner or reactive manner.

Figure 2 relates the energy-efficient rules required at each layer. As illustrated in Figure 2, the top-down communication (network layer to MAC layer) and upward communication (MAC layer to network layer) are expected to attain the energy controlling principles to increase the lifetime of WSNs. The routing protocols used for WSNs are anticipated to perform successful data communication with optimal energy-saving rules compared to other types of wireless networks. Since the internal resources of sensor nodes are more limited than other network nodes, the effective governance of suitable energy-saving mechanisms are highly needed at routing layer functions.

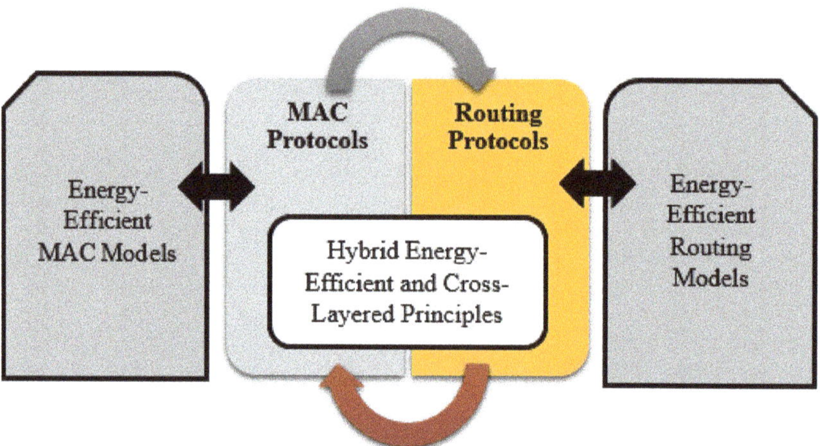

Figure 2. Cross-layered model.

The requirements of future-generation networks, IoT systems, are progressively specified by different researchers. In this manner, Dogra et al. [53] identified the importance of energy-efficient future-generation networks and fifth-generation IoT systems and wireless technologies. The reliability, scalability and lifespan of future-generation IoT systems are sorely needed to manage multiple-input and multiple-output channels. This work mainly focuses on the development of fifth-generation routing principles with energy optimization techniques to handle multi-channel data transmission practices. Notably, the distributed load balancing mechanisms, energy sharing principle and heterogeneous channel allocations are considered major technical problems in this domain. In another way, the assumptions of this proposed model lead to clustered IoT-based WSNs. Similarly, the prospects of flat IoT models, coverage liability and appropriate energy-saving routing models cannot be ignored around the domain of wireless technologies [54,55].

Kumar et al. [56] proposed an optimized zonal energy-balancing technique and adaptive Dijkstra's technique for improving routing principles. This method evaluates the node's transmission distance and residual energy in the network. In this manner, the entire wireless network (personal network) has been divided into various local zones to ensure the even distribution of the communication load. At the same time, optimized shortest-path routing algorithms are applied to manage the energy consumptions of each sensor node in a balanced manner. This scheme justifies that the proposed technique has a confirmed packet transmission ratio, energy wastage and routing delay.

In the same vein, Navarro et al. [57] and Hajipour et al. [58] developed energy-balanced routing protocols for WSNs. Angurala et al. [59] identified the solar energy production and consumption models for sensor networks. Using this platform, this scheme proposes a modified solar-constrained AODV protocol for enriching the energy plans of each sensor node. Compared to other sensor platforms, the establishment of solar sources admits periodical recharging panels and energy-utilization principles. In this concern, the modified AODV has been trained to coordinate the events of each sensor node in terms of data production, data collection, load balancing and recharging plans. In any event, the technical benefits of solar initiative AODV have not been justified against natural disturbances.

Equally, Singla et al. [60] and Asqui et al. [61] developed multi-path energy optimization possibilities for constructing flat routing protocols in small scale WSNs. Yun et al. [62] proposed a deep-learning model for improving the routing quality to reduce the energy depletion rate.

Particularly, reinforcement learning (RL) networks are trained with Q-matrix computations for observing the practices of routing protocols in order to reduce the energy consumption rate. In this case, the observation reveals that the energy of each sensor node has been spent excessively during irregular data aggregation and transmission periods. Accordingly, the Q-learning model computes event-based Q-values or weights for classifying the sensor node's issues.

In association with previous routing protocols, this method initiates data aggregation and routing processes based on the computed Q-values. On the other hand, the limitations of deep-learning procedures in tiny sensor nodes are not significantly explored in this effort.

Sharma et al. [63] implemented energy optimization techniques for achieving efficient routing policies in order to operate agricultural sensor networks. Agricultural sensor networks are widely used for automating irrigation systems, plant monitoring systems, surveillance systems and soil monitoring systems. In this field, various types of sensor nodes are used according to the sensing role. This can be considered as a heterogeneous sensor network where the nodes are oddly deployed and configured. Simultaneously, these nodes are completely different in terms of internal components and other communication tools attached with sensor nodes. This field-based sensor network has serious vulnerabilities against uniform energy plans. Under this circumstance, this work proposes LEACH principles, deterministic energy clustering principles, sensor information collection modules and stable node selection models in a hierarchical fashion. The network assumptions are made with clustered sensor communities, base stations and gateway points.

The contribution of this work is commendable, yet the computation complexity of proposed techniques is ignored for contemplation. In the same way, a related set of techniques are contributed for energy optimization problems and clustered communities in WSNs [64–66]. The considerations of energy limitations and optimization parameters in WSNs are extensively noted with regard to WMAC and routing policies. From the vast analysis, the clarifications are derived on the basis of the contributions and limitations of current technologies [67–69]. In this sense, several existing techniques only consider WMAC irregularities as the causes for the fastest downtime of sensor nodes and energy wastage. On the other hand, a few notable research works consider the energy problems against routing functions and data aggregation functions [70–73]. The perspectives of each research work vary with respect to the authors' own research motivations.

The research contributions are extended for different types of networks, such as generic sensor networks, body sensor networks, personal sensor networks, health sensor networks, surveillance sensor networks, agricultural sensor networks and IoT-based sensor networks [74]. Each existing technique is specifically formed with static assumptions under WMAC and routing policies [76–78]. More insights into energy-aware routing in WSN, MIMO systems, 4G and 5G networks are also briefly discussed in [79–83]. However, the need to integrate irregular cross-layer functions in order to monitor individual energy consumptions is a more important quality.

These cross-layer solutions and integrated energy plans are not effectively taken into consideration for research practices. With this in mind, the proposed experimental survey has been executed for justifying the solutions against energy issues. The reason for continuous energy depletion is related to listening (active and passive) activities, data transmission policies, data aggregation policies, network deployment strategies, routing policies, WMAC practices and physical problems in WSNs.

3. Experiments, Comparative Investigations and Results

Experimenting the notable technical contributions is a challenging yet useful practice to observe the practical abilities and limitations of the developed frameworks. As illustrated in Table 4, the related articles are taken under considerations such as energy-efficient WMAC policies and routing strategies [84–86]. The successful development of the earlier research frameworks leads to noteworthy benefits for energy-saving plans in WSNs [87,88].

Table 4. Related contributions on energy-efficient WMAC and routing protocols.

No.	Existing Techniques	Energy Optimization Solutions			
		WMAC Energy Optimizer/Balancer	WMAC Scheduler and Time Divider	Energy-Efficient Wireless Routing Protocol	Energy Balancer/ML-Based Routing Protocol
1	Richert et al. [28]	✓	✓	✗	✗
2	Chawra et al. [30]	✗	✓	✗	✗
3	Alzahrani et al. [31]	✗	✓	✗	✗
4	Alablani et al. [37]	✓	✗	✓	✗
5	Samal et al. [38]	✗	✓	✗	✗
6	Sakib et al. [39]	✓	✗	✓	✗
7	Darabkh et al. [40]	✗	✓	✗	✗
8	Gowda et al. [42]	✓	✗	✗	✗
9	Ajmi et al. [43]	✓	✗	✓	✗
10	Sah et al. [50]	✓	✓	✗	✗
11	Yang et al. [51]	✗	✗	✓	✗
12	Dogra et al. [53]	✗	✗	✓	✗
13	Hao et al. [54]	✗	✗	✓	✗
14	Kumar et al. [56]	✗	✗	✓	✓
15	Navarro et al. [57]	✗	✗	✓	✓
16	Sharma et al. [63]	✗	✗	✓	✗
17	Huamei et al. [65]	✗	✗	✓	✓
18	Almalki et al. [66]	✗	✗	✓	✓
19	Goswami et al. [69]	✗	✗	✓	✓
20	Mohan et al. [72]	✗	✗	✓	✗
21	Yao et al. [73]	✗	✗	✓	✗
22	Gayathri et al. [75]	✗	✗	✓	✓
23	Han et al. [79]	✗	✗	✓	✗
24	Senthil et al. [80]	✗	✗	✓	✓
25	Mir et al. [81]	✗	✗	✓	✓

✓—Techniques Implemented, ✗—Techniques Not Implemented.

At any rate, this article was motivated by the intention to conduct an experimental comparison between crucial existing works developed based on energy-efficient WMAC polices, energy-efficient routing policies and multi-layer optimization policies (WMAC and wireless routing protocols), as shown in Table 5.

Table 5. Related experiments.

No.	Existing Techniques	Energy Optimization Strategies
1	Richert et al. [28]: E1	Modified WMAC and channel allocation
2	Chawra et al. [30]: E2	Memetic-based sleep and wakeup scheduling
3	Alablani et al. [37]: E3	Integrated WMAC and routing policies using timeline management
4	Sakib et al. [39]: E4	Data priority computation and QoS modelling
5	Sah et al. [50]: E5	Load balancing and aggressive WMAC scheduling
6	Navarro et al. [57]: E6	Balanced routing for low-powered WSNs
7	Gayathri et al. [75]: E7	Cooperative authentic routing protocol and feedback system
8	Han et al. [79]: E8	Adaptive and hierarchical routing models

Based on the illustration of Table 5, the technical cohesions between each work deliver the implementation details and findings. In this experimental section, E1 [28] is denoted for modified MAC protocols for managing energy distributions in the critical WSN environment. Notably, this work analyzed weak signal detection through CSMA/CA principles (CSMA/WSD). In this regard, this work found the variants of MAC models and the MiXiM-OMNet++ environment. On this basis, the proposed CSMA/WSD was implemented to classify weak signals and channel collisions. Under this mechanism, each packet loss event was evaluated, as given in Figure 3.

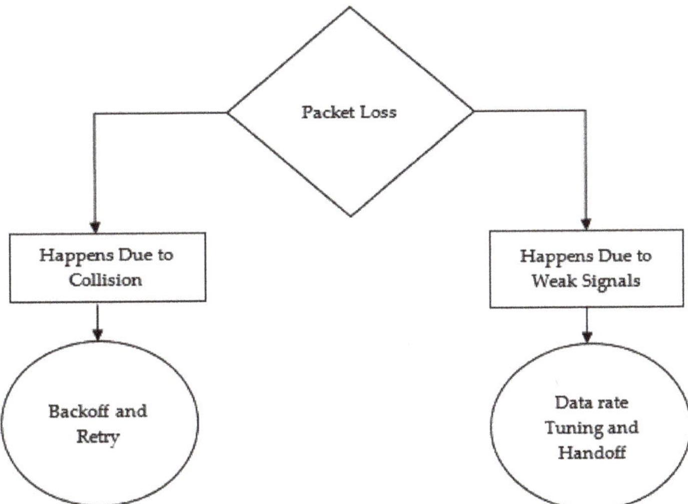

Figure 3. E1-carrier sense and weak channel detection.

In this process, the random determination of packet loss under collision state is evaluated for weak signal conditions. On successful detection of weak signals, the proposed CSMA/WSD helps to re-tune the data rate and finds handoff possibilities. Thus, the systems reduce the number of collision-based retransmissions and save the sensor node's energy. In the same manner, E2 [30] implemented a memetic algorithm–meta-heuristic suit for improving the ability of MAC-based sleep and wake operations. The proposed memetic algorithm implemented five steps to identify the possible number of active nodes in sensor networks to avoid the failure rate of data transmissions. The optimal selection of an active node on the wireless channel minimizes excessive consumption of energy.

According to the model, the active sensor node selection process is illustrated in Figure 4. Notably, the determination of this memetic approach considers the sensor node's coverage quality and connectivity factors, the remaining energy in the sensor node and the duration of sleep–wake periods.

Figure 4 shows the modified genetic algorithm-based memetic node selection approach through population vectors and solution vector computations. In this case, population vectors are computed based on successful sensor node identification marks. To an extent, the solution vectors are computed on the basis of available nodes around consecutive neighbors. In this solution vector, each node's neighbors are updated as child entries to create possible active channels. Under this MAC-based memetic approach, coverage costs, connectivity costs and energy costs are identified with maximum quantity to wake up the nodes from sleep mode. In this connection, each sensor node has been identified with possible solution vector entries (covering sleep nodes), coverage crossover points and sleep–wake mode factors. With this in mind, sensor nodes are searched to enable active node communication channels regularly. This work stated that the modified memetic approach reduces channel allocation time and energy consumption rate for channel allotment.

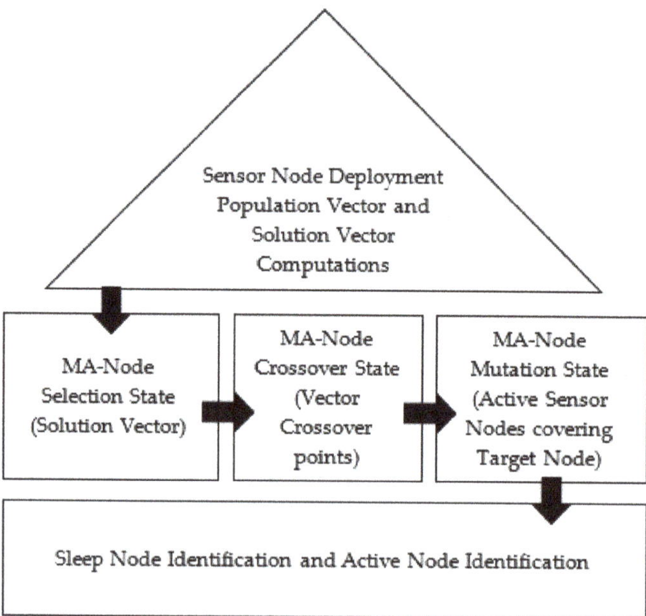

Figure 4. E2-heuristic.

Comparing E1 and E2, both works minimize the cost of packet retransmission through the proper handling of collision cases and node availability issues, respectively. In the same manner, E3 [37] provided the integrated channel management and routing solutions using energy-optimized underwater sensor networks. Compared to other types of WSNs (radio frequency), underwater WSNs use acoustic communication models. Acoustic signaling models are vulnerable to significant data loss, maximum propagation delay, restricted bandwidth, limited energy provisions and channel distortions. In comparison with E1 and E2, E3 takes critical channel (water medium) characteristics to establish energy optimization solutions. In the implementation phases, E3 found depth modelling procedures for numerous acoustic sensor nodes. In this work, the AUVNet simulator toll was used to observe the benefit of energy-efficient underwater MAC and routing protocols against other techniques such as the focused-beam routing protocol, distance-aware collision avoidance protocol and cluster-based protocol. In this scheme, multi-path clear-to-send (CTS) and request-to-send (RTS) packets are shared among the acoustic sensor nodes. On the basis of acoustic sensor placement, the score of any node was computed with respect to node distances (D) and energy levels (E). The score values are calculated based on Equation (1).

$$C = Dr + Er \tag{1}$$

$$Dr \rightarrow \frac{\text{Distance between node and sink}}{\text{maximum distance}}$$

$$Er \rightarrow \frac{\text{Available Energy in node}}{\text{maximum energy.}}$$

To a certain extent, the appropriate node-level computations are useful for finding multi-path channels to route the packets. In the multi-layer network protocol tuning process, E4 [39] developed quality of service (QoS)-MAC assisted multi-path routing protocols and data priority computation schemes. This work produced cross-layered packet analysis procedures, priority evaluation procedures, channel listening activities and flexible routing principles according to channel quality metrics. On this basis, this scheme used AODV protocol and QoS-MAC principles under a single point of concern. According to the system design, this work embedded a packet priority field in WMAC frames. This field was pointed with four priority levels, as shown in Figure 5.

MAC Frame Control	Source Node Address	Destination Node Address	Data Priority Vector	Network Vector	Frame Check Sequence

1-Generic
2-Special
3-Urgent
4-Most Urgent

Figure 5. E4-QoS and MAC and data priority analysis.

In this concern, E5 [50] and E6 [57] proposed load balanced sequence scheduling and a routing protocol, respectively, for reducing the sensor node's energy consumptions. In E5, each sensor node adaptively used the internal buffer to process the data on demand. In multi-hop communication, each sensor node is activated through aggressive scheduling-based MAC models to hold the time division multiple access (TDMA) slots. The even distribution of load among sensor nodes shall forward the channel data using minimal requirement margins. At the same time, these TDMA slots were occupied by multiple sensor data streams. Similarly, E5 proposed an energy-efficient collection tree protocol for low-powered WSNs. The protocol functions executed in each sensor node collected and distributed the data streams based on energy-sensitive tree-like paths. In particular, WSNs are categorized under a low-powered wireless personal area network standard due to their resource-limited environment. In this regard, both works suggested restricted channel allocation policies and low-powered routing mechanisms in WSNs. Notably, E6 compared energy-efficient collection tree protocol with lossy routing protocols and conventional routing protocols. Figure 6 illustrates the functions of E5.

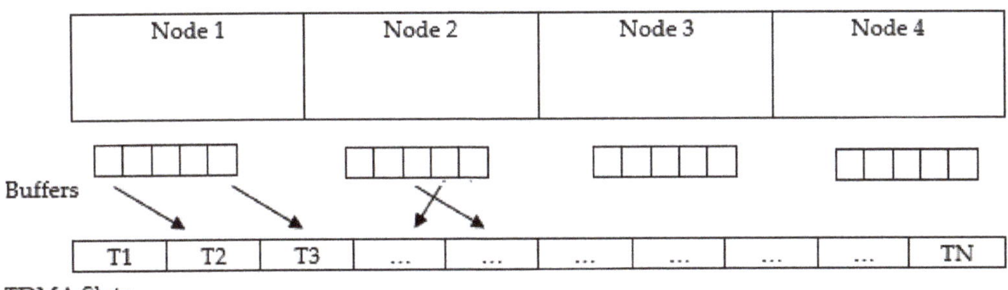

Figure 6. E5-Aggressive TDMA scheduling and buffer management.

In addition, E7 [75] and E8 [79] produced notable solutions on energy-aware wireless routing protocols. Particularly, E7 generated trust value computation techniques using an average packet delivery ratio, route reply ratio, residual energy rate and number of retransmissions on the channel. In this scheme, local trust cost and global trust cost were computed for each sensor node. Local trust values of the sensor nodes were computed in the node itself. At the same time, global trust values were computed at border router points to optimize the data transmission rate. Under this trust evaluation scheme, each trusted sensor node formed a channel to avoid any discrepancies during the data transmission period. This practice reduced the number of retransmissions and packet drops on the channel. Thus, the cooperative routing methodology manages all active sensor nodes under trusted communities (Figure 7). At the end, E8 discussed the particulars of routing protocol issues, connectivity problems and energy control mechanisms in detail. In the experimental sections, this work compared the AODV routing protocol and stateless real-time routing protocol for ensuring energy optimization qualities.

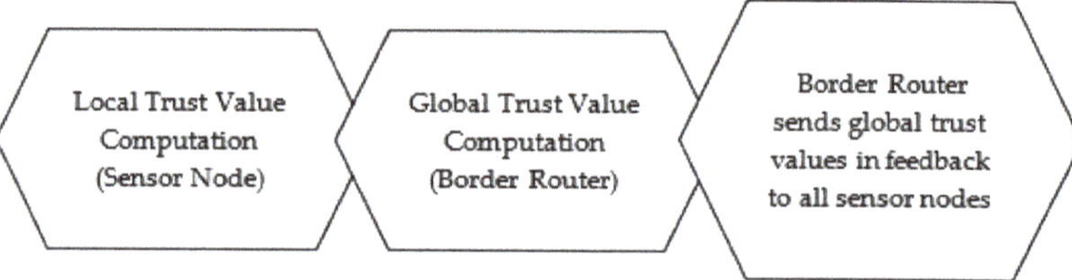

Figure 7. E7-Cooperative feedback trust values for energy-efficient routing principles.

According to the technical observations, WMAC customization and channel allocation strategies [28], sleep and wake strategies [30], cross-layered implementations [37], priority-based channel allocation [39], balanced WMAC/routing solutions [50,57] and adaptive routing models [75,79] are observed for effective comparison practices. The diversified nature of the restricted choices from existing energy-saving policies provides the best understandings under experimental testbed conditions. On the execution of experimental study practices, this work was implemented using tools such as Network Simulator (NS-3.35) and the tool command platform.

The scenario of the WSN is created with a maximum of 300 sensor nodes around the geographical area (1000 m $\times$ 1000 m). Generally, NS-3.0 supports deploying more than 300 sensor nodes. At any rate, this experimental survey sets an assumption of 300 sensor nodes around a 1000 m^2 area, which is significant in terms of network population. Additionally, the network configuration sets a mobility model for the sensor nodes in the prescribed region. The real-time deployment for this type of sensor network (300 sensor nodes with mobility features around a 1000 m^2 area) provides enough challenges for data transmissions, energy harvesting schemes, channel management and route establishment tasks. NS-3.0 is a network scenario creator with simulated configurations of WSNs.

The network scenario has been assumed with the heterogeneous nature of sensor nodes where the internal components of nodes vary in terms of energy, transmission range and mobility constraints. In this regard, node characteristics such as initial energy (joules), transmission energy (joules), receiving energy (joules) and node coverage abilities (meters) are differently configured for each sensor node in the network (Table 6).

Table 6 illustrates the implementation details of WSNs in the NS-3.0 environment. Consequently, the supportive packages of Python and C++ were used to implement the existing techniques. The performance metrics for evaluating the exiting techniques illustrated in Table 4 (E1, E2, E3, E4, E5, E6, E7 and E8) are average energy consumption rate (joules), liveliness rate, successful data delivery rate, number of retransmissions (count), computational overhead (%), routing delay (milliseconds), scheduling time (milliseconds), packet drops due to downtime (count) and energy optimization rate. The definition of each performance metric is given as follows:

- Average energy consumption rate (AECR): The average amount of joules spent by a sensor node throughout data transmission, collection and idle listening modes.
- Liveliness rate: The availability rate of active appearance (data transmission, collection and idle listening) made by each sensor node against expected lifetime.
- Successful data delivery rate (SDDR): The ratio between the quantity of packets delivered successfully by a sensor node against the packets dropped by the node.

- Number of retransmissions: Total number of packets retransmitted by a sensor node during a simulation cycle.
- Computational overhead: The excessive amount of packets (control messages, retransmitted data and other recovery messages) processed against the average number of network packets processed.
- Routing delay: The time taken by the routing protocol to find the optimal route and deliver the data to the destination.
- Scheduling time: Time taken by WMAC scheduler to make the unique channel period for sending the data through the multiple access medium.
- Packet drops due to downtime (PDD): Number of packets dropped by an inactive node due to failure.
- Energy optimization rate (EOR): The ratio between the amount of joules spent using the energy optimization policies and the amount of joules spent without using optimization policies.

Table 6. The environment of WSN.

No.	Configuration Parameters	Magnitudes
1	Initial energy (joules(J))	2.56–2.67 (variable)
2	Transmission energy (J)	1.06–1.16 (variable)
3	Receiving energy (J)	0.87–0.99 (variable)
4	Number of sensor nodes	100, 200, 300 (variable)
5	Antenna type	Omnidirectional
6	Channel and propagation	Wireless (air/water), two-ray
7	Throughput level (Kbps)	200–250 (variable)
8	Node's coverage ability (meters (m))	30, 40, 50, 60 (variable)
9	Routing protocol	AODV and AODV-LS
10	Signaling modes	Electromagnetic and acoustic
11	Mobility ranges (meters/second (m/s))	15, 25, 35, 45 (variable)
12	Simulation cycle time	100 s
13	WMAC	IEEE 802.11-CSMA-CA

The existing works are investigated under variable constraints as given in Table 5. According to that base, the routing protocols are chosen as AODV and AODV–link-state (LS) models. Similarly, the traffic characteristics, mobility, coverage and energy levels of sensor nodes are configured as variable at different sensor nodes. In addition, the signaling models configured in this experiment are built with the functionalities of both electromagnetic and acoustic nature (underwater/underground).

The experiment starts with the performance validations of the existing techniques (Table 4) using the metric AECR of each sensor node. In this experiment, the cross-layered techniques (WMAC and routing) consume minimal AECR compared with other single-layered solutions. In this concern, the observation has been conducted with the minimal AECR of E3, E4 and E5. The existing works E3, E4 and E5 consider the impactful factors of medium, power limitations and equilibrium in load distribution. As these works are developed to consider multi-layered network functions (MAC and routing issues) with evenly distributed load management policies, they produce optimal AECR between 1.75 J and 1.85 J. At the same time, the AECR of E2, E6 and E7 fall closely under E1. The reason behind this observation is that the specified approaches consider energy optimization as the main problem. In this regard, E2 used heuristic MAC management strategies to effectively organize the data transmission slots (sleep and wake up node selection). In contrast, E6 and E7 focused on balanced load management on routing procedures to reduce the AECR.

In this observation, each existing technique initiated various energy optimization solutions regarding MAC principles or routing protocols. At any rate, the successful engagement of MAC and routing protocol principles of E3 assures minimal AECR. At the same time, other techniques experience a slight hike in AECR where the number of sensor nodes is 300. The variations among the techniques are not huge in AECR, yet the range between 1.57 J and 1.95 J shows significant impacts in resource-limited sensor nodes and network lifespan reduction (Figure 8).

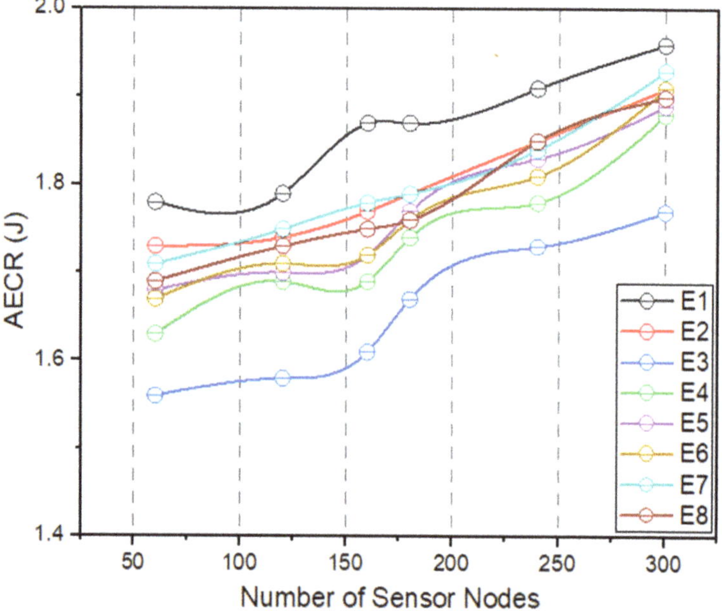

Figure 8. AECR and number of sensor nodes.

Figure 9 describes the average liveliness (active) rate of sensor nodes around the WSN. As the network population is increased in the prescribed geographical area (1000 m × 1000 m) due to the increasing number of nodes, the frequency of a node's activity increases to manage the neighbor discovery process, data transmission, data collection, route updating process and idle listening process. Hence, each sensor node consumes more energy to accomplish the requested tasks and gradually falls at the critical stage of residual energy.

At this point, the need for energy optimization techniques is essential to keep the node live to handle the data transmissions in a dense network field. The illustration given in Figure 9 implicates the fall of the average active conditions of sensor nodes in the network. The proven performance of cross-layered techniques (E3, E4 and E5) shows a better liveliness rate from 0.93 to 0.85 as the number of sensor nodes increases from 50 to 300.

In this experiment, E3, E4 and E5 diversely manage their MAC principles by considering channel quality metrics and network dynamics. Significantly, E3 managed both timeline-based channel allocation and route consistencies throughout the increasing number of sensor nodes. In the same way, E5 developed the load distribution and aggressive scheduling procedures. Based on these reasons, E3 and E5 compactly maintained the overall node liveliness rate better than other works. In the next level, E4 achieved an even better liveliness rate (0.83) under a highly populated WSN (300 sensor nodes). In this experiment, other techniques found active node selection procedures (E2-0.81 and E6-0.82) and load-optimized energy control procedures, respectively. This kind of practice improves network lifetime and the active state of nodes (liveliness rate). By contrast, other existing techniques such as E1, E7 and E8 hold the sensor nodes in an active condition for a more limited period of time than the expected case (20 to 25% of limited downtime). These existing techniques mainly concentrate on the WMAC-based energy efficiency than the node's overall behaviors (routing, advertising, discovery and listening).

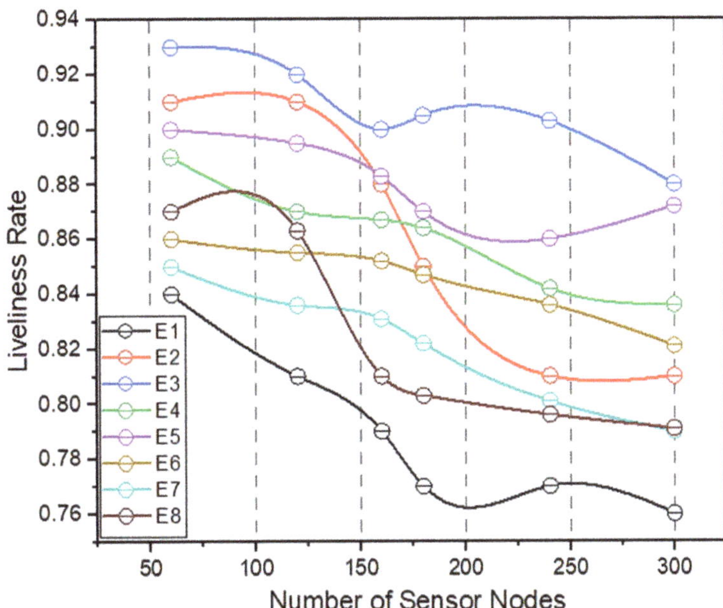

Figure 9. Average liveliness rate of sensor nodes.

From the observations, this article classifies the cross-layered energy-efficient strategies from other techniques, as given in Figure 10. The performance of each cross-layered technique is evaluated using SDDR for multiple data sessions. As denoted, the data communication sessions are populated (20 to 220) and executed for multiple test cycles. Let us assume the number of sessions increase as the number of sensor nodes increases to handle the increasing rate of throughput. In this comparison, the integrated principles of WMAC, routing assistance and scheduler policies of E3 ensure a better SDDR than other works. In any event, E4 manages the SDDR at a higher rate than E3 (94.7%) during initial sessions. The priority calculation and QoS management tasks of E4 give optimal results in SDDR around the network. In contrast, the stability in data delivery needs proper organization of routing strategies at any cost. Accordingly, E3 attains optimal SDDR during the moments of more populated sessions than other works.

During this moment, the performance of E5 starts at the lowest SDDR (94.4%) during initial sessions and manages with the average performance between E3 and E4 as it is modelling both balanced power assumptions and balanced scheduling possibilities. Finally, E5 achieves the SDDR of 93.5% against densely populated sessions. Consequently, the changes in the number of retransmissions are crucially noted for the existing techniques such as E3, E4 and E5. The number of retransmissions is closely connected with SDDR and the node's liveliness rate. The SDDR is indirectly proportional to the retransmission rate and directly proportional to the liveliness rate. According to that, E3 produces the minimal number of data retransmissions from 45 to 95 (number of packets) as the increasing number of sessions at each test cycle. In the same manner, the retransmission rate of E4 reaches a higher point for the maximum number of sessions (116). Similarly, E5 produces a moderate load in data retransmission compared to other works (between 55 and 105). The implications of data retransmission for varying number of transmission sessions are given in Figure 11. In the same manner, E6 shows notable contributions in both routing and WMAC policies. E6 achieves balanced routing and medium management policies in WSN. In this case, the observations of E6 performance using AECR, liveliness rate and SDDR are denoted in Figures 8 and 9. The effort of E6 on these metrics is crucial after the implications of E3, E4 and E5.

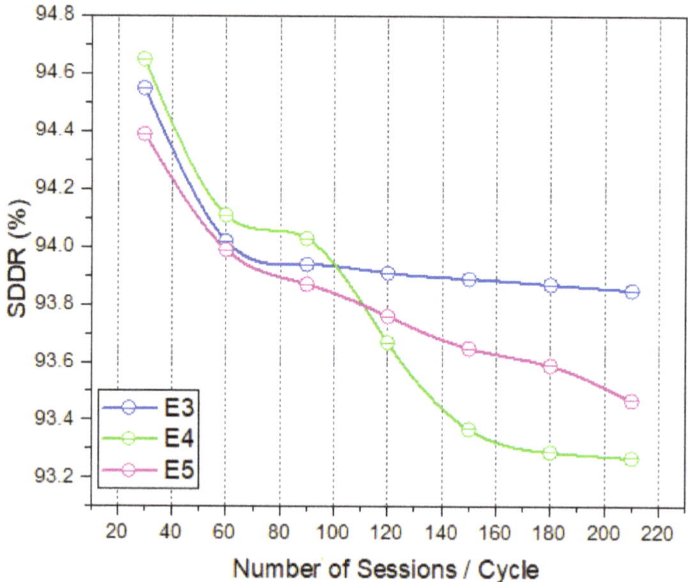

Figure 10. SDDR during the sessions.

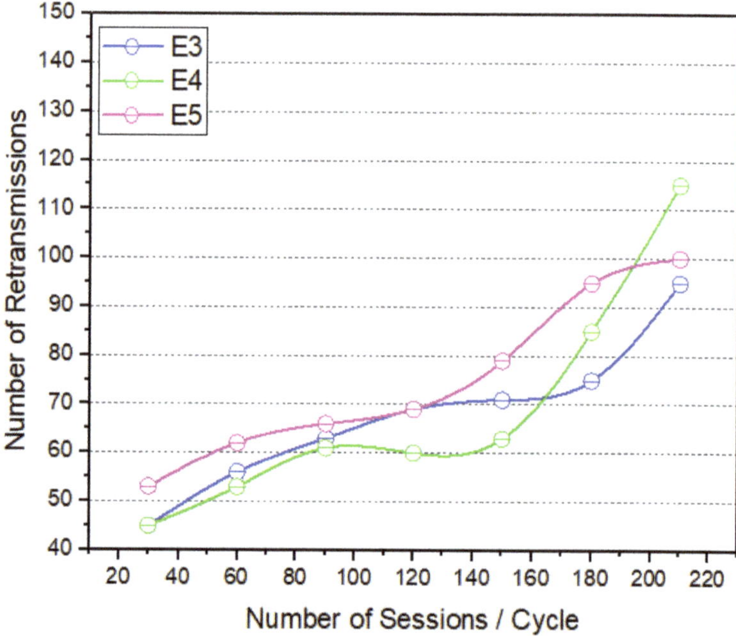

Figure 11. Total number of retransmissions.

With this in mind, Figure 12 shows the comparative cases of E3, E6, E7 and E8 with respect to computation overhead. In this association, the assumption on the computation overhead of the existing techniques E1, E2, E4 and E5 are directly mapped with the number of retransmissions taken for each session. In any event, the comparison given in Figure 12 relates the existing techniques E3, E6, E7 and E8 in terms of routing behaviors. These techniques commonly achieve energy-optimized routing solutions in WSNs. The efficient and energy-optimized routing models need to produce minimal computation overhead irrespective of the node's location. Additionally, the processes involved in each sensor node vary depends upon the mobility of sensor node.

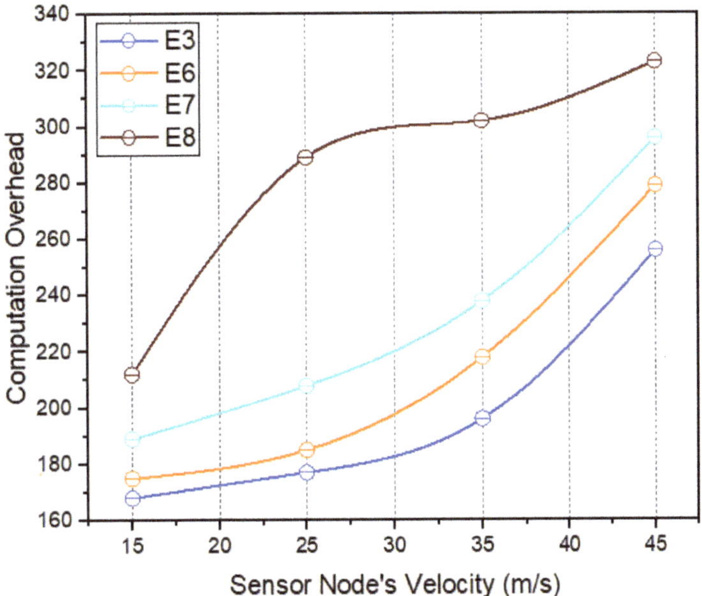

Figure 12. Overall computation overhead.

The challenges taken under mobile sensor nodes lead us to solve new problems such as link breaks, node failures, data retransmission, frequent route updates and energy loss. In this preference, the computation overhead of related routing principles are denoted in Figure 12 against the changing velocity of sensor nodes. As the sensor nodes' velocity changes from 15 m/s to 45 m/s, the excessive load initiated in the sensor node's processor slowly increases and it increases energy AECR definitively. The comparison with energy-optimized routing practices is noted in terms of the excessive number of processes raised in the distributed environment. As discussed, the additional processes in each sensor node are created as a result of managing real-time issues such as node failures, link breaks, excessive route updates, etc.

At this point, the computation overhead of E8 varies from 210 packets to 321 packets over the changing velocity of sensor nodes. Similarly, other techniques such as E7 have 290 packets as the computation load in the front end for a maximum velocity of 45 m/s. On the other hand, E6 produces a rate between 175 and 270 of overhead. Among these techniques, E8 was developed for the hierarchical routing model and it is not flexible with random networks. Hence, an excessive load in E8 is observed. In this case, E6 and E7 target load balance and cooperative routing procedures, respectively. The comparison between E6 and E7 shows the better contribution of the cooperative routing technique (E7) in computation reduction. As E7 uses an authentic and distributed cooperative model for computing twin-trust costs (local trust value and global trust value), the elimination of irregular nodes is easy in the network. In addition, the excessive packet transmission to unethical nodes or inefficient nodes is ignored in the network. At any rate, the performance of E6 directly deals with low-powered computation procedures in order to limit transmissions.

In this manner, the technical pitches and purposes used for establishing these existing techniques make notable performance variations. In this comparison, E3 falls in both channel and efficient routing practices in order to reduce the overloaded tasks at different layers. Accordingly, the production of computation overhead in E3 is maintained between 170 and 250 which is the minimum compared to other techniques. The experimental contributions of this article are extended to analyze the practical betterment of routing delay (routing protocol) and scheduling time (channel allotment). The motive of understanding the routing-based research techniques and channel slot management techniques leads to separate comparative observations.

Hence, the implications are noted as given in Tables 7 and 8, respectively. Table 7 shows the performance of energy-efficient routing strategies (E3, E6, E7 and E8). As discussed, E3, E6, E7 and E8 mainly focus on the implementation of energy-efficient routing protocols, and routing delay calculation is an important task. With this in mind, the successful elimination of overhead and retransmission leads to minimal routing delay (milliseconds (msec)). Thus, the performance of E3 is optimal in terms of routing delay production (maximum 553 msec) during the change in the node's velocity. In this experiment, this article takes the node's velocity as changing parameter to validate the routing delay. Since the dynamic velocity rate (meter/seconds, m/s) increases the possibilities of node failures and link breaks continuously in the network, the routing delay produced from each protocol varies rapidly. In this case, E6 produces a moderate routing delay compared to other techniques such as E7 and E8. The existing techniques E7 and E8 deal with secure and hierarchical routing models; therefore, the production of routing delay is higher than E6 (load balanced and low-powered scheme).

Table 7. Routing delay.

Sensor Node's Velocity	E3	E6	E7	E8
15	458	472	499	523
25	499	518	539	557
35	514	544	569	589
45	553	573	592	603

Table 8. Scheduler time.

Sensor Node's Velocity	E1	E2	E3	E5
15	101	101	104	102
25	122	117	128	126
35	130	134	148	129
45	139	145	155	142

Similarly, Table 8 gives the identifications of channel scheduling strategies (E1, E2, E3 and E5) and their performance. As E3 is noted, it is a cross-layered solution for energy optimization in WSN, and it is not effective in scheduling processes. At the same time, the development of E1, E2 and E5 are channel allocation and timeline scheduling tasks to reduce frame allocation time on the channel. In this regard, E1 and E2 effectively process the timeline slot for scheduling through modified MAC policies and memetic–heuristic scheduling policies, respectively. These techniques perform better than aggressive scheduling policies in dynamic WSNs (E5).

Table 9 illustrates the routing benefits of using a standard AODV protocol and hybrid AODV protocol with link-state features. This work identifies the limitations of standard AODV as frequent updates and overhead during network changes. Generally, an AODV protocol performs optimally for reactive updates yet takes maximum overhead for a more dynamic network. At the same time, AODV takes maximum time to update the route information for the whole network regularly. The implementation of both link-state routing models and distance vector models gives fast route updates and reactive route updates, respectively. Especially, the varying velocity of sensor nodes affects the established wireless links frequently. In this case, the uni-protocol system struggles to initiate either global updates or frequent neighbor updates. However, the idea behind AODV-LS supports both reactive and proactive updates (global updates and frequent neighbor updates).

Table 9. Routing strategies.

Sensor Node's Velocity	R1 [AODV]	R2 [AODV-LS]
15	169	125
25	190	144
35	215	151
45	246	179

Figure 13 compares the efforts of all existing techniques using the metric as the number of packets dropped due to energy shortage (downtime) at a sensor node. This is an important evaluation that relates the amount of residual energy maintained by the node and the active participation of that node itself in data communication. Apart from the continuous oscillations in packet drops produced by each technique, the optimal results are observed for E1, E3, E4 and E5 under various test cycles. As noted in Figure 13, the existing techniques E3 and E4 attain the drops between 35 and 60 as optimal compared to other techniques. At the end, the EOR was taken for performance analysis of all experimented techniques. According to this observation, the higher rates of E2, E3 and E4 show the suitable nature of energy optimization in WSN. On the scope of energy-efficient routing methodologies, many recent works are developed in the research arena [89,90].

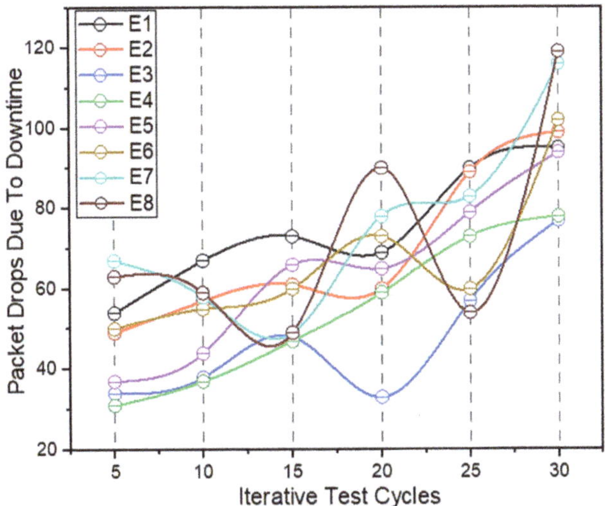

Figure 13. Total number of packets dropped due to sensor node's downtime.

Finally, the practical evaluations are crucially mounting on the considerations of packet drops due to the node's downtime and EOR (Table 10). EOR is the definite attainment of each work towards an energy optimization goal. In this experiment, better energy harvesting (saving) solutions are produced from E3, E4 and E5. Particularly, the effective reduction in computation overhead and retransmission rate gives optimal attainment in EOR. Thus, the EOR of E3 varies between 0.365 and 0.467 during various iterative simulations. In the same manner, E4 attains an EOR from 0.321 to 0.563 as it is using a channel adaptive quality evaluation procedure to initiate data transmission compared to E3. On the next level, E5 attains a better EOR (0.443) due to its load balancing principles compared to other works. In addition, the practical comparison between more recent energy-efficient routing protocols gives diversified solutions to the research community. In this regard, this proposed review article extends the evaluation of E9 [51], E10 [54], E11 [55] and E12 [56] under the considerations of energy optimization and multi-hop routing principles in WSNs.

Table 10. EOR.

Iterative Test Cycles	E1	E2	E3	E4	E5	E6	E7	E8
5	0.278	0.361	0.385	0.563	0.443	0.359	0.333	0.309
10	0.299	0.318	0.365	0.478	0.389	0.303	0.301	0.328
15	0.372	0.404	0.467	0.321	0.435	0.388	0.361	0.371
20	0.396	0.412	0.434	0.367	0.366	0.401	0.408	0.401
25	0.354	0.399	0.411	0.401	0.301	0.382	0.357	0.366
30	0.267	0.358	0.397	0.387	0.318	0.298	0.284	0.299

The most recent ideologies on novel energy-optimized routing protocol development and relevant discussions lead to future-generation WSN energy models [91,92]. In this concern, E9 developed swarm-intelligence-based chimp optimization solutions and hunger game searching principles to find energy-efficient multi-hop routing paths. This work followed the natural habits of chimps to optimize path-finding problems with minimal overhead. Particularly, the first phase of the chimp optimization algorithm initiated the network formation under a hierarchical structure (base station and clusters). The clusters were formed as driving nodes, chaser nodes, barrier nodes and attacker nodes using chaotic cost computations. In the same manner, the second phase of this work provided a hunger search-based path selection approach for initiating an energy-efficient multi-hop routing process in underwater sensor networks.

E10 proposed energy classification and channel assessment techniques using a greedy approach for minimizing the overload of the node's energy resources. According to the strategies, each sensor node's energy levels are monitored with adaptive internal buffer management policies on the reception of data packets. Similarly, the routing protocol used for this mechanism found the greedy-based route selection with sufficient node resources to avoid packet losses. In this connection, E11 and E12 considered coverage problems and locality problems, respectively. Particularly, E11 proposed link stability evaluation protocols and grid-level stimulated network models to achieve coverage optimization in WSNs. In this concern, Figure 14 illustrates the functions of link stability evaluation and coverage problem analysis models (holes or inactive nodes).

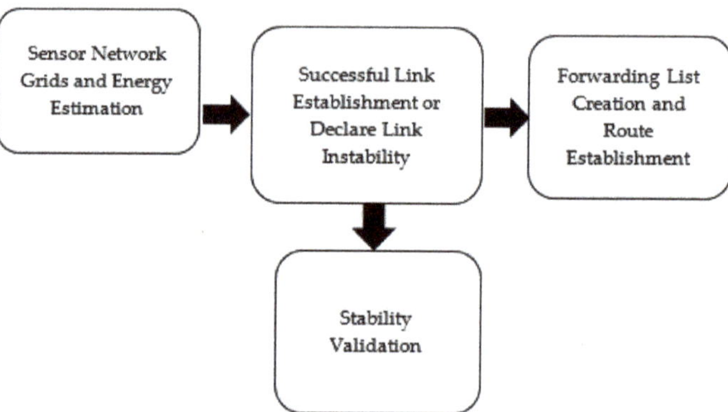

Figure 14. E11-Link stability management and routing.

Finally, E12 proposed zone-based routing protocols and energy optimization policies throughout the WSNs. In this concern, each sensor node was constructed under various zonal locations with allotted energy resources to be associated with other zonal sensor nodes. The proposed routing protocol running in each sensor node evaluated residual energies to proceed data routing into other zones. Thus, various recent routing protocols were proposed under energy consideration platforms. These works are compared as illustrates below.

Figure 15 depicts the performance comparison between E9, E10, E11 and E12 in terms of AECR against changing number of sensor nodes. In this experiment, four different types of energy-efficient routing protocols are compared. E9 has been experimented for underwater sensor networks. Compared to other sensor networks, underwater sensor networks use acoustic sensors with minimal energy resources. In this concern, E9 developed hierarchical energy optimization solutions and chimp optimization policies in order to search active nodes to enable flawless communication. In this case, the network clustering process and node searching processes consume significant energy (1.93 J). On the other hand, protocol implemented in E12 consumes maximum energy as it is related to zonal computational policies. At any rate, E10 and E11 are optimally designed for improving the quality of energy-saving mechanisms via the greedy approach and link stability validation approach.

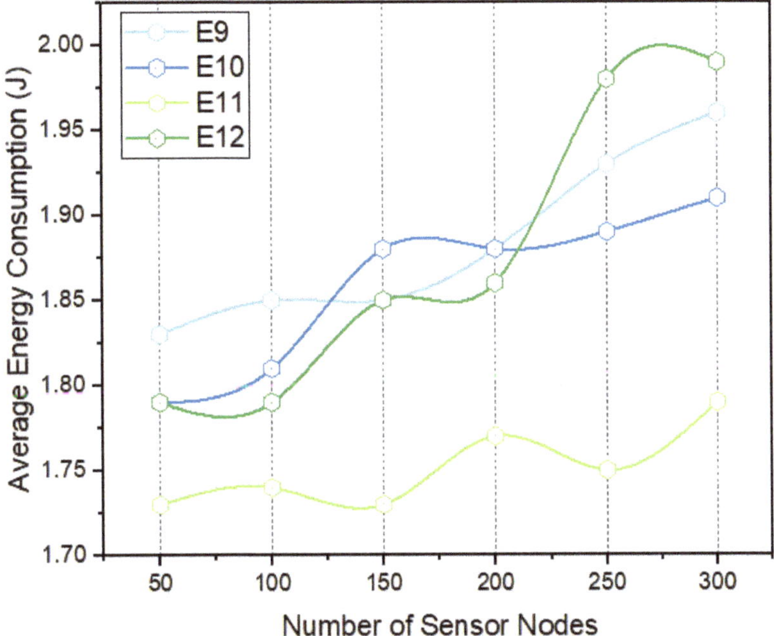

Figure 15. Energy-efficient routing schemes and AECR.

In this comparison, E11 ensures network coverage optimality and link stability concerns to operate successful data transmission from source to destination. In this analysis, E11 provides more stable and optimized channel circumstance to reduce packet drops and retransmissions. Thus, the AECR of E11 attains a minimal value (1.77 J). At the same time, E10 secures 1.88 J which is better than other works such as E9 and E12. Similarly, Figure 16 shows the calculation of the average routing delay in milliseconds.

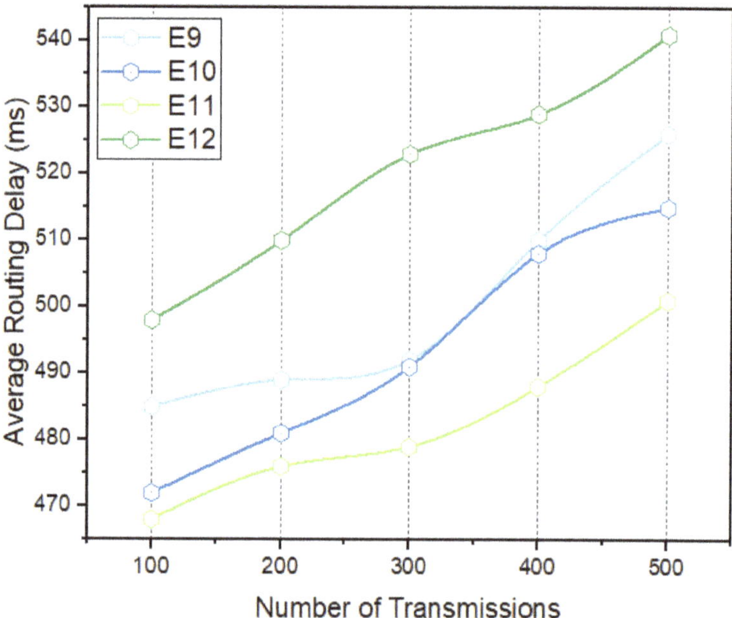

Figure 16. Energy-efficient routing schemes and Delay.

The routing delay of E10 and E11 are minimal compared to E9 and E12. The category of E9 and E12 fall under clustered or zonal network architecture. Under these network management policies, the route construction to deliver the data from source to destination has to follow clustered or zonal rules. This makes the routing delay a little higher than flat network architectures (E10 and E11). In this regard, E11 strongly builds stable links and coverage assurance in the WSN for data transmission. Once this process has been successfully developed, it reduces the routing delay during data transmission. Thus, it minimizes the delay (495 ms). In the same manner, E10 produces 510 ms of routing delay, which still better than E9 and E12. Figure 17 illustrates the computation overhead of E9, E10, E11 and E12. As discussed, E9 and E12 produce more computation overhead (additional packet transmission) than E10 and E11. The observed results of E9 (145) and E12 (170) are closely related to AECR produced by each technique.

In contrast, E10 and E11 managed the computation overhead with minimal rates compared to E9 and E12. The reason for the optimized overhead of E10 and E11 is the energy stability and link stability in the network. In this manner, E10 produces 140 additional transmissions and E12 produces 110 additional transmissions in order to stabilize network communications. The observations are gathered against the number of actively participating sensor nodes for each iteration. Finally, Figure 18 illustrates the average packet delivery ratio (PDR) attained by each work. In this experiment, E10 and E11 obtained 0.93 to 0.94 of average PDR for a maximally populated network (300 sensor nodes). On the other hand, E9 and E12 sent packets at the PDR of 0.9. In general, the difference in PDR produced by each system is not crucial, yet the delay and energy consumptions significantly vary due to network changes. Hence, the comprehensive experimental analysis and technical discussion described in this article clarify the crucial efforts of energy-efficient WMAC policies and routing strategies to extend the lifetime of sensor nodes.

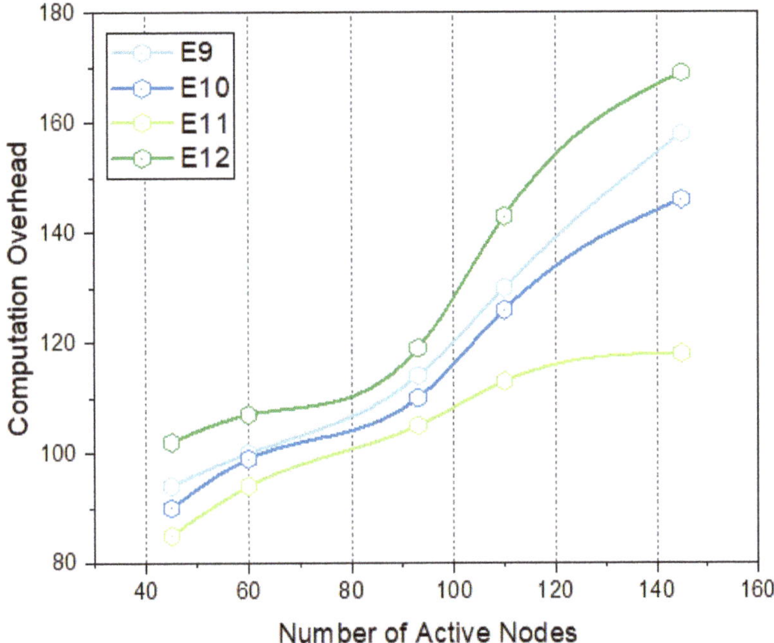

Figure 17. Energy-efficient routing schemes and Overhead.

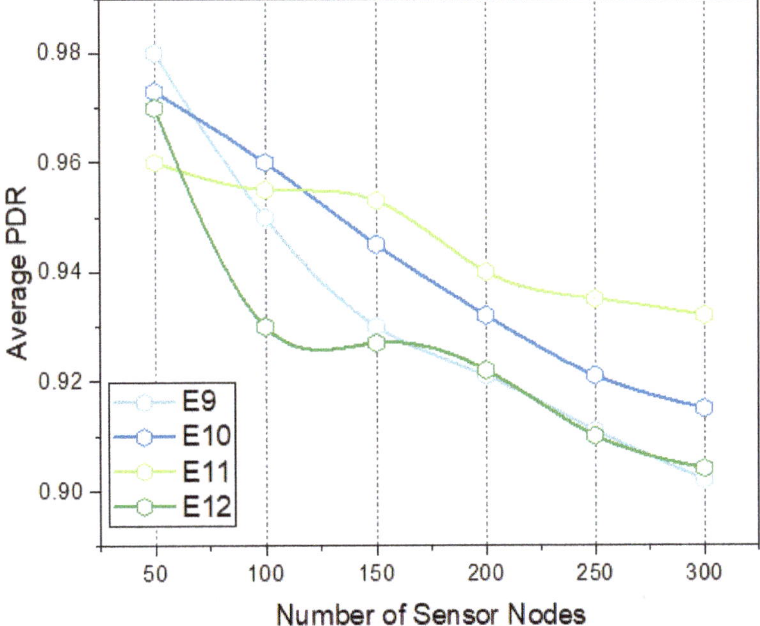

Figure 18. Energy-efficient routing schemes and Average PDR.

4. Conclusions

Wireless sensor nodes are known for their limited resources such as local memory, processor and other internal components. The deployment strategy of the WSN changes the need for unique energy-saving plans and lifetime saving plans. On this basis, this article has conducted a deep study on WMAC policies, routing protocols and energy optimization solutions. This article found the importance of energy-efficient processes under MAC policies, channel allotment policies and network routing protocols to attain novel growth towards the green computing era. As the applications of WSNs and IoT environments are widely growing around the world, energy consumption and feasible data communication practices are surely expected. Accordingly, this article technically and practically compared the recent contributions of various articles to ensure optimal energy consumption plans in WSNs. This article reviewed various energy-optimized routing protocols, load balancing approaches, variants of the WMAC protocol and data scheduling algorithms under an experimental testbed. The experimental evaluation of the existing energy optimization solutions and load balancing solutions (E1, E2, E3, E4, E5, E6, E7, E8, E9, E10, E11 and E12) against network dynamics (number of nodes, mobility, session, etc.) clarified the contributions of each work. Notably, this experimental survey found a better performance of multi-layer energy optimization policies (MAC and routing protocol) and routing policies than single-layer optimization solutions. From this experimental survey, the research community can understand the practical limitations and benefits of the mentioned techniques. In addition, the future world is in need of tiny yet efficient network participants (nodes) with the hope of safe energy plans. In particular, this article gives a future direction for researchers to build energy-optimized sensor nodes.

Author Contributions: Conceptualization, D.D.; Supervision, R.S. and S.B.G.; Validation, A.S.R. and S.B.G.; Writing—original draft, P.S. and M.S.B.; Writing—review & editing, M.S.R., T.C.M., C.V. and G.S. All authors have read and agreed to the published version of the manuscript.

Funding: This research received no external funding.

Data Availability Statement: Not applicable.

Acknowledgments: This paper was partially supported by UEFISCDI Romania and MCI through BEIA projects NGI-UAV-AGRO, SOLID-B5G, AISTOR, IPSUS, iPREMAS, ENTA, STACK, IMMI-NENCE and by European Union's Horizon 2020 research and innovation program under grant agreement No. 883522 (S4ALLCITIES) and No. 101008060 (RRREMAKER).

Conflicts of Interest: The authors declare no conflict of interest.

References

1. Cardei, M.; MacCallum, D.; Cheng, M.X.; Min, M.; Jia, X.; Li, D.; Du, D.-Z. Wireless sensor networks with energy efficient organization. *J. Interconnect. Netw.* **2002**, *3*, 213–229. [CrossRef]
2. Ekici, E.; Gu, Y.; Bozdag, D. Mobility-based communication in wireless sensor networks. *IEEE Commun. Mag.* **2006**, *44*, 56–62. [CrossRef]
3. Zhang, S.; Zhang, H. A review of wireless sensor networks and its applications. In Proceedings of the 2012 IEEE International Conference on Automation and Logistics, Zhengzhou, China, 15–17 August 2012; pp. 386–389.
4. Yick, J.; Mukherjee, B.; Ghosal, D. Wireless sensor network survey. *Comput. Netw.* **2008**, *52*, 2292–2330. [CrossRef]
5. Hou, Y.T.; Shi, Y.; Sherali, H.D. Rate allocation and network lifetime problems for wireless sensor networks. *IEEE ACM Trans. Netw.* **2008**, *16*, 321–334. [CrossRef]
6. Wei, X.; Guo, H.; Wang, X.; Wang, X.; Qiu, M. Reliable data collection techniques in underwater wireless sensor networks: A survey. *IEEE Commun. Surv. Tutor.* **2021**, *24*, 404–431. [CrossRef]
7. Boubiche, D.E.; Athmani, S.; Boubiche, S.; Toral-Cruz, H. Cybersecurity issues in wireless sensor networks: Current challenges and solutions. *Wirel. Pers. Commun.* **2021**, *117*, 177–213. [CrossRef]
8. Bashar, A.; Smys, S. Physical layer protection against sensor eavesdropper channels in wireless sensor networks. *IRO J. Sustain. Wirel. Syst.* **2021**, *3*, 59–67. [CrossRef]
9. Cao, L.; Yue, Y.; Cai, Y.; Zhang, Y. A novel coverage optimization strategy for heterogeneous wireless sensor networks based on connectivity and reliability. *IEEE Access* **2021**, *9*, 18424–18442. [CrossRef]

10. Kotiyal, V.; Singh, A.; Sharma, S.; Nagar, J.; Lee, C.C. ECS-NL: An enhanced cuckoo search algorithm for node localisation in wireless sensor networks. *Sensors* **2021**, *21*, 3576. [CrossRef] [PubMed]

11. Nain, M.; Goyal, N. Energy efficient localization through node mobility and propagation delay prediction in underwater wireless sensor network. *Wirel. Pers. Commun.* **2022**, *122*, 2667–2685. [CrossRef]

12. Ullah, F.; Khan, M.Z.; Mehmood, G.; Qureshi, M.S.; Fayaz, M. Energy Efficiency and Reliability Considerations in Wireless Body Area Networks: A Survey. *Comput. Math. Methods Med.* **2022**, *2022*, 1–15. [CrossRef]

13. Mishra, P.; Kumar, N.; Godfrey, W.W. An evolutionary computing-based energy-efficient solution for IoT-enabled software-defined sensor network architecture. *Int. J. Commun. Syst.* **2022**, *35*, e5111. [CrossRef]

14. Nweye, K.; Nagy, Z. MARTINI: Smart meter driven estimation of HVAC schedules and energy savings based on Wi-Fi sensing and clustering. *Appl. Energy* **2022**, *316*, 118980. [CrossRef]

15. Chandra, S.; Gupta, R.; Ghosh, S.; Mondal, S. An intelligent and power efficient biomedical sensor node for wireless cardiovascular health monitoring. *IETE J. Res.* **2022**, *68*, 456–466. [CrossRef]

16. Gupta, S.; Gupta, S.; Goyal, D. Wireless Sensor Network in IoT and Performance Optimization. *Recent Adv. Comput. Sci. Commun.* **2022**, *15*, 14–22. [CrossRef]

17. Ijemaru, G.K.; Ang, L.M.; Seng, K.P. Transformation from IoT to IoV for waste management in smart cities. *J. Netw. Comput. Appl.* **2022**, *204*, 103393. [CrossRef]

18. Dhaya, R.; Kanthavel, R. Energy Efficient Resource Allocation Algorithm for Agriculture IoT. In *Wireless Personal Communications*; Springer: Berlin/Heidelberg, Germany, 2022; pp. 1–23.

19. Sahu, S.; Silakari, S. Energy Efficiency and Fault Tolerance in Wireless Sensor Networks: Analysis and Review. In *Soft Computing: Theories and Applications*; Springer: Berlin/Heidelberg, Germany, 2022; pp. 389–402.

20. Humayun, M.; Alsaqer, M.S.; Jhanjhi, N. Energy Optimization for Smart Cities Using IoT. *Appl. Artif. Intell.* **2022**, *36*, 1–7. [CrossRef]

21. Zhu, S.; Ota, K.; Dong, M. Energy-Efficient Artificial Intelligence of Things With Intelligent Edge. *IEEE Internet Things J.* **2022**, *9*, 7525–7532. [CrossRef]

22. Vashisht, G. ML Algorithms for Smart Sensor Networks. In *Smart Sensor Networks*; Springer: Cham, Switzerland, 2022; pp. 73–103.

23. Nayak, P.; Swetha, G.K.; Gupta, S.; Madhavi, K. Routing in wireless sensor networks using machine learning techniques: Challenges and opportunities. *Measurement* **2021**, *178*, 108974. [CrossRef]

24. Bhargava, D.; Prasanalakshmi, B.; Vaiyapuri, T.; Alsulami, H.; Serbaya, S.H.; Rahmani, A.W. CUCKOO-ANN based novel energy-efficient optimization technique for IoT sensor node modelling. *Wirel. Commun. Mob. Comput.* **2022**, *2022*, 1–9. [CrossRef]

25. Haseeb, K.; Rehman, A.; Saba, T.; Bahaj, S.A.; Lloret, J. Device-to-device (D2D) multi-criteria learning algorithm using secured sensors. *Sensors* **2022**, *22*, 2115. [CrossRef] [PubMed]

26. Geetha, B.T.; Kumar, P.S.; Bama, B.S.; Neelakandan, S.; Dutta, C.; Babu, D.V. Green energy aware and cluster based communication for future load prediction in IoT. *Sustain. Energy Technol. Assess.* **2022**, *52*, 102244. [CrossRef]

27. Ren, X.; Vashisht, S.; Aujla, G.S.; Zhang, P. Drone-edge coalesce for energy-aware and sustainable service delivery for smart city applications. *Sustain. Cities Soc.* **2022**, *77*, 103505. [CrossRef]

28. Richert, V.; Issac, B.; Israr, N. Implementation of a modified wireless sensor network MAC protocol for critical environments. *Wirel. Commun. Mob. Comput.* **2017**, *2017*, 1–23. [CrossRef]

29. Jain, D.; Shukla, P.K.; Varma, S. Energy efficient architecture for mitigating the hot-spot problem in wireless sensor networks. *J. Ambient Intell. Humaniz. Comput.* **2022**, 1–8. [CrossRef]

30. Chawra, V.K.; Gupta, G.P. Memetic algorithm based energy efficient wake-up scheduling scheme for maximizing the network lifetime, coverage and connectivity in three-dimensional wireless sensor networks. *Wirel. Pers. Commun.* **2022**, *123*, 1507–1522. [CrossRef]

31. Alzahrani, E.; Bouabdallah, F.; Almisbahi, H. State of the Art in Quorum-Based Sleep/Wakeup Scheduling MAC Protocols for Ad Hoc and Wireless Sensor Networks. *Wirel. Commun. Mob. Comput.* **2022**, *2022*, 1–33. [CrossRef]

32. Uddin, M.N.; Rahman, M.O.; Kazary, S. A predictive schedule based energy efficient MAC protocol for wireless sensor networks. In Proceedings of the 5th International Conference on Computing and Informatics (ICCI), New Cairo, Egypt, 9–10 March 2022; pp. 424–429.

33. Khan, M.N.; Rahman, H.U.; Khan, M.Z.; Mehmood, G.; Sulaiman, A.; Shaikh, A.; Alqhatani, A. Energy-Efficient Dynamic and Adaptive State–based Scheduling (EDASS) Scheme for Wireless Sensor Networks. *IEEE Sens. J.* **2022**, *22*, 12386–12403. [CrossRef]

34. Rana, B.; Singh, Y. Duty-Cycling Techniques in IoT: Energy-Efficiency Perspective. In *Recent Innovations in Computing*; Springer: Singapore, 2022; pp. 505–512.

35. Ranjan, R.; Debasis, K.; Gupta, R.; Singh, M.P. Energy-Efficient Medium Access Control in Wireless Sensor Networks. *Wirel. Pers. Commun.* **2022**, *122*, 409–427. [CrossRef]

36. Villordo-Jimenez, I.; Torres-Cruz, N.; Menchaca-Mendez, R.; Rivero-Angeles, M.E. A Scalable and Energy-Efficient MAC Protocol for Linear Sensor Networks. *IEEE Access* **2022**, *10*, 36697–36710. [CrossRef]

37. Alablani, I.A.; Arafah, M.A. EE-UWSNs: A Joint Energy-Efficient MAC and Routing Protocol for Underwater Sensor Networks. *J. Mar. Sci. Eng.* **2022**, *10*, 488. [CrossRef]

38. Samal, T.; Kabat, M.R. Energy-Efficient Time-Sharing Multichannel MAC Protocol for Wireless Body Area Networks. *Arabian J. Sci. Eng.* **2022**, *47*, 1791–1804. [CrossRef]

39. Sakib, A.N.; Drieberg, M.; Sarang, S.; Aziz, A.A.; Hang, N.T.; Stojanović, G.M. Energy-Aware QoS MAC Protocol Based on Prioritized-Data and Multi-Hop Routing for Wireless Sensor Networks. *Sensors* **2022**, *22*, 2598. [CrossRef] [PubMed]
40. Darabkh, K.A.; Zomot, J.N.; Al-qudah, Z.; Ala'F, K. Impairments-aware time slot allocation model for energy-constrained multi-hop clustered IoT nodes considering TDMA and DSSS MAC protocols. *J. Ind. Inf. Integr.* **2022**, *25*, 100243. [CrossRef]
41. Subramanyam, R.; Bala, G.J.; Perattur, N.; Kanaga, E.G. Energy Efficient MAC with Variable Duty Cycle for Wireless Sensor Networks. *Int. J. Electron.* **2022**, *109*, 367–390. [CrossRef]
42. DR, V. Novel approach for hybrid MAC scheme for balanced energy and transmission in sensor devices. *Int. J. Electr. Comput. Eng.* **2022**, *12*, 1003.
43. Ajmi, N.; Msolli, A.; Helali, A.; Lorenz, P.; Mghaieth, R. Cross-layered energy optimization with MAC protocol based routing protocol in clustered wireless sensor network in internet of things applications. *Int. J. Commun. Syst.* **2022**, *35*, e5045. [CrossRef]
44. Dhanvijay, M.M.; Patil, S.C. Energy aware MAC protocol with mobility management in wireless body area network. *Peer Peer Netw. Appl.* **2022**, *15*, 426–443. [CrossRef]
45. Ubrurhe, O.O.; Nlerum, P.A. Energy Efficient Stable-MAC Protocol for Wireless Body Area Network. *Acad. J. Comput. Inf. Sci.* **2022**, *1*, 50–57.
46. Famitafreshi, G.; Afaqui, M.S.; Melià-Seguí, J. Enabling Energy Harvesting-Based Wi-Fi System for an e-Health Application: A MAC Layer Perspective. *Sensors* **2022**, *22*, 3831. [CrossRef]
47. Udoh, E.; Getov, V. Layered-MAC: An Energy-Protected and Efficient Protocol for Wireless Sensor Networks. In *Mobile Wireless Middleware, Operating Systems and Applications*; Springer: Cham, Switzerland, 2022; pp. 45–61.
48. Sadeq, A.S.; Hassan, R.; Sallehudin, H.; Aman, A.H.; Ibrahim, A.H. Conceptual Framework for Future WSN-MAC Protocol to Achieve Energy Consumption Enhancement. *Sensors* **2022**, *22*, 2129. [CrossRef]
49. Lakshmi, M.; Prashanth, C.R. Throughput Improvement in Energy Efficient Heterogeneous Wireless Sensor Network. In *InICDSMLA 2020*; Springer: Singapore, 2022; pp. 17–34.
50. Sah, D.K.; Nguyen, T.N.; Cengiz, K.; Dumba, B.; Kumar, V. Load-balance scheduling for intelligent sensors deployment in industrial internet of things. *Clust. Comput.* **2022**, *25*, 1715–1727. [CrossRef]
51. Yang, Y.; Wu, Y.; Yuan, H.; Khishe, M.; Mohammadi, M. Nodes clustering and multi-hop routing protocol optimization using hybrid chimp optimization and hunger games search algorithms for sustainable energy efficient underwater wireless sensor networks. *Sustain. Comput. Inform. Syst.* **2022**, *35*, 100731. [CrossRef]
52. Zagrouba, R.; Kardi, A. Comparative study of energy efficient routing techniques in wireless sensor networks. *Information* **2021**, *12*, 42. [CrossRef]
53. Dogra, R.; Rani, S.; Babbar, H.; Krah, D. Energy-efficient routing protocol for next-generation application in the internet of things and wireless sensor networks. *Wirel. Commun. Mob. Comput.* **2022**, *2022*, 1–10. [CrossRef]
54. Hao, S.; Hong, Y.; He, Y. An Energy-Efficient Routing Algorithm Based on Greedy Strategy for Energy Harvesting Wireless Sensor Networks. *Sensors* **2022**, *22*, 1645. [CrossRef] [PubMed]
55. Juneja, S.; Kaur, K.; Singh, H. An intelligent coverage optimization and link-stability routing for energy efficient wireless sensor network. *Wirel. Netw.* **2022**, *28*, 705–719. [CrossRef]
56. Kumar, D.S.; Sundaram, S.S. Associative Zone Based Energy Balancing Routing for Expanding Energy Efficient and Routing Optimization Over the Sensor Network. *Wirel. Pers. Commun.* **2022**, *124*, 2045–2057. [CrossRef]
57. Navarro, M.; Liang, Y.; Zhong, X. Energy-efficient and balanced routing in low-power wireless sensor networks for data collection. *Ad Hoc Netw.* **2022**, *127*, 102766. [CrossRef]
58. Hajipour, Z.; Barati, H. EELRP: Energy efficient layered routing protocol in wireless sensor networks. *Computing* **2021**, *103*, 2789–2809. [CrossRef]
59. Angurala, M.; Singh, H.; Grover, A.; Singh, M. Testing Solar-MAODV energy efficient model on various modulation techniques in wireless sensor and optical networks. *Wirel. Netw.* **2022**, *28*, 413–425. [CrossRef]
60. Singla, R.; Kaur, N.; Koundal, D.; Lashari, S.A.; Bhatia, S.; Rahmani, M.K. Optimized energy efficient secure routing protocol for wireless body area network. *IEEE Access* **2021**, *9*, 116745–116759. [CrossRef]
61. Asqui, O.P.; Marrone, L.A.; Chaw, E.E. Multihop Deterministic Energy Efficient Routing Protocol for Wireless Sensor Networks MDR. *Int. J. Commun. Netw. Syst. Sci.* **2021**, *14*, 31. [CrossRef]
62. Yun, W.K.; Yoo, S.J. Q-learning-based data-aggregation-aware energy-efficient routing protocol for wireless sensor networks. *IEEE Access* **2021**, *9*, 10737–10750. [CrossRef]
63. Sharma, D.; Tomar, G.S. Energy Efficient Multitier Random DEC Routing Protocols for WSN: In Agricultural. *Wirel. Pers. Commun.* **2021**, *120*, 727–747. [CrossRef]
64. Suresh Kumar, K.; Vimala, P. Energy efficient routing protocol using exponentially-ant lion whale optimization algorithm in wireless sensor networks. *Comput. Netw.* **2021**, *197*, 108250. [CrossRef]
65. Huamei, Q.; Chubin, L.; Yijiahe, G.; Wangping, X.; Ying, J. An energy-efficient non-uniform clustering routing protocol based on improved shuffled frog leaping algorithm for wireless sensor networks. *IET Commun.* **2021**, *15*, 374–383. [CrossRef]
66. Almalki, F.A.; Ben Othman, S.; AAlmalki, F.; Sakli, H. EERP-DPM: Energy efficient routing protocol using dual prediction model for healthcare using IoT. *J. Healthc. Eng.* **2021**, *2021*, 9988038. [CrossRef] [PubMed]
67. Rawat, P.; Chauhan, S. A survey on clustering protocols in wireless sensor network: Taxonomy, comparison, and future scope. *J. Ambient Intell. Humaniz. Comput.* **2021**, 1–47. [CrossRef]

68. Nguyen, N.T.; Le, T.T.; Nguyen, H.H.; Voznak, M. Energy-efficient clustering multi-hop routing protocol in a UWSN. *Sensors* **2021**, *21*, 627. [CrossRef] [PubMed]

69. Goswami, P.; Mukherjee, A.; Hazra, R.; Yang, L.; Ghosh, U.; Qi, Y.; Wang, H. AI based energy efficient routing protocol for intelligent transportation system. *IEEE Trans. Intell. Transp. Syst.* **2021**, *23*, 1670–1679. [CrossRef]

70. Khan, R.A.; Mohammadani, K.H.; Soomro, A.A.; Hussain, J.; Khan, S.; Arain, T.H.; Zafar, H. An energy efficient routing protocol for wireless body area sensor networks. *Wirel. Pers. Commun.* **2018**, *99*, 1443–1454. [CrossRef]

71. Srikanth, N.; Ashok, B.; Chandini, B.; Chandole, M.K.; Jyothi, N. Intelligent Routing Protocol for Energy Efficient Wireless Sensor Networks. In *International Conference on Electrical and Electronics Engineering*; Springer: Singapore, 2022; pp. 387–396.

72. Mohan, P.; Subramani, N.; Alotaibi, Y.; Alghamdi, S.; Khalaf, O.I.; Ulaganathan, S. Improved metaheuristics-based clustering with multihop routing protocol for underwater wireless sensor networks. *Sensors* **2022**, *22*, 1618. [CrossRef] [PubMed]

73. Yao, Y.D.; Li, X.; Cui, Y.P.; Wang, J.J.; Wang, C. Energy-Efficient Routing Protocol Based on Multi-Threshold Segmentation in Wireless Sensors Networks for Precision Agriculture. *IEEE Sens. J.* **2022**, *22*, 6216–6231. [CrossRef]

74. Kumar, A.; Sharma, S.; Goyal, N.; Gupta, S.K.; Kumari, S.; Kumar, S. Energy-efficient fog computing in Internet of Things based on Routing Protocol for Low-Power and Lossy Network with Contiki. *Int. J. Commun. Syst.* **2022**, *35*, e5049. [CrossRef]

75. Gayathri, A.; Prabu, A.V.; Rajasoundaran, S.; Routray, S.; Narayanasamy, P.; Kumar, N.; Qi, Y. Cooperative and feedback based authentic routing protocol for energy efficient IoT systems. *Concurr. Comput. Pract. Exp.* **2022**, *34*, e6886. [CrossRef]

76. Kardi, A.; Zagrouba, R. ESHARP: Energy-Efficient and Smart Hierarchical Routing Protocol Based on Smart Slumber for Wireless Sensor Networks. *Wirel. Pers. Commun.* **2022**, *123*, 1809–1824. [CrossRef]

77. Kim, W.; Umar, M.M.; Khan, S.; Khan, M.A. Novel Scoring for Energy-Efficient Routing in Multi-Sensored Networks. *Sensors* **2022**, *22*, 1673. [CrossRef]

78. Hussein, W.A.; Ali, B.M.; Rasid, M.F.; Hashim, F. Smart geographical routing protocol achieving high QoS and energy efficiency based for wireless multimedia sensor networks. *Egypt. Inform. J.* **2022**, *23*, 225–238. [CrossRef]

79. Monir, M.B.; Mohamed, A.A. Energy aware routing for wireless sensor networks. *Int. J. Commun. Netw. Inf. Secur.* **2022**, *14*, 545–550. [CrossRef]

80. Aljarrah, I.A.; Alshare, E.M. Improved Residual Dense Network for Large Scale Super-Resolution via Generative Adversarial Network. *Int. J. Commun. Netw. Inf. Secur.* **2022**, *14*, 118–124. [CrossRef]

81. Osama, I.; Rihan, M.; Elhefnawy, M.; Eldolil, S.; Abd El-Azem Malhat, H. A review on Precoding Techniques For mm-Wave Massive MIMO Wireless Systems. *Int. J. Commun. Netw. Inf. Secur.* **2022**, *14*. [CrossRef]

82. Tilwari, V.; Maheswar, R.; Jayarajan, P.; Sundararajan, T.V.P.; Hindia, M.N.; Dimyati, K.; Ojukwu, H.; Amiri, I.S. MCLMR: A Multicriteria Based Multipath Routing in the Mobile Ad Hoc Networks. *Wirel. Pers. Commun.* **2020**, *112*, 2461–2483. [CrossRef]

83. Paliwal, R.; Khan, I. Design and Analysis of Soft Computing Based Improved Routing Protocol in WSN for Energy Efficiency and Lifetime Enhancement. *Int. J. Recent Innov. Trends Comput. Commun.* **2022**, *10*, 12–24. [CrossRef]

84. Malathy, S.; Rastogi, R.; Maheswar, R.; Kanagachidambaresan, G.R.; Sundararajan, T.V.P.; Vigneswaran, D. A Novel Energy-Efficient Framework (NEEF) for the Wireless Body Sensor Network. *J. Supercomput.* **2019**, *76*, 6010–6025. [CrossRef]

85. Degambur, L.-N.; Mungur, A.; Armoogum, S.; Pudaruth, S. Resource Allocation in 4G and 5G Networks: A Review. *Int. J. Commun. Netw. Inf. Secur.* **2021**, *13*, 401–408. [CrossRef]

86. Han, B.; Ran, F.; Li, J.; Yan, L.; Shen, H.; Li, A. A Novel Adaptive Cluster Based Routing Protocol for Energy-Harvesting Wireless Sensor Networks. *Sensors* **2022**, *22*, 1564. [CrossRef] [PubMed]

87. Thirumoorthy, P.; Kalyanasundaram, P.; Maheswar, R.; Jayarajan, P.; Kanagachidambaresan, G.R.; Amiri, I.S. Time-critical energy minimization protocol using PQM (TCEM-PQM) for wireless body sensor network. *J. Supercomput.* **2019**, *76*, 5862–5872. [CrossRef]

88. Senthil, G.A.; Raaza, A.; Kumar, N. Internet of Things Energy Efficient Cluster-Based Routing Using Hybrid Particle Swarm Optimization for Wireless Sensor Network. *Wirel. Pers. Commun.* **2022**, *122*, 2603–2619. [CrossRef]

89. Mir, M.; Yaghoobi, M.; Khairabadi, M. A new approach to energy-aware routing in the Internet of Things using improved Grasshopper Metaheuristic Algorithm with Chaos theory and Fuzzy Logic. *Multimed. Tools Appl.* **2022**, 1–27. [CrossRef]

90. Altowaijri, S.M. Efficient Next-Hop Selection in Multi-Hop Routing for IoT Enabled Wireless Sensor Networks. *Future Internet* **2022**, *14*, 35. [CrossRef]

91. Shafi, I.; Ashraf, M.J.; Choi, G.S.; Din, S.; Ashraf, I. Intelligent autonomous underwater vehicle mobility with energy efficient routing in sensor networks. *Environ. Dev. Sustain.* **2022**, 1–3. [CrossRef]

92. Jayarajan, P.; Kanagachidambaresan, G.R.; Sundararajan, T.V.P.; Sakthipandi, K.; Maheswar, R.; Karthikeyan, A. An Energy Aware Buffer Management (EABM) Routing Protocol for WSN. *J. Supercomput.* **2018**, *76*, 4543–4555. [CrossRef]

Article

Energy-Aware UAV Based on Blockchain Model Using IoE Application in 6G Network-Driven Cybertwin

Atul B. Kathole [1], Jayashree Katti [1], Dharmesh Dhabliya [2], Vivek Deshpande [3], Anand Singh Rajawat [4], S. B. Goyal [5,*], Maria Simona Raboaca [6,7,8,*], Traian Candin Mihaltan [7], Chaman Verma [9] and George Suciu [10,*]

[1] Department of Information Technology, Pimpri Chinchwad College of Engineering, Pune 411044, India
[2] Department of IT, Vishwakarma Institute of Information Technology, Pune 411048, India
[3] Computer Engineering Department, Vishwakarma Institute of Information Technology, Pune 411048, India
[4] School of Computer Sciences & Engineering, Sandip University, Nashik 422213, India
[5] Faculty of Information Technology, City University, Petaling Jaya 46100, Malaysia
[6] ICSI Energy Department, National Research and Development Institute for Cryogenics and Isotopic Technologies, 240050 Ramnicu Valcea, Romania
[7] Faculty of Building Services, Technical University of Cluj-Napoca, 400114 Cluj-Napoca, Romania
[8] Doctoral School, University Politehnica of Bucharest, Splaiul Independentei Street, No. 313, 060042 Bucharest, Romania
[9] Faculty of Informatics, University of Eötvös Loránd, 1053 Budapest, Hungary
[10] R&D Department Beia Consult International Bucharest, 041386 Bucharest, Romania
* Correspondence: sb.goyal@city.edu.my (S.B.G.); simona.raboaca@icsi.ro (M.S.R.); george@beia.ro (G.S.)

Citation: Kathole, A.B.; Katti, J.; Dhabliya, D.; Deshpande, V.; Rajawat, A.S.; Goyal, S.B.; Raboaca, M.S.; Mihaltan, T.C.; Verma, C.; Suciu, G. Energy-Aware UAV Based on Blockchain Model Using IoE Application in 6G Network-Driven Cybertwin. *Energies* **2022**, *15*, 8304. https://doi.org/10.3390/en15218304

Academic Editors: R. Maheswar, M. Kathirvelu and K. Mohanasundaram

Received: 19 September 2022
Accepted: 28 October 2022
Published: 7 November 2022

Abstract: Several advanced features exist in fifth-generation (5G) correspondence than in fourth-generation (4G) correspondence. Centric cloud-computing architecture achieves resource sharing and effectively handles big data explosion. For data security problems, researchers had developed many methods to protect data against cyber-attacks. Only a few solutions are based on blockchain (BC), but are affected by expensive storage costs, network latency, confidence, and capacity. Things are represented in digital form in the virtual cyberspace which is the major responsibility of the communication model based on cybertwin. A novel cybertwin-based UAV 6G network architecture is proposed with new concepts such as cloud operators and cybertwin in UAV. Here, IoE applications have to be energy aware and provide scalability with less latency. A novel Compute first networking (CFN) framework named secure blockchain-based UAV communication (BC-UAV) is designed which offers network services such as computing, caching, and communication resources. The focus of the blockchain was to improve the security in the cloud using hashing technique. Edge clouds support core clouds to quickly respond to user requests.

Keywords: 6G; cloud-computing; cybertwin; blockchain; UAV; IoE; CFN; BC-UAV

1. Introduction

Efficient and effective options for developing intelligent cities can be found in unmanned aerial vehicles (UAVs). They are broadly employed in civil and military domains, including data collecting, distribution of data, audio and video monitoring, aerial images, crop surveys, and real-time medical services. [1]. Due to the COVID-19 pandemic situation, several businesses are forced to do online. Ericsson predicted that 5Gwill be rapidly commercialized as numerous individuals embrace this change. As a result, using the Internet has become essential for better connectivity in order to meet the demands and requirements of a more stringent network. This is essential to simplify the emerging techniques to extend the reality, tactile, related independent systems [2], telemedicine, and Industrial Internet of Things (IIoT) [3], which has an impact on latency and data speed has to be high. Low latency and higher data speed minimize collision rates and provide more secure autonomous vehicles.

These applications must necessarily provide smart autonomous life, multisensory virtual experience, smart agriculture, smart cities, and even more. Unfortunately, 5G networks are not efficient enough to meet these emerging demands [4]. Thus, there has aroused a necessity to develop efficient 6G wireless communication networks which can upgrade social needs, thereby enabling Sustainable Development Goals (SDGs) [5].

Networks of the next generation may have a demand with people and devices connected. Hence, the network architecture in the future has to overcome the increasing traffic in mobile Internet along with the services and applications by the use of heterogeneous networks. It is considered that the future Internet of Everything (IoE) can intellectually connect humans, data, processes, and things [6]. For this, techniques related to artificial intelligence (AI) and mobile communication are used which makes the connections in the network more relevant, reliable, and valuable than the ones that is existing. This IoE architecture due to its revolutionary feature supports ubiquitous data collection, processing, aggregation, fusion, distribution, and service. There arouse several challenges and issues due to the disruptive change in designing IoE network architecture, providing mobility, scalability, availability, and security.

Security, dependability, flexibility, and extensibility are a few services offered by Nebula for Internet architecture with the use of more reliable routers and extendable control planes. Moreover, it enforces arbitrary policies using multiple paths. However, still, there exist limitations in scalability and performance as processing ability at the network edge is ignored. Moreover, issues related to the growing demands for resources and services are not considered. The CloNet is in a multi-administrative domain setup which helps network and data center domains to interact, thereby providing an elastic dynamic network to serve the customers on the cloud. Additionally, resources utilized for computing and storage are deployed for better end-user experience and to reduce network dependency.

A cloud-centric network architecture built on cybertwin is proposed to deal with scalability, availability, mobile, and safety for future networks. As an IoE helper, data recorder, and digital asset owner, Cybertwin operates. This architecture is designed using blockchain technology and fog computing. For device-devices, devices-to-FN, and FN-to-FN components, the resource provisioning model was used [7].

The main contribution of this research is sensitive applications and applications connected to end-users. There are several data security and data security problems, and worldwide researchers have provided many methods to protect data against cyber-attacks. Many of them have offered very computational cryptography solutions. Very few academics provide solutions based on blockchain (BC); however, their solutions may be affected by expensive storage costs, as well as problems in network latency, confidence, and capacity. This issue can be overcome by a novel cybertwin-based 6G network. The architectural design is proposed by introducing new concepts such as cloud operators and cybertwin. In the cybertwin-based UAV, IoE applications have to be energy aware and must provide scalability with less latency. A novel CFN framework is designed securely and blockchain-based UAV communication (BC-UAV).

In Section 2, the related works are presented. Section 3 shows the proposed model for CFN-based BC-UAV. Evaluation criteria are discussed in Section 4. Finally, the conclusion is presented in Section 5.

2. Literature Review

A peer-to-peer network utilized in many different types of technology was in charge of managing UAV communication based on the blockchain. The uses of DLT are mentioned to mitigate multiple cyber-threats which classified major cyber-attacks into four categories, namely scanning, power of root, local to remote, and denial of service [8]. The security issues in fog-enabled IoT applications were examined. Blockchain was considered to address these issues [9]. Moreover, this work failed to consider the ability of blockchain with AI. A detailed survey on reinforcement learning applications based on blockchain used in industrial IoT networks was presented [10]. It was revealed that the Q-learning algorithm, one of the machine learning strategies, improved the network performance. A two-way convergence of blockchain with ML was discussed [11]. Blockchain with ML features provided security and reliability. ML was the tool used for optimizing blockchain networks.

The present blockchain based on AI techniques was examined for energy-cloud management, where the security and privacy issues were listed out [12]. A comprehensive study on blockchain-AI applications was presented along with their relationship in the IoT-enabled ecosystem. However, MEC which is a key technology of evolving 5G networks were not concentrated [13]. Blockchain techniques in 5G networks were discussed and a short survey was presented [14]. Three major challenges were identified, namely identity authentication, privacy, and trust management along with a few blockchain-based solutions. A short survey on the blockchain-enabled federated model was described [15]. In motivation to integrate MEC with blockchain was presented [16]. Edge computing was employed to enable mobile blockchains. Edge computing integrated with blockchain was examined. It was found that blockchain extended the ability of edge computing [17]. With this as the basis, the present work in this paper focuses on the blockchain-enabled distributed and decentralized ML approach. Anti-phishing approaches were used at different levels and has to meet a few predefined conditions. Random forest (RF) classifier was used for classifying the mails received on the basis of commonly known features set and the possible threats made known for which phishing incoming mails were required [18]. When an illegal mail received was redirected to a webpage, the level of similarity between legitimate and suspicious webpages was identified and classification was based on a content-based anti-phishing approach.

IoT devices performed edge computation on the resource owner nodes when required. A framework was developed in which in corporate blockchain technology in the applications based on IoT [19]. A Logchain system based on oneM2M which integrated IoT and blockchain technology was utilized which ensured block integrity. However, still, several security issues were not taken into account. To handle the inhibited nature of IoT, blockchain technology was integrated with Edge Computing in [20] which minimized the needs of the IoT device such as memory capacity. Moreover, the performance was improved and was satisfactory. The issues of this approach were addressed and one major issue evaluated was resource optimization which was not achieved. In [21], for enabling propitiation, techno-economic factors and normative assumptions were considered. However, data privacy was still low.

3. Cybertwin Based Network Architecture for 6G with Cloud-Fog Based Network (CFN) Block Chain Based UAV Communication (BC-UAV)

Cybertwin offers a few functions such as communication assistance, network behavior logger, and digital asset which satisfies several new design needs of the network. In end-to-end communication, an end-to-end connection to the server has to be established by the end devices to provide services. As depicted in Figure 1, in this cybertwin-based communication model, the things available in the physical network have to be initially connected with cybertwin which in turn obtains the necessary service from the network and then is delivered to the end-user. This is the most basic function termed as communication assistance function. During the network behavior logger function, while digital representation, cybertwin obtains and logs every data for users. In the digital asset function, cybertwin, after the removal of sensitive information, converts the behavior data of the user to a digital asset for sale. For obtaining improved performance with minimized latency, the technique for distribution assists in offering services based on demand. The standard of life for the citizens will improve as well as the residential expectations. The data processing can be speed-up by a pattern of fog computing which assists the elements of IoE through minimized latency [22].

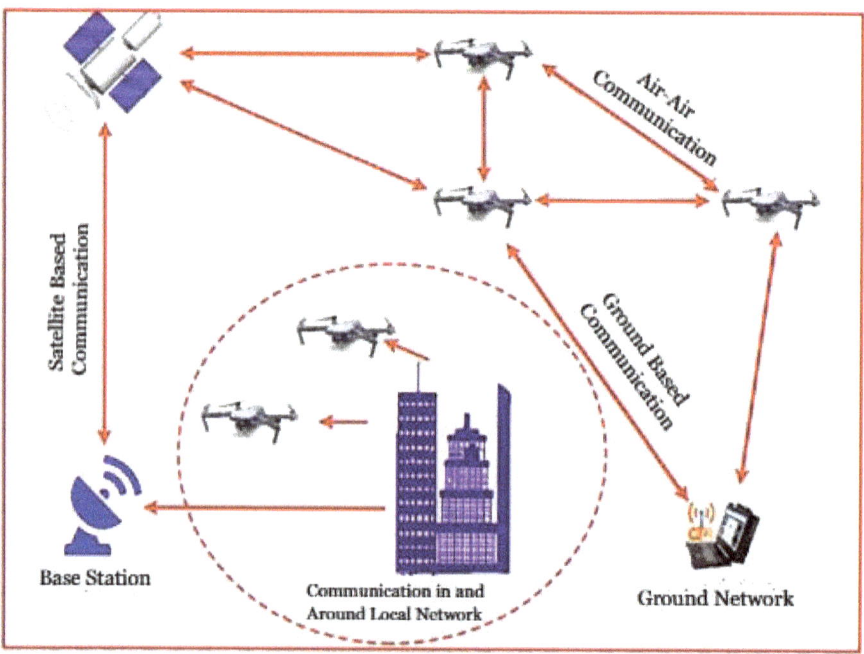

Figure 1. Cybertwin UAV-based communication.

3.1. Cybertwin Communication Model

It is already known that the recent Internet paradigm has no ability to satisfy the needs of the mobile network in the future. The IP address of the present Internet indicates the identity as well as the location information of the device. Due to this reason, the Internet faces difficulties in handling the growing needs of mobile devices as well as services thus causal ability challenges [23]. The network to be trustworthy and secure, the present Internet depends on the security measures of the end-to-end connection and adopts trustworthy users. No procedure is utilized for authenticating the users; hence several security issues are experienced. For ensuring the quality of service (QoS), the Internet considers the way communication resources are managed alone while the other entities are responsible for managing computing and caching resources. Thus, coordination among resources is not easily obtained. Hence, a novel cloud-centric Internet architecture following a cybertwin-based communication model is proposed which is better than using an end-to-end communication model [24].

3.2. Cloud-Centric Internet Architecture

This novel cloud-centric Internet architecture introduced here is depicted in Figure 2. The IP layer is still used as the "thin waist" of the stack in order to permit the developments in other layers and the continuous updates made in the infrastructure of legacy networks. In this architecture, there exist two new components in the network infrastructure which is the fog cloud above the IP layer. Many fully connected core clouds are present to offer network services such as computing, caching, and communication resources. In between core clouds and end users, edge clouds are present which support core clouds to respond to the requests of the end users quickly [25,26].

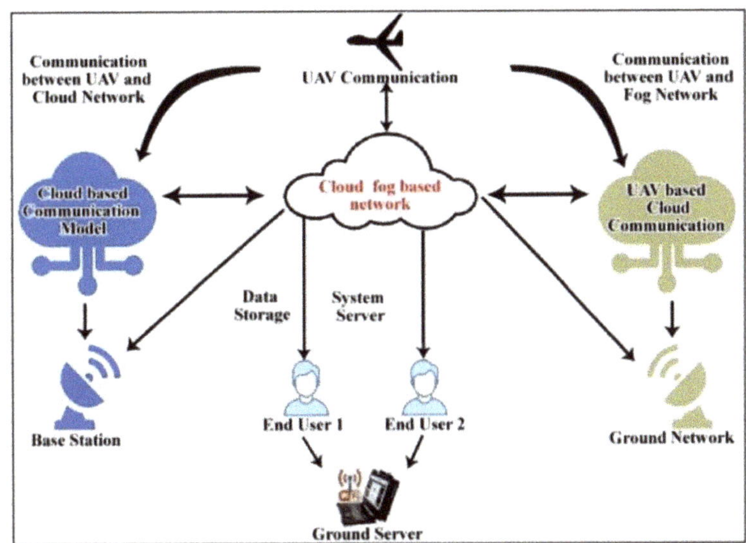

Figure 2. Cloud-fog architecture with IOT based on UAV.

In this Internet architecture, a new cloud operator is introduced to help in constructing an operating system for a cloud network, as illustrated in Figure 3. Services such as computing, caching, and communication resources are scheduled and coordinated accordingly. Moreover, a real-world trading environment for the end users is also established which helps in deploying more multi-purpose resources due to scarcity [27,28].

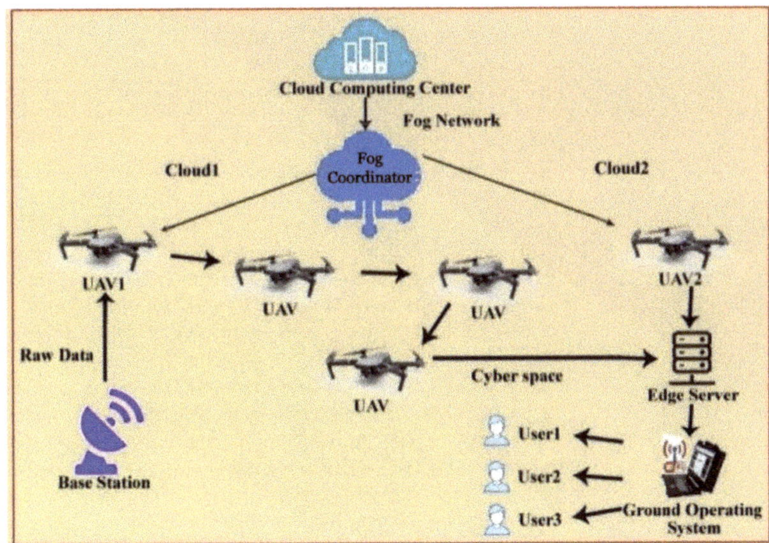

Figure 3. Cloud-fog network operating system with UAV.

According to the services provided by cloud operators in the network, the service providers of the application offer services to core clouds and edge thereby forming a service network. Due to this, the operational cost is reduced and no dedicated servers are required, thus offering the best QoS to the end users [29].

The proposed block chain-fog based network (BFN) connects to the Internet of Everything (IoE) and fog nodes of UAV. The distributed technology provides on-demand services thereby higher performance and less latency rate are achieved. From the citizen's point of view, their quality of life is improved and their expectations are met. Fog computing supports the components of IoE for quick data processing as latency is low [30]. By utilizing its unique features, such as immutability, effective cryptography, and distributed decentralized storage systems, the introduction of blockchain innovation will address these issues [31–36]. The authors [37] evaluated and organized all cryptographic ideas currently employed in blockchain technology in-depth. We also provide a list of cryptographic ideas that have not been used yet but could greatly enhance state-of-the-art blockchain solutions.

This BFN architecture comprises fog nodes. The fog nodes layer reduces the latency by processing the data received by fog nodes from the traffic of IoE. Expectations of the users are satisfied by providing faster services. In this architecture, the fog node layer denotes the devices connected with one another and with FN. The communication among the connected devices is secured as blockchain technology is employed.

3.3. Fog Node Layer

The users demand services that are met with IoE devices connected using fog computing for smart cities. Blockchain technology provides more reliability with the added new component. Several physical servers are integrated to form FN covering certain regions.

4. Block Chain Based UAV Communication (BC-UAV)

The physical UAVs fly in this layer to sense the data application/situation is used in the data sensing, that is, {S1, ... Sn, ... SN} for which UAV is used. For various applications with differing weights, capability of payload, elevation of the flight as well as choices to fly which has enormous dimensions of UAV. The UAVs are supposed to fly in a three-dimensional space having coordinates at time τ can be [x(τ), y(τ), z(τ)], are parallel to the ground and z(μ) are in a vertical height where UAVs can fly [23–26]. When the UAV is projected at an angle, the equations are given from the ground at the initial speed V at the following time as given in Equation (1):

$$\varnothing_{MN}^{ACn} = \varphi MN + \xi MN \sum_{f=1}^{Nf} \lambda_f \tag{1}$$

where λ_f is the bit rate of the flow f.

Range R of UAV attained has been represented by Equation (2)

$$\phi_{ABC}^{ACn} = N_s \cdot \varphi_{ABC} + \xi_{ABC} \sum_{f=1}^{Nf} \lambda_f \tag{2}$$

where R in particular is one of the elements that determines UAV deployment. For instance, the R should be high in the event of military use. In order to secure and reliable communication, communication of UAVs as well as data sharing between the UAVs in the B_{public} network UAV layer. A UAV can only store data in B public when the rules and criteria of SC are fulfilled. On the other hand, UAVs can also use B_{public} network to detect data from D_{SL} to assure confidentiality as well as secrecy of data. Every single UAV has a common ledger replica, thus reducing the latency of access to stored information. To reduce latency and boost system reliability, communication between UAVs and ground stations through a communication network of 6G. The blockchain-based UAV can be represented by Equations (3)–(6),

$$x_m = \frac{x}{a} \tag{3}$$

$$t = \frac{x_t}{x} \tag{4}$$

$$s_m = \frac{s}{a} \tag{5}$$

$$IDS = \frac{\sum a\epsilon X m_y}{x} \tag{6}$$

For exchange between UAVs, base stations, level of BC, and the x-layer, the communication system is a 6G system that delivers an enormously large range of information, ideal for latency-conscious applications. The features of the network layer of 6G communication include highly reliable (10−9), highly minimized latency (1 TB), and in elevation spectrum performance (3–10 P/M over 5G). The complete list of 6G functionalities compared to 5G can be found in Table 1. 6G provides software-defined networking (SDN) which isolates the control plane (CP) and the data plane (DP), allowing a centralized entity (controller) in the CP to configure the forwarding devices in the DP, enabling programmable and dynamic network setup, and network function virtualization (NFV) is a paradigm for network architecture where NFs that previously utilized specialized vendor-specific hardware [38]. SDN is a software prototype that splits the level of data from the controller to simplify and efficiently manage the UAV network. NFV virtualizes network infrastructure, including computer, memory as well as machinery components of the channel, in order to create a channel of UAV with more economic, effective and resistant. Since some of the UAV applications are crucial, even a 1 ms interval cannot be tolerated. The military and healthcare applications would be such. More latency of the network, the more chances of failure. Thus, latency (l) in UAV communication is a key parameter as given in Equation (7):

$$l_{M \rightarrow D}(d_{M,l}) = \mathbb{P}_{LOS}(d_{M,l,h_D}) L_0 d_{M,l}^{-\alpha A} + \left(1 - \mathbb{P}_{LOS}(d_{M,l,h_D})\right) L_{NLOS} L_0 d_{M,l}^{-\alpha A} \tag{7}$$

Table 1. Network efficiency.

Number of Internet Things	CT-CNN [10]	DLT [8]	BAI_ECM [14]	CT-6G_CFN
10	92	92.1	92.3	92.4
20	92.5	93	94	95
40	93.8	93.8	95.8	96
60	94.3	94.2	96	97
80	95	95.1	96.5	98
100	96.5	96.8	97.1	98.5

Therefore, the energy for communication latency of UAV is given as Equations (8) and (9):

$$E_{tx}(l,d) = E_{tx-elec}(m) + E_{tx-amp}(l,d) \tag{8}$$

$$E_{tx}(l,d) = \left\{ \begin{array}{c} m.E_{elec} + m \\ m.E_{elec} + d_{crossover} \end{array} \right\} \tag{9}$$

Value l changes from channel to channel (end-to-end delay), i.e., from LTE-A to 6G, as shown in Equation (10):

$$\begin{array}{c} \ell_{L\pi E-A} \leq 20 \text{ ms} \\ \ell_{5G} \leq 5 \text{ ms} \\ \ell_{B_{Fg}md-5G} \leq 1 \text{ ms} \\ \ell_{6G} \leq 0.1 \text{ ms.} \end{array} \tag{10}$$

4.1. IoE Layer

This layer helps the users to deploy the application in a real-time environment with no limits. Clusters are formed based on the location and function of the IoT devices thus improving throughput and reducing energy consumption, time, and cost overheads. There is an increase in the workload of data centers as hardware and software services require processing by integration. Peer-to-peer (P2P) TCP/IP communication among IoT devices takes place at a shorter distance. For longer distances, these devices make use of FN and communicate using technologies such as WIFI, ZigBee, and Bluetooth.

4.2. Blockchain for IoE

The proposed system uses blockchain for IoE due to its decentralized and tamper-proof nature. Thus, billions of devices in the network can be easily tracked. Moreover, the cost of managing and deploying the server is also reduced.

4.3. Data Transfer

This proposed CFN architecture helps in improving the mobility of the users in the applications based on IoE with fog nodes and cloud computing. Moreover, security is achieved to a greater extent using the blockchain technique where anonymous users are restricted to access IoE devices.

4.4. Cloud Network Layer Security

Assuming that the whole network is an R-circular region, where the characteristics of the network are analyzed in a huge R-based network. The received power in the Prx(xi, xj) main recipient, xj, should rise when increasing the primary transmitter P power and the Íh(xi, xj)Í amplitude of the complex fade coefficient of the primary link transmitter and the primary receptor in a quasi-static wireless environment. In addition, if the distance between primary transmitter xi and primary receiving xj rises, the received power would decrease. In addition, wireless transmission has to do with the trajectory loss exponent which, due to varying communication environments, fluctuates between 0.8 and 4. We explore situation a > 2 here to estimate the interference of primary users and eavesdroppers from the secondary users. The interference power on the primary recipient from the secondary user assumes that WP is the noise power introduced by the main receivers and IP. Now, we analyze a specific situation in the wireless environment where there is simply path loss h(xi,xj), and that is normalized to be one for everyone and not equal to j. Various antenna of the secondary receiver along with the eavesdropper have been prepared while the information of channel state is a channel of eavesdropper's and it is not available in the secondary transmitter [25]. To improve the security in the cloud the hash function is introduced. A hashing operation H is a function that converts an input of any size to an output of a specific size. There are some more characteristics of cryptographic hash functions, such as: Collision resistance: It is challenging to find two inputs a and b such that $H(x^i) = H(y^i)$; preimage resistance: It is challenging to find an input a such that $H(x^i)) = Y$ for a given output Y; and second preimage resistance: It is challenging to find an additional input yi such that $H(y^i) = Y$ for an input x^i and output $= H(x^i)$. The thermal power of the primary users and wafers is considered to be the same because this noise power can be presumed to be independent of a secondary user's location and they are both W. The powers received by the primary users and by the eavesdroppers can all be determined by wireless transmission propagation laws. This simplifies the secrecy capacity as shown in Equation (11):

$$C_s(x_i, x_j) = max \left\{ \begin{array}{c} \log_2\left(1 + \frac{P}{\|x_i - x_j\|^T (w + t_p)}\right) \\ -\log_2\left(1 + \frac{P}{\|x_i - e^*\|^\alpha (W + I_E)}\right), 0 \end{array} \right\} \tag{11}$$

The probability density function of IP and IE can first be derived. Then, the secrecy capacity Cs(i,j) is the probability density function from Equation (12):

$$f_{C_z(i,j)}(c) = \begin{cases} f_{C_P(i,j)}(c) * f_{C_E}(-c), & c > 0, \\ Pr_{0,j} \cdot \delta(c), & c = 0, \\ 0, & c < 0. \end{cases} \tag{12}$$

where I is the primary user's transmitter, j is the nearest neighbor j of transmitter I, and $f_{CP(i,j)}(c), f_{CE}(c), d(c)$, and Pr0,j denotes the primary user probability density, eavesdropper capacity probability function, and Dirac delta function and zero capability.

Depending upon the probability of secrecy, first-order expansion of $F_{\gamma_M|\{X\}}(\gamma)$ on X is represented by Equation (13),

$$F_{\gamma_M|\{X\}}(\gamma) = \begin{cases} \left(\frac{\gamma}{\overline{\overline{\gamma}}_1}\right)^{n_B}, & X \le \frac{\overline{\gamma}_p}{\overline{\overline{\gamma}}_0} \\ \left(\frac{X}{\overline{\overline{\gamma}}_1\sigma}\gamma\right)^{n_B}, & X > \frac{\overline{\overline{\gamma}}_p}{\overline{\overline{\gamma}}_0} \end{cases} \tag{13}$$

the binomial expansion, the cloud network outage probability of secrecy is computed from Equation (14),

$$\begin{aligned} P_{\text{out}}^{\infty} &= \left(-1 - e^{-\frac{\overline{\overline{\gamma}}_p}{\overline{\gamma}_0}}\right) \sum_{i=0}^{n_B} \binom{n_B}{i} \left(\frac{2^{R_s}-1}{\overline{\overline{\gamma}}_1}\right)^{n_B-i} \left(\frac{2^{R_S}}{\overline{\overline{\gamma}}_1}\right)^{i} \\ &\quad \sum_{j=0}^{n_E-1} \binom{n_E-1}{j} \frac{n_E}{\overline{\overline{\gamma}}_2}(-1)^j \int_0^{\infty} (\gamma_E)^i e^{-\frac{(j+1)\sim E}{\overline{\gamma}_2}} d\gamma_E \\ &\quad + \sum_{i=0}^{n_B} \binom{n_B}{i} \left(\frac{2^{R_S}-1}{\overline{\overline{\gamma}}_1\sigma}\right)^{n_B-i} \left(\frac{2^{R_S}}{\overline{\overline{\gamma}}_1\sigma}\right) \sum_{j=0}^{in_E-1} \binom{n_E-1}{j} \\ &\quad \frac{n_E}{\overline{\overline{\gamma}}_2\sigma}(-1)^j \frac{1}{\Omega_0} \int_{\frac{\overline{\pi}_p}{\gamma_0}}^{\infty} e^{-\frac{x}{\overline{\Omega}_0}} \int_0^{\infty} x^{n_B+1}(\gamma_E)^i e^{-\frac{(j+1)\gamma_E}{\overline{\gamma}_2\sigma}} d\gamma_E dx \end{aligned} \tag{14}$$

Employing Equation (14) given by $\int_0^{\infty} x^n e^{-\mu x} dx = \frac{1(n+1)^*}{\mu^{n+1}}$ integrally evaluated and the cloud network outage probability of secrecy is derived by Equation (15)

$$P_{\text{out}}^{\infty} = (G_a\overline{\gamma}_1)^{-G_i} + O\left(\overline{\gamma}_1^{-G_d}\right) \tag{15}$$

where the diversity order for secrecy is given by Equation (16):

$$zG_d = n_B \tag{16}$$

and the gain of secrecy array is given by Equation (17):

$$\begin{aligned} G_a &= \left[\left(1 - e^{-\frac{\sigma}{n_0}}\right) \sum_{i=0}^{n_B} \binom{n_B}{i} (2^{R_S} - 1)^{n_B-i} 2^{R_s i} \right. \\ &\quad \sum_{j=0}^{n_E-1} \binom{n_E-1}{j} n_E \overline{\gamma}_2^{\,i}(-1)^j \frac{\Gamma(i+1)}{(j+1)^{i+1}} + \sum_{i=0}^{n_B} \binom{n_B}{i} \\ &\quad (2^{R_S} - 1)^{n_B-i} \sigma^{-n_B} 2^{R_s i} \sum_{j=0}^{n_E-1} \binom{n_E-1}{j} n_E (\overline{\gamma}_2\sigma)^i \\ &\quad \left. (-1)^j (\Omega_0)^{n_B-i} \frac{\Gamma(i+1)}{(j+1)^{i+1}} \Gamma\left(n_B - i + 1, \frac{\sigma}{\Omega_0}\right)\right]^{-\frac{1}{n_B}} \end{aligned} \tag{17}$$

where incomplete gamma function is represented by $\Gamma(\cdot, \cdot)$.

The CDF and PDF of Y are written by Equations (18) and (19):

$$F_Y(y) = \sum_{n=0}^{N} \binom{N}{n}(-1)^n e^{-\frac{nv}{n_Y}} \tag{18}$$

$$f_Y(y) = \sum_{n=0}^{N-1} \binom{N-1}{n} \frac{N}{\Omega_Y}(-1)^n e^{-\frac{(n+1)y}{n_Y}} \tag{19}$$

Let $u(X) = \min\left(\frac{\overline{\gamma}_p}{X}, \overline{\gamma}_0\right)$. Using the probability theory, for $RV\gamma = u(X)Y$ the conditional CDF and PDF of γ can be obtained as (19) and (20), respectively.

Hence, the cloud network layer security in Equations (20)–(23) can be calculated as:

$$
\begin{aligned}
P_{\text{out}} = \; & \underbrace{\int_{0}^{\frac{\overline{\gamma_P}}{\gamma_0}} \int_{0}^{\infty} F_{\gamma_M|\{X=x\}}\big(\epsilon(\gamma_E)\big) f_{\gamma_E|\{X=x\}}(\gamma_E) f_X(x)\, d\gamma_E\, dx}_{\mathcal{J}_1} \\
& + \underbrace{\int_{\frac{\overline{\gamma_P}}{\gamma_0}}^{0} \int_{0}^{\infty} F_{\gamma_M|\{X=x\}}\big(\epsilon(\gamma_E)\big) f_{\gamma_E|\{X=x\}}(\gamma_E) f_X(x)\, d\gamma_E\, dx}_{\mathcal{J}_2}
\end{aligned}
\tag{20}
$$

$$
F_{\gamma u|\{X=x\}}\big(\epsilon(\gamma_{\mathrm{E}})\big) = \sum_{i=0}^{n_{\mathrm{B}}} \binom{n_{\mathrm{B}}}{i}(-1)^{i} e^{-\frac{i\omega(\gamma_{\mathfrak{F}})}{\gamma_0^{n_1}}}, \; f_{\gamma E|\{X=x\}}(\gamma_{\mathrm{E}}) =
$$
$$
\sum_{j=0}^{n\varepsilon-1} \binom{n_{\mathrm{E}}-1}{j} \frac{n_{\mathrm{E}}}{\overline{\gamma_0}\Omega_2}(-1)^{j} e^{-\frac{(j+1)\sim_2}{F_0^{l2}}}
\tag{21}
$$

$$
\text{For } X > \frac{\overline{\gamma_P}}{\gamma_0}, \text{ we have}
$$

$$
F_{\gamma_M|\{X=x\}}\big(\epsilon(\gamma_{\mathrm{E}})\big) = \sum_{i=0}^{n_{\mathrm{B}}} \binom{n_{\mathrm{B}}}{i}(-1)^{i} e^{-\frac{i_k(\gamma_5)}{\overline{v}_p^{n_1} x}},
\tag{22}
$$

$$
f_{\gamma Z|\{X=x\}}(\gamma_E) = \sum_{j=0}^{n_{\mathrm{E}}}-1 \binom{n_{\mathrm{E}}-1}{j}\frac{n_{\mathrm{E}}}{\overline{\gamma}_p\Omega_2}(-1)^{j} xe^{-\frac{(j+1)\gamma_1}{T_p^{n_2} x}}
\tag{23}
$$

5. Performance Analysis

The performance of this proposed CFN model with its numerical simulation results from the iFogSim simulator is relatively examined with the existing ones. The real-time situation of smart city fog network is involved and considers traffic of web applications. The workload needs CPU and network resources.

The parametric analysis is given by graphs below.

The above Figures 4 and 5 show the average power consumption, and latency comparison between CFN with fog and cloud frameworks. In Figure 4 for 1s the power consumption stays constant for time without any minimal oscillations because of small variations vertically. Figure 5 shows fog computing architecture latency in which the event has been notified earlier for final users of cloud computing architecture. Here, the latency extends to second for cloud computing.

The above Tables 1 and 2 show a comparison of network efficiency, communication delay, and average power consumption, and Figures 6–12 show their graphical representation in comparison between existing and proposed techniques. From Figures 6–9 shows the network efficiency as well as the cloud computing layer delay has been initiated due to a delay in communication in cloud servers with edge servers. On one hand, this is on the grounds that the correspondence with a significant distance from end clients to the cloud server center may create high postponement. Then again, the limit of organization data transfer capacity builds the transmission delay from edge devices to cloud workers incredibly. As the measure of information increments consistently, the correspondence defers increments quicker. Figures 10–12 show communication latency, UAV mobility, and connectivity of UAVs. The bends turn 100% of responsibility in the cloud. This shows that the transmission rate acts simultaneously with the mist handling rate introduced. The dormancy at the mist will be improved as more work is moved to the cloud.

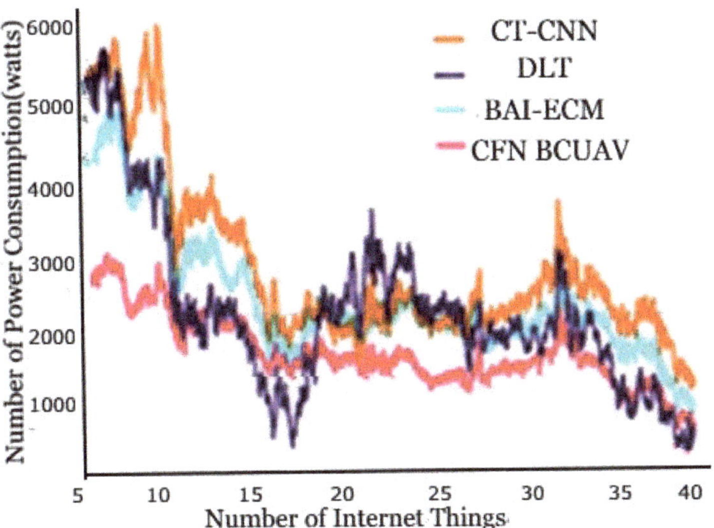

Figure 4. Average power consumption of UAV.

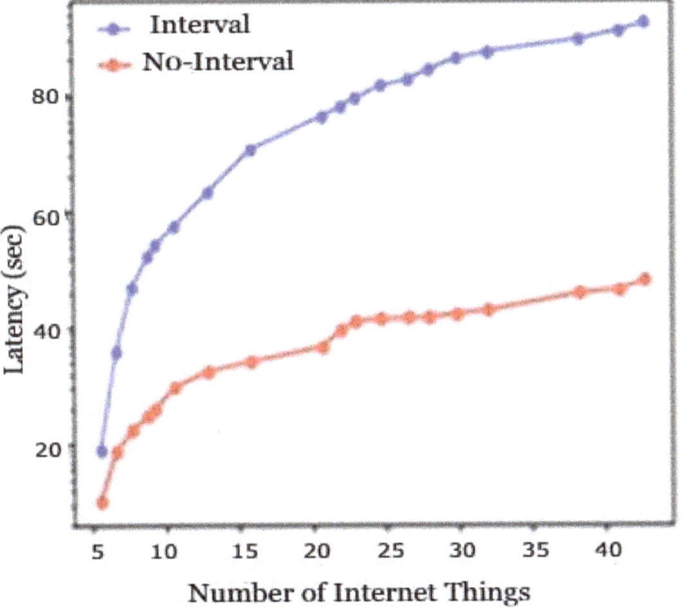

Figure 5. Comparison for latency of the proposed CFN with fog and cloud frameworks.

Table 2. Communication delay.

Number of Internet Things	CT-CNN [10]	DLT [8]	BAI_ECM [14]	CT-6G_CFN
20	4.2	3.9	3.5	3.2
40	6	6.3	6.5	5.7
60	7.3	7.5	7.8	6.8
80	8.2	6.7	7.9	8.2
100	12	11.5	11	10.8

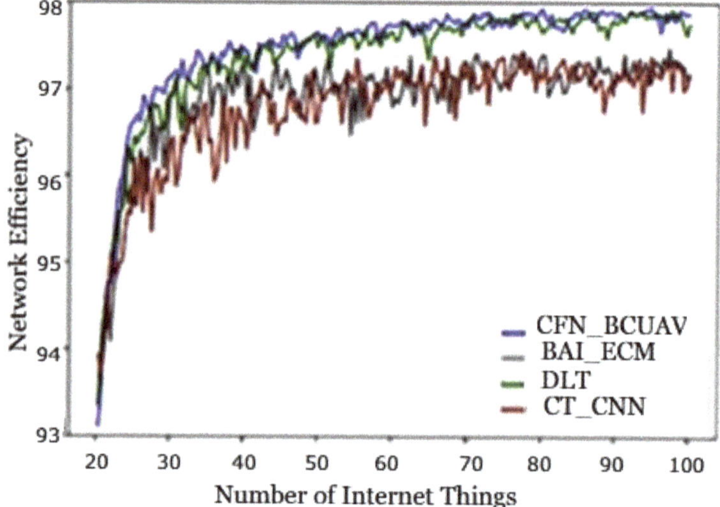

Figure 6. Network efficiency.

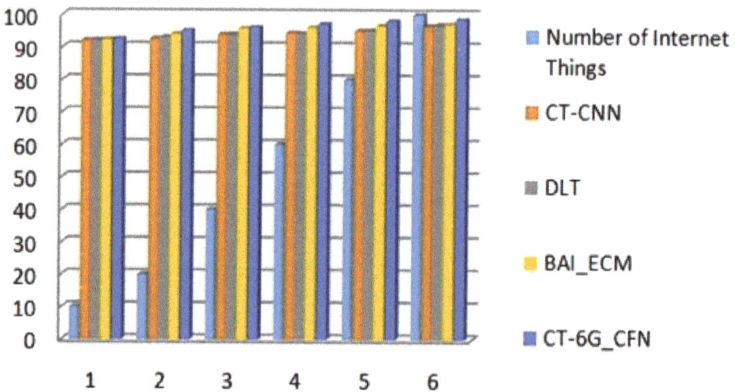

Figure 7. Comparison chart—network efficiency.

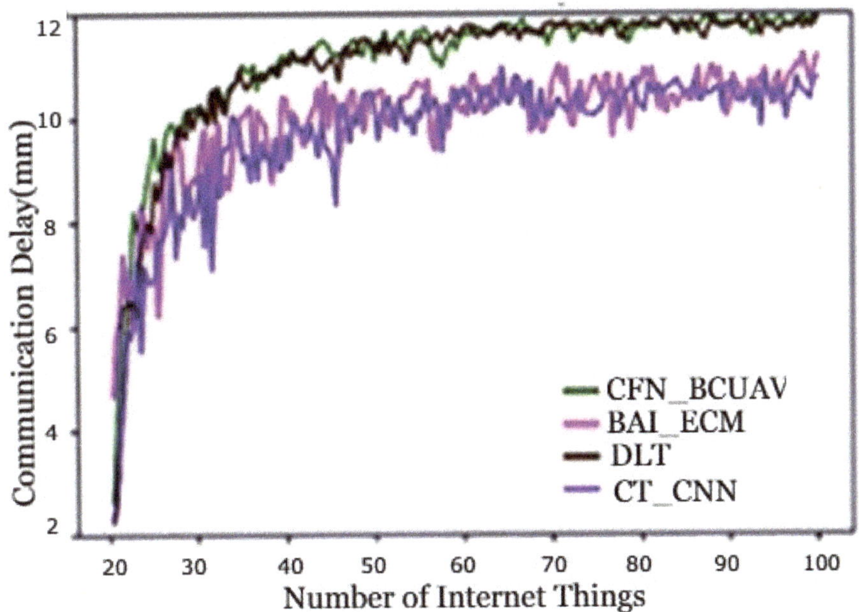

Figure 8. Communication delay.

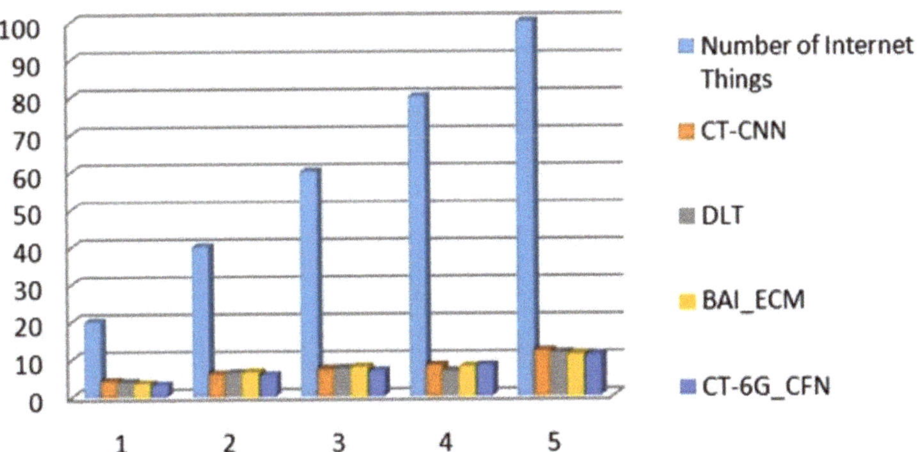

Figure 9. Comparison chart—communication delay.

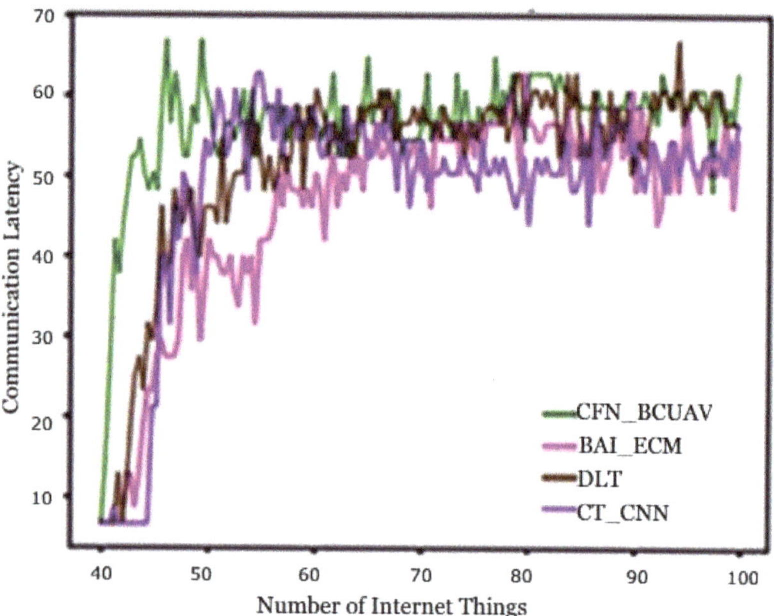

Figure 10. Communication latency.

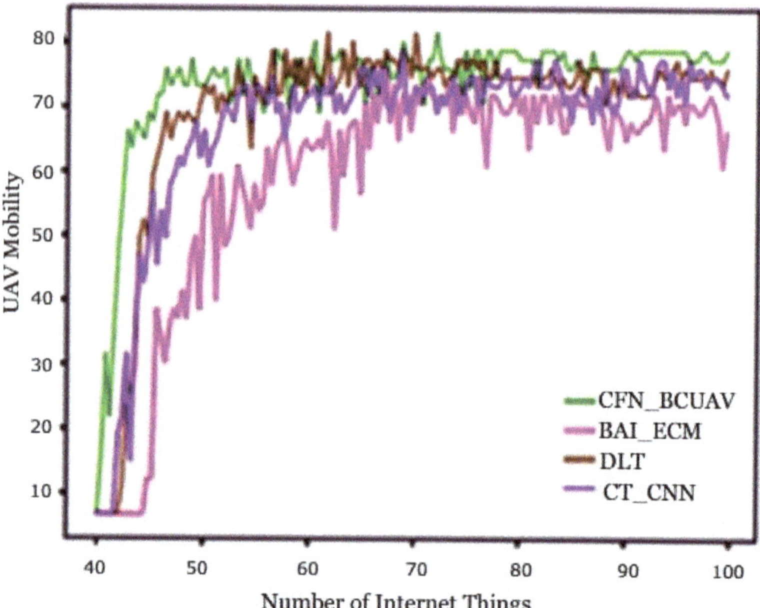

Figure 11. UAV mobility.

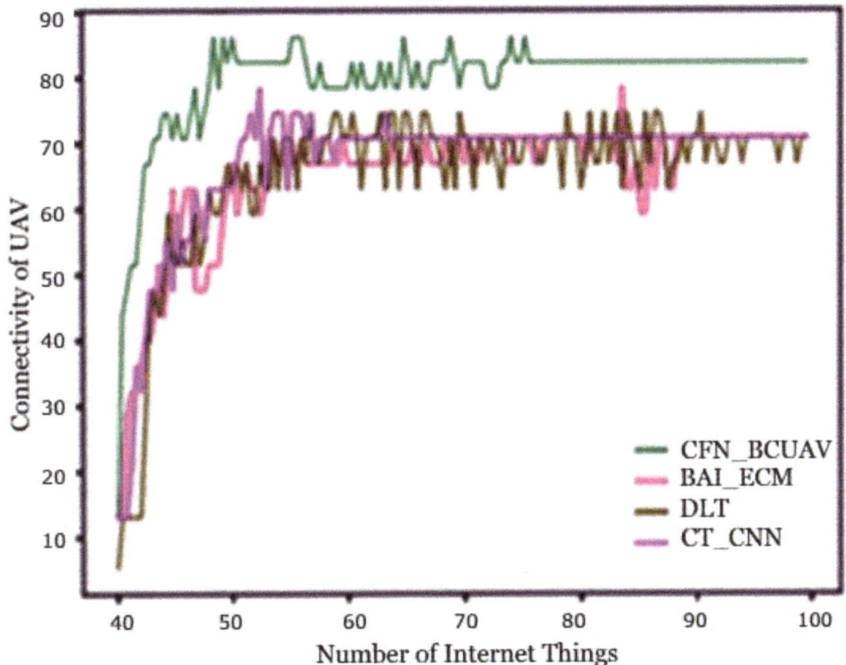

Figure 12. Connectivity of UAV.

6. Conclusions

The concept of blockchain technology, cybertwins (CTs) with UAV assumes a protuberant part in business use, convoluted security basic missions alongside numerous other assorted scopes of uses. UAVs ought to have the option to give wide inclusion and availability to distant regions under all conditions. Besides, conventional UAV correspondence is not satisfactory to manage the high-versatility and dynamic highlights of UAVs. Along these lines, there is a requirement for an effective and secure organization of UAVs as they have been broadly utilized in antagonistic conditions. So, this paper proposed that cybertwin-based UAVs and IoE applications have to be energy aware and must provide scalability with less latency. A novel CFN framework is designed securely and blockchain-based UAV communication (BC-UAV). The overall results reveal that the performance of this proposed CFN_BC-UAV architecture is better for cyber-based security in 6G techniques. Since applications were designed based on the fog computing environment which provides scalability, energy efficiency as well as security. In some current works, the information privacy and trustworthiness levels additionally stay low. The examination difficulties and open issues in joining blockchain with the 6G correspondence network are investigated. Then, at that point, the future exploration rules toward blockchain-empowered IoT with 6G correspondence are given. Security and protection issues of 5G advancements are to be diminished contingent upon requests and prerequisites.

Author Contributions: Conceptualization, A.B.K.; supervision, J.K. and S.B.G.; original draft and review and editing, D.D. and V.D.; validation, A.S.R. and S.B.G.; writing—review and editing, A.S.R., S.B.G., M.S.R., T.C.M., C.V. and G.S.; proposed the new method or methodology, D.D. and A.B.K. All authors have read and agreed to the published version of the manuscript.

Funding: This research received no external funding.

Data Availability Statement: Data will be shared for review based on the editorial reviewer's request.

Acknowledgments: This paper was partially supported by UEFISCDI Romania and MCI through BEIA projects NGI-UAV-AGRO, SOLID-B5G, I-DELTA, STACK, ENTA, IMMINENCE and by European Union's Horizon 2020 research and innovation program under grant agreement No. 883522 (S4ALLCITIES). This work is supported by Ministry of Research, Innovation, Digitization from Romania by the National Plan of R & D, Project PN 19 11, Subprogram 1.1. Institutional performance-Projects to finance excellence in RDI, Contract No. 19PFE/30.12.2021 and a grant of the National Center for Hydrogen and Fuel Cells (CNHPC) — Installations and Special Objectives of National Interest (IOSIN).

Conflicts of Interest: The authors declare no conflict of interest.

References

1. Yu, Q.; Ren, J.; Zhou, H.; Zhang, W. A Cybertwin based Network Architecture for 6G. In Proceedings of the 2020 2nd 6G Wireless Summit (6G SUMMIT), Levi, Finland, 17–20 March 2020; IEEE: Piscataway, NJ, USA, 2020.
2. Yu, Q.; Zhou, H.; Chen, J.; Li, Y.; Jing, J.; Zhao, J.J.; Qian, B.; Wang, J. A fully-decoupled RAN architecture for 6G inspired by neurotransmission. *J. Commun. Inf. Netw.* **2019**, *4*, 15–23. [CrossRef]
3. Yu, Q.; Ren, J.; Fu, Y.; Li, Y.; Zhang, W. Cybertwin: An origin of next generation network architecture. *IEEE Wirel. Commun.* **2019**, *26*, 111–117. [CrossRef]
4. Fernandez-Caram, T.M.; Fraga-Lamas, P. A Review on the Use of Blockchain for the Internet of Things. *IEEE Access* **2018**, *6*, 32979–33001. [CrossRef]
5. Nawaz, S.J.; Sharma, S.K.; Wyne, S.; Patwary, M.N.; Asaduzzaman, M. Quantum machine learning for 6G communication networks: State-of-the-art and vision for the future. *IEEE Access* **2019**, *7*, 46317–46350. [CrossRef]
6. Shafi, M.; Molisch, A.F.; Smith, P.J.; Haustein, T.; Zhu, P.; de Silva, P.; Tufvesson, F.; Benjebbour, A.; Wunder, G. 5G: A tutorial overview of standards, trials, challenges, deployment, and practice. *IEEE J. Sel. Areas Commun.* **2017**, *35*, 1201–1221. [CrossRef]
7. Zhao, J. A survey of reconfigurable intelligent surfaces: Towards 6G wireless communication networks with massive MIMO 2.0. *arXiv* **2019**, arXiv:1907.04789.
8. Yang, R.; Yu, F.R.; Si, P.; Yang, Z.; Zhang, Y. Integrated Blockchain and Edge Computing Systems: A Survey, Some Research Issues and Challenges. *IEEE Commun. Surv. Tutor.* **2019**, *21*, 1508–1532. [CrossRef]
9. Nguyen, D.C.; Pathirana, P.N.; Ding, M.; Seneviratne, A. Blockchain for 5G and beyond networks: A state of the art survey. *J. Netw. Comput. Appl.* **2020**, *166*, 102–693. [CrossRef]
10. Kumari, A.; Gupta, R.; Tanwar, S.; Kumar, N. Blockchain and AI amalgamation for energy cloud management: Challenges, solutions, and future directions. *J. Parallel Distrib. Comput.* **2020**, *143*, 148–166. [CrossRef]
11. Fernández-Caramés, T.M.; Fraga-Lamas, P. Towards next generation teaching, learning, and context-aware applications for higher education: A review on blockchain, IoT, fog and edge computing enabled smart campuses and universities. *Appl. Sci.* **2019**, *9*, 4479. [CrossRef]
12. Liu, Y.; Yu, F.R.; Li, X.; Ji, H.; Leung, V.C.M. Blockchain and Machine Learning for Communications and Networking Systems. *IEEE Commun. Surv. Tutor.* **2020**, *22*, 1392–1431. [CrossRef]
13. Gui, G.; Liu, M.; Tang, F.; Kato, N.; Adachi, F. 6G: Opening New Horizons for Integration of Comfort, Security and Intelligence. *IEEE Wirel. Commun.* **2020**, *27*, 126–132. [CrossRef]
14. Wang, M.; Zhu, T.; Zhang, T.; Zhang, J.; Yu, S.; Zhou, W. Security and privacy in 6G networks: New areas and new challenges. *Digit. Commun. Netw.* **2020**, *6*, 281–291. [CrossRef]
15. Kim, H.; Park, J.; Bennis, M.; Kim, S. Blockchained On-Device Federated Learning. *IEEE Commun. Lett.* **2020**, *24*, 1279–1283. [CrossRef]
16. Mao, Y.; You, C.; Zhang, J.; Huang, K.; Letaief, K.B. A Survey on Mobile Edge Computing: The Communication Perspective. *IEEE Commun. Surv. Tutor.* **2017**, *19*, 2322–2358. [CrossRef]
17. Jiang, X.; Yu, F.R.; Song, T.; Ma, Z.; Song, Y.; Zhu, D. Blockchain-Enabled Cross-Domain Object Detection for Autonomous Driving: A Model Sharing Approach. *IEEE Internet Thing. J.* **2020**, *7*, 3681–3692. [CrossRef]
18. Rahim, R.; Murugan, S.; Mostafa, R.R.; Dubey, A.K.; Regin, R.; Kulkarni, V.; Dhanalakshmi, K.S. Detecting the Phishing Attack Using Collaborative Approach and Secure Login through Dynamic Virtual Passwords. *Webology* **2020**, *17*, 524–535. [CrossRef]
19. Fourati, M.; Najeh, B.; Idriss, A. Blockchain Towards Secure UAV-Based Systems. In *Enabling Blockchain Technology for Secure Networking and Communications*; IGI Global: Hershey, PA, USA, 2021; pp. 149–174.
20. Gupta, R.; Shukla, A.; Anwar, S. BATS: A Blockchain and AI-empowered Drone-assisted Telesurgery System towards 6G. *IEEE Trans. Netw. Sci. Eng.* **2020**, *8*, 2958–2967. [CrossRef]
21. Pokhrel, S.R. Federated learning meets blockchain at 6g edge: A drone-assisted networking for disaster response. In Proceedings of the 2nd ACM MobiCom Workshop on Drone Assisted Wireless Communications for 5G and Beyond, London, UK, 25 September 2020; pp. 49–54.
22. Singh, P.; Nayyar, A.; Kaur, A.; Ghosh, U. Blockchain and fog based architecture for internet of everything in smart cities. *Future Internet* **2020**, *12*, 61. [CrossRef]

23. Rametta, C.; Schembra, G. Designing a softwarized network deployed on a fleet of drones for rural zone monitoring. *Future Internet* **2017**, *9*, 8. [CrossRef]
24. García-Magariño, I.; Lacuesta, R.; Rajarajan, M.; Lloret, J. Security in networks of unmanned aerial vehicles for surveillance with an agent-based approach inspired by the principles of blockchain. *Ad Hoc Netw.* **2019**, *86*, 72–82. [CrossRef]
25. Hattab, G.; Cabric, D. Energy-efficient massive IoT shared spectrum access over UAV-enabled cellular networks. *IEEE Trans. Commun.* **2020**, *68*, 5633–5648. [CrossRef]
26. Monir, M.B.; Mohamed, A.A. Energy aware routing for wireless sensor networks. *Int. J. Commun. Netw. Inf. Secur. IJCNIS* **2022**, *14*, 70–75. [CrossRef]
27. Aljarrah, I.A.; Alshare, E.M. Improved Residual Dense Network for Large Scale Super-Resolution via Generative Adversarial Network. *Int. J. Commun. Netw. Inf. Secur. IJCNIS* **2022**, *14*, 118–125. [CrossRef]
28. Osama, I.; Rihan, M.; Elhefnawy, M.; Eldolil, S.; Abd El-AzemMalhat, H. A review on Precoding Techniques For mm-Wave Massive MIMO Wireless Systems. *Int. J. Commun. Netw. Inf. Secur. IJCNIS* **2022**, *14*, 26–36. [CrossRef]
29. Paliwal, R.; Khan, I. Design and Analysis of Soft Computing Based Improved Routing Protocol in WSN for Energy Efficiency and Lifetime Enhancement. *Int. J. Recent Innov. Trends Comput. Commun.* **2022**, *10*, 12–24. [CrossRef]
30. Degambur, L.-N.; Mungur, A.; Armoogum, S.; Pudaruth, S. Resource Allocation in 4G and 5G Networks: A Review. *Int. J. Commun. Netw. Inf. Secur. IJCNIS* **2021**, *13*, 401–408. [CrossRef]
31. Arumugam, S.; Shandilya, S.K.; Bacanin, N. Federated Learning-Based Privacy Preservation with Blockchain Assistance in IoT 5G Heterogeneous Networks. *J. Web Eng.* **2022**, *21*, 1323–1346. [CrossRef]
32. Ramanan, M.; Singh, L.; Suresh Kumar, A.; Suresh, A.; Sampathkumar, A.; Jain, V.; Bacanin, N. Secure blockchain enabled Cyber-Physical health systems using ensemble convolution neural network classification. *Comput. Electr. Eng.* **2022**, *101*, 108058. [CrossRef]
33. Sampathkumar, A.; Murugan, S.; Elngar, A.A.; Garg, L.; Kanmani, R.; Malar, A.C.J. A Novel Scheme for an IoT-Based Weather Monitoring System Using a Wireless Sensor Network. In *Integration of WSN and IoT for Smart Cities. EAI/Springer Innovations in Communication and Computing*; Rani, S., Maheswar, R., Kanagachidambaresan, G., Jayarajan, P., Eds.; Springer: Cham, Switzerland, 2020. [CrossRef]
34. Ugochukwu, N.A.; Goyal, S.B.; Arumugam, S. Blockchain-Based IoT-Enabled System for Secure and Efficient Logistics Management in the Era of IR 4.0. *J. Nanomater.* **2022**, *2022*, 7295395. [CrossRef]
35. Bedi, P.; Goyal, S.B.; Rajawat, A.S.; Shaw, R.N.; Ghosh, A. Application of AI/IoT for Smart Renewable Energy Management in Smart Cities. In *AI and IoT for Smart City Applications. Studies in Computational Intelligence*; Piuri, V., Shaw, R.N., Ghosh, A., Islam, R., Eds.; Springer: Singapore, 2022; Volume 1002. [CrossRef]
36. Sharma, S.; Rani, M.; Goyal, S.B. Energy Efficient Data Dissemination with ATIM Window and Dynamic Sink in Wireless Sensor Networks. In Proceedings of the 2009 International Conference on Advances in Recent Technologies in Communication and Computing, Kottayam, India, 27–28 October 2009; pp. 559–564. [CrossRef]
37. Raikwar, M.; Gligoroski, D.; Kralevska, K. SoK of Used Cryptography in Blockchain. *IEEE Access* **2019**, *7*, 148550–148575. [CrossRef]
38. Esmaeily, A.; Kralevska, K. Small-Scale 5G Testbeds for Network Slicing Deployment: A Systematic Review. *Wirel. Commun. Mob. Comput.* **2021**, *2021*, 6655216. [CrossRef]

Article

Energy Efficient Routing and Dynamic Cluster Head Selection Using Enhanced Optimization Algorithms for Wireless Sensor Networks

I. Adumbabu [1,*] and K. Selvakumar [2]

1 Department of Electronics and Communication Engineering, Annamalai University, Chidambaram 608 002, India
2 Department of Information Technology, Annamalai University, Chidambaram 608 002, India
* Correspondence: adumbabu401@gmail.com

Citation: Adumbabu, I.; Selvakumar, K. Energy Efficient Routing and Dynamic Cluster Head Selection Using Enhanced Optimization Algorithms for Wireless Sensor Networks. *Energies* **2022**, *15*, 8016. https://doi.org/10.3390/en15218016

Academic Editors: R. Maheswar, M. Kathirvelu and K. Mohanasundaram

Received: 26 September 2022
Accepted: 25 October 2022
Published: 28 October 2022

Publisher's Note: MDPI stays neutral with regard to jurisdictional claims in published maps and institutional affiliations.

Abstract: A large number of spatially dispersed nodes on the wireless network create Wireless Sensor Networks (WSNs) to collect and analyze the physical data from the environment. The issues that affected the network and had an impact on network energy consumption were cluster head random selection, working node redundancy, and cluster head transmission path construction. Consequently, this energy constraint also has an impact on the network lifetime and energy-efficient routing. Therefore, the primary goals of this research are to decrease energy consumption and lengthen the network's lifespan. So, using improved optimization algorithms, this paper presents a dynamic cluster head-based energy-efficient routing system. The Improved Coyote Optimization Algorithm (ICOA), in this case, consists of three phases setup, transmission, and measurement phase. The Improved Jaya Optimization Algorithm with Levy Flight (IJO-LF) then determines the route between the BS and the CH. It selects the most effective course based on the distance, node degree, and remaining energy. The proposed approach is compared with traditional methods and the routing protocols Power-Efficient Gathering in Sensor Information Systems (PEGASIS) and Threshold sensitive Energy Efficient Sensor Network protocol (TEEN) during implementation on the MATLAB platform. Performance indicators for the suggested methodology are evaluated based on data packets collected by the BS, energy usage, alive nodes, and dead nodes. The outputs of the suggested methodology performed better than the conventional plans.

Keywords: IOCA; IJO-LF; WSN; energy efficient routing

1. Introduction

Hundreds of thousands of sensor nodes are deployed across the WSN to sense, analyze, and retrieve information. Such sensor nodes are less affordable and offer greater detection, computing, and communication capabilities [1]. Medical applications, defensive performance, weather prediction, and a variety of industrial and commercial uses are just a few of the applications that use WSN [2]. WSN sensors are small and run on a small battery [3]. The sensor is utilized an Analog-to-Digital Converter (ADC) to capture data, which it then processes before sending it to the primary hub, called the Base Station (BS). Across different applications, the data is evaluated at BS to make decisions. WSN sensor nodes act as a repeater, sending information to additional sinks and sensors [4]. Due to the placement of the sensor in a hostile and no-land environment, the WSN's power supply can sometimes be recharged or switched, man's it must be handled properly [5]. Various characteristics, such as energy efficiency, fault tolerance, scalability, and so on, have an impact on the construction of WSNs [6]. The energy used by the sensors in the WSN has been used in two ways: (1) environmental parameter sensing and (2) data transmission to the BS via the nodes. WSN data transmission ingests more energy than environmental sensing and data processing [7].

In the WSN, CHs are used to save energy. CH is chosen based on a set of restrictions, including the estimated distance between the receiver and fusion center, as well as residual energy [8]. For sending energy-efficient data, the CH selection of CH is significant. During the specific iterations, the CH in WSN varies to improve performance. The CH uses the Time Division Multiple Accesses (TDMA) technology to collect sensed information. This strategy eliminates redundancies by reducing and gathering information. For capturing information from multiple WSNs, dynamic clustering is the best option [9]. As a result of some aspects, such as energy-saving capability and efficient scalability, dynamic clustering has attracted the interest of various academics. Dynamic clustering is homogeneous and adapts to variations within the transmission range of the sensor. The CH operates in a variety of hop networks, in this case [10]. The periodic rearranging of clusters aids the specialist in enhancing the scalability & energy of WSNs. The spinning of CH is employed in numerous means of transport and IoT applications [11] where dynamic clustering has been used to sustain data transmission between nodes [12].

Energy constraint, in particular, is seen to be a major issue. Every one of the sensing elements requires energy to work. As a result, maximizing the network's longevity requires managing a node's energy usage [13]. One of the most well-known strategies for effectively reducing energy usage in WSNs is the hierarchical arrangement of the network in clusters.CH, the cluster's head node, is in charge of each cluster. In this regard, several unique methods have been presented, all of which have been shown to be effective [14]. Furthermore, within those processes, the direction of finding is ignored, and the heuristic function's essential choice for determining a relay node for data transmission is based solely on a single parameter [15].

Monitoring and control, surveillance systems, healthcare systems, and intelligent space have all benefited from the deployment of WSNs [16]. Furthermore, in WSNs, how to optimally employ network node energy is a worthwhile subject to investigate [17]. Cluster routing is an effective energy maintenance solution for WSNs [18], with the LEACH algorithm being the most well-known. By clustering, this algorithm splits network nodes into cluster members &cluster heads, allowing network node resources to be completely utilized and the network's life cycle to be successfully extended. To limit the number of nodes, the TEEN and PEGASIS algorithms used the clustering concept that connects directly with the base station. The energy consumption of network nodes is balanced by changing cluster heads regularly; this is utilized to extend the life cycle of the network [19].

The foremost contribution of this research work is as follows:

- For the low computational complexity and excellent stability, the ICOA is utilized in the WSN to choose the CH. Based on numerous objective values, ICOA selects the CH, such as node degree, residual energy, node centrality, distance to the BS, and distance to neighbors.
- In WSN, IT can enable the rapid discovery of solutions. By utilizing the IJLFA, the shortest route between CH and BS is discovered.
- Because of the optimal route generation and energy-efficient CH selection for the transmission of data, the network's lifetime is extended. Furthermore, by reducing the energy utilization of nodes while transferring packets, the overall packets expected by the BS are enhanced.

The research paper is organized as follows: Section 2 presents a review of recent research in the field of cluster head selection and routing algorithms. In Section 3, the proposed methodology is elaborated. The simulation results of the proposed methodology are described in Section 4. Finally, the research is concluded in Section 5.

2. Related Work

Some of the recent research works related to efficient CH selection and routing were reviewed in this section.

Moussa and El Alaoui [20] proposed an Energy-efficient Cluster-based Routing Protocol with Enhanced Ant Colony Optimization (ACO) and Unequal Clustering (ECRP-UCA).

To effectively balance the load across Cluster Heads: Based on residual energy, ECRP-UCA splits the network into uneven clusters; how far away the sink is; several nodes in the same neighborhood; and an additional factor called prior cycle's number of backward relay nodes. The next hop sensor node's energy is factored into the heuristic function, the distance between the current and the next sensor nodes.

To address these challenges, Sankar et al. [21] suggested a different cluster formation and Cluster Head (CH) selection technique. There are two stages to the procedure. Initially, by the Sailfish Optimization Algorithm (SOA),the CH is chosen to be called a Swarm Intelligence Algorithm. Next, the Euclidean distance forms the cluster. Previous clustering techniques include issues with the network's short lifespan, imbalanced load among network nodes, and end-to-end delay.

Loganathan et al. [22] introduced an Energy Enhanced Routing Protocol (EERP), which is a novel routing protocol utilizing high-efficiency data transfer rules. For creating a fault-free communication model for the WSN environment, the new EERP technique incorporates the scheme's sophisticated clustering logic. With this method, a standard routing structure concerning the base station and sensor nodes is created. The management of CH is the most significant component of a cluster-based wireless technology, as it must be chosen based on specific communication rules, including distance estimation, node capacity, positioning of the base station, and cluster region position. For analyzing the CH and improving the estimation of the pathway process, these restrictions are necessary. To provide an effective CH election process, the presented EERP method makes use of the efficient CH election algorithm Firefly.

To increase network energy, Yong et al. [23] introduced a multi-hop routing scheme depending on the path tree. To build a cluster headset, initially, several nodes that are closer to the base station and then have a lot of leftover energy are chosen. After that, the whole cluster is partitioned into smaller regions, and the nodes with higher remaining energy than the cluster's regular remaining energy are chosen as working nodes in each region. Eventually, the CHs are ordered based on how far they are from the base station.

The optimal route creation and CH selection presented by Maheshwari et al. [24] are considered tough challenges in WSNs. To improve the network's lifetime and minimize total energy usage, a mixture of ACO and BOA was employed in this study. The node centrality, neighbors' distance, nodes' residual energy, the Base Station's (BS) distance, and node degree were all used in the BOA-based CH selection. To identify the best CH, through the node groups, this fitness function was utilized. By optimizing ACO with three distinct factors, the most energy-efficient path was found. The Advanced Configuration and Power Management Interface (ACPI) and JEHDO are both independently battery- and energy-efficient, according to exploratory results by Sampathkumar et al. [25–27]. An enhanced cuckoo search algorithm is used for node localization to obtain optimal energy efficiency routing by Vaibhav Kotiyal [28].An improved version swarm intelligence approach of the whale optimization algorithm was used for node localization to improve the energy efficiency of the network by Bacanin [29].

3. Dynamic CH Selection and Energy-Efficient Routing Design

3.1. Problem Statement

The current challenges with WSNs are as follows: For WSNs to be energy efficient, an adequate selection of optimal solutions must be addressed in the network. Only when the appropriate technique is prioritized, both distance and energy, the network energy usages decrease. Both large- and small-scale WSN applications could benefit from the energy-efficient WSN. In WSN, rapid transmission of data from CH to BS uses more energy. It causes the network's hot spot problem, which results in packet loss. Because of the node's deployment in a hostile and unmanaged situation, the sensor node becomes inefficient and malfunctioning. Furthermore, while transferring data packets to their destination, the nodes' energy consumption is a major concern. As a result of the nodes' lack of energy, data transmission packets are dropped.

Because the energy consumption within every node is primarily determined by the node's distance, energy utilization is openly proportional to the node's distance. Furthermore, packet loss is prevented in WSN by taking into account each node's energy. By establishing multi-hop routing among the network, the routing challenge can be overcome. The path from the source to the BS is created using the IJLFA approach. The routing considers the nodes' residual energy, the distance between them, and the hop number in each cluster as optimal solutions. The packet loss across networks is decreased when these parameters are taken into account. As a result, an energy efficient WSN is modeled to perform well in two large and small-scale WSNs.

3.2. Network Model

Figure 1 displays the WSN schematic structure. Depending on the following conditions, the model of the network is formulated:

- In terms of processing time and initial energy, the entire nodes are the same as every other in WSN.
- Depending on the Euclidean distance, the distance of the sensor is evaluated.
- The node's position is constant after the deployment, and the nodes are casually positioned in the sensing situation.
- From the nodes, the BS obtains the distance and residual energy information. CHs are selected by an effective CH selection algorithm based on this information. Hence, the route among the CHs to the BS is obtained by the routing process.

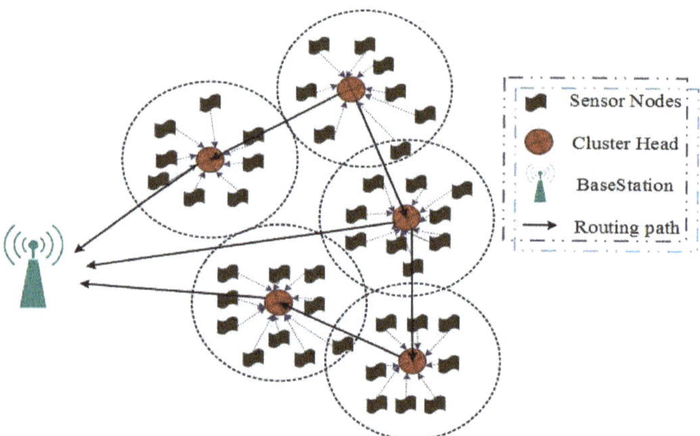

Figure 1. Structure of WSN network model.

3.3. Energy Model

A typical first-order radio paradigm is used to estimate the transmission & reception energy. Equations (1) and (2) represent the amount of energy used to broadcast and capture packets of bits at a given distance.

$$En_{txer}(\delta,d) = \begin{cases} \delta \times E_{dis} + \delta \times \varepsilon_{fs} \times d^2 & if\ d \leq d_0 \\ \delta \times E_{dis} + \delta \times \varepsilon_{mp} \times d^4 & if\ d > d_0 \end{cases} \tag{1}$$

$$En_{rxer}(\delta,d) = \delta \times E_{dis} \tag{2}$$

where the energy consumed at the receiver and transmitter is denoted as En_{txer} *and* En_{rxer} and the threshold distance is defined by d_0, δ represents the information bits, and d is the distance between the sender and receiver. Equation (3) is used to compute the threshold distance.

$$d_0 = \sqrt{\frac{\varepsilon_{fs}}{\varepsilon_{mp}}} \tag{3}$$

where ε_{fs} and ε_{mp} indicate the amplification energy in the multipath and free space models, respectively. The transmitter amplifier model was used to determine these ε_{fs} and ε_{mp}.

3.4. Proposed Methodology

The suggested method includes 2 algorithms: one for network routing and the other for selecting the CH. As IJLFO and CH select the ideal route between the BS and CH, ICOA is utilized to determine appropriate sensors. Via the IJLFO-generated channel, following that, the CHs send the data obtained to the BS. The overall process of the proposed methodology is shown in Figure 2.

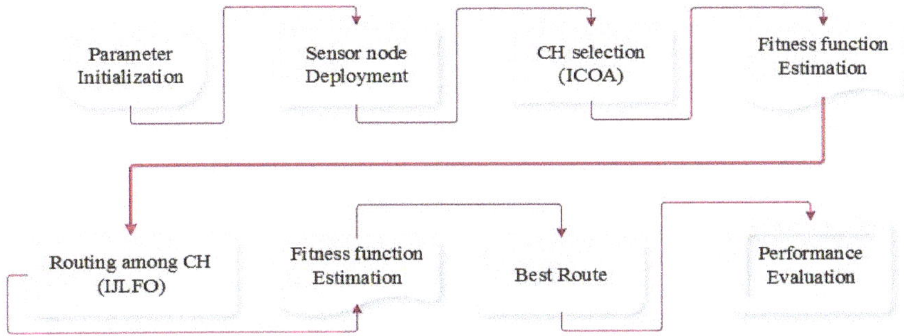

Figure 2. Schematic Block Diagram of Proposed Methodology.

3.4.1. CH Selection Using ICOA

Figure 3 represents the flow chart of selection of CH using ICOA method. The steps are as follows.

Sensor nodes parameter and random initialization: Initially, the node's global population, including Np packs, each with Ns nodes, is initialized. At a certain time, t, the pth pack population is an integer vector, and it is displayed as:

$$S_i^{p,t} = (X_1, X_2, \ldots X_d) \tag{4}$$

where the optimized problem dimension is represented as d, and the ith threshold is represented as x_i.

$$S_{i,j}^{p,t} = \text{lb}_j + \text{R}_j(\text{ub}_j - \text{lb}_j), \; j = 1, 2, \ldots, d \tag{5}$$

where lower and upper bounds are represented as ub_j and lb_j, R_j is a random value.

ICOA Fitness function: From the sensors group in the network, the optimal CH is selected by the ICOA fitness function. The residual energy is considered in the fitness function to evade a dead node as a CH. Higher centrality is used to reduce the distance of transmission among the members. The mathematical forms of fitness functions and their definitions are discussed below:

(a) CH Residual energy

CH collects data from ordinary sensor nodes and transmits it to BS in a network. Because the CH takes more energy to perform the preceding activities, the node with the most residual energy is the strongest choice to be a CH. The following Equation (6) describes residual energy (f_1)

$$f_1 = \sum_{i=1}^{m} \frac{1}{\text{E}_{\text{CHi}}} \tag{6}$$

where the ith CH's residual energy is E_{CHi}

(b)   Distance among sensor nodes:

It specifies the range among usual sensor nodes as well as its CH. The dissipation of energy for the node mostly depends on the transmission path distance. If the chosen node has a minimum transmitting distance near BS, then the node's energy consumption is small. Sensors to CH (f_2) the distance is displayed in Equation (7).

$$f_2 = \sum_{j=1}^{m} \sum_{i=1}^{I_j} D\left(s_i, CH_j / I_j\right) \tag{7}$$

where, $D\left(s_i, CH_j / I_j\right)$ denoted the sensor i and CHj distance and I_j is the sensor node's quantity belongs to CH.

(c)   CH and BS Distance:

The energy consumption of the node is evaluated based on the distance via the transmitting track. When BS is situated away from CH, for example, data transmission will require a lot of energy. As a result, the abrupt drop in CH could be related to increased energy use. Therefore, throughout information transfer, the node that is closest to BS is selected. The optimal solution of distance among the BS (f_3) and the CH is represented by the following Equation (8).

$$f_3 = \sum_{i=1}^{m} D\left(CH_j, BS\right) \tag{8}$$

where $D\left(CH_j, BS\right)$ is the distance between BS and CH_j.

(d)   Node degree

It specifies how many sensor nodes each CH has. Because CHs with more cluster members lose their energy for a shorter period, the CHs with fewer sensors are chosen. Equation (9) expresses the degree of node (f_4).

$$f_4 = \sum_{i=1}^{m} I_i \tag{9}$$

where I_i is the number of CHi sensor nodes.

(e)   Node centrality

Node centrality (f_5) is an expression that expresses how far a node is from its neighbors, represented in Equation (9).

$$f_5 = \sum_{i=1}^{m} \frac{\sqrt{\frac{\left(\sum_{j \in n}^{m} D^2(i,j)\right)}{n(i)}}}{L} \tag{10}$$

where $n(i)$ is the number of CHi's neighboring nodes and L is the network dimension.

Every objective value is assigned a weight value. Several objectives are combined into a single function in this situation. S1, S2, S3, S4, and S5 are the weighted values. Equation (10)depicts the single objective function.

$$F = S1f_1 + S2f_2 + S3f_3 + S4f_4 + S5f_5 \tag{11}$$

where, $\sum_{r=1}^{5} Sr = 1$ $Sr \epsilon (0, 1)$, the values of Sr are 0.35, 0.25, 0.2, 0.1 and 0.1 correspondingly.

Updating new solutions: In each group, the alpha node is a node indicating it has the best fitness value among the groups listed below:

$$alpha(\alpha) = min\left(S_i^{p,t}\right) \; i = 1, 2, \ldots .ni \tag{12}$$

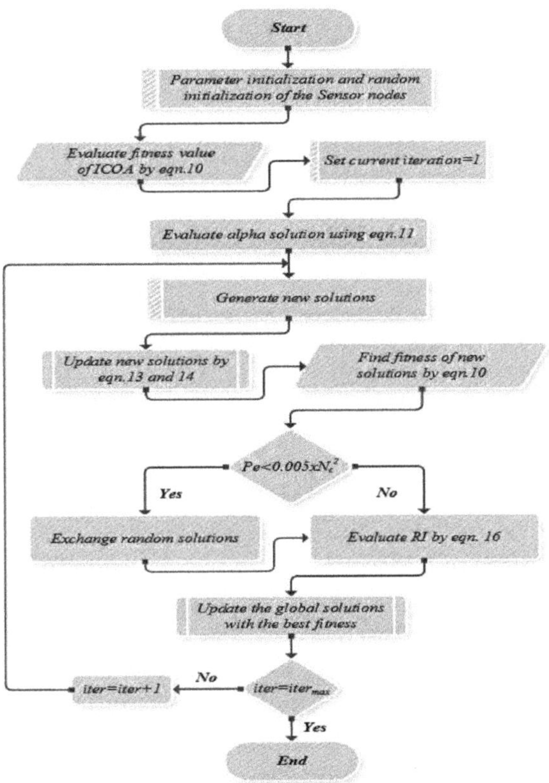

Figure 3. Flowchart for CH selection using ICOA.

The following rule is used to produce new socioeconomic problems after detecting the tendency of coyotes as well as an alpha coyote for each pack:

$$new_{S_i^P} = S_i^P + R1 \times \left(\alpha - S_{r1}^P\right) + R2 \times \left(S_{gbest} - S_{r2}^P\right) \tag{13}$$

If the fitness function improves, the previous social conditions are replaced by new ones.

$$S_i^P = \left\{new\_S_i^P, \text{ ifnew\_fit}_i^P < \text{fit}_i^P S_i^P, \text{ else} \right. \tag{14}$$

$$fit_i^P = \left\{new\_fit_i^P, \text{ ifnew\_fit}_i^P < \text{fit}_i^P \text{fit}_i^P, \text{ else} \right. \tag{15}$$

Expulsion and admission: It is possible for nodes to leave packs by becoming solitary, or they might establish a pack instead, which happens with a high probability p_e.

$$p_e = 0.005 \cdot N_C^2 \tag{16}$$

Calculate RI in two iterations: The exploiting process is used to update the best solution so far by Equation (17).

$$RI = \left| \frac{f\left(S_{gbest}(i-1)\right) - f\left(S_{gbest}(i)\right)}{f\left(S_{gbest}(i-1)\right)} \right| < toll \tag{17}$$

where $f\left(S_{gbest}(i-1)\right)$ and $f\left(S_{gbest}(i)\right)$ are the fitness values of the two most recent best solutions. The toll is a constant.

3.4.2. Clusters Formulation Using the Potential Function

Using the possible function indicated in Equation (18), the CHs are provided to sensor nodes once the CHs have been selected by ICOA. Sensor nodes with shorter transmission distances and more remaining energy are assigned to the CH. As a result, the quantity of energy consumed during the data transmission phase will be lower.

$$sn_p = \frac{z \times E\left(CH_j\right)}{D\left(s_i, CH_j\right)} \tag{18}$$

where z is the proportionality constant, sn_p is the potential of the sensor node and $D\left(s_i, CH_j\right)$ is the distance between the CH_j and sensor s_i; The sensor is assigned to a specific CH with greater potential and $E\left(CH_j\right)$ denotes the residual energy of that CH. When the distance between the two different CHs and sensor nodes is equal, the sensor node is connected to the CH with higher energy.

3.4.3. Routing Algorithm Using IJLFA

Due to the availability of local optima that can trap the search, the original Jaya method may be unable to reach the best solution. To solve this difficulty, we proposed modifying the original Jaya method by multiplying its step size by random values. Instead of using random numbers to reposition the particles, the following Equation (19) is used

$$X_{j,k}^{i+1} = X_{j,k}^i + |lf_1|\left(X_{j,best}^i - \left|X_{j,k}^i\right| - |lf_2|\left(X_{j,worst}^i - \left|X_{j,k}^i\right|\right)\right) \tag{19}$$

where lf_1 and lf_2 are 2-numbers chosen at random from the Lévy distribution, respectively. The power-law index β, which is required in Equation (20) to sample random numbers from the Lévy distribution, is the only added parameter compared to the original Jaya algorithm.

$$D_{lf} = \frac{U}{|V|^{1/\beta}} \tag{20}$$

where the power-law index is denoted by β, V is a random number drawn at N(0, 1), and U describes a random number drawn at N(0, σ2), and the standard deviation σ is given by Equation (21),

$$\sigma = \left(\frac{\Gamma(1+\beta) \times \sin\left(\frac{\pi\beta}{2}\right)}{\Gamma\left(\frac{(1+\beta)}{2}\right) \times \beta \times 2^{\frac{\beta-1}{2}}}\right)^{\frac{1}{\beta}} \tag{21}$$

where Γ denotes the gamma function. Even though this new distribution is a minor change in the algorithm, it causes significant changes in the optimization procedure, which, as we will see later, results in higher actual quality to demonstrate effectiveness.

Steps convoluted in the IJO-LF algorithm are as follows:

Step 1: Population size (N) initialization: Decision variables' numbers $D(j = 1, 2, \ldots D)$, decision variables for upper and lower bounds of $X_{max,j}, X_{min,j}$ and maximum number of iterations it_{max} as an ending stage.

Step 2: Initialization at random: The population of N solutions $X_i(1, 2, \ldots N)$ inside the search space boundary. Individually jth dimension of ith solution i.e., $X_{j,i}$, is initialized among $X_{max,j}$ and $X_{min,j}$ as per Equation (22).

$$X_{j,i} = X_{min,j} + r(0,1)\left(X_{max,j} - X_{min,j}\right) \tag{22}$$

Step 3: Fitness function evaluation: F_i for each separate solution to the specific issue and determine the worst and best solutions, i.e., X_{worst} and X_{best} correspondingly. The solution with the F greatest value is considered the worst, while the solution with the slowest value is regarded as the best when minimizing the objective function.

Step 4: Random numbers consideration: r1 and r2 among 0 and 1 with uniform distribution, Equation (18) is used to update the values of choice variables with each solution.

Step 5: Evaluate the fitness F_i^{new} for comparing the fitness F_i of each formed solution to the fitness of the prior solution.

Step 6: If $F_i^{new} > F_i$ for a maximization problem and $F_i^{new} < F_i$ for a minimization problem, replace χ_i with a fresh produced solution, i.e., χ_i^{new} new for a minimization problem, otherwise preserve the prior solution, i.e., χ_i and update the relevant fitness function value.

Step 7: Sort all of the solutions according to their fitness by,

$$\chi^{new1} = F_i^{new} > F_i^{\ new} + \alpha \oplus lf(\beta) \tag{23}$$

Step 8: Evaluate the fitness F_i^{new1} of new solutions generated.

Step 9: The worst and best solutions are determined using the updated fitness function values

Step 10: Repeating steps 4–9 until reaching the termination requirement, or else report the best solution X_{best}.

3.4.4. Cluster Maintenance

For balancing the load amongst clusters, the management of clusters is one of the most critical phases in this study. For inter-cluster traffic, clusters closer to the BS consume too much energy. As a result, the management of the cluster phase is necessary to prevent the failure of the node. As a consequence, the lifetime of data transmission from the source node to the BS increases. The ICOA is re-set to cluster the network if the CH's residual energy exceeds the threshold level. The CHs are then chosen using the clustering process, and for calculating the routing path between the BS and CHs, the IFLFA is used.

In this proposed methodology the ICOA algorithm is used to achieve an effective CH selection. The CHs are chosen based on five criteria: residual energy, distance from neighbors, node degree, node centrality, and distance from the BS. Among the nodes, these factors are utilized to choose the best CH. To avoid node failure during the transmission of data, BS constantly monitors the nodes' residual energy. From the source to BS through CH, the IJLFA algorithm is used to find the best transmission path. To minimize the nodes' energy consumption, it finds the shortest path. This IJLFA and ICOA-based route generation and optimal CH selection resulted in the development of an energy-efficient WSN. During the transmission of data, an energy efficient WSN is utilized to increase the overall packets transferred to BS, extending the network lifetime.

4. Results and Discussion

4.1. Performance Metrics

The performance metrics are as follows:

Alive nodes: The number of alive nodes in a network is defined by the number of nodes that are alive. When a network contains a large number of active nodes, the network performance improves.

Average energy consumption: During each iteration, it specifies how much energy each node uses on average

Total packets transferred to BS: The overall data packets sent to BS are proportional to the number of alive nodes and their remaining energy. When there are a lot of alive nodes, BS obtains a lot of packets.

Throughput: Over WSN, the number of bits delivered to BS is referred to as throughput. Bits per second are used to measure throughput.

Packet drops ratio: It refers to the amount of data lost from the source to BS during transmission.

Routing overhead: It is described as the ratio of the total number of packets obtained by BS to the total number of packets generated.

4.2. Simulation Setup

The ease of suitable data analyses and arithmetic operations are the major reasons for choosing MATLAB. In the detecting region of 200 m × 200 m, there are between 100 and 150 sensor nodes distributed at random. The first-order radio model is used as an energy design to compare the proposed technique to different routing protocols. The simulation parameters used in the experiment are listed in Table 1. The goal of this research is to lower the entire energy usage of each network node. As a result, IJLFO-based routing between the CHs and ICOA-based CH selection is used to provide cluster-based routing. For better CH selection, distance to BS, node centrality, residual energy, node degree, and distance to neighbors are the inputs given to ICOA. Distance, node degree, and residual energy are also inputs to the IJLFO. Because these methods are commonly employed to enhance the WSNs energy efficiency, this proposed method is contrasted to several established approaches such as the Threshold ensitive Energy Efficient Sensor Network protocol (TEEN), Power-Efficient Gathering in Sensor Information Systems (PEGASIS).

Table 1. Simulation Parameters.

Parameters	Value
Number of sensor nodes	100 and 150
Sensing range	250 m × 250 m
Initial energy	0.5 J
Base station	1
Packet size	4000 bits
Number of CH	4
Number of the source node	1

4.3. Performance Analysis

A developed methodology comparison to existing schemes is shown first. The two existing techniques used to assess the introduced methodology are TEEN and PEGASIS. This assessment is carried out in two separate circumstances relating to the base station's location. The BS is located in the center of the area used to assess short-range communications, which is referred to as the first scenario. The BS is placed outside of the region where long-range transmissions are analyzed.

The following figures show the performance analyses of the proposed method with two existing routing protocols.

Figures 4 and 5 illustrate the rate of decrease of alive nodes of proposed method compared to existing methods. Figures 6 and 7 show that the proposed method has very less average energy consumption compared with TEEN, PEGASIS methods with a network size of 100 nodes and 150 nodes. Figures 8 and 9 illustrate the proposed method network lifetime in terms of rounds with conventional methods. The graph shows a good improvement in the network lifetime of the proposed method compared with existing methods of network size 100 and 150 nodes.

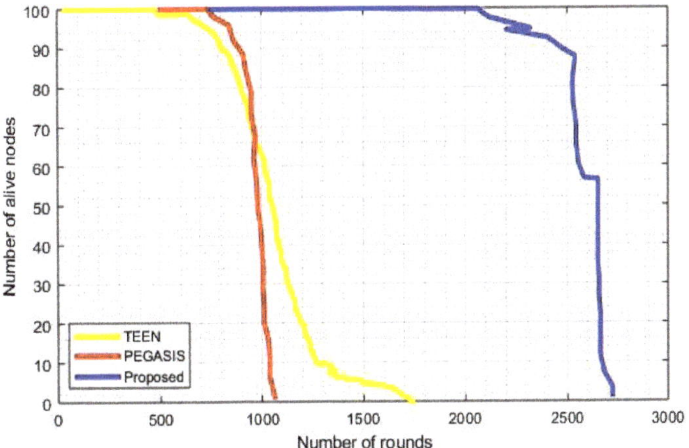

Figure 4. Number of rounds vs. number of alive nodes (100 nodes).

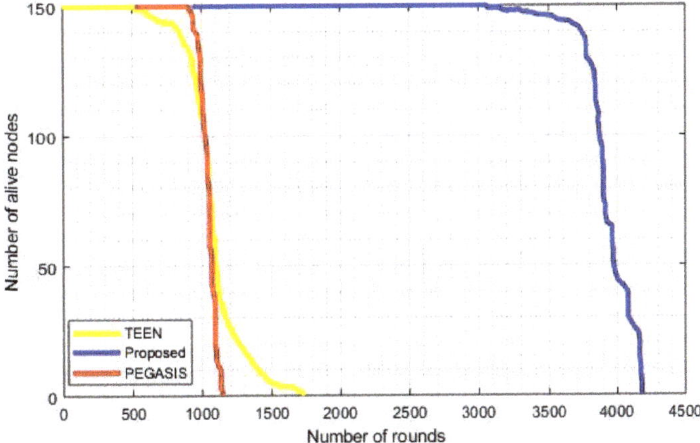

Figure 5. Number of rounds vs. number of alive nodes (150 nodes).

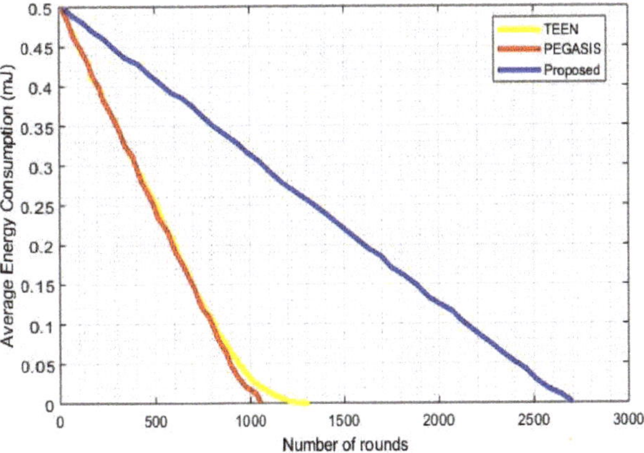

Figure 6. Energy consumption with 100 nodes.

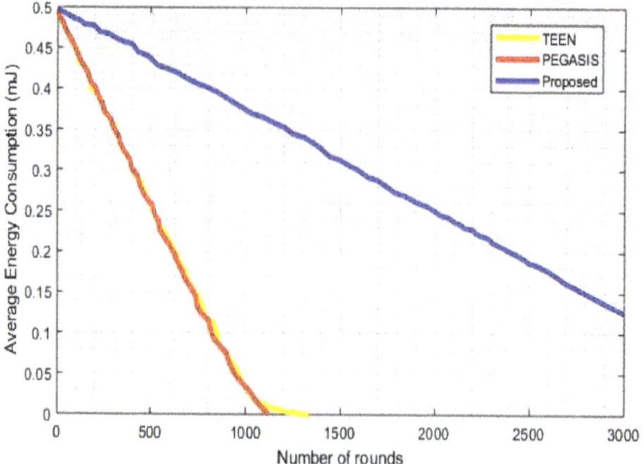

Figure 7. Energy consumption with 150 nodes.

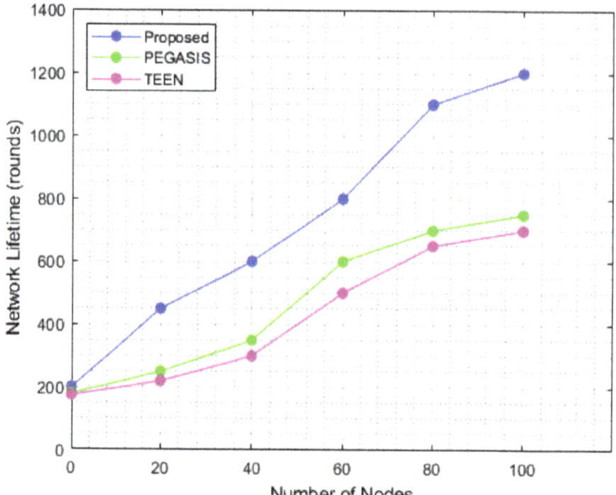

Figure 8. Number of nodes vs. network lifetime with 100 nodes.

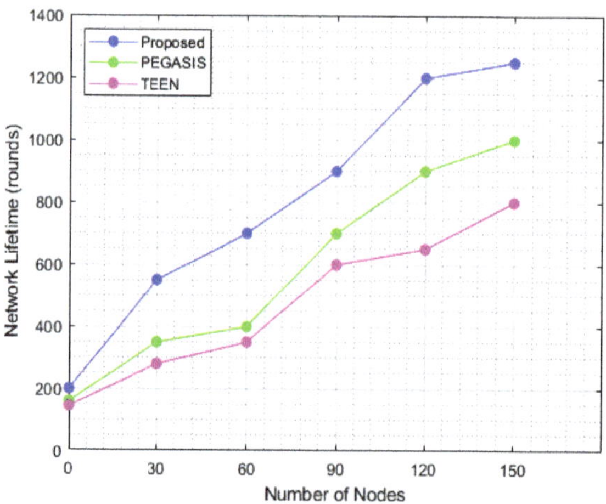

Figure 9. Number of nodes vs. network lifetime with 150 nodes.

Figures 10 and 11 illustrate the decrease in routing overhead compared to conventional methods with the proposed technique, in terms of rounds. The packet delivery ration simulation results of TEEN, PEGASIS, and proposed methods with 100 and 150 nodes are shown in Figures 12 and 13. Figures 14 and 15 depict the comparison of the proposed technique with existing techniques in terms of packet drop ratio with 100 and 150 nodes. Figures 16 and 17 show the throughput performance of the proposed method with conventional methods. The graphs show a greater improvement in throughput with the increase in sensor nodes.

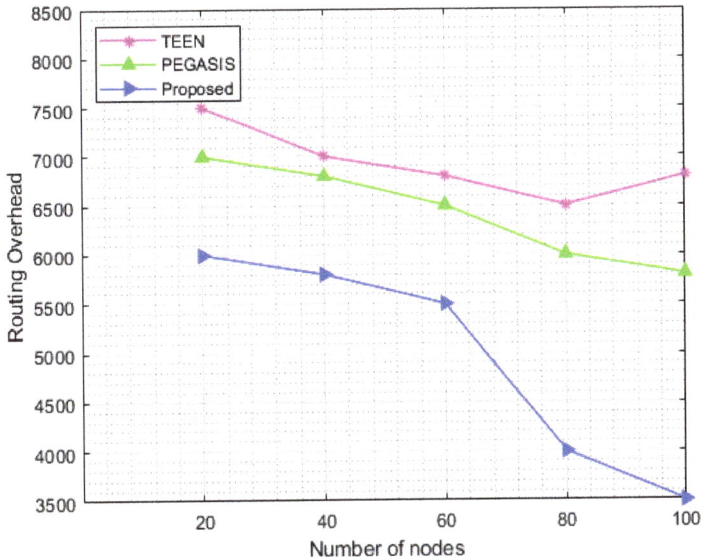

Figure 10. Number of nodes vs. Routing Overhead (100 nodes).

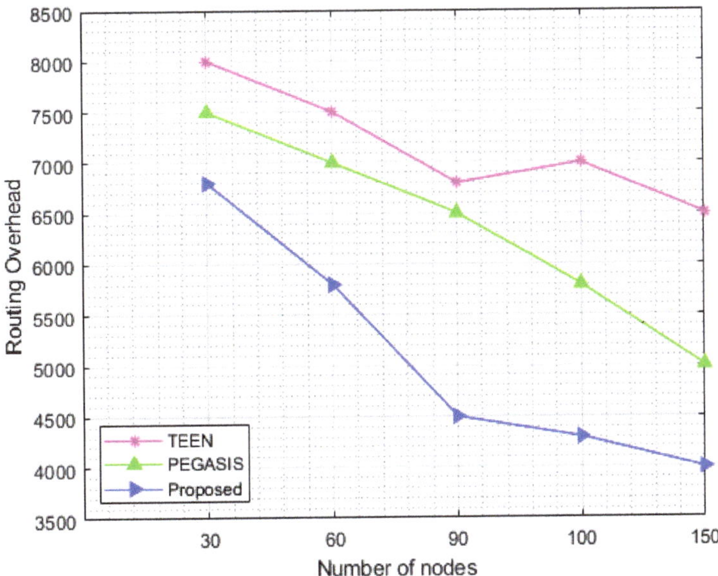

Figure 11. Number of nodes vs. Routing Overhead (150 nodes).

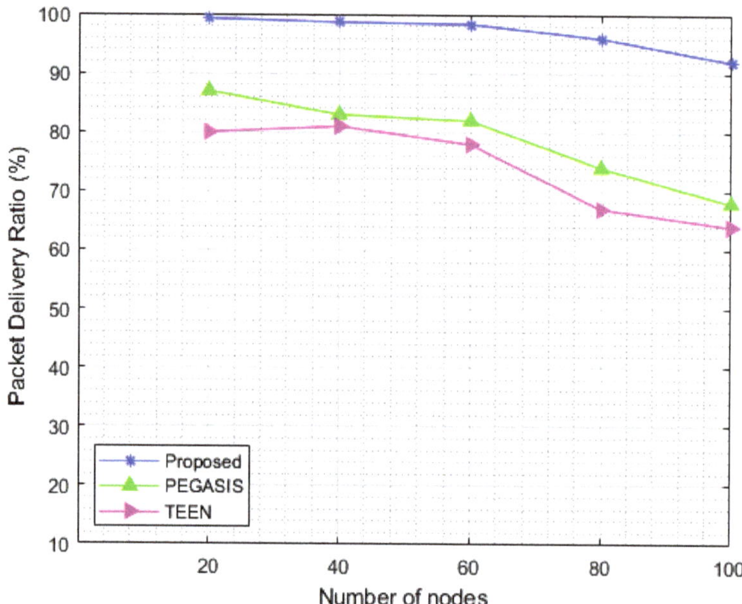

Figure 12. Number of nodes vs. packet delivery ratio (100 nodes).

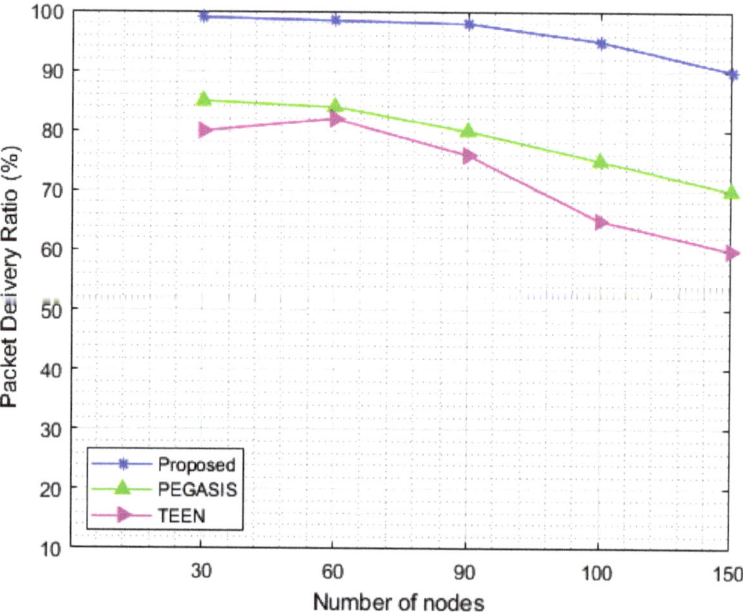

Figure 13. Number of nodes vs. Packet delivery ratio (150 nodes).

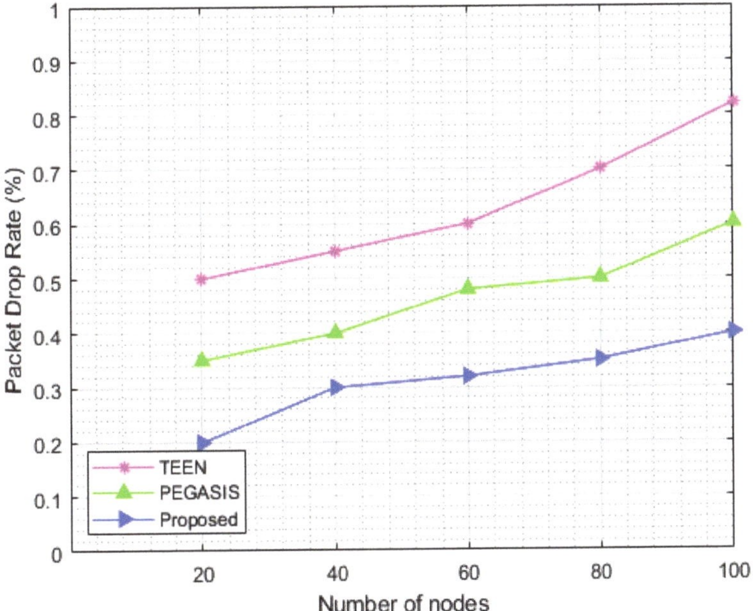

Figure 14. Number of nodes vs. packet drop ratio (100 nodes).

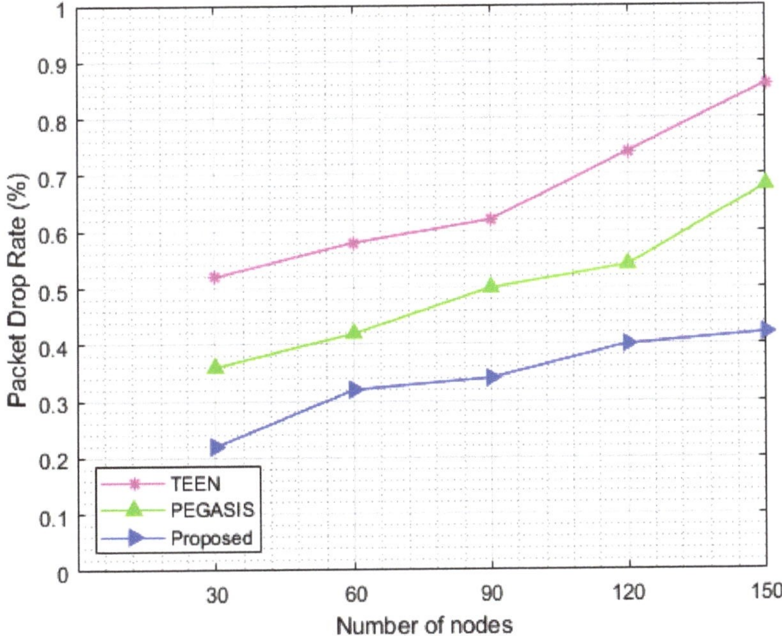

Figure 15. Number of nodes vs. packet drop ratio with 150 nodes.

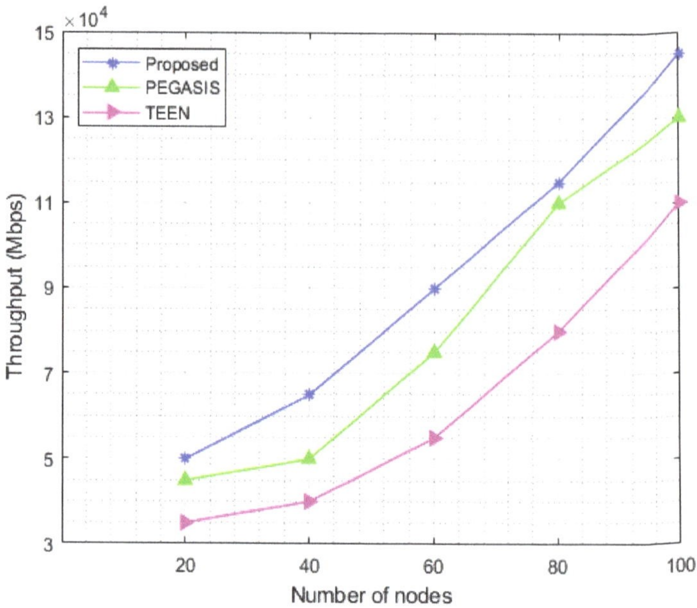

Figure 16. Number of nodes vs. Throughput (100 nodes).

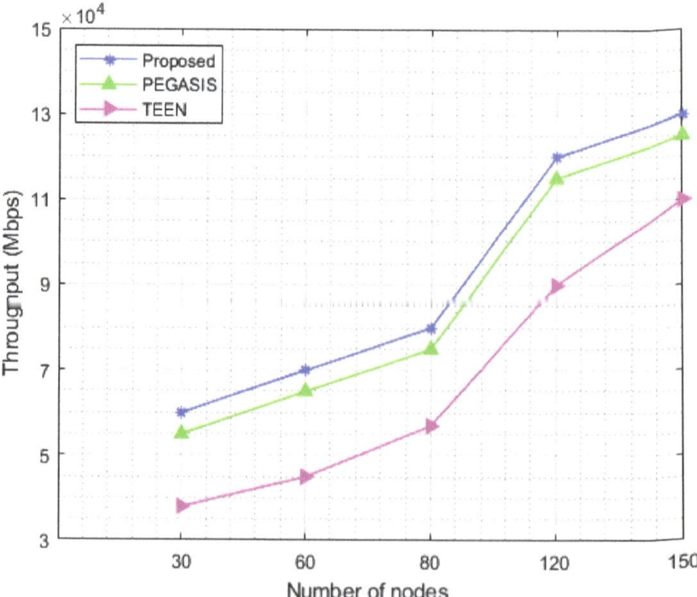

Figure 17. Number of nodes vs. throughput (150 nodes).

The simulation results indicate that the proposed strategy outperforms established methods. The average energy consumption determines the efficiency and effectiveness of the proposed method, no of alive nodes, routing overhead, packet delivery ratio, packet drop ratio, network lifetime, and throughput.

5. Conclusions

Effective route generation and CH selection are regarded as challenging problems in WSN. In this research, a combination of ICOA and IJLFA is proposed to optimize the overall energy consumption and the network lifetime is also increased. Based on five varying metrics such as node degree, base station distance, node centrality, distance to neighbors, and node residual energy. ICOA is used to choose the precise CH. With the help of the fitness values function, the best CH is selected from the cluster nodes. The energy-efficient routing is created by combining three factors such as distance, node degree, and residual energy, to improve IJLFA. Throughout the simulation of the proposed algorithms, the base station is moved from inside to outside. When compared to current methods, the proposed scheme performed better than PEGASIS and TEEN in terms of sustained network usage. Additionally, the proposed methodology is contrasted with other routing approaches. In terms of network performance, it is found that the proposed scheme performed better than PEGASIS and TEEN.

Author Contributions: Conceptualization, K.S.; Supervision; Original draft and review& editing, I.A.; validation, K.S.; All authors have read and agreed to the published version of the manuscript.

Funding: This research received no external funding.

Data Availability Statement: Data will be shared for review based on the editorial reviewer's request.

Conflicts of Interest: The authors declare no conflict of interest.

References

1. SriVenkateswaran, C.; Sivakumar, D. Secure cluster-based data aggregation in wireless sensor networks with aid of ECC. *Int. J. Bus. Inf. Syst.* **2019**, *31*, 153–169. [CrossRef]
2. Jun, Y.; Craig, A.; Shafik, W.; Sharif, L. Artificial Intelligence Application in Cybersecurity and Cyberdefense. *Wirel. Commun. Mob. Comput.* **2021**, *2021*, 3329581. [CrossRef]
3. Lee, W.K.; Schubert, M.J.; Ooi, B.Y.; Ho, S.J.Q. Multi-source energy harvesting and storage for floating wireless sensor network nodes with long range communication capability. *IEEE Trans. Ind. Appl.* **2018**, *54*, 2606–2615. [CrossRef]
4. Putra, S.A.; Trilaksono, B.R.; Riyansyah, M.; Laila, D.S.; Harsoyo, A.; Kistijantoro, A.I. Intelligent sensing in multiagent-based wireless sensor network for bridge condition monitoring system. *IEEE Internet Things J.* **2019**, *6*, 5397–5410. [CrossRef]
5. Sharma, K.; Anand, D.; Sabharwal, M.; Tiwari, P.K.; Cheikhrouhou, O.; Frikha, T. A Disaster Management Framework Using Internet of Things-Based Interconnected Devices. *Math. Probl. Eng.* **2021**, *2021*, 9916440. [CrossRef]
6. Sahu, S.; Silakari, S. Analysis of Energy, Coverage, and Fault Issues and their Impacts on Applications of Wireless Sensor Networks: A Concise Survey. *Network* **2021**, *10*, 13. [CrossRef]
7. Shu, T.; Chen, J.; Bhargava, V.K.; de Silva, C.W. An energy-efficient dual prediction scheme using LMS filter and LSTM in wireless sensor networks for environment monitoring. *IEEE Internet Things J.* **2019**, *6*, 6736–6747. [CrossRef]
8. El Khediri, S.; Fakhet, W.; Moulahi, T.; Khan, R.; Thaljaoui, A.; Kachouri, A. Improved node localization using K-means clustering for Wireless Sensor Networks. *Comput. Sci. Rev.* **2020**, *37*, 100284. [CrossRef]
9. Su, S.; Zhao, S. A hierarchical hybrid of genetic algorithm and particle swarm optimization for distributed clustering in large-scale wireless sensor networks. *J. Ambient. Intell. Humaniz. Comput.* **2017**, 1–11. [CrossRef]
10. Sangeetha, M.; Sabari, A. Genetic optimization of hybrid clustering algorithm in mobile wireless sensor networks. *Sens. Rev.* **2018**, *38*, 526–533.
11. Sennan, S.; Ramasubbareddy, S.; Balasubramaniyam, S.; Nayyar, A.; Kerrache, C.A.; Bilal, M. MADCR: Mobility aware dynamic clustering-based routing protocol in Internet of Vehicles. *China Commun.* **2021**, *18*, 69–85. [CrossRef]
12. Qureshi, K.N.; Tayyab, M.Q.; Rehman, S.U.; Jeon, G. An interference aware energy efficient data transmission approach for smart cities healthcare systems. *Sustain. Cities Soc.* **2020**, *62*, 102392. [CrossRef]
13. Arjunan, S.; Sujatha, P. Lifetime maximization of wireless sensor network using fuzzy based unequal clustering and ACO based routing hybrid protocol. *Appl. Intell.* **2018**, *48*, 2229–2246. [CrossRef]
14. Abualigah, L.M.; Khader, A.T.; Hanandeh, E.S.; Gandomi, A.H. A novel hybridization strategy for krill herd algorithm applied to clustering techniques. *Appl. Soft Comput.* **2017**, *60*, 423–435. [CrossRef]
15. Sayed, G.I.; Khoriba, G.; Haggag, M.H. A novel chaotic salp swarm algorithm for global optimization and feature selection. *Appl. Intell.* **2018**, *48*, 3462–3481. [CrossRef]
16. Venkateswarulu, B.; Subbu, N.; Ramamurthy, S. An efficient routing protocol based on polar tracing function for underwater wireless sensor networks for mobility health monitoring system application. *J. Med. Syst.* **2019**, *43*, 218. [CrossRef] [PubMed]
17. Jiang, B.; Huang, G.; Wang, T.; Gui, J.; Zhu, X. Trust based energy efficient data collection with unmanned aerial vehicle in edge network. *Trans. Emerg. Telecommun. Technol.* **2020**, *33*, e3942. [CrossRef]

18. Yuvaraj, D.; Sivaram, M.; Ahamed, A.M.U.; Nageswari, S. An efficient lion optimization based cluster formation and energy management in WSN based IoT. In Proceedings of the International Conference on Intelligent Computing & Optimization, Chongqing, China, 6–8 December 2019; Springer: Cham, Switzerland, 2019; pp. 591–607.
19. Cui, Z.; Cao, Y.; Cai, X.; Cai, J.; Chen, J. Optimal LEACH protocol with modified bat algorithm for big data sensing systems in Internet of Things. *J. Parallel Distrib. Comput.* **2019**, *132*, 217–229. [CrossRef]
20. Moussa, N.; El Alaoui, A.E.B. An energy-efficient cluster-based routing protocol using unequal clustering and improved ACO techniques for WSNs. *Peer-to-Peer Netw. Appl.* **2021**, *14*, 1334–1347. [CrossRef]
21. Sankar, S.; Ramasubbareddy, S.; Chen, F.; Gandomi, A.H. Energy-Efficient Cluster-Based Routing Protocol in Internet of Things Using Swarm Intelligence. In Proceedings of the 2020 IEEE Symposium Series on Computational Intelligence (SSCI), Canberra, ACT, Australia, 1–4 December 2020; IEEE: Manhattan, NY, USA, 2020; pp. 219–224.
22. Loganathan, G.B.; Salih, I.H.; Karthikayen, A.; Kumar, N.S.; Durairaj, U. EERP: Intelligent Cluster based Energy Enhanced Routing Protocol Design over Wireless Sensor Network Environment. *Int. J. Mod. Agric.* **2021**, *10*, 1725–1736.
23. Yong, J.; Lin, Z.; Qian, W.; Ke, B.; Chen, W.; Ji-fang, L. Tree-Based Multihop Routing Method for Energy Efficiency of Wireless Sensor Networks. *J. Sens.* **2021**, *2021*, 6671978. [CrossRef]
24. Maheshwari, P.; Sharma, A.K.; Verma, K. Energy efficient cluster based routing protocol for WSN using butterfly optimization algorithm and ant colony optimization. *Ad Hoc Netw.* **2021**, *110*, 102317. [CrossRef]
25. Sampathkumar, A.; Murugan, S.; Rastogi, R.; Mishra, M.K.; Malathy, S.; Manikandan, R. Energy Efficient ACPI and JEHDO Mechanism for IoT Device Energy Management in Healthcare. In *Internet of Things in Smart Technologies for Sustainable Urban Development*; Kanagachidambaresan, G.R., Maheswar, R., Manikandan, V., Ramakrishnan, K., Eds.; EAI/Springer Innovations in Communication and Computing; Springer: Cham, Switzerland, 2020. [CrossRef]
26. Sampathkumar, A.; Mulerikkal, J.; Sivaram, M. Glowworm swarm optimization for effectual load balancing and routing strategies in wireless sensor networks. *Wirel. Netw.* **2020**, *26*, 4227–4238. [CrossRef]
27. Sampathkumar, A.; Murugan, S.; Sivaram, M.; Sharma, V.; Venkatachalam, K.; Kalimuthu, M. Advanced Energy Management System for Smart City Application Using the IoT. In *Internet of Things in Smart Technologies for Sustainable Urban Development*; Kanagachidambaresan, G.R., Maheswar, R., Manikandan, V., Ramakrishnan, K., Eds.; EAI/Springer Innovations in Communication and Computing; Springer: Cham, Switzerland, 2020. [CrossRef]
28. Kotiyal, V.; Singh, A.; Sharma, S.; Nagar, J.; Lee, C.-C. ECS-NL: An Enhanced Cuckoo Search Algorithm for Node Localisation in Wireless Sensor Networks. *Sensors* **2021**, *21*, 3576. [CrossRef] [PubMed]
29. Bacanin, N.; Antonijevic, M.; Bezdan, T.; Zivkovic, M.; Rashid, T.A. Wireless Sensor Networks Localization by Improved Whale Optimization Algorithm. In *Proceedings of 2nd International Conference on Artificial Intelligence: Advances and Applications*; Mathur, G., Bundele, M., Lalwani, M., Paprzycki, M., Eds.; Algorithms for Intelligent Systems; Springer: Singapore, 2022. [CrossRef]

Article

A Novel Energy Efficient Threshold Based Algorithm for Wireless Body Sensor Network

Suresh Kumar Arumugam [1], Amin Salih Mohammed [2,3], Kalpana Nagarajan [4], Kanagachidambaresan Ramasubramanian [5], S. B. Goyal [6], Chaman Verma [7,*], Traian Candin Mihaltan [8,*] and Calin Ovidiu Safirescu [9]

1   Department of Computer Science and Engineering, Graphic Era Deemed to be University, Dehradun 248002, Uttarakhand, India
2   Department of Computer Engineering, Lebanese French University, Erbil 44002, Kurdistan Region, Iraq
3   Department of Software and Informatics Engineering, Salahaddin University, Erbil 44002, Kurdistan Region, Iraq
4   Department of CSE, PSNA College of Engineering and Technology, Dindigul 624622, Tamilnadu, India
5   Department of CSE, Veltech Dr. Rangarajan Dr. Sagunthala R&D Institute of Science and Technology, Chennai 600062, Tamil Nadu, India
6   Faculty of Information Technology, City University, Petaling Jaya 46100, Malaysia
7   Department of Media and Educational Informatics, Faculty of Informatics, Eotvos Lorand University, 1053 Budapest, Hungary
8   Faculty of Building Services, Technical University of Cluj-Napoca, 40033 Cluj-Napoca, Romania
9   Environment Protection Department, Faculty of Agriculture, University of Agriculture Sciences and Veterrnary Medicine Cluj-Napoca, Calea Manastur No. 3–5, 40033 Cluj-Napoca, Romania
*   Correspondence: chaman@inf.elte.hu (C.V.); mihaltantraian83@gmail.com (T.C.M.)

Citation: Arumugam, S.K.; Mohammed, A.S.; Nagarajan, K.; Ramasubramanian, K.; Goyal, S.B.; Verma, C.; Mihaltan, T.C.; Safirescu, C.O. A Novel Energy Efficient Threshold Based Algorithm for Wireless Body Sensor Network. *Energies* **2022**, *15*, 6095. https://doi.org/10.3390/en15166095

Academic Editor: Alberto Geri

Received: 29 June 2022
Accepted: 18 August 2022
Published: 22 August 2022

Publisher's Note: MDPI stays neutral with regard to jurisdictional claims in published maps and institutional affiliations.

Abstract: Wireless body sensor networks (WBSNs) monitor the changes within the human body by having continuous interactions within the nodes in the body network. Critical issues with these continuous interactions include the limited energy within the node and the nodes becoming isolated from the network easily when it fails. Moreover, when the node's burden increases because of the failure of other nodes, the energy utilization as well as the heat dissipated increases much more, causing damage to the network as well as human body. In this paper, we propose a threshold-based fail proof lifetime enhancement algorithm which schedules the nodes in an optimal way depending upon the available energy level. The proposed algorithm is experimented with a real time system setup and the proposed algorithm is compared with different routing mechanisms in terms of various network parameters. It is inferred that the proposed algorithm outperforms the existing routing mechanisms.

Keywords: wireless body sensor network (WBSN); energy; network lifetime; routing and threshold

1. Introduction

Limited medical facility and need for smart digital environments have resulted in exponential usage of WBSN. The WBSN has wide application in areas, including sports, defence, medical, smart home automation, and Internet of Things (IoT) applications. The WBSN provides mobility to patients and avoids the feeling of being monitored. However, the WBSN should also provide prodigious care like being in the hospital during critical situations [1–7]. These tiny, embedded machines are power starving and energy issue is met through battery sources. In WBSN, the sensor node monitors the physiological signals of the human body and transmits the signal to the central node, called central monitoring unit (CMU), which is connected to doctors and other health workers. The CMU is superior in computing and energy capability when compared to other nodes. The data from the nodes are communicated 24×7 to ensure the medical safety of the seniors and post-surgical patients [8–11]. Engaging full-time monitoring by a node having limited energy supply makes nodes unavailable. In some cases, the sensor nodes are implanted inside the human body to monitor the signals deep inside the subject [12–17]. The overloading and continuous monitoring of the signal from implanted node heats the node and causes tissue damage to the subject [10]. The data from each sensor node are communicated to the sink either by star or mesh topology. Figure 1 depicts the STAR topology and MESH topology of the WBSN.

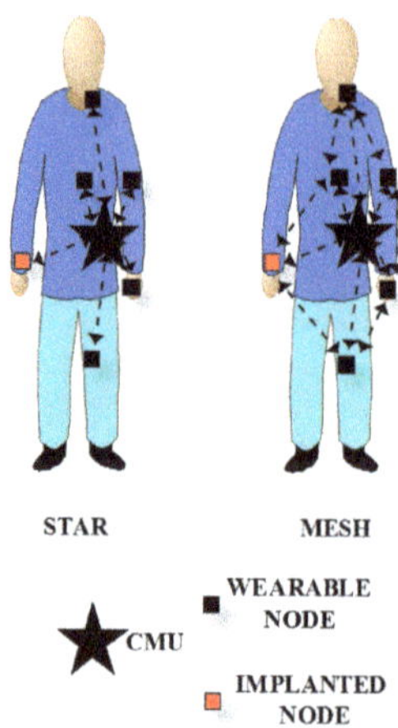

Figure 1. Wireless Body Sensor Network topology.

The star topology follows single hop communication and Mesh topology proceeds with multi hop communication. The cluster head (CH) selection or next hop selection mainly determines the lifetime of the network [18–22]. Table 1 illustrates the data rate and different physiological sensor comparison [3,4,10].

Table 1. Comparison of various sensors.

Sensor Type	Data Rate in bps	Power Consumption (H-High/L-Low)	Privacy (H-High/L-Low)	Bandwidth in Hz
Blood pressure	16	H	H	0–150
Temperature	120	L	L	0–1
EMG	300 k	L	H	0–10,000
ECG	288 k	L	H	100–1000
EEG	43.2 k	L	H	0–1

2. Related Works

Numerous algorithms on enhancing the lifetime and providing failure safe WBSN are proposed in papers [3,4,10,14,16]. These papers concentrate on providing enhanced lifetime by optimally scheduling the nodes and by selecting the next hop towards the sink. The N policy model-based scheduling of nodes in wireless sensor networks suits delay-sensitive applications. The transceiver switching energy is minimized in this N policy model [23–28]. The losses due to transceiver circuit on-off condition are taken into account. The number of the on-off condition of the transceiver circuit is reduced in N-Policy method. The packets are stored and forwarded through the N-Policy scheme. Nodes in networks are highly subjected to many failures based on depletion of energy, failure of hardware, errors in communication link and many other factors. Node failures due to communication link are common and hence the problem of mean time to failure and mean time to repair during the communication link is taken into account. Here, if any fault occurs during active state, the transmission is stopped and the fault at the node is detected. The packet transmission is continued in the active state after the faulty node recovers [10]. The Fail Safe Fault

Tolerant (FSFT) algorithm in [3] enhances network lifetime in the group based WBSN. The packets are classified based on subject status and transmitted through high energy node. However, the thermal effect and tissue damage is not considered in FSFT approach. The TA-FSFT algorithm addresses the heating issue of an implanted node. The implanted node data is routed through a high energy node [10]. The TA-FSFT algorithm fails to consider distance as a factor. However, the power consumption and heat dissipation is with respect to distance.

The Adaptive Threshold based Thermal unaware Energy-efficient Multi-hop Protocols (ATTEMPT) algorithm addresses the topology change during critical condition, the algorithm concentrates in better hop selection. The Mobility-supporting Adaptive Threshold-based Thermal-aware Energy-efficient Multi-hop Protocol (M-ATTEMPT) [4] algorithm addresses the issue of network lifetime during critical conditions the CH rotation is done to enhance the lifetime of the network. The Multihop based WBSN suggested in [8] enhances the lifetime of the network through mesh topology. However, the mesh topology is delay sensitive in nature and node with maximum load are selected as CH, resulting in the network having a shorter lifespan. The list of possible condition for sensor to provide false data or improper data is discussed in [15] that includes (a) loose connection of sensors (b) hardware failure and (c) communication failure. All the above algorithms enhance lifetime of the network, however the node availability during critical condition and safe data delivery should be ensured during critical conditions with low thermal dissipation. To manage the energy consumption of sensor nodes, uses a pseudo-random route discovery algorithm and an improved pheromone trail-based update strategy [29,30]. The routing protocols must be developed to balance traffic among the various nodes that make up a WBASN. Vital signals from the human body demand various levels of service quality for various data kinds [31,32].

The FPLE algorithm proposed improves network lifetime and also ensures availability during the critical condition of the subject. The distance between the hospital and the subjects taken into account and the amount of reserved energy required to monitor subject during critical energy is calculated. The reserved energy is utilized only during abnormal conditions. The subject status is modelled as a finite state machine (FSM) with three states, i.e., (a) normal (b) above normal, and (c) abnormal. During Above normal and abnormal states exigent care is in need. The threshold-based T* Policy scheme suits delay in sensitive applications; it stores data and forwards after N packets. However, the data from the WBSN during Normal condition are delay insensitive in nature, the T* Threshold framework scheme is adapted during this condition to save energy. During critical condition, the node provides a continuous communication to the network.

3. Fail-Proof Lifetime Enhancement (FPLE) Algorithm

The sensor node connected with the subject is classified to primary sensors and secondary sensors. The primary sensors are always made available to sense the physiological signal of the subject. The secondary sensors are essential during critical conditions and close monitoring of subject is engaged during this state of operation. Here, electro cardio graph (ECG) and pulse rate (PR) signals are considered as primary sensors and made available all time. Continuous monitoring of data from the implanted node increases its thermal dissipation causing tissue damage to the subject, hence the implanted node is activated only during the critical conditions. The subject is realized with three states, i.e., (a) normal (S1) (b) above normal and (S2), and (c) abnormal (S3). The cross-correlation coefficient of sensor data with the subject normal data is considered for state transition.In the case of S1 the cross-correlation coefficient is low and slightly deviated in case of abnormal and the deviation is high in case of the abnormal state. The transition from one state to another state depends on the present input and is memory free in nature. Since the transitions exhibit Markov nature, the probability of transition from state to another state of the FSM is predicted through the Markov approach. Figure 2 illustrates the FSM realization of the subject.

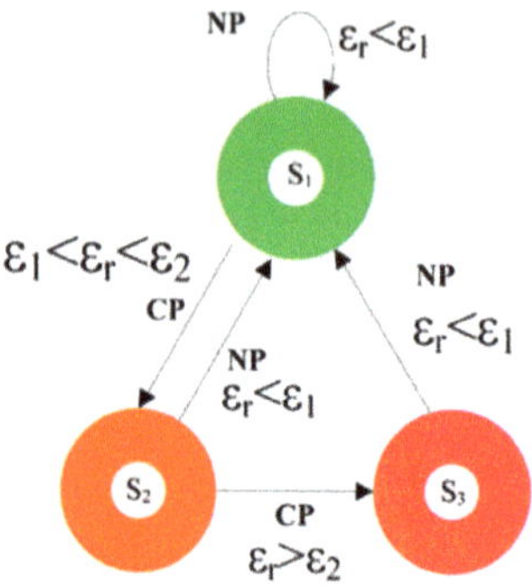

Figure 2. Finite-SM realization of subject.

3.1. Markov Model

In the case of Markov approach, the probability value of r different steps from x state to y state is given by conditional probability approach.

The probability of selecting state x to state y for n different steps is given by Equation (1).

$$P_{xy} = P_r \left(P_n = y \mid P_0 = x \right) \tag{1}$$

Equations (2)–(4) denote the next step transition in Markov chain.
The probability of one-step transition from xth to kth is given in Equation (2).

$$P_{xk} = P_r \left(P_1 = y \mid P_0 = x \right) \tag{2}$$

Equations (3) and (4) give the time homogenous transition from x state to y.
The r steps transition is determined by Equation (3).
The time-homogeneous Markov chain is given as

$$P_r \left(P_n = y \right) = \sum_{r \in s} P_{ry} P_r \left(P_{n-1} = r \right) \tag{3}$$

The general probability of choosing r steps is given in Equation (4).

$$P_r \left(P_n = y \right) = \sum_{r \in s} P_{ry} P_r \left(P_0 = r \right) \tag{4}$$

The probability P of transition x state to y state is represented by the matrix in Equation (5)

$$P = \begin{pmatrix} P_{r11} & P_{r12} & P_{r13} \\ P_{r21} & P_{r22} & P_{r23} \\ P_{r31} & P_{r22} & P_{r33} \end{pmatrix} \tag{5}$$

3.2. Battery Model

The power starving battery is modelled with the voltage decaying process. The fully charged battery shows high voltage due to high charge density and loses while discharging which results in low potential across its terminals. Figure 3 summarises the battery voltage curve of the battery in which E_0 is the initial voltage of the battery when it is fully charged. The point p_1 and p_2 are utilized for setting the threshold limits in the algorithm.

Equation (6) expresses the voltage (V) curve of the battery cell in which E is the voltage across the terminals of the battery and x_1, y_1, z_1, x_2, y_2, and z_2 are curve constants which depend on the diffusion of chemicals inside the battery.

$$E = x_{1\sin}(y_1 a + z_1) + x_2 \sin(y_2 a + z_2) \tag{6}$$

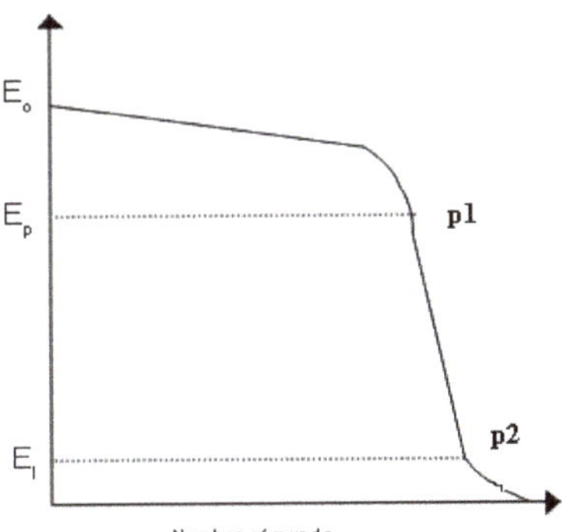

Figure 3. Voltage curve of the battery.

3.3. Radio Model

The reserved energy for a battery is computed with the data rate of the node and distance between the subject and medical help. Table 1 illustrates the data rate of the different sensor attached over the body. The energy taken by the transceiving unit for transceiving a bit of data is provided in Equations (7) and (8).

$$E_{TX}(k,d) = E_{elec}k + E_{fs}kd_2; \; d < d_0$$
$$= E_{elec}k + E_{mp}kd_4; \; d > d_0 \tag{7}$$

$$E_{RX}(k) = E_{elec}k \tag{8}$$

The residual energy for a particular physiological sensor is calculated as given in Equation (9).

$$E_{RE} = \text{Energy due to transceiving unit} \times \text{Data rate} \times \text{time} \tag{9}$$

k—umber of bits;
d—distance between the nodes;
E_{elec}—Energy expense/bit to run the transmitter (TX) or the receiver (RX) circuit;
E_{rx}—Energy expense during data reception;
E_{fs} (pJ/(bit-m^2)), E_{mp}(pJ/(bit-m^2))—Energy expense/bit to process the amplifier of the transmitter determined by the distance between the TX and RX.

The nearby medical help is assumed to be within 9–27 miles. The ambulance arrival time is considered to be 10 min in minimum and not later than 60 min [9,10].

3.4. Threshold T* Policy Framework

The threshold policy T* reduces the number of transceiver switching conditions. The store and forward strategy used reduces power consumption. The optimum number of packets to be in a hold during normal and faulty node condition is given as follows.

Terminologies

λ	Rate at which the packets are arrived
μ	Rate at which the packets are serviced
ρ	Utilization factor
T	Threshold number of packets
E_{TX}	Amount of energy consumed during transmit mode in J
E_{TR}	Amount of energy consumed due to synchronization and switching in J
E[C]	Average cycle duration
E[T]	Average energy consumption of node as a function of T in J
E[I]	Sensor node's average duration in Idle state
C_y	Mean number of cycles

L	Average number of packets
P_I	Idle-state probability

The Idle-state probability (PI) is defined as the ratio of sensor node's average duration in Idle state to the average cycle duration. Equation (10) illustrates the probability of a node to be in idle condition. The process of determining the threshold T* during normal and faulty condition is provided in (a) and (b).

(a)   T* Node during normal operation condition

$$P_I = \frac{E[I]}{E[C]} \tag{10}$$

$$P_I = \frac{E[I]}{E[C]} \tag{11}$$

$$P_I = 1 - \rho \tag{12}$$

From Equations (10) and (11).

$$E[C] = \frac{T}{\lambda(1 - \rho)} \tag{13}$$

where

$$\rho = \frac{\lambda}{\mu}$$

The mean number of cycles (C_y) is given as

$$C_y = \frac{1}{E[C]} \tag{14}$$

Hence C_y is obtained from Equation (13) is given as,

$$C_y = \frac{\lambda(1 - \rho)}{T} \tag{15}$$

The average or mean energy consumption of an sensor node E(T) is given by,

$$E(T) = E_{TX}L + E_{TR}C_y \tag{16}$$

On the basis of M/G/1 queuing model, the mean or average number of packets (L) in a sensor node is expressed as in Equation (17).

$$= \sum_{n=1}^{T-1} nP_I(n) + \sum_{n=1}^{\infty} nP_B(n) \tag{17}$$

where

L equals $\rho + \frac{\lambda^2 E[S^2]}{2(1-\rho)} + \frac{T-1}{2}$ and $E[S^2]$ is the 2nd order service time moment and L is found to be,

$$L = \frac{\rho(2 - \rho)}{2(1 - \rho)} + \frac{T - 1}{2} \tag{18}$$

Equating Equation (18) with Equation (16), E[T] is found. Here, the energy cost E[T]with reference to the average or mean number of packets is given by Equation (19).

$$E(T) = E_{TX}\left(\frac{\rho(2 - \rho)}{2(1 - \rho)} + \frac{T - 1}{2}\right) + E_{TR}\left(\frac{\lambda(1 - \rho)}{T}\right) \tag{19}$$

The optimal threshold (T*) value is the one T which corresponds to the minimal energy taken by the node and the following inequality condition is used to determine T*.

$$E(T) - E(T + 1) < 0 \tag{20}$$

T* is determined based on Equations (19) and (20) as in Equation (21),

$$T^* = \sqrt{\frac{2E_{TR}\lambda(1 - \rho)}{E_{TX}}} \tag{21}$$

(b)   T* Model under node fault condition (Communication failure)

$$E[C] = \frac{T}{\lambda(1 - \rho_{BR})} \tag{22}$$

where

$$\rho_{br} = \rho\left(1 + \frac{\alpha}{\beta}\right)$$

The mean number of cycles C_y is obtained from Equation (14) is given with α, β as follows,

$$C_y = \frac{\lambda(1 - \rho_{BR})}{T} \tag{23}$$

On basis of M/G/1 queuing model, the mean packets L in faulty condition is given in Equation (24).

$$L = \rho_{BR} + \frac{\lambda^2 \rho_{BR}^2 E\left[S^2\right]}{2\rho^2(1 - \rho_{BR})} + \frac{\lambda\alpha\rho E\left[BR^2\right]}{2(1 - \rho_{BR})} + \frac{T - 1}{2} \tag{24}$$

where $E[Br^2]$ is the second-order moment of mean repair time, failure rate follows the Poisson process with mean time to failure $1/\alpha$ and with mean repair time $1/\beta$, and the L is found in Equation (25).

$$= \rho_{BR} + \frac{\rho_{BR}^2}{2(1 - \rho_{BR})} + \frac{\lambda\alpha\rho}{2\beta^2(1 - \rho_{BR})} + \frac{T - 1}{2} \tag{25}$$

Equating Equation (25) with Equation (24), E[T] is calculated and it is given in Equation (26).

$$(T) = E_{TX}\left(\rho_{BR} + \frac{\rho_{BR}^2}{2(1 - \rho_{BR})} + \frac{\lambda\alpha\rho}{2\beta^2(1 - \rho_{BR})} + \frac{T - 1}{2}\right) + E_{TR}\left(\frac{\lambda(1 - \rho_{BR})}{T}\right) \tag{26}$$

Using Equation (26), the optimal threshold (T*) under faulty condition is given in Equation (27), and it is expressed as,

$$T^* = \sqrt{\frac{2E_{TR}\lambda(1 - \rho_{BR})}{E_{TX}}} \tag{27}$$

Figure 4 elucidates the architecture of the FPLE algorithm. The data from the primary sensors are hoped to the coordinator (or) sink through high energy cum high potential node to enhance its lifetime. The other secondary sensor transmits its I-am-alive packet directly to the CMU to ensure its presence for facing the critical conditions. The data from the implanted during above normal condition is hoped to the neighbor node nearby, under the abnormal condition the implanted node data is hoped directly to sink. As the energy dissipation is proportional to distance and number of bits transmitted, the implanted node dissipates low energy transmitting data to the node very nearby. The tissue damage to the implanted node is avoided by hoping data to the nearby node. The node acting as a transceiver follows T-threshold framework where the packets are stored and forwarded towards the sink to save energy of the nodes.

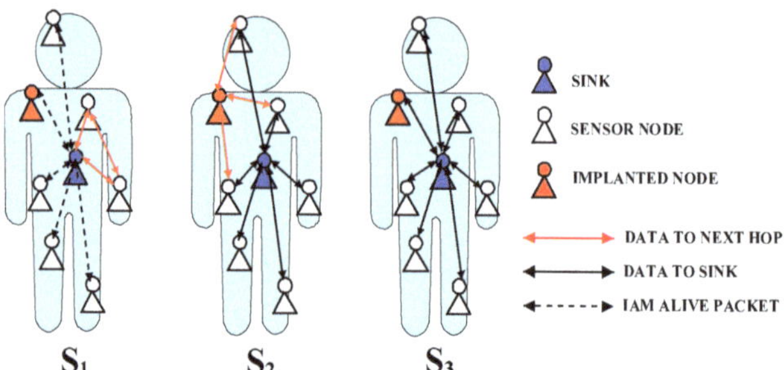

Figure 4. Architecture of the proposed FPLE algorithm.

Algorithm 1 illustrates the proposed FPLE algorithm. The reserved energy in the algorithm is calculated with subject distance and nearby medical help available as given in Equation (9). The alarm is given to the neighbor under a fault condition and low energy condition. Under fault conditions, the data arrival rate from the sensor node increases causing increased data transmission. The fault occurrence α and repair rate β are considered as from Equation (23). The threshold rate between $\beta \rightarrow \alpha$ is considered from the T-Threshold model.

When the state of the subject enters above normal or abnormal state, the sensor nodes check the data rate to detect a loose connection, value limit to find hardware failure, and also check with other primary sensor cross-correlation coefficients to avoid false computation. Algorithm 2 illustrates node fault detection of the proposed FPLE algorithm.

3.5. Proof for FPLE Being Thermal-Aware

The node chooses a high voltage node as the next hop towards the sink. The energy dissipated during the transmission and receiving of data is given in Equations (7) and (8), choosing a high voltage node as a router decreases the current consumption thereby enhancing the network lifetime. Equation (28) illustrates the energy taken from the battery.

$$E = V_B \times I_N \times t \tag{28}$$

Equation (29) provides the amount of energy consumed as distance when the node is performing the role of the router. Increase in distance increases the energy consumption thereby increasing the temperature of the node.

$$E_{CH} = E_{elec}k + E_{mp}kd^4 + E_{elec}k \tag{29}$$

Solving Equations (28) and (29)

$$V_B \times I_N \times t_d = E_{elec}k + E_{mp}kd^4 + E_{elec}k \tag{30}$$

$$I_N = \frac{E_{ELEC}k + E_{ELEC}k^4 + E_{ELEC}k}{V_B \times T_d} \tag{31}$$

Equations (30) and (31) illustrate that the increased load increases the current consumption fastening the battery decay. The voltage decaying process of the battery is illustrated in Equation (6). The node with low voltage acting as a router drains more current to compensate for the rise in load. The implanted node of the subject is only allowed as a participant. Thereby, it is awakened during above normal and abnormal condition, thereby the temperature rise in node due to overloaded is avoided.

Algorithm 1: FPLE routing

BEGIN PROCESS
While(1)
Cross-correlate the ECG and HB value with normal data
    IF cross correlation coefficient $\varepsilon_r < \varepsilon_1$
        subject under normal condition;
        reserved Energy $R_E = E_{RE}$; // compute reserved energy from Equation (9);
        While$_1$ (1)
           receive I am alive packet from all nodes; delay();
      if I am alive packet not received or $R_E < E_{RE}$
alarm;
end if
        end while$_1$
          if$_1$ $V_{ECG} > V_{PR}$
            ECG sensor works as a CH
           Route the PR & ECG data towards sink if T = T*
go to if$_1$;
          else
            PR sensor works as a CH
Routes the ECG data towards sink if T = T*;
          end if$_1$
      else IF cross correlation coefficient $\varepsilon_1 < \varepsilon_r < \varepsilon_2$
        Check node fault();
Wakeup all idle nodes subject under above normal condition;
        Reserved Energy $R_E = 0$;
        Implanted node selects high energy and high signal strength node;
        All the other nodes directly send data towards sink following star topology;
      else
        Check node fault();
Wakeup all idle nodes subject under abnormal condition;
        All nodes send data directly to sink;
end IF
end while

Algorithm 2: FPLE Node Fault Check

Begin node fault
    Check data rate;
    Check value limit
    Check   cross correlation coefficient with neighbor primary sensor node;
End

---

4. Results and Discussion

The proposed FPLE algorithm is simulated with Mat lab 2017 with 8 SNs. Tables 2 and 3 illustrates the node placement in the region of Interest (ROI), in which node 2 is considered as the implanted node as well as the network parameters. The implanted node senses the subject internal temperature during critical conditions. The status of the subject is changed with respect to the FSM and Markov model proposed. Based on the rate of failure considered, the node packet size is varied. Hence, $\alpha = 0.001$ and $(1/\beta) = 1000$ ms value for are considered, as provided by [17].

The list of prelims considered for simulation is mentioned below.

- All SNs are deployed in the ROI.
- The nodes are treated as energy starving.
- The nodes are assumed to be either a Full Function or reduced function device.
- All nodes in nature are static in their respective positions

Table 2. Sensor node deployment in the region of interest.

Node-ID	x Location	y Location
1	20.00	110.00
2	60.00	120.00
3	10.00	80.00
4	70.00	80.00
5	30.00	50.00
6	50.00	60.00
7	30.00	10.00
8	50.00	30.00
9 (CMU)	40.00	80.00

Table 3. Demonstrates the network parameters.

Network Parameters	Value
Network Area	160×80 cm^2
Number of SNs	$8 + 1$ (CMU)
E_{elec}	50 nJ/bit
E_{fs}	10 pJ/bit-m^2
Energy at time 0	0.5 Joule
Probability of being a CH	0.1
Size of normal data	2000 bytes
Size of critical data	4000 bytes
Size of header field	50 bytes

Figure 5 illustrates the lifetime comparison of FPLE, ATTEMPT, Multihop, SingleHop algorithms. The FPLE outperforms ATTEMPT algorithm by 1.94 times extended lifetime. The proposed algorithm sustains for a longer duration, whereas the first node become inactive only after 3200 rounds approximately in FPLE approach.

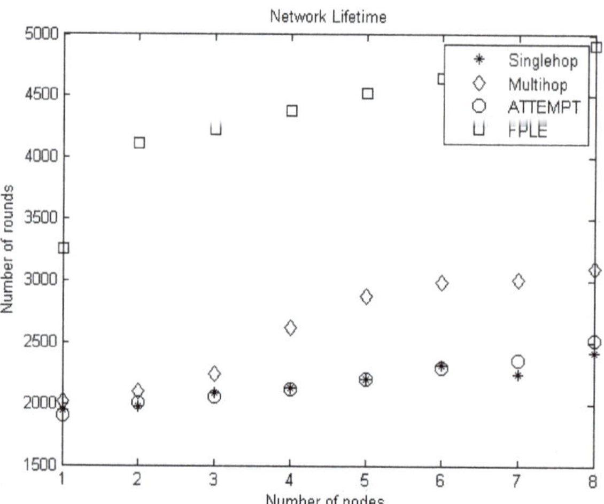

Figure 5. Network lifetime.

The number of packets sent to the coordinator (sink) by SingleHop, MultiHop, ATTEMPT, and FPLE algorithms is given in Figure 6. The FPLE algorithm provides high throughput to the network 1.1 times when compared with ATTEMPT protocol.

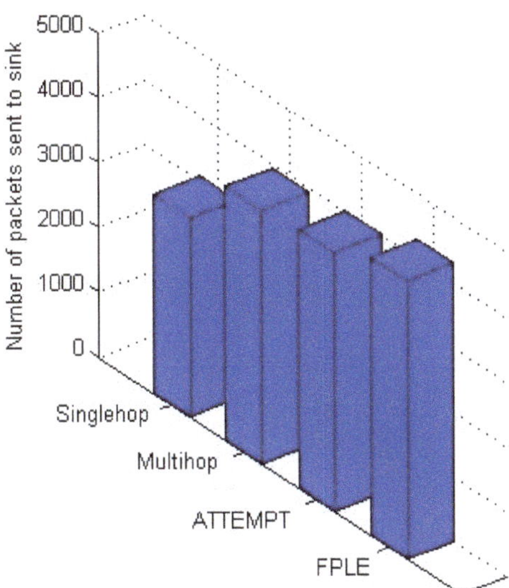

Figure 6. Network Throughput.

Figure 7 elucidates the remaining energy of the SNs after 500, 1000, 3000, and 4000 rounds. The implanted node (red node marked with the arrow) in the simulation survives longer duration. The load to the implanted node is lowest (blue node) to avoid thermal dissipation. As proof of low burdening, the node dies last in the simulation. The FPLE algorithm supports thermal-aware and emergency response during critical conditions.

The left energy of the implanted node in Figure 7 supports low energy consumption and low thermal dissipation.Figure 8 elucidates the mean energy consumed in one round in case of Single Hop, MultiHop, ATTEMPT, and FPLE algorithms.

The proposed FPLE approach consumes less power when considered with ATTEMPT, MultiHop, and SingleHop protocol. The proposed FPLE algorithm is validated with 9 Waspmote in real-time in lab condition. The nodes send the sample HB data to the sink and the battery end terminal voltage is monitored after 25 rounds and 50 rounds. Figure 9 illustrates the sample HB signal transmitted by the node to the sink. Table 4 illustrates the node specification used for validating the work.

Table 4 shows the sensor mote details used to validate the algorithm. Figure 10 illustrates the experimental setup used to validate the proposed FPLE algorithm. The node marked with pink flag is considered to be the implanted temperature node. The node transmits a temperature value to sink during critical conditions. The remaining node transmits the sample ECG signal to sink. The primary sensor nodes [33–37] are programmed to send a high data rate exhibiting noise signal randomly during the simulation time based on α and β values considered. The experimental setup is tested to send 50 cycles of ECG data. The CMU marked with white flag activates the buzzer for fault data generation and low energy.

The residual energy present in the node is proportional to the end voltage of the battery, the battery terminal voltage after every ten rounds of ECG signal transmission is listed in Figure 11. Nodes 3 and 4 serve as the primary sensors in the experimental setup in case of FPLE evaluation. The FPLE algorithm shows high voltage across the battery with respect to SingleHop, MultiHop, and ATTEMPT algorithms, supporting the energy efficiency.

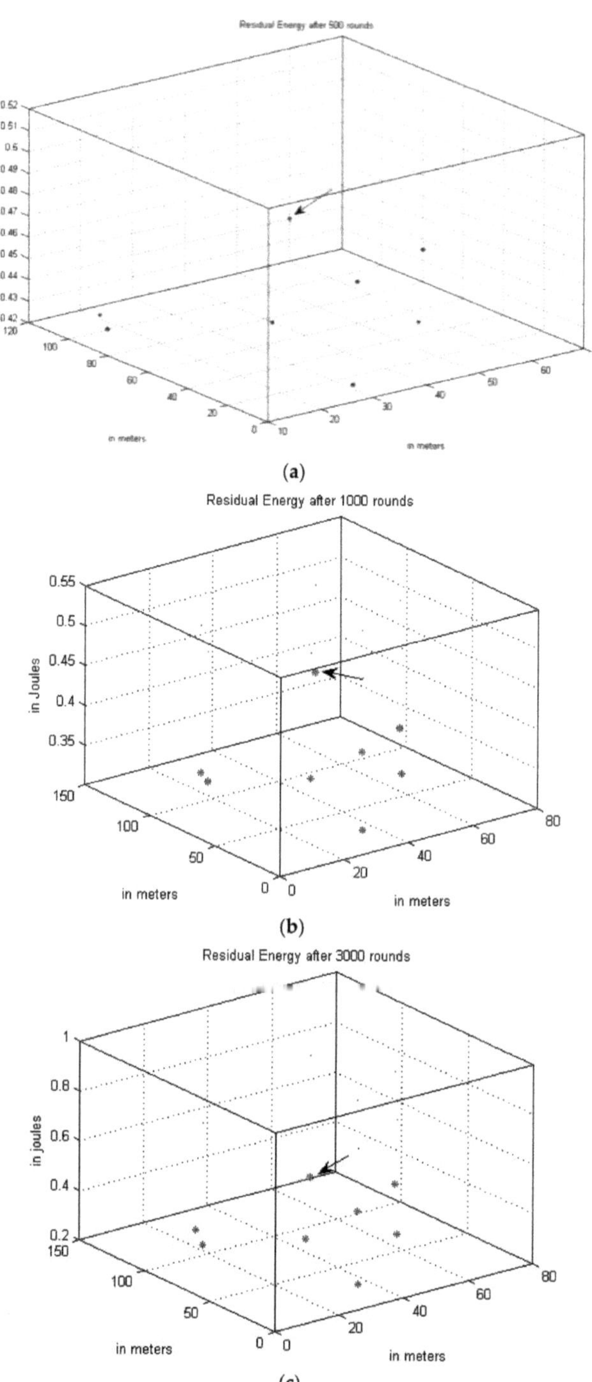

(**a**)

(**b**)

(**c**)

Figure 7. *Cont.*

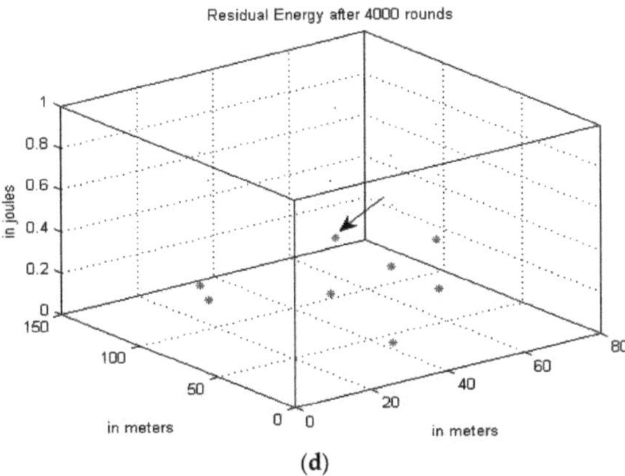

Figure 7. Residual Energy of nodes after 500, 1000, 3000 and 4000 rounds (FPLE). (**a**) Remaining Energy after 500 rounds. (**b**) Remaining Energy after 1000 rounds. (**c**) Remaining Energy after 3000 rounds. (**d**) Remaining Energy after 4000 rounds.

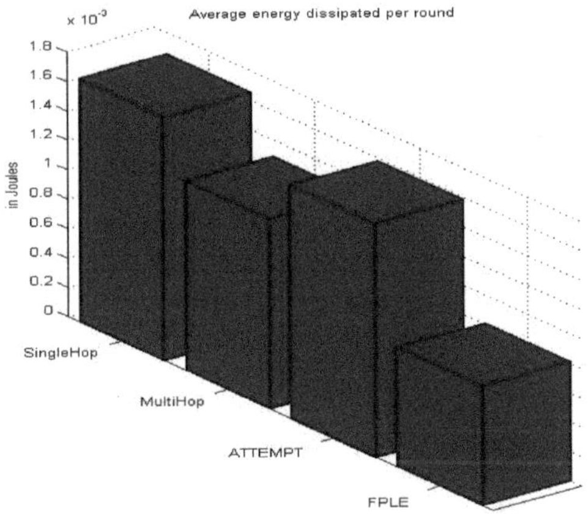

Figure 8. Average energy consumed per round.

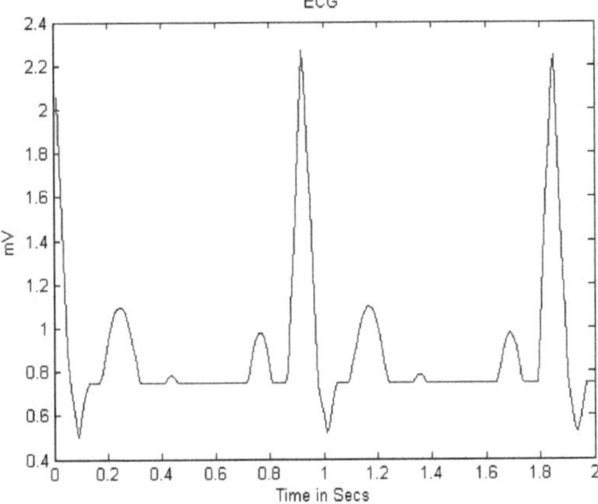

Figure 9. Sample ECG signal.

Table 4. Real-time node deployment metrics.

Parameter	Value
Network area	80 cm × 160 cm
Number of SNs	8 + 1 (CMU)
Base station place	40, 210
Battery capacity	2300 mAh, 3.3 terminal voltage
Probability to be opted as a CH	0.1
Transceiver protocol	Zigbee protocol (XBee) transceivers)
Body sensor data	ECG, Temperature (data from an implanted node)
Processing module	Arduino Uno

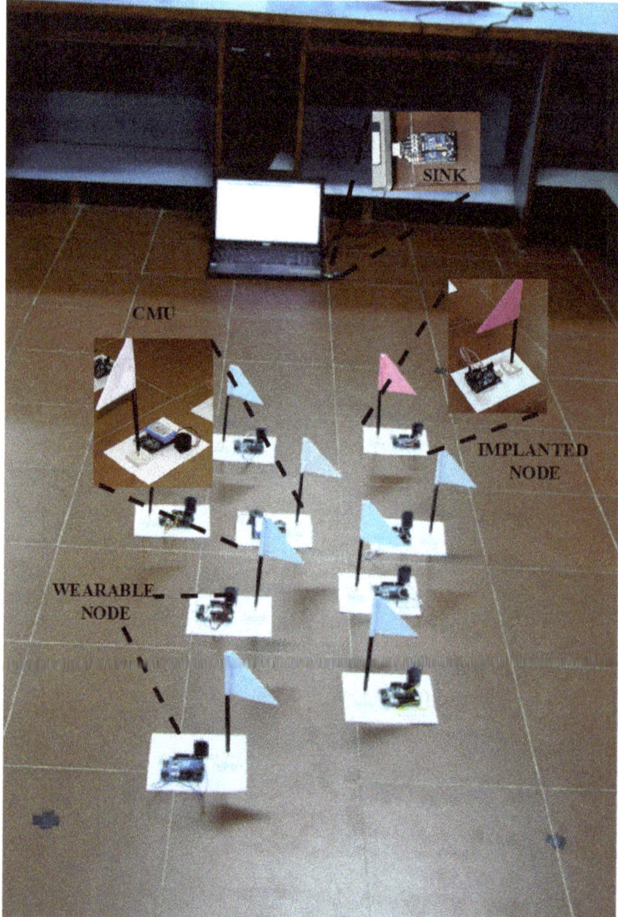

Figure 10. Experimental Setup for validation of the algorithm.

The implanted node terminal voltage after 50 rounds is high in the case of the FPLE algorithm. Table 5 illustrates the protocol comparison, that the FPLE algorithm supports lifetime enhancement, emergency situations, and exhibits thermal and fault awareness.

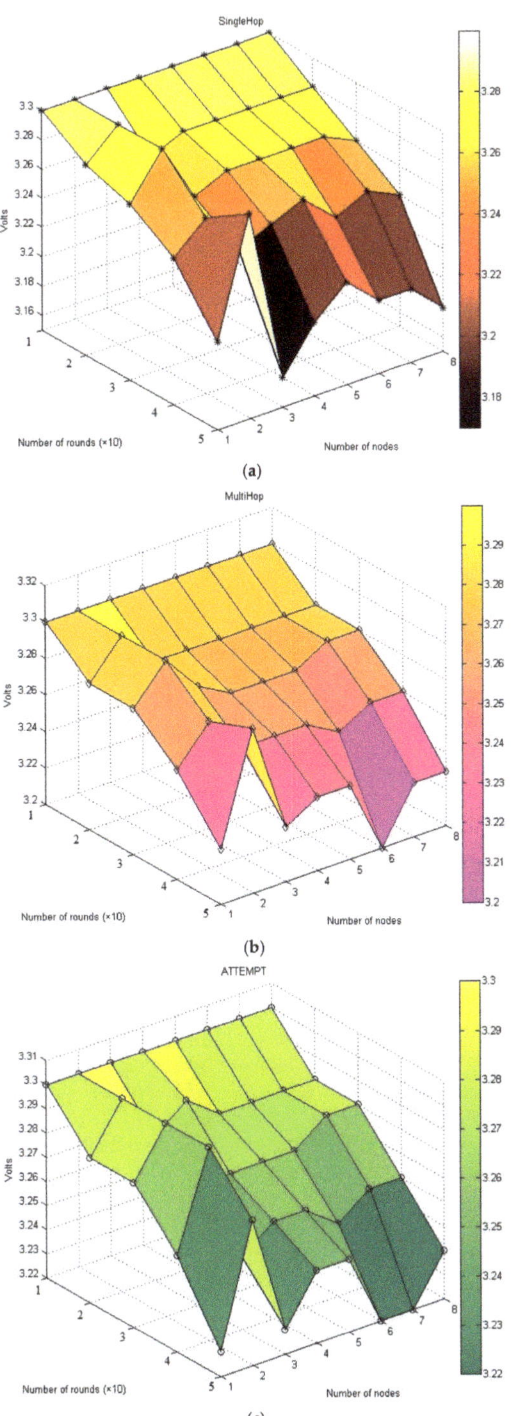

Figure 11. *Cont.*

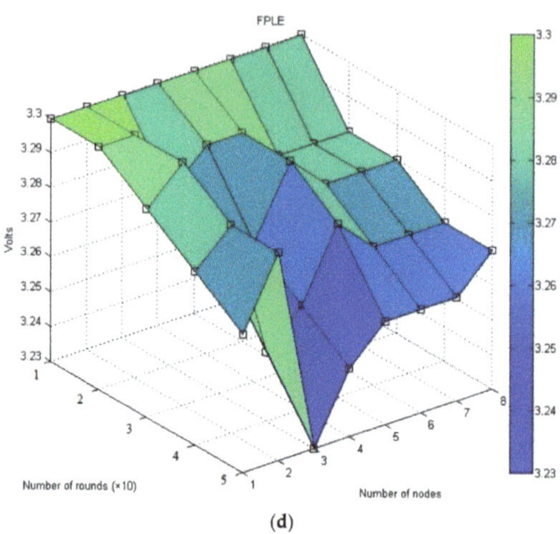

Figure 11. Battery Terminal voltage for every ten rounds of ECG signal by SingleHop, MultiHop, ATTEMPT and FPLE algorithms. (**a**) SingleHop; (**b**) MultiHop; (**c**) ATTEMPT; (**d**) FPLE.

Table 5. Protocol Comparison.

Protocol	Energy Saving	Emergency Situation Handling	Thermal-Aware	Fault Awareness
SingleHop	×	√	×	×
MultiHop	√	×	√	×
ATTEMPT	√	√	√	×
FPLE	√	√	√	√

5. Conclusions

This paper presents a novel FPLE algorithm to addresses the optimal node scheduling based on the energy level and the threshold T* and achieve better network lifetime. The objective of monitoring persons in smart digital environment is achieved by classifying packets based on their status and packets are transmitted towards the sink upon meeting a threshold value T*. A part of the energy in the sensor node is utilized during emergencies to ensure the availability of monitoring the subject during critical conditions. The FPLE algorithm is compared with SingleHop, MultiHop, and ATTEMPT routing schemes and it is inferred that the FPLE algorithm outperforms the SingleHop, MultiHop, and ATTEMPT routing schemes in terms of lifetime and throughput. The FPLE algorithm provides 1.91 times lifetime and 1.1 times throughput when compared with the ATTEMPT communication protocol. The FPLE algorithm is also tested in real-time, also providing better results when compared to ATTEMPT, SingleHop, and MultiHop protocols.

Author Contributions: Conceptualization, writing—original draft, S.K.A.; supervision, A.S.M. and S.B.G.; writing—original draft and review and editing, K.R. and K.N.; validation, S.B.G.; methodology, K.N. and S.B.G.; formal analysis, investigation, C.O.S.; resources, S.B.G. and C.O.S.; Software, T.C.M.; writing—review and editing, T.C.M. and C.V.; project administration, T.C.M., C.O.S. and C.V.; funding acquisition, T.C.M., C.O.S. and C.V. All authors have read and agreed to the published version of the manuscript.

Funding: National Research Development Projects to finance excellence (PFE)-14/2022-2024 granted by the Romanian Ministry of Research and Innovation, this paper was partially supported through BEIA projects, AISTOR, FinSESco, CREATE, I-DELTA, DEFRAUDIFY, Hydro3D, FED4FIRE—SO-SHARED, AIPLAN—STORABLE, EREMI, NGI-UAV-AGRO and by European Union's Horizon 2020 research and innovation program under grant agreements No. 872172 (TESTBED2) and No. 777996 (SealedGRID), SOLID-B5G.

Institutional Review Board Statement: Not applicable.

Informed Consent Statement: Not applicable.

Data Availability Statement: Data will be shared for review based on the editorial reviewer's request.

Acknowledgments: The work of Chaman Verma was supported by the European Social Fund under the project "Talent Management in Autonomous Vehicle Control Technologies" (EFOP-3.6.3-VEKOP-16-2017-00001).

Conflicts of Interest: The authors declare no conflict of interest.

References

1. Kanagachidambaresan, G.R.; SarmaDhulipala, V.R.; Vanusha, D.; Udhaya, M.S. Matlab based modelling of body sensor network using ZigBee protocol. In Proceedings of the International Conference on Computational Intelligence and Information Technology (CCIS 250), Pune, India, 7–8 November 2011; Springer: Berlin/Heidelberg, Germany, 2011; pp. 773–776.
2. Akyildiz, I.F.; Su, W.; Sankarasubramaniam, Y.; Cayirci, E. Wireless sensor networks: A survey. *Comput. Netw.* **2002**, *38*, 393–422. [CrossRef]
3. Kanagachidambaresan, G.R.; Chitra, A. Fail safe fault tolerant mechanism for wireless body sensor network (WBSN). *Wirel. Pers. Commun.* **2015**, *80*, 247–260. [CrossRef]
4. Javaid, N.; Abbas, Z.; Farid, M.S.; Khan, Z.A.; Alrajeh, N. M-attempt a new energy-efficient routing protocol in wireless body area sensor networks. *Procedia Comput. Sci.* **2013**, *19*, 224–231. [CrossRef]
5. Ahmad, A.; Javaid, N.; Qasim, U.; Ishfaq, M.; Khan, Z.A.; Alghamdi, T.A. RE-ATTEMPT: A new energy-efficient routing protocol for wireless body sensor networks. *Int. J. Distrib. Sens. Netw.* **2014**, *10*, 464010. [CrossRef]
6. Abdur, M.R.; Hong, C.; Lee, S. Data-centric multi objective QoS-aware routing protocol for body sensor networks. *Sensors* **2011**, *11*, 917–937. [CrossRef]
7. Kanagachidambaresan, G.R.; Chitra, A. TA-FSFT Thermal Aware Fail Safe Fault Tolerant algorithm for Wireless Body Sensor Network. *Wirel. Pers. Commun.* **2016**, *90*, 1935–1950. [CrossRef]
8. Braem, B.; Latre, B.; Blondia, C.; Moerman, I.; Demeester, P. Analysing and improving reliability in multi-hop body sensor networks. *Adv. Internet Technol.* **2009**, *2*, 152–161.
9. Fleischman, R.J.; Lundquist, M.; Jui, J.; Newgard, C.D.; Warden, C. Predicting Ambulance Time of Arrival to the Emergency Department Using Global Positioning System and Google Maps. *Prehosp. Emerg. Care* **2013**, *17*, 458–465. [CrossRef]
10. Breen, N.; Woods, J.; Bury, G.; Murphy, A.W.; Brazier, H. A national census of ambulance response times to emergency calls in Ireland. *Emerg. Med. J.* **2000**, *17*, 392–395. [CrossRef]
11. Kanagachidambaresan, G.R.; SarmaDhulipala, V.R.; Udhaya, M.S. Markovian model based trustworthy architecture. In *Procedia Engineering*; ICCTSD; Elsevier: Amsterdam, The Netherlands, 2011.
12. SarmaDhulipala, V.R.; Kanagachidambaresan, G.R.; Chandrasekaran, R.M. Lack of power avoidance: A fault classification based fault tolerant framework solution for lifetime enhancement and reliable communication in wireless sensor network. *Inf. Technol. J.* **2012**, *11*, 719.
13. Ababneh, N.; Timmons, N.; Morrison, J.; Tracey, D. Energy balanced rate assignment and routing protocol for body area networks. In Proceedings of the 26th International Conference on the Advanced Information Networking and Applications Workshops (WAINA'12), Fukuoka, Japan, 26–29 March 2012; pp. 466–471.
14. Ben Elhadj, H.; Chaari, L.; Kamoun, L. A survey of routing protocols in wireless body area networks for healthcare applications. *Int. J. E-Health Med. Commun.* **2012**, *3*, 118. [CrossRef]
15. Titouna, C.; Aliouat, M.; Gueroui, M. FDS: Fault Detection Scheme for Wireless Sensor Networks. *Wirel. Pers. Commun.* **2016**, *86*, 549–562. [CrossRef]
16. Hanson, M.A.; Powell, H.C., Jr.; Barth, A.T.; Ringgenberg, K.; Calhoun, B.H.; Aylor, J.H.; Lach, J. Body area sensor networks: Challenges and opportunities. *Computer* **2009**, *42*, 58–65. [CrossRef]
17. Liu, H.; Nayak, A.; Stojmenović, I. Fault-Tolerant Algorithms/Protocols in Wireless Sensor Networks. In *Guide to Wireless Sensor Networks*; Springer: London, UK, 2009; pp. 261–291.
18. Mhatre, V.; Rosenberg, C.P.; Kofman, D.; Mazumdar, R.; Shroff, N. A minimum cost heterogeneous sensor network with a lifetime constraint. *IEEE Trans. Mob. Comput.* **2005**, *1*, 4–15.
19. Du, X.; Guizani, M.; Xiao, Y.; Chen, H.-H. Two tier secure routing protocol for heterogeneous sensor networks. *IEEE Trans. Wirel. Commun.* **2007**, *6*, 3395–3401. [CrossRef]
20. Polastre, J.; Hill, J.; Culler, D. Versatile low energy media access for wireless sensor networks. In Proceedings of the 2nd International Conference on Embedded Networked Sensor Systems, Baltimore, MD, USA, 3–5 November 2004; pp. 95–107.
21. Maheswar, R.; Jayaparvathy, R. Performance Analysis of Cluster based Sensor Networks Using N-Policy M/G/1 Queueing Model. *Eur. J. Sci. Res.* **2011**, *58*, 177–188.
22. Maheswar, R.; Jayaparvathy, R. Performance Analysis of Fault Tolerant Node in Wireless Sensor Network. In Proceedings of the Third International Conference on Advances in Communication, Network, and Computing—CNC 2012, Chennai, India, 24 February 2012; Springer: Berlin/Heidelberg, Germany, 2012.

23. Darwish, A.; Kanagachidambaresan, G.R.; Maheswar, R.; Laktharia, K.I.; Mahima, V. Buffer Capacity Based Node Life Time Estimation in Wireless Sensor Network. In Proceedings of the 8th IEEE International Conference on Computing, Communication and Networking Technologies (ICCCNT), Delhi, India, 3–5 July 2017.
24. Jayarajan, P.; Maheswar, R.; Kanagachidambaresan, G.R. Modified Energy Minimization Scheme Using Queue Threshold Based on Priority Queueing Model. *Clust. Comput.* **2017**, *22*, 12111–12118. [CrossRef]
25. Nageswari, D.; Maheswar, R.; Kanagachidambaresan, G.R. Performance analysis of cluster based homogeneous sensor network using energy efficient N-policy (EENP) model. *Clust. Comput.* **2018**, *22*, 12243–12250. [CrossRef]
26. Jayarajan, P.; Maheswar, R.; Sivasankaran, V.; Vigneswaran, D.; Udaiyakumar, R. Performance Analysis of Contention Based Priority Queuing Model Using N-Policy Model for Cluster Based Sensor Networks. In Proceedings of the Seventh IEEE International Conference on Communication and Signal Processing (ICCSP), Chennai, India, 3 April 2018.
27. Maheswar, R.; Jayarajan, P.; Vimalraj, S.; Sivagnanam, G.; Sivasankaran, V.; Amiri, I.S. Energy Efficient Real Time Environmental Monitoring System Using Buffer Management Protocol. In Proceedings of the Ninth IEEE International Conference on Computing, Communication and Networking Technologies (ICCCNT), Bengaluru, India, 10–12 July 2018.
28. Jayarajan, P.; Maheswar, R.; Kanagachidambaresan, G.R.; Sivasankaran, V.; Balaji, M.; Das, J. Performance Evaluation of Fault Nodes Using Queue Threshold Based on N-Policy Priority Queueing Model. In Proceedings of the Ninth IEEE International Conference on Computing, Communication and Networking Technologies (ICCCNT), Bengaluru, India, 10–12 July 2018.
29. Sampathkumar, A.; Mulerikkal, J.; Sivaram, M. Glowworm swarm optimization for effectual load balancing and routing strategies in wireless sensor networks. *Wirel. Netw.* **2020**, *26*, 4227–4238. [CrossRef]
30. Sampathkumar, A.; Murugan, S.; Rastogi, R.; Mishra, M.K.; Malathy, S.; Manikandan, R. Energy Efficient ACPI and JEHDO Mechanism for IoT Device Energy Management in Healthcare. In *Internet of Things in Smart Technologies for Sustainable Urban Development*; Kanagachidambaresan, G.R., Maheswar, R., Manikandan, V., Ramakrishnan, K., Eds.; EAI/Springer Innovations in Communication and Computing; Springer: Cham, Switzerland, 2020. [CrossRef]
31. Murugan, S.; Sampathkumar, A.; Kanaga Suba Raja, S.; Ramesh, S.; Manikandan, R.; Gupta, D. Autonomous Vehicle Assisted by Heads up Display (HUD) with Augmented Reality Based on Machine Learning Techniques. In *Virtual and Augmented Reality for Automobile Industry: Innovation Vision and Applications. Studies in Systems, Decision and Control*; Hassanien, A.E., Gupta, D., Khanna, A., Slowik, A., Eds.; Springer: Cham, Switzerland, 2022; Volume 412. [CrossRef]
32. Banuselvasaraswathy, B.; Sampathkumar, A.; Jayarajan, P.; Sheriff, N.; Ashwin, M.; Sivasankaran, V. A Review on Thermal and QoS Aware Routing Protocols for Health Care Applications in WBASN. In Proceedings of the 2020 International Conference on Communication and Signal Processing (ICCSP), Chennai, India, 28–30 July 2020; pp. 1472–1477. [CrossRef]
33. Jayarajan, P.; Kanagachidambaresan, G.R.; Sundararajan, T.V.P.; Sakthipandi, K.; Maheswar, R.; Karthikeyan, A. An Energy Aware Buffer Management (EABM) Routing Protocol for WSN. *J. Supercomput.* **2018**, *76*, 4543–4555. [CrossRef]
34. Ugochukwu, N.S.; Goyal, S.B.; Arumugam, S. Blockchain-Based IoT-Enabled System for Secure and Efficient Logistics Management in the Era of IR 4.0. *J. Nanomater.* **2022**, *2022*, 7295395. [CrossRef]
35. Bedi, P.; Goyal, S.B.; Rajawat, A.S.; Shaw, R.N.; Ghosh, A. Application of AI/IoT for Smart Renewable Energy Management in Smart Cities. In *AI and IoT for Smart City Applications. Studies in Computational Intelligence*; Piuri, V., Shaw, R.N., Ghosh, A., Islam, R., Eds.; Springer: Singapore, 2022; Volume 1002. [CrossRef]
36. Sharma, S.; Rani, M.; Goyal, S.B. Energy Efficient Data Dissemination with ATIM Window and Dynamic Sink in Wireless Sensor Networks. In Proceedings of the 2009 International Conference on Advances in Recent Technologies in Communication and Computing, Kottayam, India, 27–28 October 2009; pp. 559–564. [CrossRef]
37. Olariu, S.; Goyal, S.B.; Qaimi, S. Four-Layer Architecture Model for Energy Conservation in Wireless Sensor Networks. In Proceedings of the 2009 Fourth International Conference on Embedded and Multimedia Computing, Jeju, Korea, 10–12 December 2009; pp. 1–3. [CrossRef]

 *energies*

Article

Secure Routing-Based Energy Optimization for IoT Application with Heterogeneous Wireless Sensor Networks

Regonda Nagaraju [1], Venkatesan C [2], Kalaivani J [3], Manju G [4], S. B. Goyal [5,*], Chaman Verma [6], Calin Ovidiu Safirescu [7,*] and Traian Candin Mihaltan [8]

[1] Department of Information Technology, St. Martin's Engineering College, Dhulapally, Secunderabad 500100, India; nagcse01@gmail.com
[2] Department of Electronics and Communication Engineering, HKBK College of Engineering, Bengaluru 560045, India; venkatesanc.ec@hkbk.edu.in
[3] Department of Computing Technologies, SRM Institute of Science and Technology, Kattankulathur, Chennai 603203, India; kalaivaj@srmist.edu.in
[4] Department of CSE, SRM Institute of Science and Technology, Kattankulathur, Chennai 603203, India; manju.shruthi@gmail.com
[5] Faculty of Information Technology, City University, Petaling Jaya 46100, Malaysia
[6] Department of Media and Educational Informatics, Faculty of Informatics, Eotvos Lorand University, 1053 Budapest, Hungary; chaman@inf.elte.hu
[7] Environment Protection Department, Faculty of Agriculture, University of Agriculture Sciences and Veterrnary Medicine Cluj-Napoca, Calea Manastur No. 3–5, 40033 Cluj-Napoca, Romania
[8] Faculty of Building Services, Technical University of Cluj-Napoca, 40033 Cluj-Napoca, Romania; mihaltantraian83@gmail.com
* Correspondence: sb.goyal@city.edu.my (S.B.G.); calin.safirescu@usamvcluj.ro (C.O.S.)

Citation: Nagaraju, R.; C, V.; J, K.; G, M.; Goyal, S.B.; Verma, C.; Safirescu, C.O.; Mihaltan, T.C. Secure Routing-Based Energy Optimization for IoT Application with Heterogeneous Wireless Sensor Networks. *Energies* **2022**, *15*, 4777. https://doi.org/10.3390/en15134777

Academic Editor: Jose A. Afonso

Received: 29 May 2022
Accepted: 27 June 2022
Published: 29 June 2022

Publisher's Note: MDPI stays neutral with regard to jurisdictional claims in published maps and institutional affiliations.

Abstract: Wireless sensor networks (WSNs) and the Internet of Things (IoT) are increasingly making an impact in a wide range of domain-specific applications. In IoT-integrated WSNs, nodes generally function with limited battery units and, hence, energy efficiency is considered as the main design challenge. For homogeneous WSNs, several routing techniques based on clusters are available, but only a few of them are focused on energy-efficient heterogeneous WSNs (HWSNs). However, security provisioning in end-to-end communication is the main design challenge in HWSNs. This research work presents an energy optimizing secure routing scheme for IoT application in heterogeneous WSNs. In our proposed scheme, secure routing is established for confidential data of the IoT through sensor nodes with heterogeneous energy using the multipath link routing protocol (MLRP). After establishing the secure routing, the energy and network lifetime is improved using the hybrid-based TEEN (H-TEEN) protocol, which also has load balancing capacity. Furthermore, the data storage capacity is improved using the ubiquitous data storage protocol (U-DSP). This routing protocol has been implemented and compared with two other existing routing protocols, and it shows an improvement in performance parameters such as throughput, energy efficiency, end-to-end delay, network lifetime and data storage capacity.

Keywords: WSNs; IoT; heterogeneous WSN; multipath link routing protocol (MLRP); hybrid-based TEEN; ubiquitous data storage protocol (U-DSP)

1. Introduction

The IoT and WSNs are becoming viable solutions that are widely utilized in real-time data collection and monitoring applications, which include automated irrigation, target monitoring, observing clinical records, tracking landslides and predictions for forest fire and disaster management. A WSN consists of numerous effective sensor nodes (SNs) that monitor climatic disasters found in harsh or remote areas. Furthermore, the SNs examine the atmospheric factors such as pressure, temperature, humidity, sound and moisture content representing intense symptoms. After an SN completes its sensing operation, the information is gathered and transmitted to the base station (BS). The sensor and data communication units of SNs consume more energy and once the entire energy is exhausted, it expires or is unable to process [1]. Hence, a node that is considered dead is unsuitable for either replacing with or recharging using an alternate power source. Hence, balancing the power utilization of a SN is more essential. To overcome these challenges, several developers have used clustering approaches [2], which can scale up the network lifetime and provide remarkable efficiency. Furthermore, it is supportive in maintaining power as numerous reliable clusters are formed. These clusters, based on the techniques used, can either be considered temporary or permanent. Moreover, clustering distributes the nodes that are placed jointly, which is accomplished based on similarity metrics such as the distance from the base station (BS), radius of transmission and density of the cluster. After the clusters are formed, a node from the cluster is chosen as the cluster head (CH) whose responsibility is to organize the data gathered from cluster members (CM) and transmit those data to the BS. When a single node cluster is considered, it mandatorily communicates with the sink rather than the BS, which finally results in a reduction in power utilization. In WSNs, the serious threats, along with the technical challenges, which arise due to the nature of resource constraints and its limited availability [3], have to be focused on to ensure revision and distribution. With WSNs in an open area, the sensors are more vulnerable to the unfriendly environment that arises due to humidity, increased temperature, pressure, snow, rain, dust and so on. These affect the functions of the wireless sensor network, and hence a demand arises to introduce robust and flexible SNs. Moreover, general and future challenges are constrained to the limited resources, the limited ability of communication, fault tolerance, stability, mobility, bandwidth, precision, reliability, availability, heterogeneity, accountability, uncontrollable setup and denial of service (DoS). Above all, some specific challenges have turned the attention of the researchers towards the utilization of power, network duration, throughput, security and routing protocols. In WSNs, during communication, energy consumption is a highly prominent issue that seeks attention. Energy efficiency has a greater impact on the entire performance of the network and acts as a significant function in the lifespan of a WSN [4]. The metrics that are essential while routing and estimating the cost function (CF) in a WSN are total energy, energy consumed and residual energy. Energy is one of the most significant aspects of the efficient working of wireless sensor networks. The longevity of the network is totally dependent upon the optimum use of the available energy; therefore, optimization is much needed for the efficient utilization of energy. Routing protocol, considered as an important factor, has to be selected carefully to route the data packets safely to the destination with less overhead because of the resource constraints of SNs, namely, their limited power and shorter range for communication. Numerous efforts have been made to bring out the best solutions for WSNs, yet extraordinary works are required [5]. The benefits of the IoT are as follows. An urban IoT has many facilities that may bring a number of benefits such as the management and optimization of traditional public services, including transport and parking facilities, lighting, the surveillance and maintenance of public areas, preservation of cultural heritage, garbage collection, public health care, i.e., hospitals, and schools. Therefore, the availability of different types of data that will store in the cloud or be collected by a data warehouse in the urban IoT should increase the transparency and promote the actions of the local government or municipality towards the citizens, which can increase the awareness of people about the status of their city and their life style. Therefore, the application of the IoT paradigm to the smart city is attractive and expansive for local government and administrations; however, it take times to adopt IoT technology in a wide manner.

This research is organized as follows: A detailed review of the literature related to this research is described in Section 2. The design, algorithms and functionality of the proposed protocol are elaborated on in Section 3. The simulation results of the proposed protocol are presented in Section 4, which is followed by the conclusion in Section 5.

2. Related Works

Mostly, in large scale networks such as the IoT and SNs with restricted constraints, the demanding factor is energy conservation. Generally, in networks based on clusters [6], the controlling entity is CH, whose significant role is to gather data and transmit them. Furthermore, in large scale WSNs integrated with the IoT, the crucial part is to route data securely as they involve constrained resources. Several existing approaches do not provide reliable and secure data routing in the network as they lack protection schema against threats [7]. A low-energy adaptive clustering hierarchy (LEACH) protocol consists of multi-stages [8]. The LEACH protocol's performance was improved by applying a multi-hop basis for transmitting information [9]. The design elucidated the unstable consumption of energy as the clusters were randomly created. Moreover, the multi-hop paths created were not optimized, which resulted in route breaks. In a chain–chain-based routing protocol (CCBRP) [10], the principles of LEACH as well as Power-Efficient Gathering in Sensor Information Systems (PEGASIS) were integrated, which resulted in scalable energy conservation in SNs while forwarding data. This hybrid CCBRP protocol, which was executed in two stages, reflected the drawback of a higher energy consumption in the nodes and the production of high latency. Subsequently, as the scalability is restricted, CCBRP was not suitable for larger networks. Kumar et al., proposed a data collection as well as a load balancing scheme that was designed to save energy. The network performance was improved in this scheme as the aggregating data were sequenced by data forwarding [11]. A vigorous authentication protocol for energy efficiency was developed for an industrial IoT-based WSN, which provided scalable data security. In this method, mutual authentication was followed between nodes, but the energy was consumed gratuitously, thereby compromising the network lifetime [12]. Likewise, the Shamir secret sharing method comprised two phases, namely, share generation as well as reconstruction. The generated secret was distributed among nodes and the secret key was reconstructed by the usage of any nodes of the subset. The data transmission using this method consumed additional energy leading to routing overheads [13]. If the flow entry exists, the forwarding is performed according to the flow table, and if not, the packet-in message is sent to the subset. After receiving the packet-in message, the subset will make a decision. An improved three-layer hybrid clustering method (ETL-HCM) analysed and limited the control traffic specifically while selecting CH. This method achieved an 18% improvement in the lifetime of the network as well as half of the nodes being alive compared with the hybrid hierarchical clustering approach (HHCA). However, this method was not suitable for selecting the grid head (GH) [14]. To beat hateful attacks such as Sybil, wormholes and a black hole, a novel protocol was introduced by Haseeb et al. to ensure reliable routing and data transmission [15]. The functioning of a combined GIN with GEAR protocol resulted in being too complicated to achieve data security [16]. A voting-based hybrid ensemble classification method was introduced for predicting the availability of parking lots, wherein 96% of the accuracy and 89% of the availability rate was achieved [17]. An optimal network coding backpressure routing (NCBPR) method was developed for a large-scale IoT to divert the flow of data packets from highly congested nodes to low ones, thereby balancing the load and optimizing the battery power [18]. The nodes in the network were formed as clusters and the CHs were selected based on the battery power. Moreover, an efficient data aggregation model was involved to improve the network throughput, wherein the redundant data packets were eliminated. An efficient load balancing optimization algorithm for the employment of energy effectual routing as well as a load balancing protocol were used to enhance the network lifetime [19,20]. The authors in [21] explored the cluster-based backpressure routing algorithm for IoTs, which discusses the energy and congestion issues in IoT environment. A brief discussion is also presented related to the sustainable development through the Internet of Things in [22,23].

A novel framework was proposed to handle environmental monitoring using wireless sensors connected with the internet, wherein two distinctive SNs were involved. Moreover, clients observed the information using the web application on the internet from anywhere. When the information of the SN surpassed the specified range, a notification was sent to the clients insisting on the environmental setup being altered accordingly [24–26]. Energy is a critical issue in WSNs and the IoT, specifically when deployed in smart city applications and this was discussed briefly in [27–30].

Research Objectives

1. To establish secure routing for confidential data of the IoT through sensor nodes with heterogeneous energy utilizing MLRP.
2. To optimize the energy and improve the network lifetime using a hybrid-based TEEN (H-TEEN) protocol that has load balancing capacity.
3. To improve the storage capacity using the ubiquitous data storage protocol (U-DSP).

3. Proposed Methodology

The proposed routing protocol with a storage capacity protocol has been discussed in three phases. The first phase implements data routing with the security of the sensor nodes for confidential data through the IoT, which has heterogeneous energy using a multipath link routing protocol (MLRP). The next phase is to incorporate the H-TEEN protocol, which has load balancing capacity for sensor nodes that helps in attaining the energy optimization of nodes. Then the transmission of data is carried out using the security and optimization of energy. Finally, the data are stored with improved capacity using the ubiquitous data storage protocol (U-DSP). The architecture is given in Figure 1.

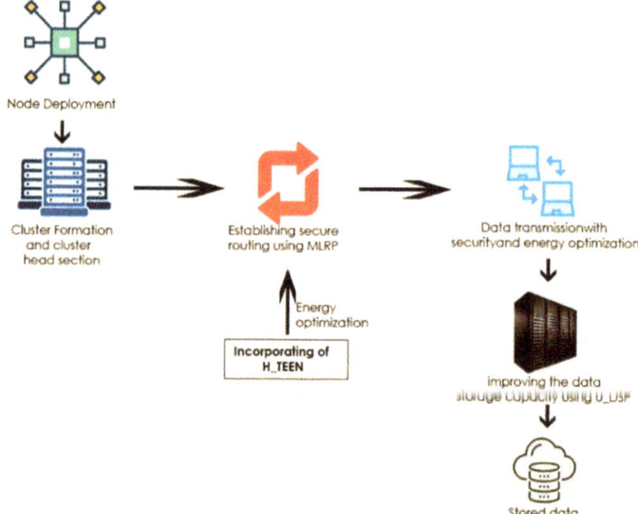

Figure 1. Proposed architecture.

3.1. Multipath Link Routing Protocol (MLRP)

MLRP comprises five various stages, namely, detecting neighbors, constructing topology, distributing pairwise keys, forming a cluster and transmitting data. Every process is described in detail below.

Detection of neighbors and construction of topology: It is considered that every node holds an ID$\{ID_x\}$, certificate $\{CERT_x\}$, public key $\{K_{bs}\}$ and unique shared key $\{K_{xbs}\}$. To identify the neighbors, node broadcasts and receives NBR DET packet containing its ID as well as its CERT as in Equation (1).

$$x \rightarrow \ast : NBR\_DET|ID_x|CERT_x \tag{1}$$

The node receiving the NBR_DET packet initially authenticates the ID of the node by verifying $CERT_x$. When authenticated, the ID is added to the neighbor list by the receiver, or else the packet is dropped such that the unauthenticated node does not participate in this process of detecting neighbors. Once broadcasting is complete, neighbor information is forwarded to BS as in Equations (2) and (3).

$$x \rightarrow BS : NBR\_INFO|ID_x|CERT_x|E(k_{xbs}, NBR_x)| \tag{2}$$

$$MAC(k_{xbs}, NBR\_INFO|ID_x|CERT_x|E(k_{xbs}, NBR_x)) \tag{3}$$

An intermediate node that is receiving the NBR_INFO packet performs few functions such as:

1. The authentication of the SN is verified by its certificate.
2. When the ID of the node is authorized, the packet is again broadcasted by the receiver node.
3. When the receiver receives a similar packet having a similar ID again, the packet is simply dropped.

Hence, a table is maintained by every node termed as a receiver packet table. Thus, the network traffic is reduced and node energy is saved to some extent. Once the NBR INFO packet is reached at BS, BS verifies MAC for authenticity and integrity; then neighbor information is authenticated using K_{xbs} between SN and BS. MAC, obtained from the data and by encrypting K_{xbs}, is used such that the intruder cannot either alter or spoof neighbor information.

3.2. Distribution of Pairwise Key

Once the information about the neighbor is obtained from the network nodes, BS analyzes the exact network topology and generates the neighbor matrix. Then the DFS algorithm is applied by which multiple paths are identified from BS to each source node. Beforehand, for each neighbor pair, BS generates the secret key, a random number, termed as pairwise key with the help of the hash function given by Equations (4)–(6).

$$k_{xy} = h\left(secret, ID_x, ID_y\right) \tag{4}$$

BS unicasts this key to the corresponding node as

$$BS \rightarrow x : PAIR\_KEY|seq_{no}|ID_{bs}|CERT_{bs}|ID_x|ID_y|E(k_{xbs}, k_{xy}\Big|E\left(k_{ybs}, k_{xy}\right))| \tag{5}$$

$$MAC(k_{xbs}, PAIR_{KEY|seq_{no}|ID_{bs}|CERT_{bs}|ID_x|ID_y|E(K_{xbs},K_{xy}|}\Big|E\left(K_{ybs}, K_{xy}\right))) \tag{6}$$

The packet holds its type, sequence number, the ID of BS, neighbor and destination, certificate of BS, pairwise key and MAC of entire data. Every intermediate node that receives the packet performs the following operations:

(1) The certificate of BS is verified with the public key.
(2) The sequence number as well as the node pair is checked in the receiver packet table. When no such value is found, sequence number, type of the packet and node pair along with packet rebroadcasting is stored or else dropped.
(3) When the ID of the destination is identical to its ID, the pairwise key is encrypted, MAC is verified and the encrypted packet of neighbor with nonce as well as its ID encrypted using a pairwise key is sent, which is given by Equation (7).

$$x \rightarrow y : CHALLENGE|ID_y|E\left(K_{yb}, K_{xy}\right)\Big|E(K_{xy}, ID_x|nonce) \tag{7}$$

The packet at node y is decrypted using unique shared key K_{yb} and the pairwise key is generated followed by decrypting the next packet using the generated pairwise key then forwards the packet to x, which is given by Equation (8).

$$y \rightarrow x : CHALLENGE\_REP|ID_x|E(k_{xy}, ID_y|nonce\_1) \tag{8}$$

Both the neighboring nodes verify one another by swapping the challenge packet as well as reducing overhead of resending pairwise key to y from BS. In end, each pair of nodes holds a pairwise key. When the CHALLENGE_REP packet is not received at node x in the expected form, node x reports about the fake node to BS.

3.3. Cluster Formation

BS initiates cluster formation and cluster head (CH) is selected based on residual energy. It is assumed that the energy level of the node does not change after the cluster is formed. In total, 5–8% of the cluster nodes are selected as CH with the criteria given below: (1) Two cluster heads should not be neighbors; (2) every CH has not less than 7–10% nodes as a neighbor. Then BS unicasts CH INT to CH along the route from CH to BS as illustrated in Figure 2. By considering that the node is next hop in route, the format of the packet CH INT is given by Equation (9).

$$BS \rightarrow CH : CH\_INT|ID_{bs}|ID_i|E(K_{ibs}, PATH|seq_{no})|MAC(k_{chbs}, CH\_INT|ID_{ch}|PATH|seq\_no) \tag{9}$$

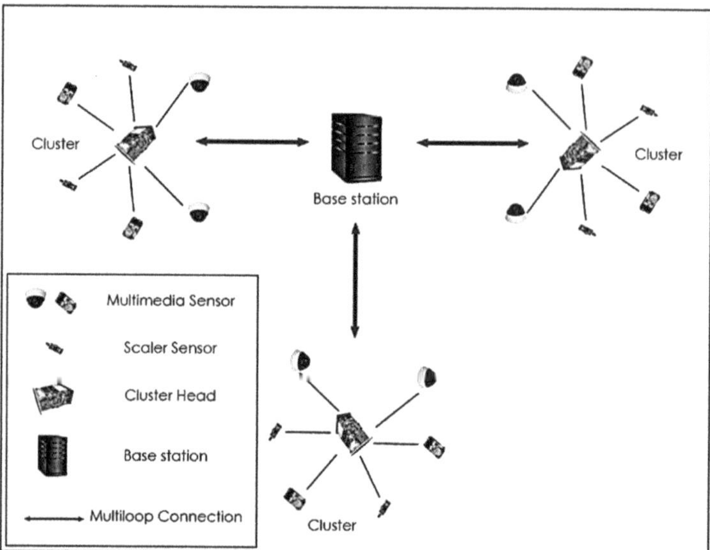

Figure 2. Clustered architecture.

Every node that receives the packet performs the following: (1) ID of the next hop is checked, if identical to its own, routing path (PATH) is decrypted and next hop from PATH is identified or else packet is dropped. (2) The sequence number is checked in the receiver packet table, then store packet type and sequence number if the sequence number did not exist in the table, and the changes needed are made, or else it is stopped. (3) Preceding hop is assigned its ID, while the subsequent one with the ID is found in PATH. (4) Routing table is stored in memory along with the previous one, as well as next hop, such that it helps in transmitting the data to BS. (5) PATH, as well as the sequence path for the subsequent hop, are encrypted with a pairwise key and the updated packet is forwarded. While CH is receiving the CH INT packet, decryption of PATH takes place and the data are verified by the MAC; then acknowledgement (ACK) is sent to BS via the same path.

When ACK is not received by BS within certain time from CH, the path is recomputed and the CH INT packet is sent again. A few criteria are involved in determining the routing path including the path having residual energy and power consumed. A path is selected with more residual energy and small hops count. To form a cluster, a CH ADV packet is sent by CHs to publicize their will.

ID and CERT are present in CH and the ADV packet, such that the receiving node ensures authentication. Nodes receiving several CH and ADV select CH based on two factors: (1) checking if the pairwise key is present in the ID advertised and (2) signal strength of advertisement forwarded. Once CH is selected, the will of the nodes is forwarded by CH JOIN packet with ID as well as MAC with pairwise key and a nonce. Once entire CH JOIN is received, information of the cluster members is sent to the BS by CH, and the TDMA schedule is created based on the total nodes in the cluster and unicast to every member. The packet has the following format by Equations (10)–(12).

$$CH \rightarrow * : CH\_ADV|ID_{ch}|CERT_{ch} \tag{10}$$

$$x \rightarrow CH : CH\_JOIN|ID_x|MAC(K_{xch}, CH\_JOIN|ID_x|nonce_x \tag{11}$$

$$CH \rightarrow x : CH\_SHED|ID_x|E(k_{xch}, t_x)MAC\left(CH_{SHED|ID_x|E(K_{xch},t_x)nocne_x} + 1\right) \tag{12}$$

3.4. Data Transmission

The data transmission phase comprises three subphases;

(1) The data sensed with encrypted and authenticated format transmitted by the member node to CH and when not linked to any route, can sleep so that energy is saved.
(2) The received data are aggregated and compressed by CH to generate a new signal, which is then transmitted to BS through the route specified. (It is considered that node j is subsequent hop in routing table.)
(3) Base station uses an exclusively shared key for decryption and authentication of the data received.

These subphases are observed with the form as given in Equations (13) and (14).

$$x \rightarrow CH : DATA|ID_x|E(K_{xch}, d_x) \tag{13}$$

$$MAC(K_{xch}, DATA|ID_x|E(K_{xch}, d_x)) \tag{14}$$

Once the data are received, they are aggregated by CH and then forwarded to BS by Equation (15).

$$CH \rightarrow BS : AGGR\_DATA|ID_{ch}|ID_j\left|E\left(K_{jch}, seq_{no}\right)\right|E(K_{chbs}, d_{ch})\left|MAC(K_{chbs}, AGGR\_DATA|seq_{no}|E(K_{chbs}, d_{ch}) \tag{15}$$

AGGR_DATA specifies the type of the packet, ID_{ch} and ID_j denote the previous and next hop in the path, respectively; the packet reply is verified with the encrypted sequence number when AGGR_DATA packet is received by any node with identical sequence number, then packet is dropped. D_{ch} represents encrypted data for BS and MAC, which supports maintenance of the packet integrity and authentication. The following operations are performed by the node receiving this packet: (1) ID of the next hop is checked; when similar to its ID, the sequence number is decrypted. (2) The packet sequence number is verified in the receiver packet table, if not found, packet type and the sequence number is entered or else left out. (3) The entry of the subsequent hop is modified with the ID of the subsequent hop nodes and that of the preceding hops with its ID. (4) The sequence number is encrypted using the pairwise key of subsequent hop as well as then forwarded again. In this manner, data reach the BS through the route specified and BS uses an exclusively shared key (K_{chbs}), which evaluates the data efficiency.

3.5. Incorporation of H-TEEN

In H-TEEN, after selecting the CHs, CHs forwards the following parameters:

1. Attributes (A): This physical set of parameters helps the user to collect the data.
2. Thresholds: This is composed of hard and soft thresholds denoted by HT and ST, respectively. HT is the particular value that triggers the node to broadcast the data. ST is a little alteration in the significance that triggers the node to rebroadcast the information.
3. Schedule: It is scheduled by TDMA that allows the slot to each node.
4. TimeCount (TC): This is the utmost duration of the pair consecutive reports forwarded by a node. This is the numerous length of the schedule that accounts for the practical component. In a WSN, closer nodes form a cluster that senses analogous data and forwards them concurrently, which leads to collisions. TDMA schedule is introduced so that every cluster member is assigned a slot for transmission.

3.6. U-DSP

In a storage system, businesses store data in the data server located remotely, and hence data authenticity is assured. When, occasionally, unauthorized users delete or modify data, the server is compromised and/or randomly leads to Byzantine failures. As this is the initial process for recovering the storage errors quickly, cloud storage systems introduce a flexible as well as effective distributed approach with explicit dynamic data support for distribution of files in a cloud server. The homomorphic token is computed with the help of a universal hash function and is integrated with verification of erasure-coded data. Moreover, servers that misbehave are also identified. At last, file retrieving and error recovering procedures based on erasure-correcting code are defined.

4. Performance Analysis and Discussion

The efficiency of the secure routing protocol with energy optimization and data storage was evaluated with several simulation experiments with randomly varying topology. NS-2 version 2.34 was the simulation tool used and considered a multi-hop network with 1000 m × 1000 m size located in a randomized grid with SNs from 50 to 200. The sink node was at the centre of the network. Traffic was CBR of 600 packets/sec and packet size was 316 bytes. The heterogeneity of the network was proven in the simulation by testing the environment with different sets of nodes. The simulation parameters are shown in Table 1.

Table 1. Simulation parameters.

Parameters	Value
Network size	1000 m × 1000 m
Sensor nodes	200
Transmission Rate	50 to 250 Kbps
Number of Nodes	10 to 500
Data Flows	2 to 10
MAC Protocol	IEEE 802.11
Initial Energy	14.0 Joules
Packet Size	512 bytes
Receiving Power	0.4 Watts

The performance measures considered were throughput, end-to-end delay, energy efficiency, the lifetime of the network and data storage capacity.

End-to-end delay, an important metric, is considered to deal with real-time traffic and transmit data packets within the stipulated time. End-to-end delay is the difference in the time taken between the source node transmitting data and the sink receiving it. It is the sum of the delay in transmission, propagation, queuing and processing at every hop.

Table 2 shows a comparison of the end–end delay. Figure 3 depicts the end–end delay for the proposed MLRP-HTEEN-UDSP protocol and Figure 4 compares it with the other protocols such as LEACH, CCBRP and PEGASIS. It suggests that the proposed protocol performs better than other protocols. The delay is at a minimum since the hierarchical architecture of the CHs for all the times chooses the route that has fewer hops with good quality links.

Table 2. Comparison of end–end delay.

Network Size	LEACH	CCBRP	PEGASIS	NCBPR	HHCA	ETLHCM	MLRP-HTEEN-UDSP
50	42	40	38	35	31	28	25
75	75	69	63	59	48	38	28
100	78	72	67	63	52	42	32
125	81	75	70	69	58	48	38
150	85	79	77	76	65	55	45

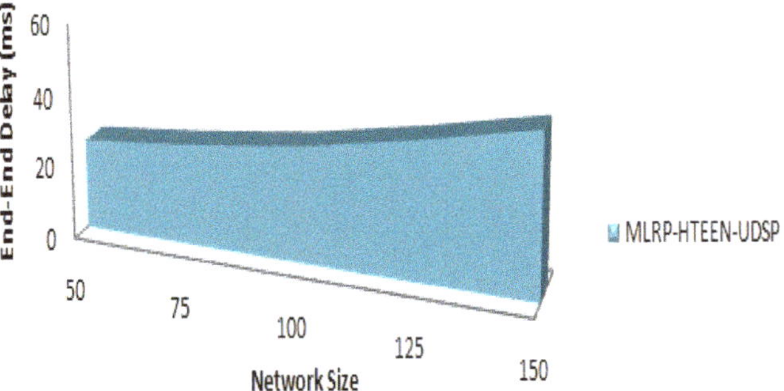

Figure 3. End-to-end delay of MLRP-HTEEN-UDSP.

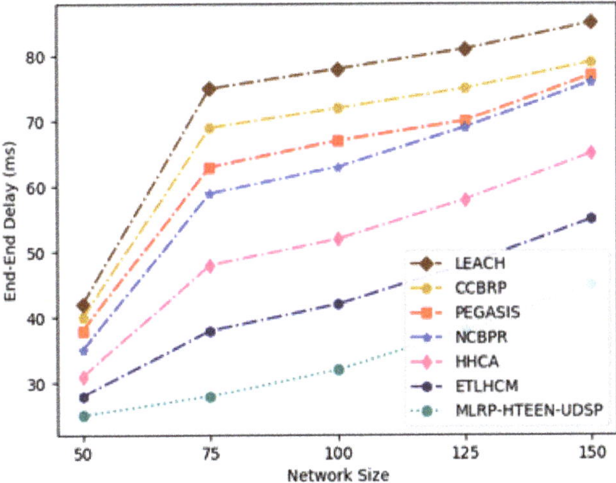

Figure 4. Comparison of end-to-end delay.

Throughput is the total packets received by the sink within a specified duration. The comparison of throughput is shown in Table 3.

Table 3. Comparison of throughput.

Network Size	LEACH	CCBRP	PEGASIS	NCBPR	HHCA	ETLHCM	MLRP-HTEEN-UDSP
50	20	28	30	35	40	45	52
75	34	36	40	44	52	55	61
100	43	47	50	52	63	65	78
125	58	62	66	69	74	79	82
150	63	68	70	72	76	80	85

Figure 5 depicts the throughput of MLRP-HTEEN-UDSP, while Figure 6 compares the throughput of the proposed protocol with other protocols. The selection of multiple paths and minimum delay balances the load and uses the wireless spectrum efficiently. Thus, it achieves a higher throughput than other protocols.

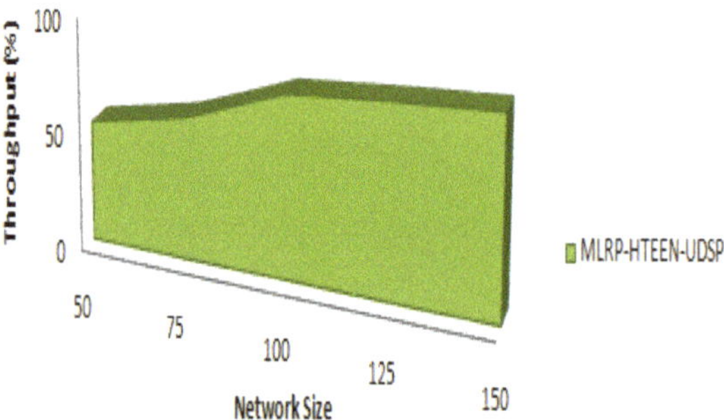

Figure 5. Throughput of MLRP-HTEEN-UDSP.

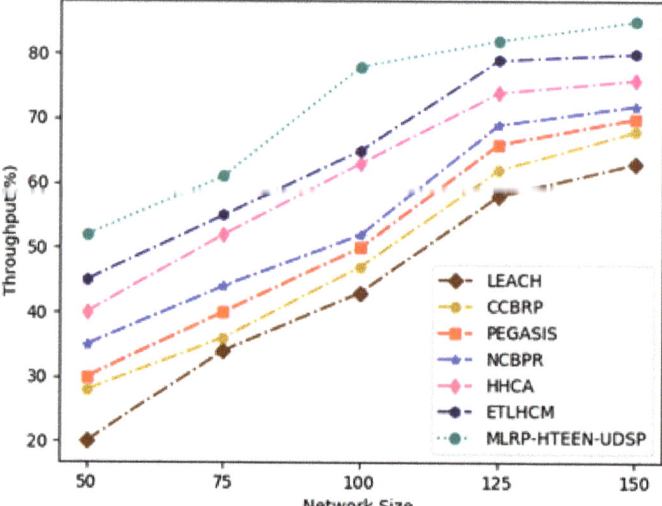

Figure 6. Comparison of throughput.

Table 4 shows a comparison of the energy efficiency. The average energy efficiency is represented in Figure 7 where we can realize that our proposed protocol MLRP-HTEEN-UDSP has less energy dissipation compared with other protocols (LEACH, CCBRP and PEGASIS) as shown in Figure 8 with various node numbers.

Table 4. Comparison of energy efficiency.

Network Size	LEACH	CCBRP	PEGASIS	NCBPR	HHCA	ETLHCM	MLRP-HTEEN-UDSP
50	20	23	25	28	32	38	44
75	35	38	40	42	46	50	51
100	38	40	43	45	51	58	61
125	50	53	55	59	63	69	72
150	52	58	60	62	68	70	75

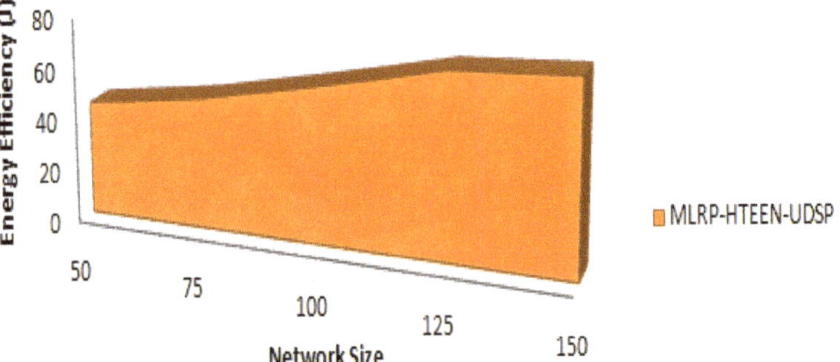

Figure 7. Energy efficiency of MLRP-HTEEN-UDSP.

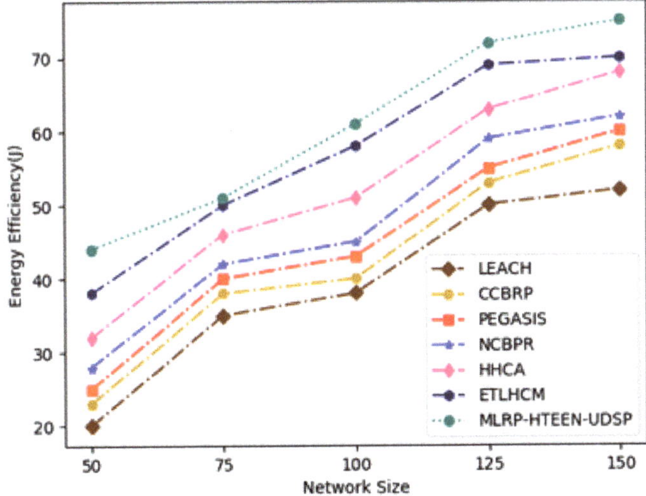

Figure 8. Comparison of energy efficiency.

Table 5 shows the comparison of the network lifetime. Figure 9 presents the network lifetime for the proposed MLRP-HTEEN-UDSP and Figure 10 shows a comparison of the network lifetime between the proposed and existing techniques.

Table 5. Comparison of network Lifetime.

Network Size	LEACH	CCBRP	PEGASIS	NCBPR	HHCA	ETLHCM	MLRP-HTEEN-UDSP
50	28	30	33	38	45	50	55
75	45	48	50	55	59	63	65
100	49	51	56	65	71	78	81
125	50	56	59	66	79	82	85
150	52	59	61	72	80	85	88

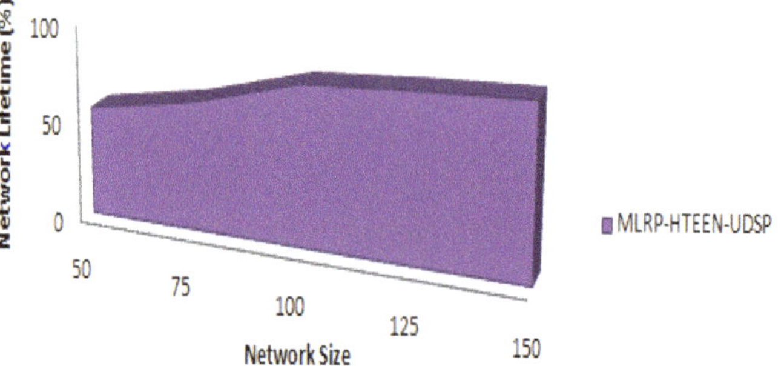

Figure 9. Network lifetime of MLRP-HTEEN-UDSP.

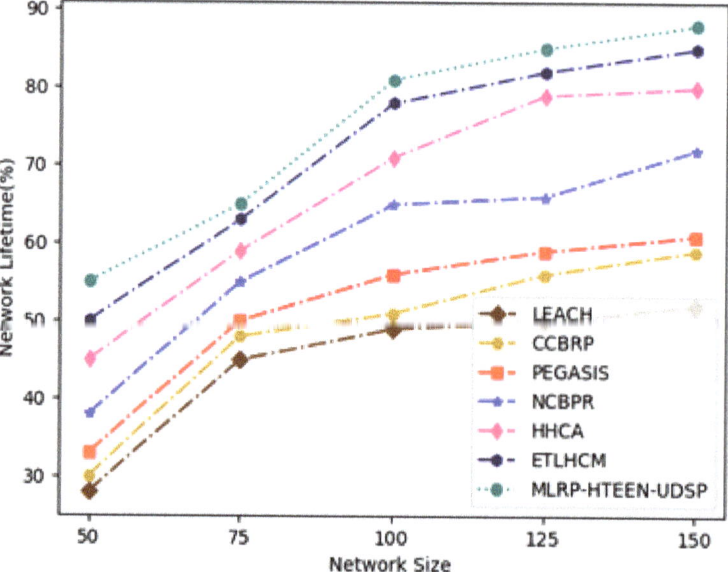

Figure 10. Comparison of network Lifetime.

Table 6 shows the comparison of data storage. Figure 11 shows the data storage for the proposed MLRP-HTEEN-UDSP and Figure 12 shows the comparison of the data storage between the proposed and existing techniques.

Table 6. Comparison of data storage.

Network Size	LEACH	CCBRP	PEGASIS	NCBPR	HHCA	ETLHCM	MLRP-HTEEN-UDSP
50	19	23	28	32	35	38	42
75	22	28	30	39	40	45	51
100	26	31	35	42	59	64	68
125	28	35	37	52	65	70	72
150	31	39	41	61	68	72	75

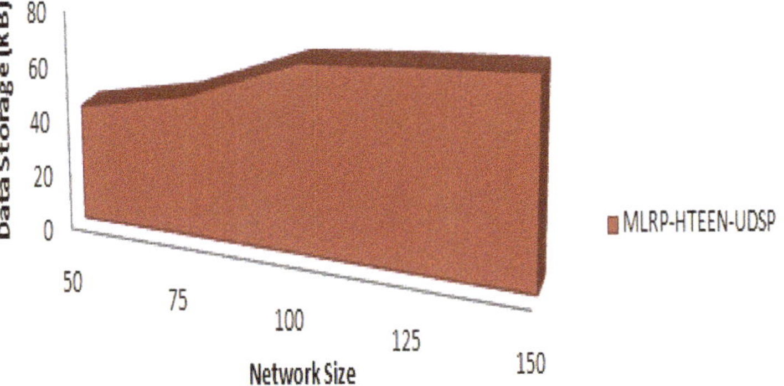

Figure 11. Data storage of MLRP-HTEEN-UDSP.

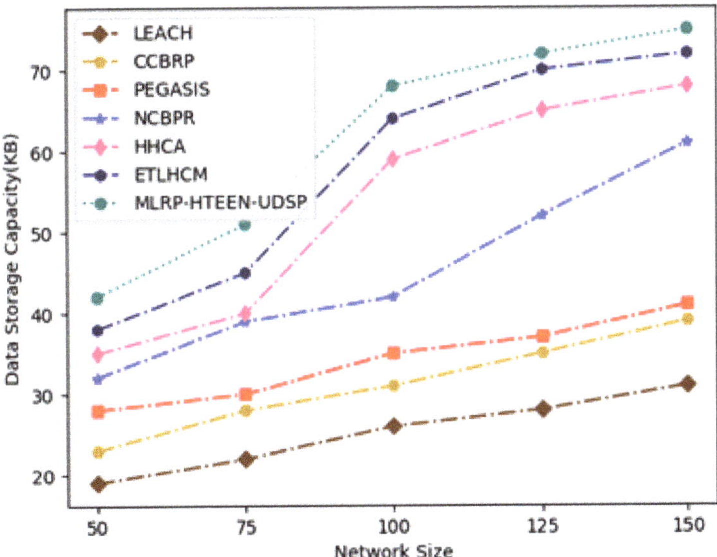

Figure 12. Comparison of data storage.

Table 7 represents the overall parameter comparison of the proposed and existing techniques, and Figure 13 is the graphical representation of the overall comparison.

Table 7. Overall comparisons of MLRP-HTEEN-UDSP.

Network Size	Throughput	End–End Delay	Energy Efficiency	Network Lifetime	Data Storage Capacity
50	52	25	44	55	42
75	61	28	52	65	51
100	78	32	61	81	68
125	82	38	72	85	72
150	85	45	75	88	75

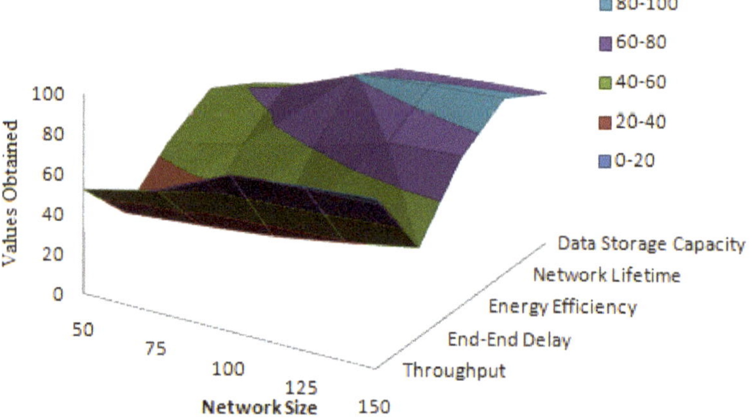

Figure 13. Overall comparison of MLRP-HTEEN-UDSP.

5. Conclusions

In this research, an energy optimization method with secure routing for IoT heterogeneous WSN applications is proposed. This secure as well as reliable routing protocol gathers data about neighbor nodes at BS, and generates the key and energy-efficient multipath for every node. CHs help in data aggregation and forward them to BS, which continuously monitors nodes for residual energy to choose some new paths and CHs. By using the complexity of multimedia processing and the aggregation process to the CHs side, as well as preventing path loops and path cycles for establishing routes, the integrated implementation of MLRP-HTEEN-UDSP produced a minimum end-to-end delay in suitable data packets and reduced the energy consumption at SNs. We also demonstrated a light-weight distributed key management method for supporting secure communication among nodes. The performance of MLRP-HTEEN-UDSP outperforms the existing ones such as LEACH, CCBRP and PEGASIS in all the performance metrics including end-to-end delay, throughput, energy efficiency, network lifetime and data storage capacity. Future work of the proposed protocol will be extended to the performance analysis in the green IoT environments with an increased network size.

Author Contributions: Conceptualization, Writing—original draft R.N.; Supervision, V.C. and S.B.G.; Writing-original draft and review & editing, K.J. and M.G.; Validation, C.V.; propose the new method or methodology, N.R and S.B.G.; Formal Analysis, Investigation C.O.S.; Resources, S.B.G. and C.O.S.; Software, T.C.M.; Writing—review & editing, T.C.M. All authors have read and agreed to the published version of the manuscript.

Funding: National Research Development Projects to finance excellence (PFE)-14/2022-2024 granted by the Romanian Ministry of Research and Innovation, this paper was partially supported by UE-FISCDI Romania and MCI through projects AISTOR, FinSESco, CREATE, I-DELTA, DEFRAUDIFY, Hydro3D, FED4FIRE—SO-SHARED, AIPLAN—STORABLE, EREMI, NGI-UAV-AGRO and by European Union's Horizon 2020 research and innovation program under grant agreements No. 872172 (TESTBED2) and No. 777996 (SealedGRID).

Institutional Review Board Statement: Not applicable.

Informed Consent Statement: Not applicable.

Data Availability Statement: Data will be shared for review based on the editorial reviewer's request.

Acknowledgments: The work of Chaman Verma was supported by the European Social Fund under the project "Talent Management in Autonomous Vehicle Control Technologies" (EFOP-3.6.3-VEKOP-16-2017-00001). The publication of C.O.S. was supported by funds from the National Research Development Projects to finance excellence (PFE)-14/2022-2024 granted by the Romanian Ministry of Research and Innovation.

Conflicts of Interest: The authors declare no conflict of interest.

References

1. Downie, J.D.; Nederlof, L.; Sutherland, J.S.; Wagner, R.E.; Webb, D.A.; Whiting, M.S. Radio Frequency Identification (RFID) Connected Tag Communications Protocol and Related Systems and Methods. U.S. Patent No. 9,652,707, 16 May 2017.
2. Koch, M.J.; Swope, C.B.; Bekritsky, B.J. System for, and Method of, Accurately and Rapidly Determining, in Real-Time, True Bearings of Radio Frequency Identification (RFID) Tags Associated with Items in a Controlled area. U.S. Patent 9,477,865 B2, 26 October 2016.
3. Pirbhulal, S.; Zhang, H.; Alahi, M.E.; Ghayvat, H.; Mukhopadhyay, S.C.; Zhang, Y.-T.; Wu, W. A Novel Secure IoT-Based Smart Home Automation System Using a Wireless Sensor Network. *Sensors* **2017**, *17*, 69. [CrossRef]
4. Sharma, N.; Sharma, A.K. Cost analysis of hybrid adaptive routing protocol for heterogeneous wireless sensor network. *Sādhanā* **2016**, *41*, 283–288. [CrossRef]
5. Wang, K.; Wang, Y.; Sun, Y.; Guo, S.; Wu, J. Green industrial Internet of things architecture: An energy-efficient perspective. *IEEE Commun. Mag.* **2016**, *54*, 48–54. [CrossRef]
6. Airehrour, D.; Gutierrez, J.; Ray, S.K. Secure routing for internet of things: A survey. *J. Netw. Comput. Appl.* **2016**, *66*, 198–213. [CrossRef]
7. Deebak, B.D.; Al-Turjman, F. A hybrid secure routing and monitoring mechanism in IoT-based wireless sensor networks. *Ad Hoc Netw.* **2020**, *97*, 102022. [CrossRef]
8. Yang, T.; Xiangyang, X.; Peng, L.; Tonghui, L.; Leina, P. A secure routing of wireless sensor networks based on trust evaluation model. *Procedia Comput. Sci.* **2018**, *131*, 1156–1163. [CrossRef]
9. Safara, F.; Souri, A.; Baker, T.; Al Ridhawi, I.; Aloqaily, M. PriNergy: A priority-based energy-efficient routing method for IoT systems. *J. Supercomput.* **2020**, *76*, 8609–8626. [CrossRef]
10. Haseeb, K.; Islam, N.; Almogren, A.; Din, I.U. Intrusion prevention framework for secure routing in WSN-based mobile Internet of Things. *IEEE Access* **2019**, *7*, 185496–185505. [CrossRef]
11. Kumar, K.; Kumar, S.; Kaiwartya, O.; Cao, Y.; Lloret, J.; Aslam, N. Cross-Layer Energy Optimization for IoT Environments: Technical Advances and Opportunities. *Energies* **2017**, *10*, 2073. [CrossRef]
12. Minoli, D.; Sohraby, K.; Occhiogrosso, B. IoT Considerations, Requirements, and Architectures for Smart Buildings—Energy Optimization and Next-Generation Building Management Systems. *IEEE Internet Things J.* **2017**, *4*, 269–283. [CrossRef]
13. Guo, X.; Lin, H.; Li, Z.; Peng, M. Deep-Reinforcement-Learning-Based QoS-Aware Secure Routing for SDN-IoT. *IEEE Internet ThingsJ.* **2020**, *7*, 6242–6251. [CrossRef]
14. Pirbhulal, S.; Wu, W.; Muhammad, K.; Mehmood, I.; Li, G.; de Albuquerque, V.H.C. Mobility enabled security for optimizing IoT based intelligent applications. *IEEE Netw.* **2020**, *34*, 72–77. [CrossRef]
15. Haseeb, K.; Almogren, A.; Islam, N.; Ud Din, I.; Jan, Z. An Energy-Efficient and Secure Routing Protocol for Intrusion Avoidance in IoT-Based WSN. *Energies* **2019**, *12*, 4174. [CrossRef]
16. Preeth, S.K.; Dhanalakshmi, R.; Kumar, R.; Shakeel, P.M. An adaptive fuzzy rule based energy efficient clustering and immune-inspired routing protocol for WSN-assisted IoT system. *J. Ambient. Intell. Humaniz. Comput.* **2018**. [CrossRef]
17. Hammi, B.; Zeadally, S.; Labiod, H.; Khatoun, R.; Begriche, Y.; Khoukhi, L. A secure multipath reactive protocol for routing in IoT and HANETs. *Ad Hoc Netw.* **2020**, *103*, 102118. [CrossRef]
18. Sampathkumar, A.; Maheswar, R.; Harshavardhanan, P.; Murugan, S.; Jayarajan, P.; Sivasankaran, V. Majority Voting based Hybrid Ensemble Classification Approach for Predicting Parking Availability in Smart City based on IoT. In Proceedings of the 2020 11th International Conference on Computing, Communication and Networking Technologies (ICCCNT), Kharagpur, India, 1–3 July 2020.

19. Sampathkumar, A.; Murugan, S.; Rastogi, R.; Mishra, M.K.; Malathy, S.; Manikandan, R. Energy Efficient ACPI and JEHDO Mechanism for IoT Device Energy Management in Healthcare. In *Internet of Things in Smart Technologies for Sustainable Urban Development*; Springer: Cham, Switzerland, 2020; pp. 131–140.
20. Sampathkumar, A.; Mulerikkal, J.; Sivaram, M. Glowworm swarm optimization for effectual load balancing and routing strategies in wireless sensor networks. *Wirel. Netw.* **2020**, *26*, 4227–4238. [CrossRef]
21. Sharma, S.; Rani, M.; Goyal, S.B. Energy Efficient Data Dissemination with ATIM Window and Dynamic Sink in Wireless Sensor Networks. In Proceedings of the 2009 International Conference on Advances in Recent Technologies in Communication and Computing, Kottayam, India, 27–28 October 2009; pp. 559–564. [CrossRef]
22. Maheswar, R.; Jayarajan, P.; Sampathkumar, A.; Kanagachidambaresan, G.R.; Hindia, M.H.D.; Tilwari, V.; Dimyati, K.; Ojukwu, H.; Sadegh Amiri, I. CBPR: A Cluster-Based Backpressure Routing for the Internet of Things. *Wirel. Pers. Commun.* **2021**, *118*, 3167–3185. [CrossRef]
23. Raut, R.; Kautish, S.; Polkowski, Z.; Kumar, A.; Liu, C.M. *Energy-Efficient Routing Protocol for Green IoT Network, Green Internet of Things and Machine Learning: Towards a Smart Sustainable World*; John Wiley & Sons: Hoboken, NJ, USA, 2021; ISBN 9781119792031. [CrossRef]
24. Kanagachidambaresan, G.R.; Maheswar, R.; Manikantan, C.; Ramakrishnan, K. *Internet of Things in Smart Technologies for Sustainable Urban Development*, 1st ed.; EAI/Springer Innovations in Communications and Computing Book Series; Springer: Cham, Switzerland, 2020.
25. Sharma, S.; Goyal, S.B.; Qamar, S. Four-Layer Architecture Model for Energy Conservation in Wireless Sensor Networks. In Proceedings of the 2009 Fourth International Conference on Embedded and Multimedia Computing, Jeju, Korea, 10–12 December 2009; pp. 1–3. [CrossRef]
26. Rajawat, A.S.; Bedi, P.; Goyal, S.B.; Alharbi, A.R.; Aljaedi, A.; Jamal, S.S.; Shukla, P.K. Fog Big Data Analysis for IoT Sensor Application Using Fusion Deep Learning. *Math. Probl. Eng.* **2021**, *2021*, 6876688. [CrossRef]
27. Rani, S.; Maheswar, R.; Kanagachidambaresan, G.R.; Jayarajan, P. *Integration of WSN and IoT for Smart Cities*, 1st ed.; EAI/Springer Innovations in Communications and Computing Book Series; Springer: Cham, Switzerland, 2020.
28. Khan, M.; Ilavendhan, A.; Babu, C.N.K.; Jain, V.; Goyal, S.B.; Verma, C.; Safirescu, C.O.; Mihaltan, T.C. Clustering Based Optimal Cluster Head Selection Using Bio-Inspired Neural Network in Energy Optimization of 6LowPAN. *Energies* **2022**, *15*, 4528. [CrossRef]
29. Goyal, S.B.; Bedi, P.; Kumar, J.; Varadarajan, V. Deep learning application for sensing available spectrum for cognitive radio: An ECRNN approach. *Peer-to-Peer Netw. Appl.* **2021**, *14*, 3235–3249. [CrossRef]
30. Rajawat, A.S.; Bedi, P.; Goyal, S.B.; Shukla, P.K.; Jamal, S.S.; Alharbi, A.R.; Aljaedi, A. Securing 5G-IoT Device Connectivity and Coverage Using Boltzmann Machine Keys Generation. *Math. Probl. Eng.* **2021**, *2021*, 2330049. [CrossRef]

 *energies*

Article

Clustering Based Optimal Cluster Head Selection Using Bio-Inspired Neural Network in Energy Optimization of 6LowPAN

Mudassir Khan [1], A. Ilavendhan [2], C. Nelson Kennedy Babu [3], Vishal Jain [4], S. B. Goyal [5,*], Chaman Verma [6], Calin Ovidiu Safirescu [7,*] and Traian Candin Mihaltan [8]

[1] Department of Computer Science, College of Science & Arts Tanumah, King Khalid University, Abha 62529, Saudi Arabia; mkmiyob@kku.edu.sa

[2] Department of Computer Science and Engineering, Veltech Rangarajan Dr. Sagunthala R&D Institute of Science and Technology, Avadi, Chennai 600062, India; ailavendhan@veltech.edu.in

[3] Department of Computer Science and Engineering, Saveetha School of Engineering, Chennai 602105, India; nelsonc.sse@saveetha.com

[4] Department of Computer Science and Engineering, School of Engineering and Technology, Sharda University, Greater Noida 201310, India; vishal.jain@sharda.ac.in

[5] Faculty of Information Technology, City University, Petaling Jaya 46100, Malaysia

[6] Department of Media and Educational Informatics, Faculty of Informatics, Eotvos Lorand University, 1053 Budapest, Hungary; chaman@inf.elte.hu

[7] Environment Protection Department, Faculty of Agriculture, University of Agriculture Sciences and Veterrnary Medicine Cluj-Napoca, Calea Manastur No 3-5, 400372 Cluj-Napoca, Romania

[8] Faculty of Building Services, Technical University of Cluj-Napoca, 400114 Cluj-Napoca, Romania; mihaltantraian83@gmail.com

* Correspondence: sb.goyal@city.edu.my (S.B.G.); calin.safirescu@usamvcluj.ro (C.O.S.)

Citation: Khan, M.; Ilavendhan, A.; Babu, C.N.K.; Jain, V.; Goyal, S.B.; Verma, C.; Safirescu, C.O.; Mihaltan, T.C. Clustering Based Optimal Cluster Head Selection Using Bio-Inspired Neural Network in Energy Optimization of 6LowPAN. *Energies* **2022**, *15*, 4528. https://doi.org/10.3390/en15134528

Academic Editors: R. Maheswar, M. Kathirvelu and K. Mohanasundaram

Received: 29 May 2022
Accepted: 16 June 2022
Published: 21 June 2022

Publisher's Note: MDPI stays neutral with regard to jurisdictional claims in published maps and institutional affiliations.

Abstract: The goal of today's technological era is to make every item smart. Internet of Things (IoT) is a model shift that gives a whole new dimension to the common items and things. Wireless sensor networks, particularly Low-Power and Lossy Networks (LLNs), are essential components of IoT that has a significant influence on daily living. Routing Protocol for Low Power and Lossy Networks (RPL) has become the standard protocol for IoT and LLNs. It is not only used widely but also researched by various groups of people. The extensive use of RPL and its customization has led to demanding research and improvements. There are certain issues in the current RPL mechanism, such as an energy hole, which is a huge issue in the context of IoT. By the initiation of Grid formation across the sensor nodes, which can simplify the cluster formation, the Cluster Head (CH) selection is accomplished using fish swarm optimization (FSO). The performance of the Graph-Grid-based Convolution clustered neural network with fish swarm optimization (GG-Conv_Clus-FSO) in energy optimization of the network is compared with existing state-of-the-art protocols, and GG-Conv_Clus-FSO outperforms the existing approaches, whereby the packet delivery ratio (PDR) is enhanced by 95.14%.

Keywords: RPL; fish swarm; bio-inspired approach; energy optimization; grid formation; convolution clustering; data transmission; cluster head; alive and dead node

1. Introduction

Every object should be smart in today's technological world. The IoT is a new paradigm that gives common objects and things a whole new dimension. Wireless sensor networks, particularly LLNs, are essential components of IoT. It has a high impact on usage in everyday life [1]. The usage at home, industry and institutions is growing exponentially every day. It is considered one of the most influential technologies of the modern era. The devices added to IoT are growing in leaps and bounds every day. Homes, classrooms and cities are becoming smart with IoT [2,3].

IoT is a new paradigm that connects computers, humans, devices and objects together for communication. There are many definitions given for IoT [4]. A popular definition is: IoT is a connection between the physical and digital worlds. Sensors and actuators are used to connect the digital and physical worlds. IoT is a concept in which computing and networking capabilities are incorporated virtually into any device. The capabilities are used to query the state of the object as well as change it, if possible. IoT is the networking of persons, things, objects and devices that communicate with each other to achieve a complex task, where a high degree of collective intelligence is required [5]. For computing and communications, IoT makes use of sensors, actuators, transceivers and processors. IoT cannot be considered as a single or standalone technology. It is a large collection of connected technology that works synchronously [6].

The network layer in IoT performs the major task of establishing connections among nodes and the server. It is the core layer that does the addressing, routing, formation and maintenance of the network. The network layer has protocols that perform the connectivity and networking tasks. The protocols of the network layer are: IPv4, IPv6, 6LoWPAN, 6TiSCH, 6Lo, IPv6 over Bluetooth Low Energy, IPv6 over G.9959, etc. [7]. IoT consists of LLNs that have low power, low energy and scarce computing resources. The conventional routing protocols for the networks may not be suitable for LLNs. The network layer protocols available for IoT are IPv6 RPL [8], Cognitive RPL (CORPL) [9], Channel Aware Routing Protocol (CARP) [10], Enhancement over CARP (E-CARP) [11] and others.

RPL has become a standard routing protocol suitable for LLNs due to the following characteristics in comparison with the rest: (i) RPL has a better packet reception ratio (PRR) and energy consumption; (ii) it has lesser churn and control traffic overhead; (iii) it had a shorter convergence time; (iv) it is independent of the link layer. Additionally, RPL has the following features: (i) self-healing; (ii) auto-configuration; (iii) Loop avoidance and detection; (iv) independence and transparency; (v) multiple edge routers [12].

Various features of RPL remain the main reason for preferring RPL over other protocols. In spite of its features, RPL also has a lot of room for improvement since the IoT is exponentially growing, and improved routing support is required [13]. Various enhancement methods have been devised for the RPL based on its function and application. Each enhancement method focuses on improving any one of the limitations of RPL or adding more effectiveness to the existing function of RPL [14].

A.   Motivation

A reliable and energy-efficient routing is one important research area. The link in LLNs is unreliable, but it is used to transmit valuable data. The data become vital in some conditions. Considering scenarios such as the healthcare sector, the physiological condition of the patients is monitored round the clock, especially in critical care units. Any little variation in their physiological health parameters would be critical. In that scenario, the communication needs to be reliable and in real time. Any delay in the communication would endanger life. Another environment would be safety, security and surveillance systems, where any behavioral anomaly has to be reported immediately to avoid serious damage. While the objective is to improve the reliability, other QoS parameters must also be satisfied. Those parameters are packet delivery ratio, throughput, convergence time, lifetime, energy consumption and so on. There are many researches focusing on this area, still, the optimum has not yet been achieved.

B.   Contribution

- The RPL due to its wide usage and popularity has become the de facto standard routing protocol for LLNs in IoT. A wide range of research is going on to enhance the RPL for various environments.
- The enhancement methods are based on various components of RPL. The research work in this article is based on energy hole rectification-based transmission enhancement and energy-efficiency improvement mechanism for RPL.

- The main focus of research is on enhancement for reliability in critical environments. A study on the research gap also suggests this as one of the focus areas of research.
- To formulate a grid across the network and generate clusters in the area of grids formed in the network.
- To select CH, a bio-inspired approach is introduced; a Graph-Grid-based Convolution clustered neural network with fish swarm optimization (GG-Conv_Clus-FSO) is utilized.

The rest of the research work is arranged as follows: the related mechanism in energy hole detection and their drawbacks are reviewed in Section 2; the proposed grid-based clustering with FSO for the detection of energy holes is illustrated in Section 3; simulation results are illustrated in Section 4; the research is concluded in Section 5.

2. Literature Review

Several reports have been published in 2009 and 2010 [15,16] to identify the routing requirements for the standardization of RPL based on its application in various routing environments. The widespread usage of RPL and its customisation has necessitated substantial study and development. The control packets in the network are necessary to establish a connection and maintain the network. The frequent change and resetting of the network in a mobile setup led to overhead in the link. An efficient way of detecting and controlling congestion is required [17]. LLNs are backbone networks of IoT. They are constrained by energy, memory and processing capacity. The traditional and popular network protocols are not suitable for LLNs due to these constraints. Among the existing routing protocols, RPL is more suitable for LLNs, due to its special features such as auto configuration, self-healing, loop avoidance, multiple edge routers and robustness [18]. RPL is also easily malleable to various environments of LLNs. This section presents an overview of RPL with the background, characteristics and various components.

The Internet Engineering Task Force (IETF) envisioned the standardization of IPv6RPL and started the Routing over Low-power and Lossy networks (ROLL) working group in 2008. The working group aimed at the standardization of RPL, which has the following implicit characteristics [19]: (i) LLNs are constituted by hundreds of nodes that are constrained by energy, size, processing power and memory. (ii) The constrained nodes of LLNs are connected to each other through lossy links, which have low data rates and are unstable. (iii) The traffic patterns of these LLNs may be point-to-multipoint, multipoint-to-point or, in some cases, point-to-point [20], such as urban settings [21], industrial settings, home automation and building automation [22].

Multi-hop WSN node restrictions are closer to BS's demand to infuse traffic from some other channel, enabling their energy to be spent quicker and possibly leading to very high remaining energy. As a consequence, Distributed Wedge Merging in Multi-Hop Access (DWMA) is presented here as a possible solution to the energy hole problem and routing. The major objective is to remove energy gaps while reducing the likelihood of them emerging in the future. To avoid energy holes from occurring, this DWMA method is combined with a nearby wedge [23].

In heterogeneous networks, violating the response and broadcast buffer specifications has resulted in uneven traffic loads, congestion and, as a result, packet loss in RPL, according to the author. This paper discusses the CBR-RPL technique, which uses a unique drop-aware Objective Function (OF) to arrange nodes into route data. The newly defined OF takes into account both queue occupancy and node transceiver drop rates [24]. The Energy Hole problem, which is common in WSN, drastically affects the lifetime of any established network. The energy diffusion required for data packet forwarding between HN is reduced when a good Head Node (HN) selection technique is used [25].

PEGASIS (Power-Efficient Gathering in Sensor Information Systems) is an energy-saving protocol that tries to extend the network's lifespan by reducing energy consumption. This research proposes a modification of the PEGASIS approach. SNs are sorted into groups, clustering is carried out using the k-means method and every group is assigned the PEGASIS label. Rechargeable sensor nodes were also used in the suggested strategy. The sensor node's Euclidean distance from the base station and the sensor node's residual energy are utilised to determine the chain leader. Every CH's datum is instantly forwarded to the BS [26]. Various surveys on energy use, energy gaps and attacks on RPL and LLP systems can be found in [27–30]. The glowworm swarm-based approach cast-off energy-based transmission strategy has been presented to decrease energy consumption caused by control overhead [31]. A least-square support vector machine (LS-SVM) based on modified particle swarm optimization (MPSO) is developed. To begin, the MPSO's inertial weight is adjusted to accomplish faster iterations, and an LS-SVM-based MPSO's prediction model is constructed. Second, the predictive simulation was performed and confirmed using the MPSO's optimised parameters, and the MPSO and PSO predicted values were compared [32]. This work introduces a resilient clustering routing mechanism for WSNs. To estimate the number of cluster heads and identify the best cluster heads, the technique employs the Locust Search (LS-II) approach. After the cluster heads have been identified, other sensor elements are allocated to the cluster heads that are closest to them [33]. Based on the Optimal Low-Energy Adaptive Clustering Hierarchy (LEACH) protocol, a methodology for an energy-efficient clustering algorithm for gathering and transferring data is developed. The new optimised threshold function is used in the selection of CH. LEACH, on the other hand, is a hierarchy routing protocol that picks cluster head nodes at random in a loop, resulting in a higher cluster headcount but higher power consumption. In order to improve the energy per unit node and packet delivery ratio with less energy use, the Centralised Low-Energy Adaptive Clustering Hierarchy Protocol is the best [34]. WSNs are designed for specialised applications, such as monitoring or tracking, in both indoor and outdoor conditions, where battery capacity is a major issue. Several routing protocols are designed to solve this problem. A sub-cluster LEACH-derived approach is also proposed in order to improve performance. The Sub-LEACH with LMNN surpassed its competitors in terms of energy efficiency, according to simulation data [35].

3. Proposed Graph-Grid-Based Clustering for Energy Hole Detection

Grid cells of equal length are used to partition the whole network. Every grid cell represents the square territory. Every grid cell has only static nodes. The Sink can be either stationary or moveable for gathering data. The grid cell CH is the one that is closest to the mid-point of the grid cell. Every grid cell has a node ID as well as an associated grid ID that identifies nodes. The sink is responsible for the initial cluster setup, which includes calculating node IDs and grid IDs, establishing the CH for each grid cell and scheduling data transmission and reception for nodes in the grid cells. The GG-Conv_Clus-FSO protocol uses double disjoint anchor group nodes for packet forwarding, and node nomination is based on the clustering method. To locate holes quickly, a grid-based hole identification method is utilised. Data packets are accurately routed to the anchor and destination nodes while consuming the least amount of energy.

Grid Formation

The whole network is divided into equal-sized rectangle-shaped grid cells. Each grid keeps track of the exact location of each cell relative to its border, which is subsequently used to determine the size of the holes. In this, $D_a \times D_b$ denotes the grid cell dimension where the length is determined by D_a and width is determined by X_b. Equation (1) [31] indicates the grid cell generation process. The grid-building procedure is completed by

$$f(g, r) = ((g_0 + g \times D_a, r_0 + r \times D_b)) \tag{1}$$

where the count of a horizontal line is indicated by g, and the count of the vertical line is indicated by r. The process of grid construction is given in Algorithm 1.

Algorithm 1: Construction of Grid

---

for g = 0 to p = s
  for r = 0 to r = t
    $f(g,r) = ((g_0 + g \times D_a, r_0 + r \times D_b))$
  end for
end for

Selection of Cluster Head (CH): In the actual world, fish can identify nutrient-rich places by searching on their own or by swimming near other fish; the region with the most fish often has the most nutrition. Artificial fish swarm optimization (AFSO) is based on mimicking fish behaviour, such as preying, swarming and tracking local fish hunts to attain global optima. The solution space and the states of other artificial fishes are generally the areas where an Artificial Fish (AF) dwells. The subsequent behaviour is determined by the current state as well as the immediate environment, such as the current quality of query responses and the status of nearby neighbours. The movements of an artificial fish, as well as the actions of its neighbours, have an impact on the ecosystem. If fish are discovered in a water area with more food, they will migrate quickly to that area. Equation (2) may be used to describe this behaviour.

$$F_v = F_i + Visual \times rand \quad i\epsilon[0,n]$$

$$F_{next} = F + \frac{F_v - F}{||F_v - F||} \times step \times rand \tag{2}$$

During preying mode, fish behaviour is represented by Equation (3):

$$Prey(F_i) = \begin{cases} f_i + step\frac{f_j - f_i}{||f_j - f_i||} & if\ y_j - y_i \\ f_i + (2rand - 1).step & else \end{cases} \tag{3}$$

where rand is the random function with range [0, 1].

The behaviour of the swarm is represented in Equation (4)

$$Swarm(F_i) = \begin{cases} f_i + step\frac{f_j - f_i}{||f_j - f_i||} & if\ \frac{y_c}{nf} > \delta y_i \\ prey\ (f_i) & else \end{cases} \tag{4}$$

In the follow stage, behaviour is given by equation

$$Follow(F_i) = \begin{cases} f_i + step\frac{f_{max} - f_i}{||f_{max} - f_i||} & if\ \frac{y_{max}}{nf} > \delta y_i \\ prey\ (f_i) & else \end{cases} \tag{5}$$

The three processes outlined above guarantee that both global and local searches are conducted, as well as a search direction that leads to the greatest food source. The suggested approach differs from the AFSA in two significant ways. The solutions are split at random and behave in one of two ways: swarming or following. The best fish are chosen via tournament selection, and preying processing begins. Fish who are very good at preying are chosen and allowed to breed amongst themselves. The best fish and the new solution are carried to the next iteration. The answers are represented as binary numbers in this study, and the distance between fish is calculated utilizing Hamming distance. The number of locations where two strings u and v differ is the hamming distance between them. The best fish are chosen by spinning the roulette wheel. The likelihood of a fish being picked on a roulette wheel is exactly proportional to its fitness. Equation (6) computes the probability of a fish,

$$pb_i = \frac{fit_i}{\sum_{j=1}^{N} fit_i} \tag{6}$$

To enhance QOS, a multi-objective function based on E2E delay as well as energy is proposed and represented by Equation (7):

$$minfit(f_i) = \frac{e^{-\left(\frac{D_{td}}{D_m}\right)}}{\left(\frac{E_{ri}}{E_i}\right)} \tag{7}$$

where d_{ri} is the E2E delay, D_{td} is the total delay to reach BS, D_m is the maximum delay, E_{ri} is the remaining energy in CH and E_i is the initial energy.

Subsequent assumptions are made:

- The nodes in the network are distributed arbitrarily;
- Starting energy of every node is the similar;
- In ecology, all fishes are unisex;
- Because fish are unisexual, mating among any two fish is feasible;
- Because the free space radio method is utilised, the energy needed to transmit one bit of data grows as distance improves.

The flowchart for FSO-based CH selection is shown in Figure 1. Because solution space is binary, a transfer function is required to fill the bit as the fish swims. In this paper, a novel transfer function for flipping the bits described by Equation (8) is,

$$transfer(F_i) = \frac{1}{\left(1 + e^{-\tan h(f_i)}\right)} \tag{8}$$

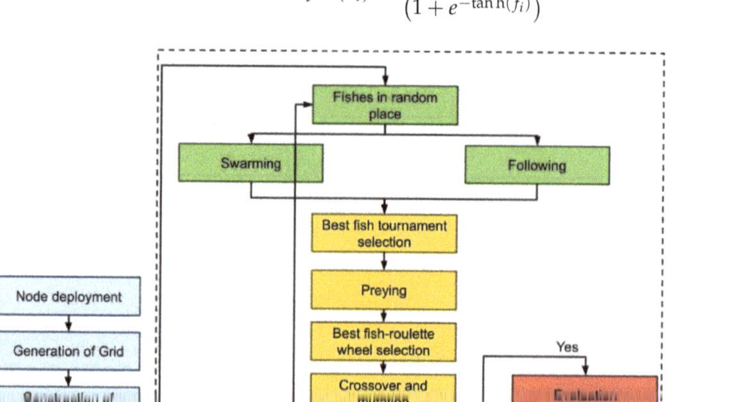

Figure 1. Flow chart of FSO based CH selection.

To attain flipping, a random number between 0 and 1 is generated, and if the random number is less than the transfer function provided by Equation (8), the bit is flipped.

$$\begin{cases} 1, & if \; p(0,1) < transfer(F_i) \\ 0, & otherwise \end{cases} \tag{9}$$

Hole Detection: The grid hole is found by comparing the cell coordinates to SNs radius as well as closest count (SN). SN count is closer to the sensor's radius, which is used to determine the hole's coverage area. The region in which a hole has developed is said to be B_i. Equations (10) and (11) show the position of the hole in the cell and the grid.

$$B_{cell} = \sum_1^i Bi \tag{10}$$

$$B_{total} = \sum_1^j Bcell \tag{11}$$

Data propagation across selected CH: Send data around the borders of the hole and transfer it to the correct location. The sensor nodes in the region are clustered, and the CH is selected as the node closest to the grid's centre. In the GBC-SS, the Static Sink is in charge of coordination, whereas in the GBC-MS, Mobile Sink is in charge of data gathering. The information is transmitted to nearby sensor nodes, and the position is recorded for future data transfer.

$$LoadintheCHnode = \left(\frac{Whole\ CH\ communication\ cost}{total\ cost\ of\ CH} \right) \tag{12}$$

$$Loadpergridcell = \left(r^2 \rho dt \right) \tag{13}$$

$$Intra-andinter-clustercommunication\ cos\ t = \left(r^2 \rho dt \right) \times \sum_{i=1}^n (2i-1)gl1 + \sum_{i=1}^n (2i-1)i \tag{14}$$

where the density is indicated by ρ, the transmission rate of data is characterized by a dot and the determined location is shown as gl.

$\mathcal{G} = (\mathcal{F}, \mathcal{E}, \mathcal{H})$ is the definition of an undirected and connected graph, where V and E are finite sets of V = N vertices and edges, and $W \in \mathbb{R}^{N \times N}$ is an adjacency matrix. The graph signals are represented by numerous variables in each vertex. The description indices are represented by the vertex variables in this study. The graph is given by its Laplacian matrix L, which is defined as $\ell = D - W$, where $D = \text{diag}\left(d_0, \ldots, d_{N-1} \right)$ is the degree matrix created by degrees $d_i = \sum_j W_{i,j}$ of vertex i. represented as $\{\chi_\ell\}_{/=0}^{N-1}$ with corresponding nonnegative eigenvalues $0 \le \lambda_0 \le \ldots \le \lambda_{N-1}$. Laplacian matrix L, is diagonalized by the eigenvector matrix $\mathcal{X} = [\chi_0, \ldots, \chi_{N-1}]$ so that $\ell = \mathcal{X} \Lambda \mathcal{X}^T$, where is the diagonal eigenvalue matrix. $\mathcal{L} = I_N - D^{-1/2} \ell D^{-1/2}$ is a normalised version [1,1].

Instead of complex exponentials, the eigenvectors $\{\chi_\ell\}_{/=0}^{N-1}$ of the Laplacian matrix L that meet the orthogonality criterion are employed as decomposition bases for graph-structured data. On a graph, the Fourier transform of a given signal $f(n)$ is defined as Equation (15):

$$\widehat{f}(\lambda_\ell) = \sum_{m=0}^{N-1} \chi_\ell^T(n) f(n) = X^T f \tag{15}$$

Inverse Fourier transformation is represented by Equation (16):

$$f(n) g(n); g(n) \{\theta_{,}\}_{=0}^{N-1} \mathcal{G}(\Lambda).f \tag{16}$$

Convolution is turned into a point-wise product in the Fourier domain as well as reconverted into the vertex domain utilizing the graph Fourier transform as well as the convolution theorem, as in Equation (17):

$$f \times g = X \left(\sum_{k=0}^K \theta_k \Lambda^k \right) \mathcal{T}^T f = \left(\sum_{k=0}^K \theta_k \left(X^T \Lambda^k X^T \right) \right) f = \sum_{k=0}^K \theta_k \mathcal{L}^k f \tag{17}$$

A convolution kernel is the graph convolution operation of two graph signals, *f(n)* and *g(n)*, and its transform, $\mathcal{G}(\Lambda)$. A set of free parameters in the Fourier domain, i.e., Laplacian eigenspace, is used to construct this kernel. Convolution is then written as Equation (18):

$$f*g = \mathcal{X}\mathrm{diag}(\theta_0,\dots,\theta_{N-1})\mathcal{X}^T f = \mathcal{X}\mathcal{G}(\Lambda)\mathcal{X}^T f\mathcal{G}(\Lambda) \tag{18}$$

$\mathcal{G}(\Lambda)$ as an eigenvalue polynomial function: As illustrated in Equation (19), a rapid localised convolution based on low-order polynomial approximation was proposed:

$$G(\Lambda) = \sum_{(k=0)}^{K} \theta_k \Lambda^k \tag{19}$$

which $\{\theta_k\}_{k=0}^{K}$ is the polynomial order, and Ki is a vector of polynomial coefficients. K is a tiny positive integer, such as 3, for example. Convolution is then rewritten as Equation (20):

$$f \times g = X\left(\sum_{k=0}^{K} \theta_k \Lambda^k\right)T^T f = \left(\sum_{k=0}^{K} {}^{K}\theta_k\left(X^{\top}\Lambda^k X^T\right)\right)f = \sum_{k=0}^{K} \theta_k \mathcal{L}^k \tag{20}$$

The convolution is performed by K multiplications of the sparse matrix L, which speeds up computation by avoiding the composition procedure.

The following is the updated version of Equation (21) for layer *l*:

$$\widehat{h}_i^{l+1} = O_h^l H_{k=1}\left(\sum_{j\in N_i} w_{i,j}^{k,l} V^{k,l} h_j^l\right)$$
$$\widehat{e}_i^{l+1} = O_e^l H_{k=1}\left(\widehat{w}_{i,j}^{k,l}\right)$$

where

$$w_{ij}^{k,l} = \mathrm{softmax}x_j\left(\widehat{w}_{i,j}^{k,l}\right) \tag{21}$$

$$\widehat{w}_{i,j}^{k,l} = \left(\frac{Q^{k,l}h_i^l \cdot K^{k,l}h_j^l}{\sqrt{d_k}}\right) \cdot E^{k,l}e_{i,j}^l$$

with $Q^{k,l}, K^{k,l}, V^{k,l}, E^{k,l} \in \mathbf{R}d_k, O_h^l, O_e^l \in \mathbf{R}^{d\times d}, k \in \{1,2,\dots,H\}$ relates the number of attention heads, and where $O_h^l \in \mathbf{R}^{d\times d}, V^{k,l} \in \mathbf{R}^{d_k\times d}, d_k H$ denotes the number of heads. Note that $h\ l\ i$ is *i*-th node's feature at *l*-th layer in Equation (22).

$$\mathrm{cut}(S_k, S_k) = \sum_{v_i\in S_k, v_j\in S_j} e(v_i, v_j), \tag{22}$$

where S_k is *k*-th set of a given graph, $\widehat{S}_k$ indicates the remaining sets, except S_k, and $e(v_i, v_j)$ is the edge between vertices v_i and v_j. When referring to multiple sets, the cut issue is represented as Equation (23):

$$\mathrm{cut}(S_1, S_2, S_3 \dots S_8) = \frac{1}{2}\sum_{i=k}^{8} \mathrm{cut}(S_k, S_k) \tag{23}$$

The issue of less cuts is extensively studied in the literature in Equation (24):

$$\mathrm{Ncut}(S_1, S_2 \dots S_g) = \sum_{k=1}^{g} \frac{\mathrm{cut}(S_k, \widehat{S}_k)}{\mathrm{vol}(S_k, V)}, \tag{24}$$

where $\mathrm{vol}(S_k, V)\sum = v_i \in S_k, v_i \in v^e(v_i, v_j)$ is the total degree of nodes from S_k in graph *g*. The normalised cut problem utilizing DL optimisation, transforming the minimum cut issue into a DL format, as in Equation (25):

$$L_{\mathrm{cut}} = \sum\nolimits_{\mathrm{reduce\_sum}}(Y\oslash\Gamma)(1-Y)^T \odot A + \sum\nolimits_{\mathrm{reduce\_sum}}\left(1^T Y - \frac{n}{g}\right)^2 \tag{25}$$

A is the adjacency matrix, and, finally, Γ is evaluated by Equation (26):

$$H_j^{[l+1]} = \sigma \left(\sum_{i=1}^{F_{in}} \left(\sum_{k=0}^{K} \theta_{i,jk} \mathcal{L}^k H_i^{[l]} \right) + b_j^{[l]} \right) \tag{26}$$

where $\sigma(\cdot)$ relates a non-linear activation function, e.g., ReLU $(\cdot)$ max$(0, \cdot) = $; Hi [ ]l indicates ith input graph; *ijk*, and *bj* [ ] *l* are trainable F F in out $\times$ vector of K-order polynomial coefficients and 1 $\times$ Fout vector of bias in *l* th layer.

4. Result and Discussion

In this section, the simulation outcome of the proposed Graph-Grid-based Convolution clustered neural network with fish swarm optimization (GG-Conv_Clus-FSO) is compared with the existing techniques such as DWMA for 6LowPAN RPL, CBR-RPL, CCS and WEMER and PEGASIS. The simulation of the above-mentioned approaches is investigated with the assistance of the number of rounds vs. alive nodes, the number of rounds vs. dead nodes, PDR, energy consumption and delivery delay. The simulation setup is given in Table 1.

Table 1. Simulation Setup.

Parameters	Values
No. of nodes	200
Simulation area	1000×1000 m^2
Routing protocol	RPL
Initial energy	100 Joule
Packet size	300 bits
Simulation time	65 ms

4.1. Energy Consumption

Every sensor node in the data transmission environment in the WSN is equipped with rechargeable batteries that consume the least amount of energy, making battery recharging difficult. The cluster and duty cycle scheduling mechanisms start the data transfer. The data transmission process is completed without interruption, and data are transmitted in the quickest way possible while consuming the least amount of energy. The transmission nodes' energy consumption is minimized as a result of this condition is given in Table 2.

Table 2. Comparison of energy consumption.

No of Nodes	DWMA	CBR-RPL	WEMER	GG-Conv_Clus-FSO
0	0	0	0	0
20	5.12	3.47	2.49	1.08
40	12.17	9.22	4.13	3.42
60	25.46	13.79	5.43	4.79
80	31.45	19.47	8.24	6.33
100	43.78	29.55	11.85	7.04

In Figure 2, energy consumption during data transmission for a different number of nodes is illustrated. The energy consumption of the proposed approach is minimal than existing approaches, namely DWMA, CBR-RPL, WEMER and PEGASIS.

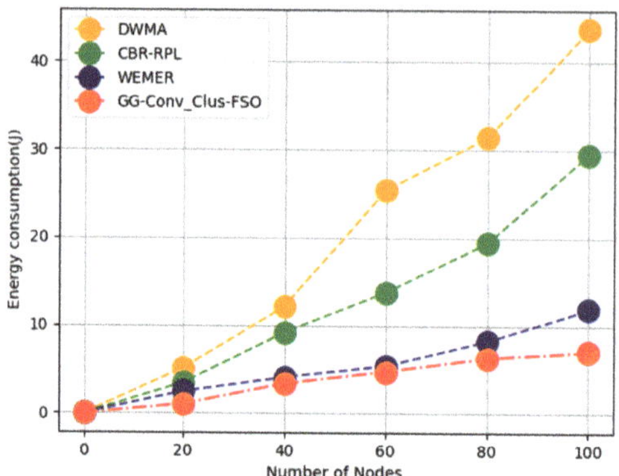

Figure 2. Comparison of energy consumption.

4.2. End to End Delay

It is time takes to transport data from source to destination node. A protocol that has the shortest transmission delay is considered to be effective, the comparisons are given in Table 3.

Table 3. Comparison of end-to-end delay.

No of Nodes	DWMA	CBR-RPL	WEMER	GG-Conv_Clus-FSO
0	0	0	0	0
20	72.49	51.34	15.46	9.13
40	125.44	71.66	43.21	29.47
60	163.27	116.77	79.34	63.06
80	255.79	166.06	134.29	103.22
100	321.44	233.45	179.11	121.19

In Figure 3, end-to-end delay during data transmission for various numbers of nodes is illustrated. E2E delay of the proposed technique is minimal than existing approaches, namely DWMA, CBR-RPL, WEMER and PEGASIS.

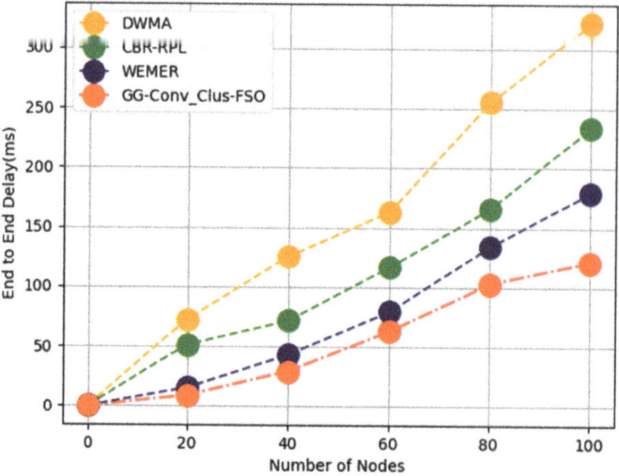

Figure 3. Comparison of end-to-end delay.

4.3. Packet Delivery Ratio

PDR is determined by dividing the total number of data packets sent from source to destination node by the number of data packets delivered. Data communication technology that delivers most packets is deemed the best. The packet delivery ratio of the different method is mentioned in Table 4.

Table 4. Comparison of the packet delivery ratio.

No of Nodes	DWMA	CBR-RPL	WEMER	GG-Conv_Clus-FSO
0	0	0	0	0
20	17.32	24.55	28.91	32.63
40	32.67	35.46	38.64	42.03
60	41.04	48.75	52.06	50.17
80	63.93	67.11	69.47	71.90
100	71.46	79.02	82.03	95.14

In Figure 4, PDR during data transmission for different numbers of nodes is illustrated. PDR of the proposed approach is higher than existing approaches, namely DWMA, CBR-RPL, WEMER and PEGASIS.

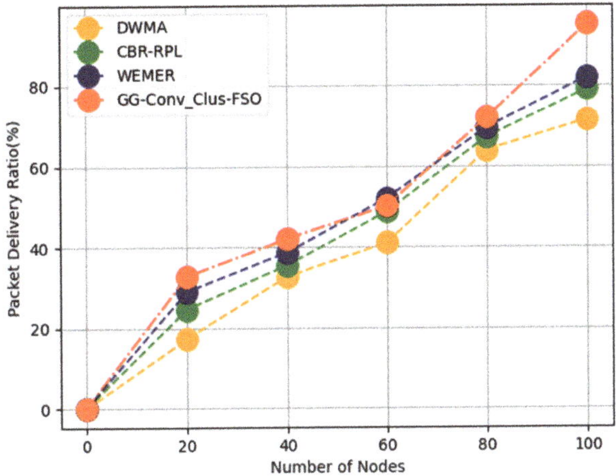

Figure 4. Comparison of the packet delivery ratio.

4.4. Packet Loss

Packet loss can be triggered by a mixture of circumstances, including signal deterioration owing to multi-path fading on the network media. In WSNs, packet loss is conceivable. In order for attackers to simply acquire the data. Packet loss occurs when one or more sent packets fail to reach their intended destination. The Packet Delivery Ratio is reduced when packets are lost. The packet loss of the different method is compared and its mentioned in Table 5

Table 5. Comparison of packet loss.

No of Nodes	DWMA	CBR-RPL	WEMER	GG-Conv_Clus-FSO
0	0	0	0	0
20	89	76	59	31
40	143	137	147	98
60	221	226	204	149
80	349	358	269	223
100	520	440	320	282

In Figure 5, packet loss during data transmission for different number of node is illustrated. The packet loss of proposed approach is minimal than existing approaches, namely DWMA, CBR-RPL, WEMER and PEGASIS.

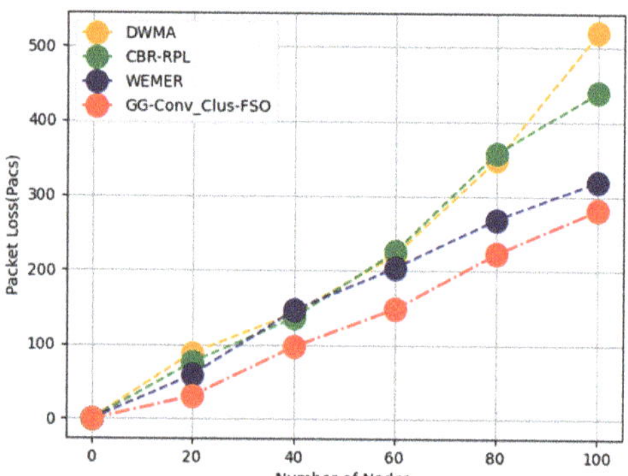

Figure 5. Comparison of packet loss.

4.5. Throughput

The amount of data that is efficiently sent/received through a communication channel is referred to as throughput. Throughput is calculated in kilobits per second, megabits per second or gigabits per second and might differ from bandwidth owing to a variety of technical issues such as packet loss, latency, jitter, and more. The quantity of data that is moved from one location to another in a given length of time is referred to as throughput. The comparisons of previous methods and proposed method is mentioned in the below Table 6.

Table 6. Comparison of throughput.

No of Nodes	DWMA	CBR-RPL	WEMER	GG-Conv_Clus-FSO
0	0	0	0	0
20	65.22	87.24	98.53	127.43
40	127.42	154.36	187.25	206.33
60	178.11	221.42	267.22	281.16
80	229.55	265.86	298.45	379.55
100	276.34	321.46	357.11	465.51

In Figure 6, the throughput during data transmission for different numbers of nodes is illustrated. The throughput of the proposed approach is minimal compared to the existing approaches, namely DWMA, CBR-RPL, WEMER and PEGASIS.

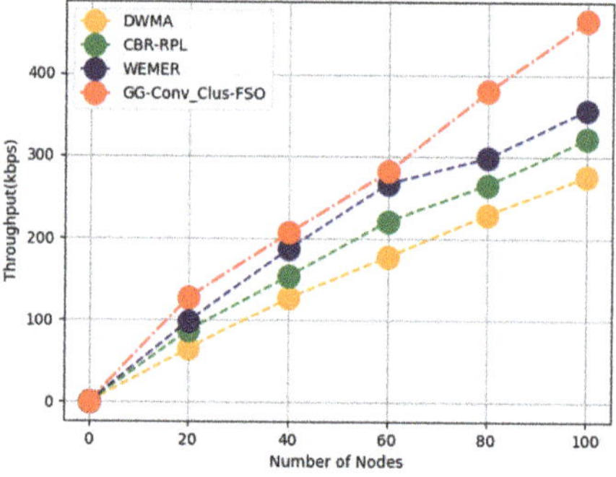

Figure 6. Comparison of throughput.

5. Conclusions

The Internet of Things has a significant influence on daily living. For IoT and LLNs, the RPL has become the standard protocol. It is not only extensively utilised, but it has also been studied by diverse groups of individuals. The widespread usage of RPL and its customisation has necessitated substantial study and development. There are certain flaws with the existing RPL mechanism, one of which is an energy hole, which is a major problem in the context of IoT. Fish swarm optimization is used to initiate Grid creation among sensor nodes, which can help in cluster formation and Cluster Head (CH) selection with energy optimization by calculating the energy consumption of the network. The performance of a Graph-Grid-based Convolution clustered neural network with fish swarm optimization (GG-Conv_Clus-FSO) is compared to existing state-of-the-art protocols, and the GG-Conv_Clus-FSObeats the existing techniques, with a 95.14 percent increase in the packet delivery ratio (PDR).

Author Contributions: Conceptualization, Writing—original draft M.K.; Supervision, A.I. and S.B.G.; Writing—original draft and review and editing, C.N.K.B. and V.J.; Validation, C.V.; propose the new method or methodology, M.K. and S.B.G.; Formal Analysis, Investigation C.O.S.; Resources, S.B.G. and C.O.S.; Software, T.C.M.; Writing—review and editing, T.C.M. All authors have read and agreed to the published version of the manuscript.

Funding: National Research Development Projects to finance excellence (PFE)-14/2022-2024 granted by the Romanian Ministry of Research and Innovation.

Informed Consent Statement: Not applicable.

Data Availability Statement: Data will be shared for review based on the editorial reviewer's request.

Acknowledgments: The work of Chaman Verma was supported by the European Social Fund under the project "Talent Management in Autonomous Vehicle Control Technologies" (EFOP-3.6.3-VEKOP-16-2017-00001).

Conflicts of Interest: The authors declare no conflict of interest.

References

1. Ul Hassan, T.; Asim, M.; Baker, T.; Hassan, J.; Tariq, N. CTrust-RPL: A control layer-based trust mechanism for supporting secure routing in routing protocol for low power and lossy networks-based Internet of Things applications. *Trans. Emerg. Telecommun. Technol.* **2021**, *32*, e4224. [CrossRef]
2. Kim, H.; Kim, H.S.; Bahk, S. MobiRPL: Adaptive, robust, and RSSI-based mobile routing in low power and lossy networks. *J. Commun. Netw.* **2022**, 1–19, early access. [CrossRef]
3. Garg, S.; Mehrotra, D.; Pandey, S.; Pandey, H.M. Network efficient topology for low power and lossy networks in smart corridor design using RPL. *Int. J. Pervasive Comput. Commun.* **2021**, *ahead-of-print*. [CrossRef]
4. Tharini, V.J.; Vijayarani, S. IoT in healthcare: Ecosystem, pillars, design challenges, applications, vulnerabilities, privacy, and security concerns. In *Incorporating the Internet of Things in Healthcare Applications and Wearable Devices*; IGI Global: Hershey, PA, USA, 2020; pp. 1–22.
5. Hua, J.; Shunwuritu, N. Research on term extraction technology in the computer field based on wireless network technology. *Microprocess. Microsyst.* **2021**, *80*, 103336. [CrossRef]
6. Mabrouki, J.; Azrour, M.; Dhiba, D.; Farhaoui, Y.; El Hajjaji, S. IoT-based data logger for weather monitoring using Arduino-based wireless sensor networks with remote graphical applications and alerts. *Big Data Min. Anal.* **2021**, *4*, 25–32. [CrossRef]
7. Lecluyse, C.; Minnaert, B.; Kleemann, M. A Review of the Current State of Technology of Capacitive Wireless Power Transfer. *Energies* **2021**, *14*, 5862. [CrossRef]
8. Mogensen, R.S.; Rodriguez, I.; Schou, C.; Mortensen, S.; Sørensen, M.S. Evaluation of the impact of wireless communication in production via factory digital twins. *Manuf. Lett.* **2021**, *28*, 1–5. [CrossRef]
9. Aijaz, A.; Su, H.; Aghvami, A.H. CORPL: A routing protocol for cognitive radio enabled AMI networks. *IEEE Trans. Smart Grid* **2014**, *6*, 477–485. [CrossRef]
10. Basagni, S.; Petrioli, C.; Petroccia, R.; Spaccini, D. CARP: A channel-aware routing protocol for underwater acoustic wireless networks. *Ad Hoc Netw.* **2015**, *34*, 92–104. [CrossRef]
11. Zhou, Z.; Yao, B.; Xing, R.; Shu, L.; Bu, S. E-CARP: An energy efficient routing protocol for UWSNs in the internet of underwater things. *IEEE Sens. J.* **2015**, *16*, 4072–4082. [CrossRef]
12. Knežević, Ž.; Beck, N.; Milković, Đ.; Miljanić, S.; Ranogajec-Komor, M. Characterisation of RPL and TL dosimetry systems and comparison in medical dosimetry applications. *Radiat. Meas.* **2011**, *46*, 1582–1585. [CrossRef]

13. Safaei, B.; Salehi, A.A.M.; Monazzah, A.M.H.; Ejlali, A. Effects of RPL objective functions on the primitive characteristics of mobile and static IoT infrastructures. *Microprocess. Microsyst.* **2019**, *69*, 79–91. [CrossRef]
14. Zhang, L.; Li, C.; Shi, H.; Xia, Y. Techniques to improve the hit rate of unicast node-to-node (n2n) delivery in channel-hopping and multi-hop low-power and lossy networks (LLNS). *Tech. Discl. Commons* **2021**, *4097*, 1–9. Available online: https://www.tdcommons.org/dpubs_series/4097 (accessed on 28 May 2022).
15. Jara, A.J.; Zamora, M.A.; Skarmeta, A.F. HWSN6: Hospital wireless sensor networks based on 6LoWPAN technology: Mobility and fault tolerance management. In Proceedings of the 2009 International Conference on Computational Science and Engineering, Vancouver, BC, Canada, 29–31 August 2009; Volume 2, pp. 879–884.
16. Islam, M.M.; Hassan, M.M.; Huh, E.N. Sensor proxy mobile IPv6 (SPMIPv6)-A mobility-supported framework IP-WSN. In Proceedings of the 2010 13th International Conference on Computer and Information Technology (ICCIT), Dhaka, Bangladesh, 23–25 December 2010; pp. 295–299.
17. Jara, A.J.; Zamora, M.A.; Skarmeta, A.F. An Initial Approach to Support Mobility in Hospital Wireless Sensor Networks Based on 6LoWPAN (HWSN6). *J. Wirel. Mob.Netw. UbiquitousComput. Dependable Appl.* **2010**, *1*, 107–122.
18. Petäjäjärvi, J.; Karvonen, H. Soft handover method for mobile wireless sensor networks based on 6lowpan. In Proceedings of the 2011 International Conference on Distributed Computing in Sensor Systems and Workshops (DCOSS), Barcelona, Spain, 27–29 June 2011; pp. 1–6.
19. Ha, M.; Kim, D.; Kim, S.H.; Hong, S. Inter-MARIO: A fast and seamless mobility protocol to support inter-PAN handover in 6LoWPAN. In Proceedings of the 2010 IEEE Global Telecommunications Conference GLOBECOM 2010, Miami, FL, USA, 6–10 December 2010; pp. 1–6.
20. Koster, V.; Dorn, D.; Lewandowski, A.; Wietfeld, C. A novel approach for combining Micro and Macro Mobility in 6LoWPAN enabled Networks. In Proceedings of the 2011 IEEE Vehicular Technology Conference (VTC Fall), San Francisco, CA, USA, 5–8 September 2011; pp. 1–5.
21. Bag, G.; Mukhtar, H.; Shams, S.S.; Kim, K.H.; Yoo, S.W. Inter-PAN mobility support for 6LoWPAN. In Proceedings of the 2008 Third International Conference on Convergence and Hybrid Information Technology, Busan, Korea, 11–13 November 2008; Volume 1, pp. 787–792.
22. Bag, G.; Raza, M.T.; Kim, K.H.; Yoo, S.W. LoWMob: Intra-PAN mobility support schemes for 6LoWPAN. *Sensors* **2009**, *9*, 5844–5877. [CrossRef] [PubMed]
23. Saravanakumar, V.; DWMA: An Energy Hole Reduction Mechanism on RPL for 6LoWPAN. EasyChair Prepr. 2020. Available online: https://easychair.org/publications/preprint/n6hC (accessed on 28 May 2022).
24. Shirbeigi, M.; Safaei, B.; Mohammadsalehi, A.; Monazzah, A.M.H.; Henkel, J.; Ejlali, A. A cluster-based and drop-aware extension of RPL to provide reliability in IoT applications. In Proceedings of the 2021 IEEE International Systems Conference (SysCon), Vancouver, BC, Canada, 15 April–15 May 2021; pp. 1–7.
25. Sharmin, N.; Karmaker, A.; Lambert, W.L.; Alam, M.S.; Shawkat, M.S.T. Minimizing the energy hole problem in wireless sensor networks: A wedge merging approach. *Sensors* **2020**, *20*, 277. [CrossRef] [PubMed]
26. Elsheikh, A.H.; Abd Elaziz, M.; Vendan, A. Modeling ultrasonic welding of polymers using an optimized artificial intelligence model using a gradient-based optimizer. *Weld. World* **2021**, *66*, 1–18. [CrossRef]
27. Abd Elaziz, M.; Elsheikh, A.H.; Oliva, D.; Abualigah, L.; Lu, S.; Ewees, A.A. Advanced metaheuristic techniques for mechanical design problems. *Arch. Comput. Methods Eng.* **2022**, *29*, 695–716. [CrossRef]
28. Bhale, P.; Dey, S.; Biswas, S.; Nandi, S. Energy-efficient approach to detect sinkhole attack using roving IDS in 6LoWPAN network. In *International Conference on Innovations for Community Services*; Springer: Cham, Switzerland, 2020; pp. 187–207.
29. Sujatha, R.; Srivaramangai, P. Performance Comparison of Black Hole Attack Detection Mechanism in 6lowpan over Manet. *Int. J. Adv. Res. Comput. Sci.* **2018**, *9*. [CrossRef]
30. Nandi, S. Energy-Efficient Approach to Detect Sinkhole Attack Using Roving IDS in 6LoWPAN Network. In *Innovations for Community Services: 20th International Conference, I4CS 2020, Bhubaneswar, India, 12–14 January 2020, Proceedings*; Springer Nature: Berlin/Heidelberg, Germany, 2019; Volume 1139, p. 187.
31. Sampathkumar, A.; Mulerikkal, J.; Sivaram, M. Glowworm swarm optimization for effectual load balancing and routing strategies in wireless sensor networks. *Wirel. Netw.* **2020**, *26*, 4227–4238. [CrossRef]
32. Liu, G.; Zhu, H. Displacement Estimation of Six-Pole Hybrid Magnetic Bearing Using Modified Particle Swarm Optimization Support Vector Machine. *Energies* **2022**, *15*, 1610. [CrossRef]
33. Rodríguez, A.; Pérez-Cisneros, M.; Rosas-Caro, J.C.; Del-Valle-Soto, C.; Gálvez, J.; Cuevas, E. Robust Clustering Routing Method for Wireless Sensor Networks Considering the Locust Search Scheme. *Energies* **2021**, *14*, 3019. [CrossRef]
34. Bharany, S.; Sharma, S.; Badotra, S.; Khalaf, O.I.; Alotaibi, Y.; Alghamdi, S.; Alassery, F. Energy-Efficient Clustering Scheme for Flying Ad-Hoc Networks Using an Optimized LEACH Protocol. *Energies* **2021**, *14*, 6016. [CrossRef]
35. Mittal, M.; De Prado, R.P.; Kawai, Y.; Nakajima, S.; Muñoz-Expósito, J.E. Machine Learning Techniques for Energy Efficiency and Anomaly Detection in Hybrid Wireless Sensor Networks. *Energies* **2021**, *14*, 3125. [CrossRef]

MDPI
St. Alban-Anlage 66
4052 Basel
Switzerland
www.mdpi.com

Energies Editorial Office
E-mail: energies@mdpi.com
www.mdpi.com/journal/energies